Alcohol and its Role in the Evolution of Human Society

Alcohol and its Role in the Evolution of Human Society

Ian S. Hornsey
Founder, Nethergate Brewery Co. Ltd.
Email: *hornsey.brews@virgin.net*

RSC Publishing

ISBN: 978-1-84973-161-4

A catalogue record for this book is available from the British Library

Published by The Royal Society of Chemistry,
Thomas Graham House, Science Park, Milton Road,
Cambridge CB4 0WF, UK

Registered Charity Number 207890

For further information see our website at www.rsc.org

Dedication

This book is dedicated to the memory of Dr. Denis Hide; tutor and friend.

Foreword

Hats off to Ian for producing such an informative and enlightening book.

Farming in the nutrient deficient soils of the Suffolk Brecks as I do, plant variety selection is of prime importance, as it is to farmers everywhere. In the aftermath of this dry early summer we face uncertainty as regards barley quality and yield. What is striking is that without the great advances in science and artificial selection, it would be doubtful whether we would have any crop to harvest.

This seminal book explains why we have selected the few plants that we have and why we have put so much effort into enhancing their sought-after qualities; matters of prime concern to human-kind – our survival depends on our continued success.

These labours of ours can now for the first time be tracked to pre-history thanks to cutting-edge technology and this subject area will remain of utmost importance given our resource-scarce world as we ever increase the demands made upon it.

This book, Ian's fourth to be published by the Royal Society of Chemistry, unites archaeology and anthropology, plant breeding and industrial process, together with so many other disciplines besides. It is nothing short of revelatory and thoroughly up-to-date in our fast-moving world; this represents a Herculean effort on the part of the author.

Only through understanding the fundamentals of where we have come from can we achieve bountiful crops into the future and we should all raise a glass to that!

Arthur Edward Guinness, Earl of Iveagh

Alcohol and its Role in the Evolution of Human Society
Ian S. Hornsey
© Ian S. Hornsey 2012
Published by the Royal Society of Chemistry, www.rsc.org

Preface

It is almost impossible to imagine the world without ethanol as a component of many beverages, but one should not forget that the fact that we are able to enjoy such drinks as beer, malt whisky, wine and cider is primarily due to a single-celled fungus, *Saccharomyces cerevisiae*, the yeast. Many microbes are capable of fermenting sugars to obtain cellular energy, liberating, as they do, ethanol as a by-product. The extent of use of the pathway is, however, limited by the very toxicity of ethanol itself, which destabilizes cell membranes and interferes with other aspects of cellular metabolism. What deserves our eternal gratitude is the fact that, over the millennia, yeast has evolved mechanisms that allow it to generate and tolerate levels of ethanol that are sufficiently high (7–9% in many fermentations) to inhibit the growth of most competing microbes, and to be of beneficial use to man. From our point of view, the critical adaptation evolved by yeast concerns its ability first to synthesize ethanol (by fermentation, which can occur even under aerobic conditions) and later to consume it (by respiration). This is achieved by the differential regulation of two duplicated alcohol dehydrogenase genes, *ADH1* and *ADH2*, which encode for two alcohol dehydrogenases (ADHs), ADH1 and ADH2, that interconvert ethanol and acetaldehyde.

Today, in much of the developed world, the production of alcoholic beverages is manifested as a massive, globalized industry that relies upon ongoing advances in our knowledge of the biochemistry, physiology and molecular biology of yeasts.

Alcohol and its Role in the Evolution of Human Society
Ian S. Hornsey
© Ian S. Hornsey 2012
Published by the Royal Society of Chemistry, www.rsc.org

Concomitant with this is a plethora of smaller, more artisanal industries doing essentially the same thing, but with very little cognisance of the scientific niceties of that important micro-organism.

Ethanol is a relatively simple chemical compound, yet it is astounding to see the many different ways in which people use, perceive and react to the substance. Most anthropologists and ethnologists would agree that ethanol has long been one of the most widespread chemical compounds encompassed by mankind. Its potential roles are many-fold, since it can act as a disinfectant, an anaesthetic, a tranquilizer, an appetizer, a food, a solvent and an economic commodity, as well as being a potent symbol in many cultures.

The present work represents an attempt to provide a basis for further research into the significance of alcoholic beverages in the advancement of mankind. It attempts to integrate some of the botany, ethnobotany and history of important providers of fermentable material with the role of the microbial world in the production of some widely used, non-distilled, alcoholic drinks. Due regard has been paid to the relevant information we have gleaned from the relatively new field of genomics, and it will, hopefully, give an indication of how genome-related work has transformed the scientific community.

The book deals solely with non-distilled beverages, and pressure on space demanded that this should be the case. Important beverages though whisky, gin, vodka, *etc.* are, products of the art of distillation were one of man's more recent (albeit very effective) means of achieving insobriety. I trust that this omission does not detract from the usefulness of the book. While on the subject of omissions, for similar, spatial, reasons I have not been able to do justice to the role of the fig and the date palm (and several other plants) as sources of fermentable sugar. Another, possibly contentious, point is my decision to include *chicha* in the section dealing with cassava rather than maize.

Regarding dating, I have used the dating systems adopted by various authorities when quoting their work. For an erudite account of how the various systems are interrelated, see Zohary and Hopf (reference 16, page 15).

I have been lucky to have had the use of two of the greatest libraries in the world: the British Library and the Cambridge

University Library (at both the Central Science and West Road sites). Staff at the latter have, as always, been particularly helpful, and I am particularly indebted to Bill Noblett who responded to my emergency calls with diligence and alacrity.

I am also grateful for the fact that numerous authorities – in varying subject areas – have given helpful assistance, and, in this regard, especial thanks are due to Robin Allaby, John Arthur, Robert Carlson, Alan Clapham, Frances Hayashida, Elizabeth Kellogg, Max Nelson, Mark Nesbitt, Roger Putman, Ian Roberts, José Sampaio, Janet Spitz and Ryan Williams. All of this help was invaluable, but, as ever, any errors are entirely down to me. Finally, I wish to thank my wife for tolerating the strange ways of an author!

Contents

Alcohol and its Role in the Evolution of Human Society
Ian S. Hornsey
© Ian S. Hornsey 2012
Published by the Royal Society of Chemistry, www.rsc.org

CHAPTER 1

Introductory Matter

THERE IS ONLY ONE HISTORY OF LIFE ON EARTH

Steven A. Benner

There is, and always has been, a never-ending struggle between man and micro-organisms for the opportunity to consume food supplies. These food supplies are ultimately the legacy of the photosynthetic activity of plants. The study of the history of alcoholic fermentation is necessarily multi-faceted, involving basic biological science, the life and culture of ancient peoples, the study of races and their customs and the interactions of plants and humans. Understanding the process of plant domestication is fundamental to our comprehension of the rise of agriculture, and knowledge of the latter is a prerequisite for understanding the importance of alcoholic beverages in the development of mankind.

Before the modern era, only the Eskimos, the peoples of Tierra del Fuego at the southern tip of South America and the Australian aborigines apparently lived out their lives without the medical benefits and mind-altering effects of alcohol. While polar regions generally lacked resources for fermentable monosaccharides, honey and sugar-rich fruits and other plants are plentiful in temperate parts of the globe and the tropics. In the New World, most of North America was devoid of indigenous alcoholic beverages prior to Columbus, while, further south, maize, the juice of the century plant and the saguaro cactus were sweet enough to

Alcohol and its Role in the Evolution of Human Society
Ian S. Hornsey
© Ian S. Hornsey 2012
Published by the Royal Society of Chemistry, www.rsc.org

ferment directly into *chicha*, pulque and cactus wine, respectively. As a rule of thumb, the plant species-rich areas of the globe, such as the tropics were/are home to the greatest variety of alcoholic drinks. In Europe, ancient northern peoples were somewhat starved of "available sugar" sources and relied on grains to make their alcoholic drink (beer), while those in southern Europe had recourse to berries (for wine), which would yield fermentable material with far less persuasion.

The major raw materials (fermentable sugars) available to prehistoric man for fermentation purposes would have been the monosaccharides glucose and fructose and the disaccharides sucrose, lactose and maltose. The latter could be obtained from sprouting grain, and lactose is present in milk. The two monosaccharides, and their dimer, sucrose, are invariable components of fruits and honey. It is likely that, where they grew, juice from the fruits of the date palm (*Phoenix dactylifera* L.) and the fig (*Ficus carica* L.) would have been important sources of fermentable sugar.

The date palm was one of the first trees to be taken into cultivation in the Old World.[1] Fossil evidence suggests that fermentable fruits became prominent around 80 million years ago (Mya), in the Cretaceous, during the age of the dinosaurs. The end of the Cretaceous saw, in addition to the emergence of fruits, the extinction of the dinosaurs and the emergence of mammals and fruit flies. Provided that the relevant microbes were available, these sugar-containing fruits could have undergone spontaneous fermentation and thus, in essence, become the "first alcoholic drinks". According to Dilcher,[2] the strongest selection pressure in the angiosperms was directed toward the flower, fruits and seeds.

It is a long-held, almost universal, human belief (albeit in various forms) that alcoholic drinks, particularly wine and beer, were bestowed upon mankind by a deity that took pity on his plight. In many ancient cultures beer was a gift to women from a female deity, and brewing was for centuries intimately connected with the fairer sex (Figure 1.1). Certainly, the promotion of amylolysis in grain by chewing seems to have been (and still is) largely a female occupation (Figure 1.2). There seems to have been scant regard for gender equality in ancient times, for beverage consumption was largely the preserve of males (Figure 1.3)!

Figure 1.1 Nineteenth-century lithograph showing women brewing *chicha* in the town of Arequipa, Peru. After Marcoy, 1873.[3]

Increasing amounts of archaeological, ethnographic and evidence from around the world suggest that alcoholic beverages have been integral to the social, economic, religious and political aspects of many cultures. Feasting activities were especially important,[5] and often marked critical events in the lives of individuals and communities, and frequently involved public rituals.

Figure 1.2 Women chewing maize for *chicha* production. After Marcoy, 1873.

Figure 1.3   Men drinking *chicha*. From Hield, 1883.[4]

1.1   CROP DOMESTICATION

As for the study of the history of alcoholic beverages *per se*, "it was Robert Braidwood who started it all!" The University of Chicago anthropologist, who died in 2003 aged 95, was a leading light in the field of Near Eastern prehistory, and a pioneer in the study of crop domestication. Among other projects, Braidwood (and his wife, Linda) worked on the important southern Turkish mound site at Çayönü (literally, "beside the stream"), and obtained extensive evidence for the notion that there was a shift from hunter-gathering to agriculturalism in that area between 8000 and 12,000 years ago. The Braidwoods[6] had previously been working at Jarmo, in the foothills of the Zagros Mountains in northern Iraq, the earliest Neolithic farming village in western Asia, or, in Braidwood's words "the world's oldest food producing community". The work at Jarmo tested Braidwood's "Hilly Flanks" hypothesis and, a little later, he proposed that there was a direct relationship between humans adopting a sedentary lifestyle and the domestication of wild barley. In his opinion, barley bread was the driving force for the "Neolithic Revolution", and this stimulated other workers to enter the field. One of Braidwood's major achievements was to assemble a group of scientists to identify and interpret archaeologically retrieved floral and faunal remains, thus giving birth to archaeological studies of agricultural origins.

Braidwood's articles "From cave to village" in the October 1952 *Scientific American*[7] and his use of the new "symposium-by-mail" section in the *American Anthropologist* in 1953,[8] which fostered "Did man once live by beer alone?" were responsible for preparing the ground for all subsequent work relating to man and alcohol. It is testimony to his scholarship that, 60 years on, his work is still held in high regard, and the "archaeology of alcohol" is now seen as a credible area of study (see Dietler's excellent review[9]).

In my opinion, in order to exhaustively investigate the evolution of alcoholic beverages, and to appreciate their significance in the development of *Homo sapiens*, it is necessary to have much more than a passing acquaintance with the plants that provide fermentable material. Familiarity with the domestication profiles of these plants will also help us to integrate man and his most widely used intoxicant.

Archaeologists, anthropologists and others generally agree that the emergence of agriculture, together with the domestication of animals for food and labour, has brought about the most important transformation in man's culture since the last Ice Age. Despite the fundamental role of plant domestication in human history and the seminal nature of relatively few crop plants to modern man, we still have a paucity of information on how plants adapt under the influence of domestication.

In his 1859 *On the Origin of Species by Means of Natural Selection*,[10] Charles Darwin devoted his first chapter to "Variation under Domestication", and he expanded on this theme nine years later in *Variation in Animals and Plants under Domestication*.[11] The fact that one of the founders of evolutionary theory paid such attention to domestication, and the selection processes associated with it, is testimony to the value of crops (and animal breeds) in the study of natural selection. One of the major observations made by Darwin was that the morphological modifications selected during domestication have been of such magnitudes that most crop plants cannot survive in the wild any more without human assistance.

In addition, he pointed out that selection by breeders could lead to a wide array of variation in domesticated plants and animals when compared with their wild progenitors. He also suggested that selection under human cultivation happened unconsciously or inadvertently (*i.e.* without deliberate human action). He argued that crops are so different morphologically from their wild progenitors that humans could not have possibly identified target traits so different from those existing in the wild progenitor.

For Darwin, one of the benefits of considering selection under domestication was that he was able to demonstrate that selection had heritable effects, and, remember, this in the absence of any information about the histological, biochemical or genetic foundations of heredity. Darwin's elucidation of natural selection was one of the most monumental intellectual achievements in the history of science. No longer was it permissible to explain the innumerable adaptations of living organisms as being the result of the supernatural.

As Avise and Ayala[12] have said: "Natural selection is an inevitable process of nature whenever organisms show heritable variation in their capacity to survive and reproduce in particular environments, but the operation has no more consciousness or intelligence than do natural physical forces such as gravity or the weather." Darwin's legacy, therefore, was not that evolution occurs, but rather that it results from a natural event rather than a supernatural one. The adaptation of plants to cultivation was vital to man's change from hunter-gatherer to agricultural societies, and ultimately spawned the rise of civilization. Paul Gepts[13] makes an eloquent case for regarding crop domestication as being a long-term selection experiment, and for an explanation of "plant domestication", I can do no better than to quote Barbara Pickersgill:[14]

"Domestication is generally considered to be the end point of a continuum that starts with exploitation of wild plants, continues through cultivation of plants selected from the wild but not yet genetically different from wild plants, and terminates in fixation, through human selection, of morphological and hence genetic differences distinguishing a domesticate from its wild progenitor. These differences constitute the domestication syndrome and generally render the domesticate less capable of survival in the wild, thus dependent on man for its growth and reproduction. Features of the domestication syndrome include loss of dispersal, increase in size (especially of the harvested part of the plant), loss of seed dormancy and loss of chemical or mechanical protection against herbivores."

Until fairly recently, domestication had been interpreted as a rapid process with scant pre-domestication cultivation and a fairly rapid rise of domestication characteristics (the "rapid-transition" model), and this has been central to the way in which biologists have organized their research into the origins of our crops.[15] To substantiate this widely held premise, it has been assumed that artificial selection pressures were much stronger than natural selection pressures, resulting in genetic patterns of diversity that reflect genetic independence of geographic localities (such as the Near East). The rapid-transition model supported the concept of the "Neolithic Package", which implies that different crops were brought into domestication by the same group of early farmers.[16] Zohary,[17] in a paper discussing the monophyletic/polyphyletic origin of the ancient Near Eastern crops, argued that, if the Neolithic Package is an actuality then domesticated crops should be monophyletic, since there would have been little need for the same group of farmers to domesticate crops.

As Bruce Smith[18] pointed out, it was between 5000 and 10,000 years ago, that humans domesticated virtually all major crop species used by modern agricultural societies. This feat was accomplished through artificial selection for traits that improved agronomic qualities. As a result of this process, favourable alleles at loci controlling agronomic traits were brought to fixation in the population during the domestication period. After the initial domestication, the continued practice of selective breeding allowed additional favourable alleles to sweep through the crop species, while diversifying selection in response to the different environments encountered during the expansion of the crop caused regional fixation of distinct favourable alleles. As a consequence of this complex history of selection, only a limited portion of the population contributed to each subsequent generation. Some anticipated consequences are a genome-wide loss of diversity at unselected genes because of the genetic bottleneck effect, a severe reduction in diversity at genes under directional selection during domestication and artificially high diversity at genes under diversifying selection.

These two processes – selection targeted on agronomic genes and drift due to the domestication bottleneck affecting the entire genome – are the principal factors that influence the amount and distribution of genetic variation in crop genomes as compared to their wild progenitors.

Jared Diamond[19] is in no doubt about the relevance of plant and animal domestication; he calls it "the most important development in the past 13,000 years of human history ... it was prerequisite to the rise of civilization, and it transformed global demography". As Diamond says, food production could not possibly have arisen through a conscious decision, because the world's first farmers had no model of farming around them to observe, hence they could not have known that there was a goal of domestication to strive for, and could not have guessed the consequences that domestication would bring for them. He argues that if they had actually foreseen the consequences, they would surely have outlawed the first steps towards domestication, because the archaeological and ethnographic record throughout the world shows that the transition from hunting and gathering to farming eventually resulted in more work, lower adult stature, worse nutritional condition and heavier disease burdens.[20] The only peoples who could make a conscious choice about becoming farmers were hunter-gatherers living adjacent to the first farming communities, and they generally disliked what they saw and rejected farming, for the good reasons just mentioned and others.

If plant and animal domestication was such a boon for mankind, why did it only arise in a few areas on the planet, and why did it occur earlier in some areas than others? Diamond reckons that the wild animal species that most plausibly could have yielded useful domesticates, all those years ago, were large terrestrial mammalian herbivores and omnivores, of which the planet holds 148 species weighing 45 kg or more. Only 14 of these were actually domesticated, which prompts us to ask what prevented the domestication of the other 134? Similarly, of the 200,000-odd wild species of higher plants worldwide, why did only around 100 give us valuable domesticates? Food for thought, indeed!

Diamond argues that those peoples who through "biogeographic serendipity" first produced domesticates acquired "enormous advantages over other peoples and expanded". According to Diamond,[21] domestication ultimately yielded agents of conquest; his "guns, germs, and steel". It is evident that the ability to produce one's own food conferred huge demographic, technological, political and military advantages over neighbouring hunter-gatherers. As Diamond put it, "the history of the past 13,000 years consists of tales of hunter-gatherer societies becoming driven out,

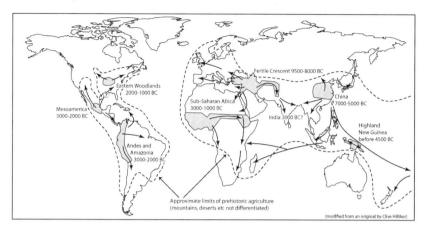

Figure 1.4 Areas where agriculture originated (from Bellwood, 2005[53]).

infected, conquered or exterminated by farming societies in every area of the world suitable for farming. One might, therefore, have anticipated that, in any part of the world, one or more of the local hunter-gatherer societies would have stumbled upon domestication, become farmers, and thereby out-competed the other local hunter-gatherer societies". The reality is that food production arose independently in, at most, nine areas of the world: the Fertile Crescent, China, Mesoamerica, Andes/Amazonia, eastern United States, Sahel, tropical West Africa, Ethiopia and New Guinea. From archaeological evidence, the areas of the world where agriculture originated are shown in Figure 1.4.

1.2 PLANT ADDITIVES

As a species, *Homo sapiens* seems to have an innate desire to alter its state of consciousness and, to this end, narcotics, hallucinogens and other mind-altering compounds have been ingested in the form of vegetable matter, as have mild stimulants such as tea and coffee. I recommend Richard Rudgley's book[23] as an introduction to intoxicants in society. The use of psychoactive plants goes back to prehistoric times and, therefore, there has been a long association between alcoholic beverages and plants with preservative, medicinal, psychtropic and flavouring properties. Some plants, and we may use the hop (*Humulus lupulus* L.) as an example, possess more than one of the above attributes – in this case having flavouring, preservative and medicinal properties.

Most lower-strength (<5% ABV) alcoholic beverages are inherently unstable, and have to be consumed soon after production. In particular, many palm wines are particularly susceptible to microbial spoilage, and we find that a number of plants have been used in an attempt to improve their shelf-life (*e.g. Saccoglottis gabonensis*, *Vernonia amygdalina* and *Euphorbia* spp.). The taste of newly made palm wines is appealing to devotees, and attempts to confer enhanced stability are often hampered by the fact that some herbal additives have undesirable organoleptic effects.

The situation regarding beers is rather different, and an unflavoured product is rather uninteresting to the drinker. Accordingly, many plants have been successfully used over the years for flavouring and preserving beers.[24–27] Some plants, the aforementioned *V. amygdalina* being one, have multiple benefits as beer additives and have been extensively studied.[28] *V. amygdalina* Del. (bitter leaf; Figure 1.5), a soft-wooded shrub of tropical Africa, also has remarkable medicinal properties, including being antibacterial, antifungal, antimalarial, antioxidant and

Figure 1.5   *Vernonia amygdalina* Del.

hepato-protective. It also produces beneficial cytotoxic effects.[29] It is one of a number of African medicinal plants that have been investigated for their potential as hop substitutes[30,31] and, when blended with *Garcinia kola* E. Heckel and *Gongronema latifolium*, produces an acceptable hop substitute for sorghum lager beer.[31] It should be pointed out that the search for hop alternatives in beer-drinking countries where the plant is difficult (or impossible) to grow, and thus expensive, is never ending. Research into the history of plant additives in alcoholic beverages is hampered by the fact that, in ancient times, little information was ever written down. Add to this the fact that each region on the planet has its own, centuries-old, traditions and that thousands of plants, worldwide, have been used for medicinal purposes, and one can appreciate the problem.

To date, the earliest biochemical evidence for plant material being incorporated into fermented drinks comes from ancient China[32,33] and the ancient Near East.[34] The latter work concerns wine jars found at the Neolithic village of Hajji Firuz Tepe, in the northern Zagros mountains, and was significant because the wine had manifestly been infused with resin from wild pistachio (*Pistacia atlantica* Desf.), which is the most economically important tree species in many rural areas in western Iran.[35] Also known as the terebinth tree, *P. atlantica* produces a resin that is useful for a wide variety of industrial and traditional uses, including food and medicine. Terebinth resin acts as a preservative and a flavouring, and resinated wine was later to become highly popular in ancient Greece.

The Chinese material came from deposits in pottery jars recovered from the early Neolithic village of Jiahu in Henan province, the material being dated to the seventh millennium BC. Chemical analysis indicated that these vessels contained a mixed fermented drink consisting of rice, honey and grape and/or hawthorn (*Crataegus* sp.). Whether all of these plants were used as a source of sugar, or whether some were for flavouring, is a matter of conjecture. "Mixed" alcoholic beverages were almost certainly widely used throughout many parts of the world, and we find a similar drink (grape wine, barley beer and mead) to be the "Phrygian grog" identified from the Gordion site (*ca.* 700 BC) in Turkey,[36] and a mixed fermented honey, wheat and cranberry drink from a Bronze Age grave in Egtved, Denmark.[37]

Alcoholic beverages have regularly been used as a vehicle for medicinal herbs, and we have plentiful written evidence of this from Roman (Pliny the Elder), Anglo-Saxon[38–40] and contemporary times. Much of the necessary information concerning the role of plants in man's milieu is covered in the excellent *Cultural History of Plants.*[41] A glance at Ann Van Arsdall's *Medieval Herbal Remedies*[42] will indicate the role of wine and beer in the medicine of that era.

Archaeological evidence that wine was used in ancient Egypt as a carrier for botanicals has recently been documented.[43] Vessels (which had evidently been used as wine containers) from the beginning of advanced ancient Egyptian culture (*ca.* 3150 BC) right up to the early Byzantine period (fourth to early-sixth century AD) were shown to have contained a range of infused natural products (Figure 1.6).

Folkloristic accounts of the medicinal attributes accorded to some plants suffused into alcoholic drinks are now being substantiated by ethnopharmacological research,[29] and it is evident that the choice of many plants was much more than serendipity. The *Journal of Ethnopharmacology* should be consulted for papers in this field. Toxic plants that have been suffused into alcoholic drinks include wormwood, henbane, opium poppy and mandrake.

1.3 THE GENOMIC AGE

When, over 50 years ago, Watson and Crick[44] elucidated the structure of the double helix, the consequences of this precise knowledge were unknown. It was hoped, however, that there would be possibilities for conquering disease, a greater understanding of the mechanisms of the origin of life and a better knowledge of the interactions between heredity and the environment. This was to be the start of a "new world". Forty-three years were to elapse before the first nucleotide sequence of a eukaryotic genome, the budding yeast *Saccharomyces cerevisiae*, was published.[45] This information enabled the scientific community to ask an entirely new class of questions. Perhaps the most remarkable change has been the redirection of emphasis from individual genes and functionalities to a more all-embracing aim of how cellular networks and systems interact and function, thus producing the highly evolved organisms that we witness today. With the yeast

Figure 1.6   Wine amphora from tomb 217 of the cemetery 4 at Gebel Adda, dated to early Byzantine times (fourth–sixth century AD). Height 67.3 cm. With permission of the Royal Ontario Museum; courtesy of W. Pratt (museum number: 973.24.1217).

genome being the pioneer eukaryotic genome, it was the yeast research community, which had come together with a huge collaborative effort to achieve its goal, that was the first to benefit from knowledge of the complete genome sequence. Hundreds of genomes have now been completely sequenced. The changing perspectives of yeast research since 1996 have been discussed.[46] For treatments of this broad topic, see Smith[47] and Atkinson, et al.[48]

Plant classification has changed significantly over the last 150 years, and enormously so over the past dozen, or so, years when major advances in DNA sequencing have provided a completely new approach to taxonomy. Each of numerous schemes that have been published for many plant groups is varied and, as a result, families, orders and sub-classes are not generally compatible between any of them. Molecular studies are by no means infallible, and modern schemes are just as likely to be scrutinized. Fortunately, the combined work of numerous botanists, under the umbrella of the Angiosperm Phylogeny Group (APG), has arrived at a number of agreed evolutionary trees,[49–51] and made the work of taxonomic botanists that much easier. The amount of phylogenetic work that has been undertaken in the last decade is enormous. Not only has genomics provided new insights into plant classification, it has also helped us to establish the origins of many important crops.

In contrast to the information that we have about crop domestication, we know relatively little about the processes of microbe domestication. With the onset of the genomic age, that situation is changing, and a recent study shows that, by combining microbial ecology with comparative genomics, the events that led to the establishment and domestication of the cryotolerant lager yeast could be established. It was suggested that the yeast in question, *Saccharomyces pastorianus*, was created by fusion of the ale-yeast, *S. cerevisiae*, and a newly identified cryotolerant species, *S. eubayanus*, which is found in the *Nothofagus* (Southern beech) forests in Papagonia.[52]

REFERENCES

1. D. Zohary and P. Spiegel-Roy, *Science*, 1975, **187**, 319.
2. D. Dilcher, *Proc. Nat. Acad. Sci. USA*, 2000, **97**, 7030.
3. P. Marcoy, *A Journey Across South America, from the Pacific Ocean to the Atlantic Ocean*, 2 vols, Blackie & Son, London, 1873.
4. M. Hield, *Glimpses of South America: or, the Land of the Pampas*, Cassell, Petter, Galpin, New York, 1882.

5. M. K. Jones, *Feast: Why Humans Share Food*, Oxford University Press, Oxford, 2007.
6. R. J. Braidwood and L. S. Braidwood, *Antiquity*, 1950, **24**, 189.
7. R. J. Braidwood, *Sci. Am.*, October 1952, 62.
8. R. J. Braidwood, *Am. Anthropol.*, 1953, **55**, 515.
9. M. Dietler, *Annu. Rev. Anthropol.*, 2006, **35**, 229.
10. C. Darwin, in *On the Origin of Species*, John Murray, London, 1859, p. 7.
11. C. Darwin, *Variation of Plants and Animals Under Domestication*, 2 vols, John Murray, London, 1868.
12. J. C. Avise and F. J. Ayala, *Proc. Nat. Acad. Sci. USA*, 2007, **104**, 8563.
13. P. Gepts, *Plant Breed. Rev.*, 2004, **24**, 1.
14. B. Pickersgill, *Ann. Bot.*, 2007, **100**, 925.
15. R. G. Allaby, D. Q. Fuller and T. A. Brown, *Proc. Nat. Acad. Sci. USA*, 2008, **105**, 13982.
16. D. Zohary and M. Hopf, *Domestication of Plants in the Old World: The Origin and Spread of Cultivated Plants in West Asia, Europe and the Nile Valley*, Clarendon Press, Oxford, 3rd edn, 2000.
17. D. Zohary, *Genet. Res. Crop Evol.*, 1999, **46**, 133.
18. B. Smith, *Proc. Nat. Acad. Sci. USA*, 2001, **98**, 1324.
19. J. Diamond, *Nature*, 2002, **418**, 700.
20. M. N. Cohen and G. J. Armelagos, *Palaeopathology at the Origins of Agriculture*, Academic Press, London, 1984.
21. J. Diamond, *Guns, Germs, and Steel: The Fates of Human Societies*, Jonathan Cape, London, 1997.
22. P. Bellwood and M. Oxenham, in *The Neolithic Demographic Transition and its Consequences*, ed. J.-P. Bocquet-Appel and O. Bar-Yosef, Springer Science+Business Media B.V., 2008, p. 13.
23. R. Rudgley, *The Alchemy of Culture: Intoxicants in Culture*, British Museum Press, London, 1993.
24. K.-E. Behre, *Veg. Hist. Archaeobotany*, 1999, **8**, 35.
25. S. H. Buhner, *Sacred and Herbal Healing Beers*, Siris Books, Boulder, CO, 1998.
26. H. T. Beck, in *The Cultural History of Plants*, ed. G. T. Prance and M. Nesbitt, Routledge, New York, 2005, p. 173.

27. I. S. Hornsey, *Brewer and Distiller International*, November 2009, p. 36.
28. O. O. Lasekan, W. O. Lasekan and J. O. Babalola, *Food Chem.*, 1999, **64**, 507.
29. S. K. Yeap, W. H. Ho, B. K. Beh, W. S. Liang, H. Ky, A. Hadi, N. Yousr and N. B. Alitheen, *J. Medical Plants Research*, 2010, **4**, 2787.
30. A. A. Oshodi, I. A. Amoo and A. F. Eleyinmi, *MBAA Technical Quarterly*, 2004, **43**, 398.
31. A. F. Eleyinmi, I. A. Amoo, A. A. Oshodi and A. Hezekiah, *MBAA Technical Quarterly*, 2004, **41**, 403.
32. P. E. McGovern, J. Zhang, J. Tang, Z. Zhang, G. R. Hall, R. A. Moreau, A. Nuñez, E. D. Butrym, M. P. Richards, C. S. Wang, G. Cheng, Z. Zhao and C. Wang, *Proc. Nat. Acad. Sci. USA*, 2004, **101**, 17593.
33. P. E. McGovern, A. P. Underhill, H. Fang, F. Luan, G. R. Hall, H. Yu, C.-S. Wang, F. Cai, Z. Zhao and G. M. Feinman, *Asian Perspect.*, 2005, **44**, 249.
34. P. E. McGovern, D. L. Glusker, L. J. Exner and M. M. Voigt, *Nature*, 1996, **381**, 480.
35. M. Pourreza, J. D. Shaw and H. Zangeneh, *Forest Ecol. Manag.*, 2008, **255**, 3667.
36. P. E. McGovern, *Uncorking the Past*, University of California Press, Berkeley, CA, 2009.
37. T. Thomsen, *Nordisk Fortidsminder*, 1929, **2**, 165.
38. O. Cockayne, *Leechdoms, Wortcunning and Starcraft of Early England*, Longmans, Green, Reader and Dyer, London, 1864, vol. I.
39. O. Cockayne, *Leechdoms, Wortcunning and Starcraft of Early England*, Longmans, Green, Reader and Dyer, London, 1865, vol. II.
40. O. Cockayne, *Leechdoms, Wortcunning and Starcraft of Early England*, Longmans, Green, Reader and Dyer, London, 1866, vol. III.
41. G. T. Prance and M. Nesbitt, ed., *The Cultural History of Plants*, Routledge, New York, 2005.
42. A. Van Arsdall, *Medieval Herbal Remedies*, Routledge, London, 2002.
43. P. E. McGovern, A. Mirzoian and G. R. Hall, *Proc. Nat. Acad. Sci. USA*, 2009, **106**, 7361.

44. J. D. Watson and J. H. C. Crick, *Nature*, 1953, **171**,737.
45. A. Goffeau, B. G. Barrell, H. Bussey, R. W. Davis, B. Dujon, H. Feldmann, F. Galibert, J. D. Hoheisel, C. Jacq, M. Johnston, E. J. Louis, H. W. Mewes, Y. Murakami, P. Philippsen, H. Tettelin and S. G. Oliver, *Science*, 1996, **274**, 546.
46. K. Dolinski and D. Botstein, *Genome Res.*, 2005, **15**, 1611.
47. G. Smith, *The Genomics Age*, AMACOM Books, New York, 2005.
48. P. Atkinson, P. Glasner and M. Lock, ed., *Handbook of Genetics and Society: Mapping the New Genomic Era*, Routledge, London, 2009.
49. APG, *Ann. Mo. Bot. Gard.*, 1998, **85**, 531.
50. APGII, *Bot. J. Linn. Soc.*, 2003, **141**, 399.
51. APGIII, *Bot. J. Linn. Soc.*, 2009, **161**, 105.
52. D. Libkind, C. T. Hittinger, E. Valério, C. Gonçalves, J. Dover, M. Johnston, P. Gonçalves and J. P. Sampaio, *Proc. Nat. Acad. Sci. USA*, 2011, **108**, 14539.
53. P. Bellwood, *First Farmers*, Blackwell, Oxford, 2005.

CHAPTER 2

Ethanologenic Organisms

Ethanol produced by microbial fermentations is used by man in a number of ways, but by far the greatest proportion is used for potable purposes. When mentioning "fermentation", one's mind comes up with "yeast" (*Saccharomyces cerevisiae* and related spp.), but there are a number of microbes that are able to produce significant (>1% w/v) quantities of ethanol, including the yeasts *Zygosaccharomyces rouxii, Kluyveromyces lactis, K. marxianus, Candida tropicalis, Pachysolen tannophilus, Debaryomyces occidentalis* and *Saccharomycopsis fibuligera*, and the bacteria *Zymomonas mobilis, Escherichia coli, Clostridium* spp. and some thermophiles.

Ethanol formation occurs *via* a number of established metabolic pathways, depending on which microbe is being employed. Between them, the above-mentioned can utilize a wide range of substrates, including sucrose, lactose, maltose, mannose, glucose, fructose and galactose, and some strains of *S. cerevisiae* (ex-*S. diastaticus*) can assimilate dextrins and starch because of their propensity for producing glucoamylase. The EMP pathway, operative in *S. cerevisiae*, and the Entner-Doudoroff pathway, responsible for ethanol production in *Z. mobilis*, are described in Chapter 3.

Microbes utilized for the production of ethanol should ideally have the following characteristics: 1) rapid rate of fermentation, 2) adequate flocculation and sedimentation properties, 3) genetic stability, 4) osmotolerance, 5) ethanol tolerance, 6) temperature tolerance, 7) good cell viability (to enable recycling).

Alcohol and its Role in the Evolution of Human Society
Ian S. Hornsey
© Ian S. Hornsey 2012
Published by the Royal Society of Chemistry, www.rsc.org

2.1 THE YEAST (*SACCHAROMYCES CEREVISIAE* MEYEN *EX* E.C. HANSEN)

2.1.1 Taxonomy

Yeasts were the first microbes to be extensively studied scientifically, mainly because of their relatively large cell size, and inherent industrial importance. As microscopic fungi, yeasts are the least complicated of the eukaryotes, and the very name "yeast" encompasses a wide range of unicellular ascomycete and basidiomycete genera, which are known to undergo some form of sexual reproduction during their life-cycle, which culminates in the formation of ascospores and basidiospores, respectively. Also included are a number of forms that reproduce only by asexual means, and these are known as "imperfect yeasts", and are classified in the Deuteromycetes. Over the years, many of the latter have been shown to be non-sexual stages of the former two types. In a major, somewhat controversial, taxonomic work, Lodder[1] defined yeasts as "those fungi whose predominant growth form is unicellular", a statement that implies, as is the case, that some species can produce transient filamentous (hyphal) outgrowths.

Emil Hansen's first classification of yeasts, which appeared around the dawn of the twentieth century, only made a distinction between those that produced sexual spores, and those that did not (referred to as "sporogenous" and "asporogenous", respectively). We now know that Hansen's sporogenous yeasts belong to the phyla Ascomycota, and Basidiomycota, whilst the asporogenous forms are now classified as Fungi Imperfecti. Yeasts have been traditionally classified by means of conventional cultural techniques that determine a range of morphological, biochemical and physical properties. Criteria and methods for conducting such tests have been reviewed by Kreger-van Rij,[2] and include: cell shape and size; sporulation; fermentation and assimilation of different sugars; assimilation of a nitrogen source; growth-factor requirement, and resistance to cycloheximide. Workload for such tests is demanding and, in some cases, results for final identification may not be available for a fortnight. Progress in the development of simpler, more rapid methods was made during the late 1970s and the 1980s, and several easy-to-use diagnostic kits have been made available commercially. Such kits, however, are mainly applicable to a limited range of yeasts of industrial and medical significance.

The genus *Saccharomyces* has undergone innumerable taxonomic changes over the years, something that has not necessarily gone down well with the folk entrusted to use yeast in an industrial context (*e.g.* brewers, distillers and winemakers), but, with the development of new techniques, the science of taxonomy is constantly improving. That being said, it is not uncommon to find brewers, for example, using a long defunct taxonomic system to describe their yeasts.

A modern fungal taxonomy would have us believe that ethanolic fermentations are carried out by strains of the yeast *Saccharomyces cerevisiae*, regardless of whether the end-product is ale, lager, wine or cider. Over the years, there have been a variety of specific synonyms given to certain industrial strains of this fungus that have been used to produce specific alcoholic beverages, and some of these names have elevated the organism concerned to species status. Examples are *Saccharomyces carlsbergensis*/*Saccharomyces uvarum* for the bottom-fermenting yeast used in lager beer production, and *Saccharomyces ellipsoideus* for the wine yeast. Such names have now largely disappeared from the literature, and are only used by microbiologists in the specialized fields in which they have been historically relevant.

Bearing the above in mind, Naumov,[3] in a novel approach to the taxonomy of *S. cerevisiae*, proposed six cultivars:

- "Cerevisiae", for brewer's top-fermenting yeasts, and baker's and distiller's strains;
- "Ellipsoideus", for strains from primary wine fermentations;
- "Cheresanus", for strains from secondary wine fermentations (*e.g.* those forming a film on sherry during oxidation of ethanol);
- "Oviformis", for wine yeasts that do not ferment galactose, and that are resistant to high concentrations of ethanol and sulfites;
- "Diastaticus", for strains able to metabolize dextrins (*e.g.* soluble starch);
- "Logos", for strains able to ferment melibiose.

In their monumental study, Barnett *et al.*[4] report 678 recognized yeast species (16 *Saccharomyces* spp.) for which cultures and the results of standard physiological tests were available. The same work provides a register of almost 4000 taxonomic names that have been used at various times, since Hansen's work, together with their provenances and synonyms. The register is important, because some yeast names have changed so regularly that it is difficult for non-systematists to know which species is being referred to in the various publications. Barnett[5] has also written a succinct appraisal of yeast taxonomy. Barnett *et al.*[4] used the following scheme for classifying *Saccharomyces cerevisiae* Meyen *ex* E.C. Hansen (it should be remembered that differing suffixes can be applied to these taxonomic groups):

KINGDOM: Fungi
PHYLUM: Ascomycota
CLASS: Hemiascomycetes
ORDER: Saccharomycetales
FAMILY: Saccharomycetaceae
GENUS: Saccharomyces
SPECIES: Saccharomyces cerevisiae

The application of gene sequence analyses to yeast systematics has highlighted a conflict between placement of species on gene trees and their classification based on phenotype. According to Kurtzman and Robnett,[6] there is now is now a widespread pattern of disparity between phenotype and genotype as a means for classifying yeasts, and these differences have been demonstrated from analyses of 18S ribosomal DNA (rDNA),[7] internal transcribed spacer (ITS) rDNA,[8] 26S rDNA,[9] and cytochrome oxidase II (COX II),[10] leaving little doubt that phenotype is a poor predictor of genetic relationships among species. They also made the point that, at that time, most molecular comparisons had been made from single genes and it was unclear whether the relationships portrayed were representative of the evolutionary history of the organisms themselves. In addition, Kurtzman and Robnett maintained that basal branches in phylogenetic trees derived from single genes were often only weakly supported, resulting in uncertain relationships among more divergent species. This was particularly evident in studies of the "*Saccharomyces* complex", where 18S rDNA and 26S rDNA comparisons showed that many genera were incorrectly circumscribed. Accordingly, these two workers subjected 75 yeasts in the "*Saccharomyces* complex" to multi-gene sequence analyses, and placed them into 14 clades. In many cases, these clades did not correspond to the then circumscribed genera. The *S. cerevisiae* clade (Clade 1) comprised the *Saccharomyces sensu stricto* species proved to be phylogentically separate from the *Saccharomyces sensu lato* species.

According to late-twentieth century taxonomic reviews,[7,11] there are 14 species in the genus *Sacharomyces*, and these are still classified into the three groups previously established by van der Walt.[12] The *S. cerevisiae sensu lato* complex includes *S. barnettii*, *S. castellii*, *S. dairenensis*, *S. exiguus*, *S. rosinii*, *S. servazzii*, *S. spencerorum*, *S. transvaalensis* and *S. unisporus*. The second group comprises *S. kluyveri*. The third group, which comprises species of biotechnological interest, consists of the *Saccharomyces sensu stricto* complex, comprising four different taxa: the domesticated *S. bayanus*, *S. cerevisiae* and *S. pastorianus*, all associated with the fermentation industry, and *S. paradoxus*, the only *Saccharomyces sensu stricto* species typically isolated from natural habitats (insects, tree exudates, *etc.*). At the start of this century, three new species isolated from natural habitats, *S. cariocanus*, *S. kudriavzevii* and *S. mikatae*, were established as new taxa included in the *Saccharomyces sensu stricto* group.[13]

As Scannell *et al.*[14] reported, the number of extant fungal species is thought to be in the millions, although only around 80,000 have thus far been described. The known species are normally divided into five phyla: Ascomycota, Basidiomycota, Glomeromycota, Zygomycota and Chytridomycota, although some of these may not be monophyletic.[15] The largest phylum, Ascomycota, is defined by the production of the ascus, a specialized fruiting structure that surrounds the spores (ascospores) formed during meiosis.

The Ascomycota diverged from the Basidiomycota somewhere in the region of 741–1195 Mya,[16] and the monophyly of the Ascomycota is well supported by molecular data. There are two main subphyla within the Ascomycota: the Pezizomycotina (which include s hyphal fungi such as *Neurospora crassa*) and the Saccharomycotina, also referred to as Hemiascomycetes, which includes yeasts such as *Candida albicans* and *Saccharomyces cerevisiae*. These two subphyla diverged about 798–1166 Mya,[16] shortly after the above-mentioned split from the Basidiomycota. Recent molecular evidence showing that *Schizosaccharomyces pombe* is an outcrop to both of these taxa has prompted the proposal of a third ascomycete class, the Archiascomycotina.[17] Additional molecular sequence information has supported this new classification, and indicates that unicellular yeasts have evolved from multi-cellular (hyphal) progenitors more than once.

More than 1000 species of Saccharomycotina have been described, and their greatest centre of biodiversity would appear to be in insect guts. It is likely that thousands more species remain to be discovered from this environment. All of the Saccharomycotina genomes sequenced thus far fall into three clusters (see Scannell *et al.*, 2007[14]). The first cluster is comprised primarily of species from the genera *Saccharomyces* and *Kluyveromyces*, and is now generally referred to as the "*Saccharomyces* complex". The second cluster consists of species that translate CTG codons as serine rather than leucine, a reassignment that occurred more than 170 Mya.[18,19] This cluster includes *Candida* species and yeasts such as *Debaryomyces hansenii* and *Lodderomyces elongisporus*. The only sequenced member of the third cluster is *Yarrowia lipolytica*.[20]

2.1.2 Cell Structure

S. cerevisiae cells are generally ellipsoidal in shape, ranging from 5–10 μm at the large diameter, and 1–7 μm at the small diameter, the cells increasing in size with age. Mean cell volumes are 29 μm^3 for a haploid cell, and 55 μm^3 for a diploid cell. Yeast cells show most of the structural and functional features of higher eukaryotic cells, and have been used as a useful model for eukaryotic cell biology. The main features of a "typical" cell are illustrated in Figure 2.1, whilst the molecular composition of a freeze-dried yeast is given in Table 2.1.

The macromolecular composition of a similar dried organism would be as follows: moisture, 2–5%; protein, 42–46%; carbohydrate, 30–37%; nucleic acid, 6–8%; lipids, 4–5%; minerals, 7–8%. From the data presented in Table 2.1, Rosen gave this particular yeast the approximate molecular formula of:

$$C_{4.02} H_{6.5} O_{2.11} N_{0.43} P_{0.03}$$

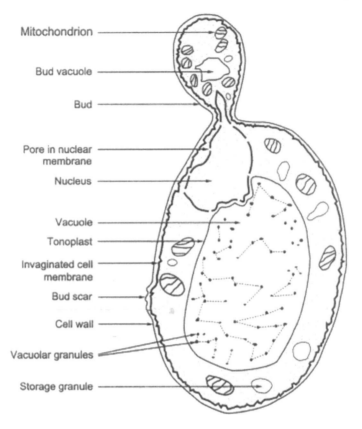

Mitochondrion

Bud vacuole

Bud

Pore in nuclear
membrane

Nucleus

Vacuole

Tonoplast

Invaginated cell
membrane

Bud scar

Cell wall

Vacuolar granules

Storage granule

Figure 2.1   Diagrammatic representation of a section of a budding yeast cell as
seen under the electron microscope (after Hornsey, 2007[21]).

Table 2.1   Molecular composition of dried wine yeast (after Rosen[22]).

Component	% of dry weight
Carbon	48.2
Hydrogen	6.5
Oxygen	33.8
Nitrogen	6.0
Phosphorus	1.0
Magnesium	0.1
Calcium	0.04
Potassium	2.1
Sulfur	0.01
Iron	0.005

The carbohydrate fraction of yeast includes structural components, such as the cell wall (majority), and compounds concerned with food storage, and/or stress resistance, such as glycogen and trehalose. Quite large differences may be observed in the concentrations of individual minerals during the various stages of the cell cycle, and, of course, the nucleic acid complement of cells will vary according to their ploidy. Some 75–80% of the actively growing yeast cell consists of water, much of which forms the basis of the cytoplasm (cytosol). The cell is surrounded by a rigid envelope, the cell wall, which comprises some 20–25% of the wet weight of the cell, and which determines the shape of the cell, provides a barrier against the high osmotic pressure in the cytoplasm and generally maintains the integrity of the cell. The wall overlies the cytoplasmic membrane (plasmalemma), and is separated from it by a discrete gap, called the periplasmic space. The cell wall, periplasmic space and plasmalemma are often referred to as the "cell envelope", and constitute some 15–25% of the total volume of the cell. The yeast cell wall is a complex extracellular organelle, capable of dynamic change in response to changes in external conditions, and to different stages of the life-cycle. Rigid the yeast cell wall might be; inert it certainly isn't. The wall can be 150–300 nm thick, according to strain and growth conditions, and is metabolically active, containing enzymes capable of facilitating the transfer of macromolecules into the cytosol. Under the electron microscope, the yeast cell membrane appears 7–10 nm in thickness, with the periplasmic space being 35–45 Å thick.

As a rule, plant cell walls are constructed of different polysaccharide moieties to which a variety of proteins are attached, and consist of two basic components: a malleable gel-like substance, and a fibrous matrix, which provides great strength. In most fungi, the fibrous component is chitin, a linear polymer of β-1,4 linked *N*-acetyl glucosamine (GlcNAc) residues. *S. cerevisiae* cells contain relatively little chitin (*ca.* 1% dry weight), most of it being concentrated at the bud neck, where it plays a role in cytokinesis (in this case, budding). In yeasts, the basic fibrous network is based upon 1,3-β-glucan, which accounts for some 40% of the cell wall dry weight, and is mainly responsible for the unusually high mechanical stability of the wall. 1,3-β-Glucan is synthesized at the cell surface, and consists of linear chains, with an average of

1500 glucose units. When viewed under the electron-microscope, it appears to exist in two forms: fibrous and amorphous. The former is responsible for the shape and rigidity of the wall, whilst the latter contributes elasticity, and acts as an anchor for the outer layer of mannoproteins (glycoproteins). The fibrous glucan component is insoluble in water, alkali and acetic acid, whereas the amorphous fraction is insoluble in water and acid, but soluble in alkali. Two other polysaccharides, 1,6-β-glucan and chitin, are linked to 1,3-β-glucan, and have functions other than preserving cellular shape and rigidity. 1,6-β-Glucan, which constitutes approximately 10% of the cell wall dry weight, consists of polymers with an average length of 350 glucose units, and plays a critical role in the cell wall architecture by linking cell wall components together, and by anchoring mannoproteins. 1,3-β-Glucan is a sparsely branched polymer, whereas 1,6-β-glucan is much branched. Most of the mannosylated proteins are located externally to the 1,3-β-glucan network, resulting in a discrete outer "mannan" layer, which may comprise 25–30% of the cell wall. The degree of glycosylation in the mannoproteins varies, one group, for example, consisting of 90% mannose and 10% peptide, and their molecular weight varies accordingly (between 20 and 450 kDa). Some mannoproteins are purely structural, and are anchored to the wall, whilst others manifest themselves as extracellular enzymes, such as invertase.

Small quantities of lipid and inorganic phosphate are also incorporated into the wall matrix. The reported lipid content in the wall of *S. cerevisiae* varies from 2–15%, but, in truth much of this may arise by contamination by membrane lipid. In a healthy yeast cell, the cell wall surface carries a net negative charge, which is attributable to phosphate chains located on the outer manno-protein layer. The charge can be qualitatively demonstrated by alcian blue staining, and quantified by ion-exchange or electro-phoretic (zeta-potential) methods. The cell surface is also hydro-phobic, a property conferred upon it by lipids, again located in the outer wall complex. As we have said, a number of important enzymes, such as invertase, have been reported as being associated with the cell wall, although some of these are actually located in the periplasmic space.

To partly illustrate the importance of the cell wall, consider, for a moment, the problems and stresses confronting yeast cells during the course of a fermentation. Initially, they find themselves in a nutrient-rich environment with a lowish pH and a high sugar content. The latter, in particular, presents a considerable osmotic challenge. By the end of the fermentation, the cells find themselves in a more osmotically friendly environment, but now have the problem of having to contend with ethanol levels, which can be as high as 15% ABV (toxic to most organisms). In addition, some of the nutrients may now start to become depleted and, depending on how well the fermentation can be attemperated, there may well have been considerable temperature fluctuations. The yeast cell wall stands between all of these environmental changes and the cytoplasm and its organelles within the cell membrane. The importance of the wall to the yeast may be gauged by the fact that some 20% of *S. cerevisiae* genes control functions that are in some way related to cell wall biogenesis.

Most of the lipids in the cell are structural components of the plasmalemma, and the intracellular membrane-bound organelles. When viewed under the electron microscope, the cell membrane has a series of invaginations on its outer surface, which become most prominent during the stationary phase of growth. In frozen-etched surface view these invaginations can be seen as elongated folds, some 4 μm in length. Also visible are clusters of hexagonally arranged particles, *ca.* 180 Å in diameter. The role of these invaginations has not yet been fully elucidated. The membrane consists of lipids and proteins in approximately equal proportions, the proteins being functional rather than structural.[23] Membrane proteins are involved in the regulation of solute transport, and include enzymes that mediate cell wall synthesis, as well as constituting the ATPase responsible for maintaining the plasma membrane proton-motive force. Intriguingly, the enzymes for synthesizing membrane lipids are located elsewhere, it being known, for instance, that the initial steps of sterol synthesis occur in the mitochondria (as do some enzymes associated with sterol degradation). The major lipid components of the membrane are phospholipids and sterols, and it is these compounds that are responsible for its hydrophobicity. The principal phospholipids are phosphotidylinositol, phosphotidylserine, phosphotidylcholine and phosphotidylethanolamine, whilst the main sterol is ergosterol. The last named can, if necessary, be added to the early stages of a fermentation as a substitute for the oxygen requirement during the synthesis of membrane components. Zymosterol is also a relatively abundant sterol in the yeast cell membrane. Sterols are orientated in perpendicular fashion within the membrane, between the chains of other lipids, and modulate its fluidity.

The periplasmic space (periplasm) is not strictly an organelle, but is more than a mere cavity between wall and membrane. Through the periplasm there is a dynamic migration of cellular components passing to the cell from the external medium (must) mixing with those moving out of the cell as a consequence of metabolism (*e.g.* ethanol). It appears that a number of enzymes are "stored" here, most significantly invertase and acid phosphatase. The protein (mannoprotein) concentration in the periplasm is very high, much higher than would be needed if the protein was for purely enzymatic purposes, and it has been suggested that the resultant gel-like consistency provides a "protective" layer for the plasmalemma.

The yeast cytoplasm (cytosol) is an acidic (pH 5.25) colloidal fluid, mainly containing ions and low/intermediate molecular weight organic compounds and soluble macromolecules (*e.g.* enzyme proteins and glycogen). The main cytosolic enzymes are those involved in glycolysis, fatty acid synthesis and protein synthesis (some). The internal stability and structural organization of the cell are guaranteed by the cytoskeleton, which is composed of microtubules (tubulin) and microfilaments (actin). These are dynamic structures which fulfil their function by regulated assembly and disassembly of protein sub-units. Thus, α- and β-tubulin monomers polymerize as heterodimers to give microtubules, while globular monomers of G-actin polymerize into double-stranded microfilaments of F-actin. Microtubules and microfilaments are important in several cell reproductive processes, including mitosis, meiosis, septa formation and organelle motility. The cytoskeleton is invisible using conventional microscopy, but can be demonstrated by specialized electron microscope techniques. In the1980s, it was shown that yeast had genes for actin and tubulin, and actin and tubulin proteins behaved like their mammalian counterparts. The main protein in the actin cytoskeleton is 86% identical to mammalian actins. It has been shown that the actin cytoskeleton consisted of two filament-based structures: the actin cortical patch and the actin cables. The actin cortical patches show a polarized distribution that changes during the cell cycle: first they appear at the incipient bud site, suggesting a role in bud emergence; soon thereafter they are found within the growing bud, indicating a role in bud growth; and late in the cell cycle they reorganize into two rings in the neck, where they are believed to be involved in septation and cytokinesis. Electron microscopy has shown actin cortical patches to be invaginations of the plasma-lemma around which actin filaments and actin-associated proteins are organized. It has also been shown that sub-sets of actin cortical patches can move at speeds of up to 1 μm/s. The actin cables, which consist of bundled actin filaments, were observed to run along the long axis of budding cells. Such information fits in well with the understanding that actin is involved in: polarized cell growth; dynamic reorganization of the cell cortex; membrane trafficking at the cell cortex; and organelle segregation at cell division (*e.g.* actin filaments direct cell growth to the emergent bud). Yeast has a single actin gene; *ACT1*. The actin cytoskeleton disassembles in response to osmotic stress, and is induced to re-assemble only after osmotic balance has been restored.

Yeast mitochondria have been the subject of much intensive research, and the literature describing the structure and function of these organelles is vast.[24] As in other eukaryotes, they contain their own ribosomes (70S), which are different from cytoplasmic (80S) ribosomes. Some mitochondria are distributed freely in the cytoplasm, whilst others are located on fragments of endoplasmic reticulum (ER) within the cytoplasm (called "rough ER"). At certain phases of growth, yeast mitochondria can comprise around 12% of the total cell volume. These organelles contain autonomous self-replicating DNA and systems for protein synthesis. Mitochondrial DNA (mtDNA) was not included in the yeast genome project, but material from the same haploid strain used in that work has since been sequenced.[25] Weighing in at 86 kb, mtDNA is much less substantial than the smallest yeast chromosome, and represents only some 0.5% of the yeast genome. The mitochondrial genome codes for just 5% of all mitochondrial proteins (in fact, it codes for only about 25 identified proteins, and 7 hypothetical proteins), the remainder being coded for by nuclear genes, some 300–400 proteins having to be imported. The mitochondrial genome is enriched, approximately ten-fold, by genes that code for transfer RNA. The significance of mitochondria is amply illustrated by the fact that some 8–10% of nuclear genes are involved in their biogenesis.

Classically, the mitochondrion is known as the site of the respiratory chain leading to ATP formation. The electron transport chain generates ATP as per the chemi-osmotic theory of Mitchell,[26] which confirms that the components of the electron transport chain are located in the inner membrane of the mitochondrion, such that protons are translocated from the inner to the outer side as electrons pass down the chain. A very readable account of the history of scientific research into the main respiratory pathway in yeast appeared in 2003.[27]

Several other enzyme systems are located within the mitochondria, in the inter-membrane space (which contains adenylate kinase) and on the outer membrane, where there are some enzymes associated with phospholipid synthesis. Unlike the mitochondria of higher eukaryotes, those in the yeast lack the facility for β-oxidation. As a consequence, the sites of fatty acid degradation in *S. cerevisiae* are the peroxisomes, small membrane-bound bodies that also exhibit catalase and glycollate cycle activity. Peroxisomes proliferate in the cytosol when oleic acid is present in the growth medium, but are repressed during growth on glucose-based media. Under the light microscope, the most obvious features of yeast cells are the vacuoles, whose size and number vary during the growth cycle. Under certain conditions of growth, there is one large vacuole present in the cell; under others they are much more dynamic. The vacuolar membrane (tonoplast) is similar in structure to the plasmalemma, but is more elastic, presumably to accommodate frequent vacuole shrinkage and expansion. Vacuoles have two main functions: firstly, they serve as stores of nutrients; secondly, they provide a site for the breakdown of certain macromolecules, particularly proteins, thus releasing intermediates for other metabolic pathways. Accordingly, amino acids are major soluble constituents of vacuoles, as are proteinases (such as carboxypeptidases and aminopeptidases). Vacuoles also serve as a food store for inorganic phosphorus, where the element exists as a linear polymer of polyphosphate linked by high-energy bonds.

The vacuoles represent only a fraction of a complex, internal system of membranes within the cytoplasm of the cell. There are other membrane-bound organelles, as well, such as the endoplasmic reticulum, the Golgi apparatus (dictyosomes) and numerous vesicles, both secretory and endocytic. The ER connects the plasmalemma with the nuclear membrane and, in effect, divides the cytoplasm of the cell into two main areas; that which is enclosed within the membrane system, and that which is exterior to it (the cytoplasmic matrix).

The Golgi complex consists of a series of stacked membranes and associated vesicles, which, like the ER, is highly dynamic. Both organelles form part of a secretory pathway, by which proteins are sorted and trafficked back and forth to the plasmalemma, or vacuole. Dictyosomes are implicated in cell wall synthesis and, at certain stages of the cell cycle, they are difficult to distinguish from ER fragments.

After the vacuole, the second largest organelle is the nucleus, which is typically double unit membrane-bound, and littered with nuclear pores (now called nuclear pore complexes, NPCs) that control the passage of molecules in and out of the nucleus. The yeast nucleus is some 0.1–2.0 μm in diameter, whilst the pores, which are dynamic entities, are in the order of 0.1 μm in diameter. Nuclear pores facilitate the exchange of low molecular weight proteins between nucleus and cytosol. The nuclear membrane, which, as we have stated, is connected to the ER, is unusual inasmuch as it does not disappear during cell division (mitosis), as it does in other eukaryotes. Within the nucleus there is a darker, more dense area called the nucleolus, which disappears during mitosis, and reforms during interphase. Although it is seen as a distinct region within the nucleus, it is not strictly delimited from the rest of the nuclear sap by anything like a membrane. The nucleolus is rich in RNA and protein, and is the site of the synthesis of ribosomal RNA, and the synthesis and organization of cytoplasmic (80S) ribosomes. High-resolution electron microscopy indicates that there are two distinct regions to the nucleolus; one particulate, one fibrillar. A concise survey of the milestones in yeast cytology, abundantly referenced, was provided by Barnett and Robinow.[28,29]

There is relatively little nuclear DNA in yeasts, as compared to other eukaryotes (*ca.* 14,000 kb in a haploid strain), and this has, inevitably, made the study of their chromosomes, under the light microscope, a difficult process. Each chromosome contains a single molecule of DNA, which is associated with an equal mass of proteins. Collectively, the DNA with its associated proteins is called chromatin. Most of the protein consists of multiple copies of five kinds of histone. These are basic proteins bristling with positively charged arginine and lysine residues. Both arginine and lysine have a free amino group on their R group, which attracts protons (H$^+$), giving them a positive charge, which enables them to bind to the negatively charged phosphate groups of DNA. Chromatin also contains small amounts of a wide variety of non-histone proteins, most of which are transcription factors, and are not permanently associated with DNA.

The number of "sets" of chromosomes is referred to as the "ploidy", and most laboratory strains of *S. cerevisiae* usually have one set of chromosomes, *i.e.* are haploid (n), or have two sets of chromosomes (diploid, 2n). Most industrial ("domesticated") yeasts, on the other hand, are polyploid, usually triploid (3n) or tetraploid (4n), although, in some strains, the number of copies of chromosomes is not necessarily a perfect multiple of the "n" number. If the latter occurs, the condition is known as "aneuploidy", and the phenomenon allows for extra, or reduced, numbers of chromosomes to be present. Fundamentally, significant differences have been shown between specific chromosomes of laboratory haploid strains and the chromosomes of commercial strains. It is often difficult to ascertain the ploidy of industrial strains but, in one study, the ploidy of wine strains was found to be 1.9n,[30] the results being obtained by using flow-cytometry to determine DNA content, rather than measurement of cellular DNA and comparison to a haploid reference strain. The copy number of each individual chromosome in polyploid (or aneuploid) strains is not necessarily identical. The recent technique of chromosomal fingerprinting, or karyotyping, has shown differences in chromosome size (called polymorphism) to be widespread. The *S. cerevisiae* nucleus is particularly suitable to karyotyping as it has 16 chromosomes, which range considerably in size from 230 kb (chromosome I) to *ca.* 1.5 Mb (chromosome IV). As we see below, *S. cerevisiae* was the first eukaryotic genome to be completely sequenced, work commencing in 1989, and ending with the publication of the sequence on 24th April 1996 by Goffeau *et al.*[31]

Several non-chromosomal genetic elements may be present in the yeast nucleus: 1) 2 μm DNA, a stably maintained circular DNA plasmid, which replicates once during the S-phase of the cell cycle; 2) double-stranded RNA and linear DNA, found in "killer" strains. They harbour the genes for toxins, which are deleterious to non-killer strains; 3) retrotransposons, mobile fragments called Ty elements (originally designated Ty1, Ty2, Ty3, Ty4).

Sexual reproduction in *S. cerevisiae*, which involves mating, or conjugation, and culminates in the production of ascospores, is relevant only to haploid or diploid cells, not generally to polyploid industrial strains, and so will not be detailed here. The interested reader is recommended to Sprague.[32]

All strains of *Saccharomyces* undergo a unique, asexual, asymmetric form of cell division called "budding", whereby one cell gives rise to two daughter cells that are genetically identical to their mother. Under suitable conditions, the daughter cells, themselves, become mother cells, and this sequence will continue until cell division is arrested. Cell division is not a never-ending process, and when nutrients become limiting, cells become senescent and eventually die. The cell cycle of yeast has been much studied, not from the point of view of its industrial significance, but because of the role that the organism plays as a "model" for eukaryotic cell research. It is now known that key elements of the eukaryotic cell cycle have been highly conserved during evolution, and so, what happens in yeast is peculiarly relevant to what happens in higher eukaryotes (man included). Add to this the fact that yeast has a small genome, is highly amenable to genetic manipulation and the cell cycle is completed in quick time, and one can appreciate its usefulness in the laboratory. A comprehensive review of the yeast cell cycle has been penned by Wheals.[33]

To appreciate fully the biological events that occur during an alcoholic fermentation it is necessary briefly to consider the cell cycle of *S. cerevisiae*, a representation of which is illustrated in Figure 2.2. In this organism, as in all eukaryotes, the mitotic cell cycle can be divided into four intervals, G1-, S-, G2- and M-phase. S-phase and M-phase are the periods when DNA replication and nuclear division (mitosis) occur, respectively, and overall control of cell division is achieved by regulating entry into these two phases. Interspersed between DNA replication and nuclear division are two "gap" periods, G1 and G2, which are of variable length and, during which, organelle production and normal cell processes, such as growth and development, take place. Phase G1 represents the period immediately after a daughter cell has separated from the mother, and is immediately prior to the onset of another cell division. As soon as the synthesis of DNA has been completed, the cell, by definition, enters the G2 phase, where it remains until mitosis commences. The major controlling event in the cell cycle, termed "START", occurs late in the G1 phase, and compels a cell to undergo division; there is no going back.[35] After the initiation of START, environmental factors, such as external stress and poor nutrient levels, are no longer able to prevent cell division. Completion of START requires that a cell grows beyond a minimum size, and commits that cell to a round of mitotic division, as opposed to alternative developmental fates, such as conjugation, or, for diploid cells, meiosis. Commitment to mitotic division sets in motion the pathways required for the initiation of DNA synthesis and the transition to S-phase. If nutrients are in sufficient supply (they are rarely a limiting factor in commercial fermentations), the S-phase is entered and, in addition to embarking upon DNA synthesis, an embryo bud is initiated. After the completion of DNA synthesis, and entry into phase G2, extensive bud growth ensues, and nuclear events culminate in mitosis (M-phase). The final steps of the cell cycle are septation, which separates the cytoplasm of mother and daughter cells (cytokinesis), and total separation of the two cells, although complete separation is not a prerequisite for continued reproduction. Ineffective mother-daughter cell separation can result in chains of cells being formed, which is a characteristic phenotype of some individual strains. Total separation of the bud leaves a permanent mark on the mother cell, the bud scar, and a temporary birth mark on the daughter cell. Some workers recognize a fifth phase to the cell cycle, G0, which represents cells in a state of "suspended animation" (see below).

The genesis of a bud begins with the selection of a locus on the surface of the mother cell, where outgrowth will later take place. Placement of the bud is governed by a set of rules, which can be predicted from the genotype; for example: haploid "α" and "a" bud axially, as do homozygous diploids, but heterozygous diploids ("αa") bud in a bipolar manner. When the markers are activated, they induce a localized assembly of an annular structure that marks the new bud initial; this includes a ring of chitin, another of 10 nm neck filaments and several additional proteins. The daughter bud normally emerges within this ring. A new-born yeast cell grows and deposits cell wall polysaccharides uniformly all over its surface until it attains a critical size, whence it initiates a bud;

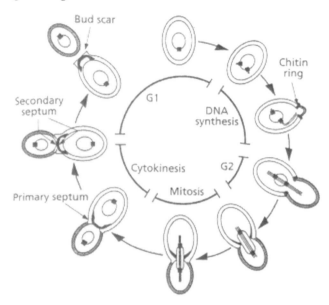

Figure 2.2   The cell cycle of *Sacch. cerevisiae* (after Harold[34]). (Reproduced by kind permission of the Society for General Microbiology.)

thereafter, deposition of new wall material is confined to the growing bud. Expansion of the bud wall is patterned in space and in time: to begin with, growth takes place chiefly at the tip of the forming bud, then it becomes uniformly distributed, and eventually expansion halts. Expansion of the wall is a secretory process, membrane-bound vesicles, which seem to originate in the Golgi apparatus, playing a prominent role. It is generally understood that these vesicles carry precursors and enzymes, such as mannosylated proteins, chitin synthase and β-glucan synthase, for wall biosynthesis. When the vesicles reach the plasmalemma, in the growth area, they fuse and undergo exocytosis. It appears that the prominent cables of actin microfilaments that course through the mother cell cytoplasm and reach the bud serve as "tracks" upon which wall precursor vesicles move toward the site of exocytosis. The cables seem to steer the vesicles to specific fusion sites, which appear as prominent plaques on the plasmalemma that contain actin and actin-binding proteins.

Figure 2.3 identifies processes known, or surmised, to contribute to bud enlargement, and represents a pictorial summary of the findings of a number of laboratories.

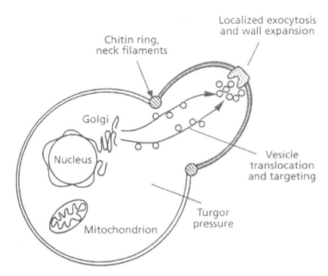

Figure 2.3   Diagrammatic representation of bud enlargement in *Sacch. cerevisiae* (after Harold[34]). (Reproduced by kind permission of the Society for General Microbiology.)

The nuclear events accompanying budding are now well known. Just prior to the formation of the bud primordium, the spindle-pole body (SPB), a centriole-equivalent situated on the nuclear membrane, undergoes duplication. The products separate and migrate, and this is followed by rotation of the nucleus so bringing the spindle-pole bodies into line with the axis of the emerging bud. The two spindle poles then nucleate development of the mitotic spindle, which guides separation of the duplicated chromosomes. Another set of microtubules sprouts from one spindle pole to reach deeper into the bud, and these are needed to guide the nucleus into the neck of the developing bud, and to guide other cytoplasmic organelles into the bud as well. Upon completion of mitosis, the maturing bud separates from its mother by ingrowth of the septum. Unlike the bulk of the cell wall, which is basically composed of β-glucan and mannoproteins, the septum is composed of chitin. Its formation is a lengthy process, requiring two distinct chitin synthases, and actually begins prior to the emergence of a visible bud. Septation culminates with cytokinesis, and splitting of the septum to liberate the daughter cell. At the end of the cell cycle, after septation, daughter cells are invariably smaller than their mother cells. As a consequence a daughter cell has to increase its cell volume to a certain level (called the "critical cell size") before it is capable of budding. As a result, the G1-phase of a daughter will be longer than that of its mother. As Figure 2.3 intimates, the growth (enlargement) of the bud is driven from within by the turgor pressure of the mother cell.

2.1.3 The Yeast Genome Project

With the completion of the genome sequence of *S. cerevisiae*, a finite number of about 6000 genes was shown to be sufficient for encoding all the proteins from a eukaryotic cell. Writing four years after the event, André Goffeau[36] summed up progress as follows:

"Less than a decade ago, the full molecular description of any eukaryotic cell seemed unreachable ... The large scientific community, combined with the relative simplicity of the cell, its unique physiological and genetic assets, the emergence of powerful global tools such as DNA hybridization arrays, proteomics and two-hybrid interaction analyses should rapidly make available millions of new biological data points. Within the next decade, the yeast cell, this humble servant of mankind, will be upgraded to the status of the first eukaryotic cell from which informatics Boolean networks will compile all molecular interactions between genes and gene products and all metabolic fluxes in hundreds of different physiological conditions."

The first chromosome ever sequenced was yeast chromosome III,[37] and then, between 1994 and 1996, all 15 other yeast chromosomes were completed and the details made publicly available by 16 different teams of the international community of "yeast scientists". The first complete nucleotide sequence of a eukaryotic genome, that of the yeast, was put on the web by MIPS (Munich Information Center for Protein Sequences) in a fully organized and annotated form, on 24th April 1996 – the first example of such data being published in this form. This was the start of, the now trendy, "publication on the web", and was quickly followed by a more traditional, peer-reviewed paper in *Nature*.[31] By the start of the new millennium, around 5000 scientists in more than 1000 laboratories (60% in the USA; 35% in Europe) had published nearly 7000 papers on yeast genes and genomics. This effort was only possible because of an unprecedented degree of co-operation between scientists worldwide.

The whole operation was stimulated by a consensus, reached in the USA in the mid-1980s, that there should be an extraordinary 15-year effort to sequence the human genome. A particularly significant feature of the US National Academy of Sciences report that announced this consensus was the recommendation that the genome sequences of a few other eukaryotes should be determined first. The genomes chosen were those of leading "model organisms", because their genomes are significantly smaller than that of the human, and because substantial and creditable molecular genetics research communities already existed to study them. Largely because of the efforts of these communities, it was already known that many of the proteins carrying out basic cellular functions are highly conserved among all the eukaryotes, suggesting that knowing the sequences of both the model genomes and the human genome would be an important path to understanding them both. Apart from the yeast, other organisms used were a nematode worm (*Caenorhabditis elegans*) and the fruit fly, *Drosophila melanogaster*. Subsequent events deemed that the yeast genome would become the pioneer eukaryotic genome, and the yeast research community would be the first to profit from knowledge of the complete genome sequence.

According to Dujon,[38] who summarized the lessons learned from the project, the rationale put forward in the late 1980s for sequencing the genome of *S. cerevisiae* was two-fold. Firstly, the organism has a high economic value (*e.g.* the fermentation industries), and is one of the primary model organisms for fundamental research into eukaryotic genetics and molecular biology. Secondly, its genome has many properties that made it an attractive target for early genome sequencing efforts: it has one of the smallest genome sizes among well-studied eukaryotes (only 14 million base pairs); a high gene density (72% of the genome codes for protein); few introns and little repetitive DNA – a combination of factors that made the sequencing relatively easy and cost-effective. The genome project was, therefore, both doable and worth doing.

When the sequencing was commenced, there were some 37 genetic markers on the map of chromosome III, whereas the complete sequence revealed the presence of some 170 genes specifying proteins of $\geqslant 100$ amino acids. The project involved the sequencing of some 12 million nucleotide bases, which encode about 6217 potential proteins (called "open reading frames" – ORFs), which account for almost 70% of the total sequence. By the beginning of the new millennium, only about half of these proteins were "known", *i.e.* were well characterized, biochemically or genetically. Of the remainder, some 30% had no recognized function, and are described as "orphan genes".

The yeasts used in the genome project were closely related haploid strains derived from the progenitor strain S288C. Strain EM93, the main progenitor strain of S288C, which is estimated to constitute 88% of the genome of this strain, was originally isolated from a rotting fig in California in the late 1930s, and is, therefore, arguably of some relevance to winemaking strains of yeast. Some genes that have been found in industrial strains cannot be found in the genome sequence, and it is assumed that these have become redundant in the "unreal" laboratory environment, and it is likely that there are many more apparently redundant genes that are necessary to cope with "natural" environments, such as the skin of the grape. The answer will only be known when (if) the genome of a natural winemaking strain of yeast is sequenced. It seems to be generally accepted that polyploidy is advantageous for domesticated industrial strains (since such strains are, at the very least, diploid), even though there is still little concrete evidence to that effect. The most extensive study on yeast thus far has been by Galitski *et al.*,[39] who suggest that there is a selective advantage in maintaining a number of copies of most chromosomes that are of utmost benefit to the cell.

The complete yeast genome can be searched and browsed at the Saccharomyces Gene Database (SGD) at Stanford University: http://genome-www.stanford.edu/Saccharomyces/. A list of useful addresses for yeast researchers has been provided by Brown.[40]

Elsewhere (p.111), the difference between respiration and fermentation is outlined, and the fact that some yeasts favour fermentation – even in the presence of oxygen – is mentioned. These yeasts, which include *S. cerevisiae* among their number, are now known to have undergone a process called whole-genome duplication (WGD) millions of years ago. This phenomenon was first reported by Wolfe and Shields,[41] who searched systematically for duplicated regions by using BLAST[42] amino-acid sequence similarity searches of all yeast proteins against one another, and plotted the results on dot matrices. Duplicate regions were visible as a diagonal series of "hits" with conserved gene orientation. The Dublin group then updated their map of duplicated chromosomal segments,[43] their new analyses being based on the more sensitive Smith-Waterman search method.[44]

Yeast contains several hundred duplicated genes, and these duplications probably occurred as a single event and, according to Langkjaer *et al.*,[45] that event probably took place before the *Saccharomyces* and *Kluyveromyces* lineages diverged from each other. Further evolution of each duplicated gene pair – such as specialization or differentiation of the two copies, or deletion of a single copy – has taken place independently throughout the evolution of these species.

Most genomes show a high degree of redundancy, which can arise from single-gene duplications, duplications of short chromosomal segments or of entire chromosomes or by duplication of the entire genome. All of these events are believed to play an important part in biological evolution. That gene duplication is an important source of evolutionary novelty was first propounded in the canonical work of Ohno in 1970,[46] who made the case for the importance of gene duplication and considered the various types of duplications and their potential for yielding novel functions.

Most duplications are of just a single gene, in which case the two genes derived from a duplication are said to be paralogous. One of the paralogues is often subsequently deleted from the genome, and it has been found that only rarely are both paralogues preserved because they differentiate and become functionally specialized. In some organisms (*e.g.* the fruit fly and fission yeast), most paralogous genes are dispersed. In contrast, duplicated genes in yeast (and in man) often occur in large segmental duplications in which members of the homologous gene pairs are located in the same order along the two distinct segments and are sporadically interspersed with unique genes. Owing to its small size, high gene density and a paucity of introns,[31] the *S. cerevisiae* genome is ideal for studying the origins of genome redundancy.

Whole-genome duplication (WGD) followed by massive gene loss and specialization has also long been postulated as a significant evolutionary tool, because duplicated genes can supply plentiful genetic raw material for the emergence of new functions through the force of mutation and natural selection. In principle, co-ordinate duplication of an entire genome may allow for large-scale adaptation to new environments, but polyploidy comes at the cost of major genomic instability, which persists until the genome returns to functionally normal ploidy through mutation, gene loss and genomic rearrangements. On the frequency of gene duplications, and how frequently they evolve new functions, Lynch and Conery[47] offered the following:

"Observations from the genomic databases for several eukaryotic species suggest that duplicate genes arise at a very high rate, on average 0.01 per gene per million years. Most duplicated genes experience a brief period of relaxed selection early in their history, with a moderate fraction of them evolving in an effectively neutral manner during this period. However, the vast majority of gene duplicates are silenced within a few million years, with a few survivors subsequently experiencing strong purifying selection. Although duplicate genes may only rarely evolve new functions, the stochastic silencing of such genes may play a significant role in the passive origin of new species."

Many duplicate genes have been found in *Saccharomyces cerevisiae*, and these often seem to be phenotypically redundant.[48] Subtle analysis of the genomic locations of paralogues has revealed the existence of ancestral duplication blocks, but their origin is controversial. Wolfe and Shields[41] interpreted the presence and distribution of such regions in the *S. cerevisiae* genome as supporting a model for WGD, but were criticized for relying on a small set of genes (*ca.* 8%) on which to base their model. Wolfe and Shields postulated that *S. cerevisiae* is a degenerate tetraploid resulting from a WGD that occurred after the divergence of *S. cerevisiae* and *Kluyveromyces*. Only a small fraction of the genes were subsequently retained in duplicate (most were deleted), and gene order was rearranged by many reciprocal translocations between chromosomes, and these workers suggested that tetraploidy may have facilitated the evolution of anaerobic fermentation in *Saccharomyces*.

We now know that the ancestral genome of the *Saccharomyces sensu stricto* complex has been subject to both whole-genome duplication, followed by massive sequence loss and divergence, and segmental duplication. In addition, the subtelomeric regions are subject to further duplications and rearrangements *via* ectopic exchanges. For a review of yeast evolution and comparative genomics see the paper by Liti and Louis.[49]

Jiang *et al.*[50] showed that the rate of evolution for the nuclear-encoded mitochondrial genes was greater in post-WGD species than in pre-WGD species. In addition, they found that codon usage bias was relaxed for these genes in post-WGD species. The codon usage pattern and the distribution of a particular transcription regulatory element suggested that the change to an efficient aerobic fermentation lifestyle in this lineage might have emerged after WGD, and between the divergence of *Kluyveromyces polysporus* and *Saccharomyces castellii* from their common ancestor. It was proposed that this new energy production strategy could have led to the relaxation of mitochondrial function in the yeast species concerned. The metabolic adaptations in organisms after WGD was the subject of study by van Hoek and Hogeweg.[51] Their model confirmed the hypothesis that WGD has been important in the adaptation of yeast to the new, glucose-rich environment that arose after the appearance of the angiosperms. The hypothesis had been put forward by Conant and Wolfe,[52] who, using a kinetic model of glycolysis and assuming a dosage effect for duplicated genes, showed that the duplicated genes could enhance glycolytic flux. They also showed that increased glycolytic flux could generate a fitness advantage in a glucose-rich environment.

WGD in yeast is estimated to have occurred in the *Saccharomyces sensu stricto* species complex around 100 Mya,[41,53–55] and has been proposed to have led to the evolution of an efficient fermentation system in this lineage.[56] This hypothesis is based on the observation that with plentiful glucose, most post-WGD yeast species, such as *S. cerevisiae* and *Schizosaccharomyces pombe*, primarily carry out fermentation to generate energy even under aerobic conditions. Yeasts, which accumulate ethanol, even in the presence of dioxygen, are called Crabtree-positive, whereas those yeasts that degrade sugars to CO_2, such as *Kluyveromyces lactis* and *Candida albicans*, are designated as Crabtree-negative. Merico *et al.*[57] studied the fermentative capacity, the ability to grow in anaerobic conditions and the occurrence of the petite phenotype (all characteristic of *S. cerevisiae*) of over 40 strains in the *Saccharomyces* complex, with a view to providing a basis for assessing how these traits might have evolved in the 150 million years since these yeasts emerged. In 2008, a team from Cornell University[58] provided new evidence for the WGD-fermentation hypothesis and new insights into the relationship between the genome duplication and the evolution of new lifestyles in eukaryotes.

2.1.4 Yeast Genetics

It was during the 1930s and 1940s that Øjvind Winge (1886–1964), working in the Carlsberg Laboratory in Copenhagen, pioneered research on the genetics of yeast. Often called the "Father of Yeast Genetics", he was director of the Physiology Department when he carried out his ground-breaking research on single cells and spores. Although his scientific career extended back to 1907, Winge actually started working on yeast in 1933, and used strains that had been stored in 10% sucrose many years earlier (some dating back to 1886) by Emil Hansen.[59] Along with Lausten and Roberts, Winge elegantly elucidated the life-cycle of several yeasts, and commenced the genetic analyses of sucrose and maltose fermentation.[60] In 1949, the same workers had characterized the first gene for homothallism.[61] Even in those far-off days, Winge was able to create genetic crosses that combined industrially important fermentation characters into a single strain.

Life-cycles, cytology and hybridization of yeast, in relation to genetics, were reviewed by Carl Lindgren in 1945.[62] The early days of yeast genetics were the subject of an excellent review by Roman[63] and, more latterly, the foundations of the subject have been reviewed by Barnett.[64] The importance of this aspect of zymology has fostered a huge volume of literature, and the following are of relevance: Hall and Linder;[65] Johnston;[66] Wheals *et al.*;[67] Brown and Tuite;[68] Weissman *et al.*[69] There are two types of geneticists – those whose aim is to study the nature of inheritance and those who use genetics as a tool to understand general biological problems. Yeasts are of vital importance in both of these research areas.

2.1.5 "Tame" Yeasts

When introduced into the laboratory, wild micro-organisms often undergo selection for "easier" growth requirements. This adaptation, and loss of selective pressures normally present in the wild, may have wide-ranging effects, such that the biology of a laboratory organism may no longer reflect that of wild populations. As such, these microbes become less relevant as biological models. The genome sequence of one laboratory strain, a derivative of S288c, as we have seen, was the first genome of a free-living eukaryote to be sequenced and, over the last dozen or so years, this genome has served as the reference for the species *S. cerevisiae*, and has catalysed the development of whole-genome approaches to biology. Despite frequent laboratory use of alternative strains, sequence information for *S. cerevisiae* beyond the domesticate S288c is, at best, fragmentary.

S288c was chosen because it possesses easy-to-work-with properties, such as minimal colony morphology switching, and consistent growth rates in glucose media. It is also non-flocculent. At several loci, S288c contains polymorphisms not found in natural isolates, and these have been seen as hallmarks of domestication.[70,71]

The S288C laboratory yeast strain was bred in the 1950s from wild and commercial strains, its genealogy being constructed by Mortimer and Johnston.[72] It apparently had six progenitor strains, with around 88% of its gene pool being contributed by one of them (EM93). Strain EM93 was isolated by Emil Mrak of the University of Califonia, Davis, in 1938, from rotting figs. Another progenitor strain, EM126, was isolated from the same environment the following year. Whether these two strains were part of the figs' flora or were spoilage organisms originating as commercial baking and/or brewing yeasts is unknown. Another S288c progenitor, NRRL-210, was isolated from rotting bananas from Costa Rica in 1942, and was originally classified (erroneously) as *S. microellipsoideus*.

Using the maximum likelihood (ML) method, Gu *et al.*[70] made a genome-wide comparison of the evolutionary rates in the lineages leading to strain S288C and a wild strain YJM 789, a clinical isolate from the lung of a patient with pneumonia.[73] They found that genes in the laboratory strain had tended to evolve faster than in the wild strain. The pattern of elevated evolution suggested that relaxation of selection intensity was the dominant underlying reason for this. Supporting this conclusion were the following observations: (1) the increases in non-synonymous evolutionary rate occurred for genes in all functional categories; (2) most of the synonymous evolutionary rate increases in S288C occurred in genes with strong codon usage bias; (3) genes under stronger negative selection had a larger increase in non-synonymous evolutionary rate; and (4) more genes with adaptive evolution were detected in the laboratory strain, but they did not account for the majority of the increased evolution. Overall, the results suggested that experimental and, possibly, the industrial manipulations of the laboratory strain of yeast could have had a significant effect on the genetic make-up of this model organism. Furthermore, it was implied that the evolution of model organisms away from their wild counterparts puts into question the relevance of "models", especially when extensive laboratory cultivation has occurred (the results also shed light on the evolution of livestock and crop species that have been under human cultivation for years).

2.2 ALCOHOL DEHYDROGENASES

Alcohol dehydrogenases (ADHs) are oxidoreductases that catalyse the reversible oxidation of alcohols to aldehydes or ketones, with the concomitant reduction of NAD^+ or $NADP^+$. These enzymes have been identified not only in yeasts, but also in several other eukaryotes, including vertebrates,[74] and even in some prokaryotes. Before the sequencing of the *S. cerevisiae* genome, five ADHs were known, but since then, *ADH6* and *ADH7* have also been identified and their enzyme products have been characterized. ADH enzymes can be sub-divided into at least three distinct enzyme superfamilies: medium chain (MDR), short chain dehydrogenases/reductases and iron-activated ADHs.

ADH1 (systematic name YOL086C) encodes for ADH1 (standard name Adh1p), which was the first pyridine nucleotide-dependent dehydrogenase to be crystallized. Determination of the primary structure of Adh1p in *S. cerevisae* marked the first case of gene cloning by functional complementation, and physico-chemical methods revealed that the protein had a molecular weight of 150 kDa. As long ago as 1964, Harris[75] found that the active enzyme contained four identical reactive sites and that, in all probability, it consisted of four similar, if not identical, polypeptide chains. Later on, it was shown that Ca^{2+} stabilizes Adh1p by preventing the dissociation of the reduced form of the enzyme and by preventing the unfolding of the oxidized form of the enzyme.

Adh2p, like Adh1p, is located in the yeast cytosol, and is encoded by *ADH2* (YMR303C). By the early 1980s, the structural genes *ADH1* and *ADH2* were identified genetically, cloned and their DNA sequences determined. High homology was evident at both the nucleotide sequence level (90%) and in their amino acid sequences (95%). The latter for Adh1 and Adh2 had only 22 differences out of 347 residues, with no differences in the groups directly involved in catalysis.[76]

The presence of Adh3p in respiratory deficient mutants proved that nuclear DNA encodes the synthesis of this mitochondrially bound enzyme[77] and, when the *ADH3* (YMR038W) gene was isolated and sequenced some ten years later,[78] the nucleotide sequence indicated a 73% and 74% identity with *ADH1* and *ADH2*, respectively. Wiesenfeld *et al.*[77] had assigned Adh3p its tetrameric structure, and the amino acid similarity of the predicted Adh3p polypeptide to Adh1p and Adh2p was 79% and 80%, respectively. Bakker *et al.*[79] reported that Adh3p is involved in a redox shuttle.

ADH4 (YGL256W) is the most distal marker on the left arm of chromosome VII, and both restriction and genetic analysis of the chromosome copy of *ADH4* indicated that it was situated near a telomere. Sequencing analysis of the gene revealed a long ORF that was not homologous to other yeast ADHs, and only distantly related to other characterized eukaryotic ADHs.[80] *ADH5* (YBR145W) was first identified through the sequencing of yeast chromosome II, and shares a 76%, 77% and 70% sequence identity with *ADH1*, *ADH2* and *ADH3*, respectively.[81] The *ADH6* (YMR218C) gene product was the first NADPH-dependent medium chain ADH to be characterized in yeast, and was also the first cinnamyl ADH and member of the MDR superfamily. The hetero-dimeric enzyme Adh6p consists of two 40 kDa subunits; one in the *apo* conformation and one in the *holo* conformation.[82] Adh7p has a 64% sequence identity with Adh6p and the purified *ADH7* (YCR105W) gene product is a homodimer whose reductase activity is about five-fold that of the dehydrogenase activity. Work carried out thus far places Adh7p in a family of enzymes structurally related to the cinnamyl ADHs.[83,84]

Initial biochemical data for Adh1p and Adh2p showed that the kinetic properties of both enzymes favoured ethanol production. Working on the sub-cellular localization of ADH activity in baker's yeast, Heick *et al.*[85] found that under conditions of high ethanol concentration and the efficient removal of acetaldehyde, both enzymes could function in the oxidation of ethanol. A few years later, Wills[86] reported that Adh1p was normally constitutive under laboratory conditions.

Wills produced mutants of yeast ADH that protected the cell against the poisonous aldehyde acrolein by increasing the NADH-NAD ratio. His specific selective procedure was based on two observations. The first was that petite yeasts (those unable to respire aerobically) synthesized only Adh1p, which is normally constitutive under laboratory conditions, has a high Michaelis–Menten constant (K_m) for ethanol (17,000–20,000 µM) and seems to be chiefly responsible for the production of ethanol from acetaldehyde in cells grown anaerobically. Adh2p has a much lower K_m for ethanol (600–800 µM), and is found only in aerobically grown grande yeast cells. The second observation was that grande cells lacking cytoplasmic ADH activity cannot be induced to become petite. They die instead, indicating that ADH activity is essential for the survival of petite cells. Reinforcing this conclusion, Wills found it impossible to select petite strains to yield cells lacking ADH activity and, in his view, it seemed likely that glycolysis could not proceed without the presence of ADH to catalyse the production of ethanol as a final step, although grande cells lacking ADH activity could survive by means of oxidative phosphorylation alone.

The kinetic properties of Adh1p optimize it as a catalyst to make ethanol from acetaldehyde[87] and, as the sugar concentration drops, Adh2p becomes de-repressed, and starts to oxidize ethanol to acetaldehyde (the paralogue has appropriate kinetic parameters for ethanol being its substrate).

2.2.1 Resurrection of Ancestral ADHs, and What This Can Tell Us

Dating events in the molecular record has always presented problems, mainly because, unlike radiochemical dating, morpho-logical change cannot follow exponential kinetics (radioisotopes decay *via* a first-order process, and we use *f* to represent the fraction of the initial atoms remaining), as it does not approach an end point. No known chemical process has rate properties that are comparable to those displayed by radioactive decay. Many chemical processes display first-order (or pseudo-first order) kinetics, but the rate constants for nearly all of these are much influenced by environmental factors (temperature, pH, *etc.*). Given this, it seemed unlikely at one time that one would be able to identify a chemical process in living systems that had sufficient first-order character to be useful for dating biological events, especially one reflected in DNA or protein sequences.[88]

As Thornton[89] outlined, ancient genes can now be reconstructed, expressed and functionally characterized, thanks to improved techniques for inferring and synthesizing ancestral sequences. Known as "ancestral gene resurrection", it is now possible to empirically test hypotheses about the function of genes from our evolutionary past. The practicalities of ancestral gene resurrection took some 30 years to evolve, for it was in 1963 that Linus Pauling and Emil Zuckerkandl[90] prophesied that it would be possible one day to infer the gene sequences of ancestral species, to "synthesize these presumed components of extinct organisms ... and study the physico-chemical properties of these molecules". Pauling and Zuckerkandl had simple models in mind when they proposed the construction of trees and reconstruction of ancestral sequences, which emerged as procedure of "maximum parsimony". According to the rule of parsimony, the best tree relating protein sequences to reconstructed ancestral sequences is the one that obtains the derived sequences from ancestral sequences with the smallest number of independent evolutionary events. All this, of course, was in the pre-genomic world! A number of software packages provide parsimonious reconstructions of ancestral states given a set of homologous sequences.

In the 1970s, many workers explored the possibility of constructing a molecular clock by counting replacements separating two sequences and assuming that the rate constant for amino acid replacement is invariant over time. Unfortunately, protein behaviours are too closely tied to the demands and constraints of natural selection. Amino acid replacement is faster or slower depending on how these change, making protein sequences irregular clocks at best. Thus, individual protein sequences generally serve as poor clocks, and it is difficult to correlate the molecular record for a specific protein family with the palaeontological record.

Accordingly, to minimize the effects of selective pressure as they construct molecular clocks, most workers now examine silent sites in a gene.[91] Because they do not change the coding sequence of a protein, nucleotide substitutions at silent sites cannot alter the behaviour of a protein. They are, therefore, most likely to be free of the selective pressures that cause protein clocks to "tick" irregularly.

Silent sites are of many types in the standard genetic code, and some offer better clocks than others. Most useful are silent sites in codon systems that are two-fold redundant, where exactly two codons encode the same amino acid. These codons are inter-converted by transitions: a pyrimidine replacing another pyrimidine, or a purine replacing another purine. When the amino acid itself is conserved, the divergence at such sites can be modelled as an "approach to equilibrium" kinetic process, just as radioactive decay, with the end point being the codon bias. The fraction of codon pairs that are preserved (f_2) can be calculated and, as Benner[92] explains, under certain circumstances f_2 can be treated as a molecular clock.

Then, in 1990, Benner and colleagues used the parsimony principle to infer the protein sequence of the highly conserved gene that encodes digestive ribonuclease from the last common ancestor of three organisms: swamp buffalo, river buffalo and ox, that lived 5–10 Mya.[93] At around the same time, the late Allan Wilson's laboratory at Berkeley, working independently, reconstructed ancestral lysozymes.[94] Subsequently, Benner's group resurrected a series of even more ancient ribonucleases from ancestral species dating up to 40 Mya on the artiodactyl phylogeny.[95]

The homologues ADH1 and ADH2 differ by 24 of 348 amino acids. Their last common ancestor, called ADH_A, had an unknown function, and was the subject of a study by Steve Benner's laboratory in Florida.[96] It was reasoned that if ADH_A existed in a yeast that made, but did not accumulate, ethanol, its physiological role would presumably have been the same as that of lactate dehydrogenase in mammals during anaerobic glycolysis: namely to recycle NADH generated by the oxidation of glyceraldehyde-3-phosphate. Lactate in human muscle is removed by the blood-stream, whereas ethanol would be lost by the yeast as it diffuses into the environment. If this is so, ADH_A should be optimized for ethanol synthesis, as is modern ADH1 (*i.e.* the kinetic behaviours of ADH_A should resemble those of modern ADH1 more than those of modern ADH2). Using the powerful tools then newly available, the Florida team, in effect, resurrected ADH_A.

Benner's laboratory cloned and sequenced 15 additional homologues of ADH from yeasts related to *S. cerevisiae*. They then constructed a maximum likelihood (ML) evolutionary tree using PAUP*4.0[97] to combine these with sequences already in the database. They then constructed ML sequences for ADH_A using both codon and amino acid models in the package for phylogenetic analysis by maximum likelihood (PAML).[98]

After rigorous analysis, it was inferred that the ancestral yeast did not have an ADH specialized for the consumption of ethanol, similar to modern ADH2, but rather had an ADH specialized for making ethanol (à la modern ADH1). This suggests that before the ADH1-ADH2 duplication, the ancestral yeast did not consume ethanol. This implies that the ancestral yeast also did not accumulate ethanol under aerobic conditions for future consumption and that the make-accumulate-consume strategy emerged after ADH1 and ADH2 diverged.

The assignment of a primitive function to ADH_A raised a broader historical question: did the ADH1-ADH2 duplication, and the accumulate-consume strategy that it presumably enabled, become fixed in response to a particular selective pressure? Two hypothetical situations were envisaged: the make-accumulate-consume strategy may have been driven by the emergence of the fleshy fruits of angiosperms that provided a food source worth defending – using ethanol accumulation. Alternatively, the strategy may have been driven by man selecting for yeasts that produce bulk (*i.e.* accumulated) ethanol.

By estimating the date of the ADH1-ADH2 duplication, it was possible to distinguish between the two above possibilities. We know that fleshy fruits evolved in the Cretaceous period, after the first angiosperms appeared in the fossil record 125 Mya,[99] but before the extinction of the dinosaurs 65 Mya.[100] From fossil records, Dilcher[101] puts the likely date for the prominence of fleshy fruits at around 80 Mya, during the age of the dinosaurs, and it is possible that overgrazing by these beasts could explain why flowering plants flourished.[102,103] Other genomes evidently indicate episodes of duplication at around this time, including those of the fruit-creating angiosperms and fruit flies, whose larvae feed on the yeast population in ripening fruit.[104,105]

The topology of the aforementioned ML evolutionary tree suggested that the Adh1-Adh2 duplication occurred before the divergence of the *sensu strictu* species of *Saccharomyces*, but after the divergence of *Saccharomyces* and *Kluyveromyces*. The exact date of the latter is unknown but is estimated to have occurred 80 ± 15 Mya,[106] a date that is consistent with a transition redundant exchange (TREx) clock, which exploits the fractional identity (f_2) of silent sites in conserved two-fold redundant codon systems to estimate the time since the divergence of two genes.[92]

Between pairs of presumed orthologues from *Saccharomyces* and *Kluyveromyces*, f_2 is typically 0.82, not much lower than the f_2 value separating ADH1 and ADH2 (0.85), but much lower than the paralogue pairs within the *Saccharomyces* genome that seem to have arisen by more recent duplication (~ 0.98).

Significantly, Benner's group found that ADH1 and ADH2 were not the only pair of paralogues with an f_2 value between 0.80 and 0.86.[95] Around 350 pairs of paralogues in the yeast genome shared at least 100 silent sites and diverged by less than 120 point-accepted placements per 100 sites, and the group identified 15 pairs with f_2 values between 0.80 and 0.86. These represented eight duplications that occurred around the time of the ADH1-ADH2 duplication (as long as it was assumed that f_2 values supported a molecular clock).

These eight duplications were not randomly distributed in the yeast genome, since six of them involved proteins that are concerned with the conversion of glucose to ethanol. Furthermore, the enzymes arising from the duplicates were those that seem, from expression analysis, to control flux from hexose sugar to ethanol.[107] These included proteins that import glucose, pyruvate decarboxylases that generate acetaldehyde from pyruvate, the transporter that imports thiamine for these decarboxylases and the ADHs (see Figure 2.4). As was reported, if the f_2 clock (within its expected variance) is assumed to date paralogues in yeast, this cluster suggested that several proteins other than ADH duplicated as part of the emergence of the new make-accumulate-consume strategy of yeast, at around the time that angiosperm fleshy fruits arose.

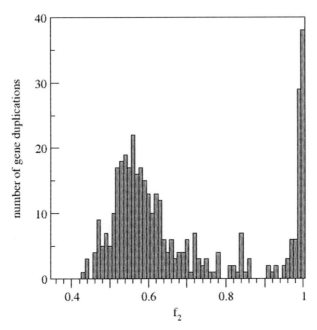

Figure 2.4 Number of duplicated gene pairs in the genome of *S. cerevisiae* vs. f_2 (from Benner *et al.*[95] - with permission).

The six duplications with f_2 values between 0.80 and 0.86, which have been proposed to be part of the make-accumulate-consume strategy, are not associated with one of the documented blocks of genes duplicated in ancient fungi, possibly as part of a WGD.[41,55]

The number of duplicated gene pairs in the genome of *S. cerevisiae versus* f_2 is shown by the histogram in Figure 2.4. The metric f_2 models divergence of silent positions in two-fold redundant codon systems *via* an approach-to-equilibrium kinetic process, and therefore acts as a logarithmic scale of time since the duplications occurred. Recent duplications are represented by bars at the right. Duplications that diverged so long ago that equilibrium at the silent sites has been reached are represented by bars where $f_2 \approx 0.55$. There are also episodes of gene duplication between the two extremes, including one at $f_2 \approx 0.84$. This represents the duplication, at ≈ 80 Mya, whereby yeast gained its ability to ferment sugars found in angiosperm fruits. Also noticeable are recent duplications of genes that enable yeast to speed DNA synthesis, protein synthesis and malt degradation, presumably representing yeast's relatively recent interaction with man.

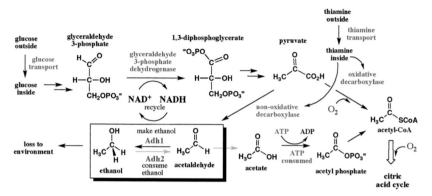

Figure 2.5 Pathway for yeast making, accumulating and consuming ethanol (from Thomson *et al.*[96] - with permission).

The biochemical pathway by which yeast makes, accumulates and then consumes ethanol is shown in Figure 2.5. Enzymes in red are associated with gene duplications that, according to the TREx clock, arose nearly contemporaneously. The make-accumulate-consume pathway is boxed. Shunting of the carbon atoms from pyruvate into (and then out of, blue arrows) ethanol is energy-expensive, consuming a molecule of ATP (green) for every molecule of ethanol generated. This ATP is not consumed if pyruvate is oxidatively decarboxylated directly to acetyl-coenzyme A to enter the citric acid directly (dashed arrow to the right). If dioxygen is available, the recycling of NADH does not need the acetaldehyde-to-ethanol reduction.

Because the energy for growth is provided by the glycolytic and fermentation pathways rather than by respiration during ethanologenesis in yeast, the organism does not need fully active mitochondria (sufficient energy being provided by the above-mentioned routes). Ihmels *et al.*[108] (2005) found that in *S. cerevisiae* the expression of genes encoding the mitochondrial ribosomal proteins (MRPs), those encoding cytoplasmic ribosomal proteins (CRPs) and those encoding the rRNA processing proteins were not co-ordinated. Instead MRP expression was correlated with the expression of stress genes induced during the slower respiratory growth in non-fermentable carbohydrate sources, such as glycerol and ethanol. In *Candida albicans*, however (aerobic and Crabtree-negative), the expression of genes encoding MRP, CRP and rRNA were all correlated.

Many different yeasts are capable of facultative fermentation. The fission yeast, *S. pombe*, for example, separated from the *Saccharomyce-Kluyveromyces* lineage at least 200 Mya and, like the yeast, is capable of aerobic alcoholic fermentation in the presence of excess sugars. It can also consume ethanol, but not as its sole carbon source, owing to the absence of a complete glyoxylate cycle rather than its inability to convert ethanol to acetaldehyde. *K. lactis*, on the other hand, is a poor producer of ethanol, but can efficiently use the compound as a sole source of carbon. It has duplicated the *ADH* gene(s) independently more than once during its evolutionary history, resulting in four modern ADHs. One of its genes, *KlADH4*, is induced by ethanol and not repressed by glucose and the corresponding enzyme is involved in the consumption of ethanol (arising from either the external medium, or previous intracellular production).

In terms of successful growth in a sugar-rich environment, *Saccharomyces* yeasts owe their competitiveness to a combination of several properties including rapid growth, efficient glucose repression, good ability to produce and consume ethanol and a tolerance for several environmental stresses (*e.g.* low oxygen levels). These properties are randomly distributed among different modern yeasts but are uniquely combined, perfected, regulated and co-ordinated through an efficient network in *S. cerevisiae* and its closest relatives. It is probably "efficient regulation" that is the most unique "invention" of *Saccharomyces* yeasts, and provides these organisms with their competitive advantages. It has also made them ideal companions for humans over the millennia, and encouraged their harnessing for industrial processes. The evolution of wine yeasts has been covered,[109] and how the yeast evolved to become a good brewer was explained by Piškur *et al.*[56] A review highlighting the achievements in genetic engineering of *S. cerevisiae* yeasts strains used in the food and beverage industry appeared in 2008.[110]

Generating ethanol from glucose in the presence of O_2, only to re-oxidize it, is energetically expensive. For every molecule of ethanol converted to acetyl-coenzyme A, a molecule of ATP is used. This ATP would not be "wasted" if pyruvate made initially from glucose were delivered directly to the citric acid cycle. It is inferred, therefore, that yeast must have a reason for rapidly converting the available sugar source in fruits to bulk ethanol in the presence of dioxygen.

One hypothesis to explain this "inefficiency" suggests that yeast, which is relatively resistant to the toxic effects of ethanol, may accumulate the latter to defend the carbohydrate resources in fruit from innumerable competing microbes. Apart from the physiological sensations resulting from ingestion, the preservative effects of ethanol have been employed by mankind for millennia. As we have seen, both the timing of *Adh* expression in *S. cerevisiae* and the properties of the expressed proteins are consistent with the above hypothesis.

2.3 *KLUYVEROMYCES MARXIANUS* (E.C. HANSEN) VAN DER WALT (1971)

This is a homothallic hemiascomycete yeast often encountered on cheese and other dairy products, notably yoghurt. As is mentioned elsewhere, the yeast is sometimes present as part of a "spontaneous fermentation" flora, especially in wineries. Occasionally, it can cause human infections, such as oesophagitis and vaginitis. The organism has been isolated from a wide variety of habitats with the resulting high metabolic diversity and fair degree of intraspecific polymorphism. The chromosome number of *K. marxianus* isolates varies from 6 to 12, with most strains containing 8. The asexual (anamorphic) state is generally accepted as being *Candida kefyr* (syn. *C. pseudotropicalis*).

The genus *Kluyveromyces* was established in 1956 by van der Walt[111] to classify a newly isolated (from soil), budding, fermentative yeast (*K. polysporus* – the type species). Characteristics of the new species were fragile asci containing multiple (up to 70) smooth spores with shapes varying from reniform to long oval. The following year, another soil species, *K. africanus*, was added (16 spores per ascus). The salient diagnostic feature of the asci of these fungi was their early evanescence at maturity, to release their ascospores. Van der Walt[112] later amended the diagnosis of the genus to include organisms producing fewer spores, usually four per ascus. As a result of this several *Saccharomyces* species were transferred to *Kluyveromyces*, and a number of taxonomic treatments subsequently appeared.

With their industrial potential, *Kluyveromyces* species have been the subject of much study since van der Walt's classic 1970 monograph.[113] The closeness of *K. marxianus* to *S. cerevisiae* can be judged by the fact that Hansen originally named it *S. marxianus*, and Campbell,[114] using serological data and numerical taxonomy, even suggested rejecting *Kluyveromyces* as a genus, and resurrected Hansen's nomen. Subsequent studies on nucleotide sequence homology, genetic analysis and chromosome number determined that *Kluveromyces* was later universally accepted, and a number of varieties of *K. marxianus* had been adopted by the time of Kreger-van Rij's tome.[2]

With some workers unhappy about the status quo, a number of other studies were undertaken. Sidenberg and Lachance,[115] for example, examined natural isolates from a variety of habitats and made a comparison on the basis of seven isoenzyme patterns. Some strains of *Candida*, considered to be *Kluyveromyces* anamorphs, were included in the study, and some relationships were confirmed. When the results of isoenzyme electrophoresis were compared with deoxyribonucleic acid reassociation data, however, the latter technique gave a more accurate view of relationships among *Kluyveromyces* species. A number of taxonomic changes were proposed as a result, including one for *K. marxianus*.

A taxonomic revision of the genus, based on nuclear DNA reassociation, has been produced,[116] and Fuson *et al.*[117] used DNA base composition and DNA base sequence relatedness for species delimitation. As a result, base composition values enabled species to be placed in three groups, and when these were further divided by comparing base sequences by using DNA/DNA renaturation experiments, two homology groups were identifiable. These two groups proved to be only distantly related to each other ($\leqslant 15\%$ DNA base sequence complementarity). The major developments in *Kluyveromyces systematics* since van der Walt's 1970 monograph were summarized by Lachance.[118]

A molecular systematic investigation of some ascomycetous genera, using 18S rRNA sequence analyses,[119] demonstrated the marked phylogenetic heterogeneity of the genus *Kluveromyces*. A phylogenetic analysis of the 17 species listed in Barnett *et al.*[4] was performed using mitochondrial cytochrome-*c* oxidase II gene sequences;[10] the authors confirmed the polyphyletic nature of the genus but found that species could be placed into four groups (as opposed to the three previously attested).

A few *K. marxianus* strains ferment xylose,[120] and some are being developed to use on specialized substrates for bioethanol production, for example whey,[121] corn silage,[122] sugar cane[123] and Jerusalem artichoke.[124]

The biotechnological potential of *K. marxianus* has been discussed by Fonseca *et al.*,[125] who noted, among other aspects of its usefulness, the significance of the organism in enzyme production (*e.g.* β-galactosidase, β-glucosidase, inulinase and polygalacturonase).

The thermotolerance of some strains allows highly efficient carbohydrate fermentations to be undertaken,[126] and this is seen as being a major advantage over *S. cerevisiae*. The effect of temperature on the kinetics of ethanol production of *K. marxianus* is well known,[127] and it has been shown that at least five strains of the organism, isolated in an Indian distillery, will grow and ferment at temperatures in the range 45–50 °C.[128] All of the strains (known as the IMB strains) produced ethanol in the range of 5.7–7.0% (w/v) at 45 °C and 5.0–5.5% (w/v) at 50 °C when grown on media containing 14% (w/v) glucose. One of the major applications envisaged for these thermotolerant strains is for the simultaneous saccharification and fermentation (SSF) of cellulosic biomass. The optimum temperature of cellulases used in SSF is 45–50 °C, which most yeasts cannot tolerate. The use of these IMB strains would permit SSF temperatures to be increased, resulting in faster cellulose hydrolysis.

An investigation into the ability of three of the IMB strains to ferment xylose under anaerobic conditions at elevated temperatures, and thus be of use for ethanol production from lignocelluloses, did not promise any industrial application.[129] Although these strains were not good candidates for anaerobic fermentation of xylose, the group suggested that small additions of oxygen would improve ethanol yield, as has been shown with other xylose fermenters.[130]

An elegant case has been made for increased research into this organism, explaining that *K. marxianus* had now lost its status as a "non-conventional yeast".[131] The paper, which is excellently referenced, describes the desirable biotechnological traits exhibited by this species, which are listed as: capacity to assimilate key sugars, namely lactose and inulin; an extremely rapid growth rate, with typical generation times of around 70 minutes; thermotolerance (can grow up to 52 °C); and a high secretory capacity.[125]

2.4 *PACHYSOLEN TANNOPHILA* BOIDIN & ADZET (1957)

First reported from a French tannery fluid in 1957,[132] this member of the family Saccharomycetaceae was the first yeast identified that had a high capacity for ethanol production from xylose.[133] Details of the general biology and physiology of this yeast have been provided by Kurtzman.[134]

Conversion of D-xylose to ethanol is relatively inefficient in batch culture, and two main reasons for this have been forwarded; 1) concurrent utilization of ethanol in the presence of considerable quantities of xylose[135] and 2) the formation of xylitol and other substances as by-products.[136,137] Thus, strains that cannot utilize ethanol were sought (*e.g.* respiration deficient, or petite, mutants) with a view to strain improvement. The first stable mutants were produced in 1986[138] by UV mutation. Three of them produced a greater yield and volumetric rate of ethanol production than the wild type. One produced less xylitol. The initial metabolism of D-xylose follows a different pathway from that of D-glucose and other hexoses, and basically consists of reduction of the xylose (aldopentose) to xylitol (pentitol) by xylose reductase. This is followed by oxidation of xylitol to xylulose (ketopentose) by NAD-requiring xylitol dehydrogenase. Xylulose is then phosphorylated and may produce ethanol *via* the pentose phosphate/ EMP pathways.[139,140]

A number of enzymes associated with xylose catabolism in *P. tannophilus* were isolated and purified during the early work on this organism, including aldose (xylose) reductase[141] and xylitol dehydrogenase.[142] The same group reported that xylose reduction under anaerobic conditions is catalysed by two separate enzymes; NADH-specific and NADPH-specific forms of xylose reductase. These two forms were the subject of a study by van Cauwenberge *et al.*,[144] who showed that both forms had a molecular weight of 36,000 daltons and may exist as a single protein that undergoes modification. Then, Xu and Taylor[145] identified the source of electrons for the production of xylitol from xylose as an oxidative pentose-hexose cycle.

Jeffries *et al.*[146] investigated the effect of glucose supplements on the fermentation of D-xylose by *P. tannophilus* and found that periodic additions to an aerobic fermentation enhanced ethanol yield. Glucose additions to anaerobic cultures did not have the same effect, and these facts were seen as being evidence for a Custers effect[147] in this yeast. The effect of pH on the kinetics of the alcoholic fermentation, and the ethanol yield, have been studied,[148] and the oxygen requirement of the process has been looked at.[149] The intricacies of the influence of temperature on the fermentation were reported by Sánchez *et al.*[150]

The genetics of this organism were initially investigated by James and Zahab,[151] who laid the foundations for subsequent work. They concluded that, although the organism has a life-cycle that is apparently unsuited to genetic studies, it is totally amenable to genetic manipulation, and might be used for investigating pentose fermentation in *S. cerevisiae*. The following year, the same group[152] found that increasing the chromosome number above the haploid level increased the yield of ethanol from D-xylose, whereas the amount of xylitol decreased. The increase of ploidy did not appreciably increase the growth rate on xylose, but it did with D-galactose.

Lignocellulose, including agricultural and industrial wastes, is the most abundant renewable raw material in nature and, theoretically, an ideal substrate for bioethanol production. The composition of lignocellulose is cellulose, hemicellulose and lignin, and the first two polymers can be hydrolysed into monosaccharides. The cellulose hydrolysate is glucose, but hemicellulose is heterogeneous and its hydrolysate contains several hexoses and pentoses, of which glucose and xylose would be the principal components. Jeffries[153] has given an account of the utilization of xylose by bacteria, yeasts and other fungi.

With the discovery of xylose-fermenting *P. tannophilus*, some 200 other yeast species capable of fermenting glucose and growing on xylose were screened for fermentation of D-xylose.[154] Most strains yielded negligible ethanol from the standard 2% xylose solution, but 19 produced between 0.1 and 1.0 g ethanol L^{-1}. The following produced more than 1.0 g ethanol L^{-1}: *Brettanomyces naardenensis*, *Candida shehatae*, *C. tenuis*, *Pichia segobiensis* and *P. stipitis*.

Unfortunately, there is no single microbe that can convert glucose and xylose sugars to ethanol, although some recombinant organisms will ferment to some degree.[155] *P. tannophilus* might have been a candidate, but it is subject to hexose catabolite repression[140,156] and has low ethanol tolerance, which prevents complete utilization of D-xylose. With a view to partially solving these problems, Wedlock *et al.*[157] produced glucose-negative mutants of *P. tannophilus*, which were able to ferment D-xylose in the presence of glucose, but these have not been used commercially.

In view of this, co-fermentation was seen as being the best way of converting such sugar mixtures to ethanol. By 1982,[158] the bioconversion of wheat straw cellulose and hemicellulose by *P. tannophilus* and *Saccharomyces uvarum* had been reported, although "deleterious substances in the wheat straw hydrolysates" prevented maximal conversion of sugars to ethanol.

As another example, Fu and Peiris[159] attempted to maximize ethanol production from a glucose-xylose mixture by co-fermenting with *Z. mobilis* and *P. tannophila*. The volumetric ethanol productivity from a 50 g L^{-1} glucose solution with their *Z. mobilis* strain was 2.87 g L^{-1} h^{-1} (an ethanol yield of 0.50 g g glucose^{-1} – 98% of theoretical), while from a similar strength xylose solution, volumetric ethanol productivity was 0.10 g L^{-1} h^{-1} (an ethanol yield of 0.15 g g xylose^{-1} – 29% of theoretical). With a sugar mixture of 60 g glucose L^{-1} and 40 g xylose L^{-1}) a total ethanol yield of 0.33 g g sugar mixture^{-1} was obtained, which is 65% of the theoretical yield. Co-fermentation increased the ethanol yield from xylose to 0.17 g g^{-1} (33% of theoretical), and it was found that both sugars were completely utilized (no residual sugar). Among the features considered necessary for optimum conversion were a fermentation temperature of 30 °C, initial inoculation with *Z. mobilis* and incubation without aeration, inactivation of *Z. mobilis* after glucose exhaustion and then inoculation with *P. tannophilus*, followed by incubation with limited aeration. An excellent review of the relevance of co-culture fermentations in biotechnology has recently appeared.[160]

A number of other Gram-negative bacteria have been successfully used to ferment sugars not normally touched by *S. cerevisiae*, the most useful being *Escherichia coli* and *Klebsiella oxytoca*. Research has tended to concentrate on engineering these organisms to selectively produce ethanol. Indeed, the construction of *E. coli* strains to selectively produce ethanol was one of the first successful applications of genetic engineering.[161] Ingram's team inserted genes encoding essential enzymes from the ethanologenic pathway of *Z. mobilis* into *E. coli* under the control of a common promoter. The genes in question, ADH II (*adhB*) and pyruvate decarboxylase (*pdc*), were expressed at high levels in their "new home", and resulted in increased cell growth and the production of ethanol as the principal glucose fermentation product in the recipient *E. coli* cells.

Its ability to ferment a wide range of sugars, the fact that it does not need complex growth factors, plus the fact that it has a number of industrial uses, render *E. coli* a potentially useful ethanologenic organism. *E. coli* ferments sugars to a mixture of ethanol and organic acids, the ethanol being produced from pyruvate *via* pyruvate formate lyase.[162] This pathway is unbalanced because one $NADH,H^+$ is generated for each pyruvate made from a sugar, and two $NADH,H^+$ are needed to convert pyruvate into ethanol. Besides ethanol, *E. coli* also produces acetic and succinic acids. Compare this to homoethanol fermenters, such as *S. cerevisiae* and *Z. mobilis*, where pyruvate decarboxylase (PDC) catalyses the non-oxidative decarboxylation of pyruvate to acetaldehyde and CO_2. This reaction consumes only one $NADH,H^+$ for each molecule of ethanol produced. When the *pdc* gene was first expressed in *E. coli*,[163] it was assumed that it would elicit the production of just ethanol, but this was not the case because the native ADH activity of the bacterium was insufficient to accommodate high ethanol yields. This led Ingram's team to isolate *adhB* from a genomic library of *Z. mobilis* using a novel indicator medium for ADH activity. When transformed with a plasmid expressing *pdc* and *adhB*, *E. coli* produced ethanol almost exclusively. They co-expressed the two genes under the control of the native lac promoter, and the system was named the "production of ethanol" (PET) operon.[161] Ingram's laboratory then reported on the efficient production of ethanol from glucose, lactose and xylose by recombinant *E. coli*.[164]

These modified *E. coli* cells, dependent as they were on a plasmid for inserted gene expression, proved too genetically unstable for any likely industrial use, and so attempts were made to stabilize the function by integrating the genes into the *E. coli* chromosome. This genetic improvement was achieved[165] when *pdc* and *adhB* were inserted into the pyruvate formate-lyase gene (*pfl*). Unfortunately, the newly constructed cell yielded much lower quantities of ethanol than the plasmid-bearing strains, and attempts to increase *pdc* and *adhB* expression were made. The integrated DNA contained a chloramphenicol resistance marker and spontaneous mutants were selected for resistance to high levels of this antibiotic (that also expressed high levels of *pdc* and *adhB*). Analogous mutants were selected for increased expression of ADH (on aldehyde indicator plates), and these proved to be equivalent to the previous plasmid-based strains in their ability to produce ethanol from glucose and xylose. The fermentation pathways of *E. coli* have been evinced by Clark.[166]

2.5   *ZYMOMONAS MOBILIS* KLUYVER AND VAN NIEL 1936

2.5.1   General Introduction

Zymomonas mobilis is a Gram-negative, facultative anaerobic, rod-shaped bacterium with a high ethanol tolerance (12%) and an efficient and very rapidly operating homoethanol fermentation pathway. Lacking an oxidative electron transport system, the organism is obligately fermentative. In the definitive description of the genus *Zymomonas* in *Bergey's Manual*, Sprenger and Swings[167] list a single species (*Z. mobilis*) with two sub-species; *mobilis* and *pomacii* (*pomaceae*). The former is described as having been isolated from "bees, ripening honey, fermenting sap of *Agave* in Mexico, fermenting palm juice, molasses and sugarcane juice, and from beer and brewery equipment in England", the last-mentioned niche being attributable to Shimwell.[168] The entry for sub-species *pomaceae* says succinctly: "isolated in England from sick cider and from apple pulp". Sick cider has long been known to be a reservoir of *Z. mobilis*, and it was Carr and Passmore[169] who first isolated it from apple pulp (as "*Z. anaerobia*"). A molecular method for quick differentiation between the *mobilis* and *pomaceae* sub-species, employing amplified ribosomal DNA restriction analysis (ARDRA), has been developed,[170] and has proved to be more reliable than conventional methods based on physiological tests. In addition this study showed that there were consistent differences in the 16S rDNA sequences between wild strains and collection strains of sub-species *pomaceae*, which suggested that French strains might constitute a new genomovar within this sub-species (see below). Swings and De Ley[171] have suggested that *Zymomonas* is either of recent evolutionary origin or is a very stable genus.

Cider sickness, a disorder of sweet, low-acid (<0.5 g malic acid 100 ml^{-1}; pH >3.7) ciders and perries, was first reported in England by Frederick Lloyd in 1903, and an organism, called "the cider sickness bacillus", was isolated from cider around a decade later,[172] although it was not accurately described. In the early stages of the disorder a slight haze develops, and this is followed by much gas production as sugar is fermented. A "fruity" aroma then develops, due to acetaldehyde production, and cider aroma and flavour become very harsh and a copious precipitate of an aldehyde-tannin complex forms. Cider thus spoiled could only ever be used for blending. In France, cider sickness is known as "framboisé" and the first unequivocal implication of *Z. mobilis* in this disorder was not presented until 2003.[173] Acetaldehyde production by *Z. mobilis* in cider can be as high as 1000 mg L^{-1}, but is normally of the order of 150–400 mg L^{-1}. Levels of this magnitude can affect the growth of the bacterium and its fermentative ability, although low levels have been found to stimulate growth.[174]

Workers studying the phenomenon have given the causative bacillus a variety of names, including *Thermobacterium mobile* (Lindner, 1928;[175] synonyms *Pseudomonas lindneri*, *Zymomonas mobile*), *Achromobacter anaerobium*[168] and *Zymomonas anaerobia*,[176] after the generic name of Kluyver and van Niel[177] had been generally accepted. Meanwhile, Shimwell[178] had proposed *Saccharomonas* as being an appropriate name, based on the ability of the bacillus to quantitatively produce ethanol from glucose, but *Zymomonas* had precedence. It was Millis, in the work for her PhD thesis (University of Bristol, 1951) who actually demonstrated that *Z. mobilis* was the cider sickness organism.

A proposal for an improved taxonomy of the genus *Zymomonas* was forwarded by De Ley and Swings.[179] They examined around 40 strains of diverse origin, mainly from fermenting palm saps, fermenting *Agave* species, spoiled British beers and sick cider. All strains barring one showed very similar DNA base composition, genome size and DNA homology,[180] which was transcribed into phenotypic similarity. The one exception was a cider sickness organism that was received as "*Z. anaerobia* var. *pomaceae*".[181]

The work concluded that *Z. mobilis* was the only valid species in the genus, and that most strains are correctly described as *Z. mobilis* subsp. *mobilis*, the type strain being ATCC 10988. They also proposed that cider sickness organisms should accurately be described as *Z. mobilis* subsp. *pomaceae* (Millis), with the type strain being ATCC 29192.

On the basis of a polyphasic study,[182] it was concluded that the causative organism of framboisé in France was different from *Z. mobilis* subsp. *pomaceae* causing cider sickness over the English Channel. The name *Z. mobilis* subsp. *francensis* was thus invoked.

Most accounts of the genus *Zymomonas* indicate that *Z. mobilis* is primarily an organism of tropical and sub-tropical regions, and cite plant saps from *Agave* spp. (Mexico), sugar cane (Brazil; Fiji) and palm wine from central Africa. Certainly, much of the early work in the field is attributable to Lindner, who, during his visit to Mexico in the 1920s, studied the fermentation of agave sap into pulque (see p.465), a report of which was written in 1928.[175] Lindner was certainly of the opinion that it was restricted to tropical regions. *Zymomonas* is now accepted as being an organism that, during evolution, has become highly specialized for growth in plant saps with high sugar content. It is prototrophic, except for strain-dependent vitamin requirements, and high rates of glycolytic flux allow this organism to compete successfully in sugar-rich natural environments.[183,184]

2.5.2 General Metabolism

A feature of *Zymomonas* metabolism is the dissociation between catabolism and anabolism, which is generally termed "metabolic uncoupling". Thus, when growing in a complex medium (peptone or yeast extract), 98% of glucose consumed is converted into ethanol, CO_2, ATP and heat, which leaves only 2% for cellular building (anabolic) materials. Having said that, in such a medium, the 2% glucose incorporated produces 48% of cellular carbon.[185] The remaining 52% is provided by the organic constituents of the medium, which are thus used as biosynthetic building blocks rather than energy sources. This energetically inefficient metabolism leads to very high fermentative fluxes with a generally very low biomass yield.[186] The highest reported biomass yield values for *Z. mobilis* are slightly above 20 g dry biomass mole glucose^{-1}, obtained for aerobic chemostat culture at high flow rate, low glucose concentration and efficient ventilation, ensuring removal of volatile inhibitory metabolic by products.[187]

It is well established that pantothenic acid is a growth factor in *Z. mobilis*,[185] and the uncoupling of growth by pantothenate starvation has been reported.[188] Starvation resulted in a simultaneous decrease of both the molecular growth yield and of the specific growth rate, when cells were grown on glucose, and in all treatments the rate of cellular catabolic activity (*i.e.* rate of glucose fermentation per unit of dry weight of organism) remained constant.[189] Such results demonstrate that in *Z. mobilis* (as in some other microbes) anabolic processes do not control catabolic activity.

2.5.3 Carbohydrate Metabolism

Carbohydrate metabolism has been studied in great detail,[171,190,191] and we know that *Z. mobilis* is unique among bacteria in fermenting sugar anaerobically by the Entner–Doudoroff glycolytic pathway and in catabolizing pyruvate to ethanol *via* pyruvate decarboxylase.[192] By using this pathway anaerobically, the organism can produce ethanol as an endproduct by an almost quantitative conversion of glucose into ethanol and CO_2 (each mole of glucose is metabolized to two moles of ethanol and of CO_2 plus a single mole of ATP). Thus, the energy yield *via* this pathway is one-half of that gained by *S. cerevisiae* (say) *via* the EMP pathway.[193] Having said this, studies have clearly demonstrated that *Zymomonas* is superior to *Saccharomyces* in its ability to produce ethanol on defined substrates. Both rate of ethanol production and the final ethanol concentration can be greater.

The rate at which *Z. mobilis* cells can ferment glucose in an exponential culture under anaerobic conditions reaches 0.75–1.0 µmole glucose mg dry weight^{-1} min^{-1} (Rogers *et al.* 1982[190]), which is three to five times faster than yeast. Each minute, the organism consumes a quantity of glucose equal to one-third of its mass.

As Sprenger[194] beautifully put it, carbohydrate metabolism in *Z. mobilis* operates as a veritable "catabolic highway, with some scenic routes". Around four decades earlier, Bauchop and

Elsden[195] had elaborated upon the quantitative aspects of the nutrition of this organism with their classic, more prosaically named "growth of micro-organisms in relation to their energy supply" – a milestone for its time.

The absence of the EMP pathway can be demonstrated by [1-^{13}C]glucose experiments, where no ^{13}C label can be detected at the C-3 position of pyruvate (owing to the missing phosophofructokinase). Quantification of glucose metabolism in *Z. mobilis*, and some other bacteria, was reported by Fuhrer *et al.*,[196] who used ^{13}C tracer experiments to identify its network topology and *in vivo* operation by quantification of intracellular carbon fluxes.

The sub-cellular locations of 11 glycolytic and fermentative enzymes in *Z. mobilis* were elucidated in 1992.[197] Glucose-fructose oxidoreductase was localized in the periplasmic region, phosphogluconate lactonase and ADH I were confined to cytoplasm adjacent to the cell membrane and the remainder were evenly distributed in the cytoplasm. Further experiments disclosed that glyceraldehyde-3-phosphate dehydrogenase, phosphoglycerate kinase and ADH I form an enzyme complex.

Because of the exclusive use of the Entner–Doudoroff pathway for ATP synthesis, and the concomitant limited energy supply, *Z. mobilis* is essentially prototrophic and grows rapidly with a doubling time of around 90 minutes. Together, the 13 glycolytic and ethanologenic enzymes comprise around one-half of soluble cellular protein in exponential-phase cells[183,184] and genes encoding most of these enzymes have been cloned and sequenced.

Being totally independent on substrate-level phosphorylation for energy production and obtaining only a single mole of ATP per mole glucose fermented, rapid carbon flux is necessary to counteract inefficient energy production, and is facilitated by the high levels of glycolytic enzymes (mentioned above). In addition, these high cytoplasmic protein levels in *Z. mobilis* are correlated with an increased stability of their transcripts.[198] Transcript stability is seen as being an important feature that may distinguish highly expressed genes (for energy generation) from biosynthetic genes, which are required at much lower levels.

In a study of the glycolytic flux during the fermentation of a high glucose (20%) solution, Osman *et al.*[183] found a peak of fermentative activity after 18 h (1.1% ethanol), and after 30 h (6.2% ethanol) this rate had halved, although cell number continued to increase. These times corresponded to the end of exponential growth and to the onset of the stationary phase. Three possible reasons were forwarded for this decline: decreased level of nucleotides; a decrease in internal pH (from 6.3 to 5.3) and a decrease in the specific activities of two glycolytic enzymes (pyruvate kinase and glyceraldehyde-3-phosphate dehydrogenase).

Dawes and Large[199] examined the effect of starvation on *Z. mobilis* cells, and found that, of all the intracellular constituents that become degraded, only RNA is significantly dissipated (but this could be reversed in the presence of Mg^{++}, which are known to stabilize ribosomes). Half of the RNA was degraded in the first 24 h of starvation after which time the decline was much slower. No endocellular carbohydrate, DNA and protein were degraded, even under prolonged starvation conditions, and ATP concentration increased (from 0.5 to 1.0 μg mg dry weight^{-1}) during growth but began to decrease exponentially in the last generation before growth ceased because of glucose exhaustion.

The ethanol yield obtained from sucrose in batch and continuous cultures is generally lower than that from glucose. This is due to the production of the fructose polymer levan and other by-products, such as sorbitol, and fructo-oligomers at higher levels during sucrose metabolism. If fermentation conditions are optimized, however, the formation of these "unwanted" compounds is inhibited, and ethanol yields from sucrose can parallel those from glucose metabolism, with levels of 0.42–0.47 ethanol g sucrose consumed^{-1} (Viikari, 1984a;[200] Doelle and Greenfield, 1985[201]). Prior to being metabolized, sucrose is broken down to glucose and fructose outside of the cell by various sucrases,[202] and the free products taken up.

In the presence of high sugar concentrations, *Z. mobilis* can synthesize substantial quantities of sorbitol[203,204] *via* the reduction of fructose and the concomitant oxidation of glucose to gluconolactone.[205] These two reactions are catalysed by a single enzyme, glucose-fructose oxidoreductase (GFOR), a tetrameric enzyme (at low pH) with a sub-unit size of 40,000 daltons and consisting of four identical sub-units.[206] Each sub-unit contains one tightly bound NADP molecule, which is not released during catalysis. As Hardman and Scopes[208] discovered, GFOR operates by a "ping-pong" mechanism, catalysing the reaction of one of its substrates to yield a product that dissociates before the other substrate binds. This means that the overall reaction consists of two half reactions, with alternate reduction of the bound $NADP^+$ (as glucose is oxidized to gluconolactone) and oxidation of NADPH (as fructose is reduced to sorbitol). Of the two end-products of these half reactions, gluconolactone is converted into ethanol, but sorbitol persists as it is not further converted in the cell. Attempts to grow *Z. mobilis* on sorbitol have been unsuccessful.[204]

GFOR is located in the perisplasmic space in bacteria,[197] and Loos *et al.*[207] proposed that its *raison d'être* was to protect the cell against osmotic stress. The protective mechanism emanates from the conversion of fructose into sorbitol, which can be accumulated in the cell without deleterious effect. Kinetic studies[208] have indicated that significant sorbitol formation will only occur in the presence of high levels of glucose and fructose, something that may be deduced from the relatively large Michaelis constants for the two substrates (10.8 ± 0.8 mM for glucose and 400 ± 30 mM for sucrose). The structure of *Z. mobilis* GFOR was elucidated in 1996.[209]

When *Z. mobilis* is grown on sucrose medium, sucrose-hydrolysing enzymes, such as intracellular sucrase (SacA or InvA), extracellular levansucrase (SacB, LevU or InvC) and extracellular sucrase (SacC or InvB), are formed. The *sacA* gene from *Z. mobilis* was cloned, sequenced and characterized in 1990[210] and the amino acid sequences of the gene product showed strong homology with yeast invertases. It was later found that the *sacB* and *sacC* genes form a gene cluster in *Z. mobilis*.[211] Molecular cloning and characterization of *sacC* was effected in 1995,[212] and a simple method for the purification of the thermostable levansucrase of *Z. mobilis* from a recombinant *E. coli* was later developed.[213] The same workers[214] found that when a 1.7 kb *Z. mobilis* gene encoding levansucrase was cloned in *E. coli*, the gene product was able to hydrolyse sucrose and synthesize levan. A sucrase mutant of *Z. mobilis*, which could not grow on sucrose, was shown to be unable to produce SacC and SacB, but when a cloned (sub-cloned into an *E. coli-Z. mobilis* shuttle vector) *sacB* gene was transferred, the recombinant strain produced extra-cellular SacB activity, and yielded quantities of levan of almost "normal" levels.[215]

Goldman *et al.*[216] showed that *Z. mobilis* levansucrase exists in two discrete and active forms. Above pH 7.0, the enzyme is soluble and is a dimer, but below pH 6.0 the protein self-assembles into long microfibrils that tend to precipitate out of solution. The enzyme is active in both forms, although its substrate and product specificity alter.

With so little energy available for sugar transport within the cell, *Z. mobilis* transports the vast amounts of glucose necessary for survival *via* a non-energy dependent, low-affinity, high-velocity diffusion system,[217] which has an approximately ten-fold higher affinity for glucose than for fructose. The kinetic differences between glucose and fructose transport have been discussed by Parker *et al.*,[218] who maintained that the kinetics of sugar transport and phosphorylation encourage the preferential consumption of glucose and the accumulation of fructose. In turn, this makes fructose available for formation of the important osmoregulator sorbitol (*via* GFOR activity).

In an attempt to understand more about the role of respiration in aerobically growing *Z. mobilis*, the effects of cyanide on culture growth were studied by Kalnenieks *et al.*,[219] who found that sub-millimolar concentrations inhibited the high respiration rates of aerobic cultures but stimulated culture growth. This stimulatory effect was attributed to decreased production of the inhibitory metabolite acetaldehyde at lower respiration rates, when more reducing equivalents are channelled through ADH. The same group[220] showed that *Z. mobilis* ADH II largely loses its sensitivity to cyanide when grown aerobically, and it was concluded that it was the membrane-associated respiratory chain, not ADH II, that is responsible for whole-cell cyanide sensitivity, while the cyanide-resistant ADH II is needed for respiration in the presence of cyanide, and represents an adaptive response of *Z. mobilis* to cyanide. Cyanide did not affect the activity of ADH I.

The two cytoplasmic isozymes of ADH known to be present in *Z. mobilis* were characterized both kinetically and physiologically by Wills *et al.*,[221] who found that one, initially called ZADH-I, a tetramer, had a sub-unit molecular weight of 34,700 daltons, very close to that of yeast ADH. The other (ZADH-II) was a dimer and had a slightly smaller sub-unit size (31,100 daltons). It was clear that these two isozymes showed more difference from each other than did the two major isozymes of yeast. Later work showed that ADH I was a zinc-containing enzyme and that ADH II was iron-containing.[222,223] Both enzymes were shown to be NAD^+-dependent. ADH I is implicated in early stages of *Z. mobilis* growth, while ADH II plays a key role later on during fermentation, when ethanol levels are higher.[224] Experiments with ADH II-negative mutants result in prolonged generation time and impaired growth and ethanol synthesis during the late exponential and early stationary phases of growth.

Some evidence suggests that ethanol synthesis may not be the sole function of iron-containing bacterial ADHs and that they are also involved in respiratory metabolism and oxidative stress response. For example, An *et al.*[225] identified ZADH-II as a stress protein that is induced both by exposure to ethanol and by elevated temperature (45 °C).

2.5.4 Genome Sequence

The complete genome sequence of ZM4 (ATCC 31821), an industrial strain of *Z. mobilis*, was reported in 2005,[226] when it was found that it consisted of 2,056,416 base pairs forming a circular chromosome with 1998 open reading frames and three ribosomal RNA transcription units. The genome lacked recognizable genes for 6-phosphofructokinase (essential in the EMP pathway), and the 2-oxoglutarate dehydrogenase complex and malate dehydrogenase, which are necessary for the Krebs Cycle. Such evidence confirms that this species can only metabolize glucose by the ED pathway. An improved genome annotation for the ZM4 genome, based on new experimental data, was forwarded in 2009,[227] and the same year the first entire genome sequence for *Z. mobilis* was published for a British ale-infecting isolate (NCIMB 11163).[228]

2.5.5 Membrane Integrity

In prokaryotes, sterols are usually absent or present in insignificant amounts. The role of sterols in prokaryotic membranes is replaced by a number of novel, complex lipids (such as carotenoids) and these appear to have the same role in prokaryotic membranes as cholesterol has in eukaryotic ones. In this respect, another notable feature of *Z. mobilis* is the high level of pentacyclic triterpenoids of the hopane family, the so-called "hopanoids" in the cell, a characteristic thought to be associated with ethanol tolerance in this species. Hopane is one of the basic skeletons of pentacyclic triterpenes, and was regarded as unimportant until found in prokaryotes.

Bringer *et al.*[229] observed that the *Z. mobilis* membrane was home to copious quantities of hopanoids. The known distribution of hopanoid triterpenes in prokaryotes at the time was presented by Rohmer *et al.*,[230] and Flesch and Rohmer[231] reported 32-oxobacteriohopane-33,34,35-triol, a novel hopanoid, among the triterpenoid fraction of *Z. mobilis*. The compound is β-linked *via* its primary hydroxyl group to glucosamine.

Hopanoids are widespread among prokaryotes and can be found in diverse taxonomic groups, and it has long been suggested that, because of their structural characteristics and rigidity, they might be equivalent to the sterols found in eukaryotes. Hopanoids are structurally similar to sterols in their molecular dimensions and amphiphilic character, and their biosynthesis is also similar, although it includes several primitive features compared to that of sterols. According to some, they qualify as phylogenetic precursors of sterols, for both groups of compounds are cyclization products of squalene, by an enzyme-mediated, acid-catalysed process, although hopanoid synthesis is regarded as being more primitive.[232] A very readable summary of the role of prokaryotic polyterpenoids and the evolution of membrane reinforcers was provided by Ourisson *et al.*[233]

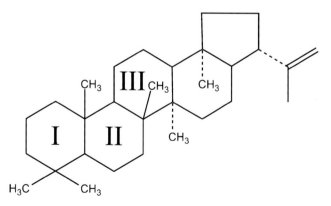

Figure 2.6 Diploptene (hopanoid).

The wide distribution and structural variety of hopanoids was initially revealed by their molecular fossils.[234] Of those found in extant prokaryotes, the "simplest" are C_{30} derivatives of hopane, diploptene and diplopterol (Figure 2.6), which are found in all hopanoid-containing prokaryotes.[233] There is also a group of C_{35} derivatives of bacteriohopane, and these C-pentosyltriterpenes are usually the major hopanoids, the most commonly encountered being bacteriohopanetetrol. Hopanoids have been found at levels similar to those of sterols in eukaryotes (0.1–2.0 mg g^{-1} dry weight).

Geohopanoids are molecular fossils ubiquitous in sediments, and have accumulated over hundreds of millions of years thanks to incomplete breakdown of polymers resistant to biodegradation. The most complex contain up to 35 carbon atoms and were discovered before their bacterial (biohopanoid) counterparts. Around 150 different structures have been identified. In typical prokaryotic hopanoids, the C_{30} hopane skeleton is linked at C-30 to a C_5 *n*-alkyl polysubstituted chain, to give C_{35} bacteriophane derivatives as mentioned above.

Several studies have suggested that exposure to ethanol decreases membrane organization, which results in increased leakage and modified activity of a number of membrane-associated protein and transport systems.[235] As Ingram[236] showed, in ethanol-tolerant microbes, adaptation to ethanol stress involves an increase in the mean chain length of incorporated fatty acids, thus thickening the hydrophobic membrane core. He also noted that the proportion of monounsaturated fatty acids (notably *cis*-vaccenic acid) increased.

In yeasts, sterols are known to be one way to enhance the membrane's insensitivity to ethanol,[237] but the absence of such compounds in bacteria means that there must be surrogates somewhere in the membrane. Studies on the membrane of *Z. mobilis*[238] showed that more than 70% of the fatty acids in the phospholipid fraction is *cis*-vaccenic acid, and that the major phospholipids were cardiolipin, phosphatidylethanolamine, phosphatidylglycerol and phosphatidylcholine. After vaccenic acid, the next most abundant fatty acids were myristic, palmitic and palmitoleic. Glucose and ethanol were shown to have no major effect on the fatty acid content of *Z. mobilis*, but both compounds caused a decrease in phosphatidylethanolamine and phosphatidylglycerol and an increase in cardiolipin and phosphatidylcholine.

Ethanol also caused a dose-dependent reduction in the lipid-to-protein ratios of crude membrane fractions.

2.5.6 Genetic Manipulation

Recent high oil prices, concern over energy security and consideration of environmental goals have rekindled interest in producing alternative fuels from industrial scale fermentations, and to achieve this goal we need to improve methods of producing ethanol from cellulosic biomass. The lack of industrially suitable microbes for converting biomass (especially lignocellulosic biomass) into fuel alcohol is often cited as being the major roadblock to progress. Lignocellulosic biomass contains complex carbohydrates that require breaking down by microbes capable of breaking down sugars not fermentable by *S. cerevisiae* (most notably the pentose sugar, xylose).

An early, erudite paper on genetic modification of *Z. mobilis* was written by Buchholz and Everleigh in 1990,[239] who prophetically attested that: "It is now practically possible to construct strains of *Z. mobilis* which are capable of efficient production of high concentrations of ethanol from inexpensive substrates such as starch and cellulosics. The transfer of *Z. mobilis* genes to other hosts in order to channel the conversion of pentoses and other substrates to ethanol has given indications of success. This all bodes well for *Z. mobilis*".

After a number of unsuccessful attempts by various workers, Zhang *et al.*[240] succeeded in genetically engineering a pentose metabolism pathway into *Z. mobilis*. With the modified organism, anaerobic fermentation of xylose (and glucose) to ethanol was achieved through a combination of the pentose phosphate and ED pathways. It was known, through experience, that genes encoding xylose isomerase (*xylA*), xylulose kinase (*xylB*), transaldolase (*talB*) and transketolase (*tktA*) would have to be introduced and expressed before a functional pathway capable of converting xylose to central intermediates of the ED pathway (thence to ethanol) could be constructed. With the help of *E. coli*, the relevant genes were inserted. The transformed strain, CP4 (pZB5), grew on xylol and ethanol yields obtained were almost theoretical (86%). The group used the same sort of strategy as was used to produce a strain capable of fermenting arabinose.[241]

Subsequently, Zhang's laboratory made some improvements to their *Z. mobilis* strains, and one, AX101, ferments both xylose and arabinose and carries the seven necessary recombinant genes as part of its chromosomal DNA. Genetic stability was confirmed when AX101 was transferred through 160 generations on glucose yet retained the ability to ferment xylose and arabinose. It was found that arabinose was fermented much slower than xylose, and its fermentations were often incomplete.[242,243]

In an attempt to make *Z. mobilis* more tolerant to abiotic stresses, Zhang *et al.*[244] inserted the *irrE* gene and found that its expression conferred protection against ethanol, acid, osmotic and thermal shocks. It also markedly improved cell viability, the expression levels and enzyme activities of pyruvate decarboxylase and ADH and the production of ethanol under both ethanol and acid stresses. In nature, the *irrE* gene encodes a regulatory protein (IrrE) in *Deinococcus radiodurans*, one of the most radioresistant organisms known, and a bacterium that can also withstand extreme cold and dehydration, as well as exposure to acid conditions and a vacuum.[245]

Acetic acid is an inhibitor produced during biomass pre-treatment processes, and at pH 5 it is in the uncharged and undissociated form. As such, it can permeate the plasmalemma and cause cytoplasmic acidification,[246] with the inevitable results. Acetate removal procedures are expensive, and so the acetate tolerant mutant (AcR) described by Joachimstahl *et al.*[247] has great potential. AcR was developed while many systems biology tools were being developed, and the molecular mechanism of AcR sodium acetate tolerance was not elucidated until 2010,[248] with the concomitant benefits for industrial strain improvement.

With a view to providing an overview of *Z. mobilis* physiology and a foundation for future comparisons of other wild-type and mutant strains, Bochner *et al.*[249] developed Phenotype MicroArray^{TM} (PM) protocol to profile cellular phenotypes. The work included a standard set of nearly 2000 assays for carbon, nitrogen, phosphorus and sulfur source utilization, nutrient stimulation, pH and osmotic stresses and chemical sensitivities with 240 inhibitory chemicals.

Dien *et al.*[162] produced a review of the bacteria used for fuel ethanol production, and Panesar *et al.*[250] have reviewed the role of *Z. mobilis* as an ethanol producer. New directions for the production of bioethanol *via* the consolidated bioprocessing of lignocelluloses have recently been described.[251]

REFERENCES

1. J. Lodder, *Yeasts: A Taxonomic Study*, North-Holland Publishing Co., Amsterdam, 1970.
2. N. J. W. Kreger-van Rij, *The Yeasts: A Taxonomic Study*, Elsevier, Amsterdam, 3rd edn, 1984.
3. G. I. Naumov, *J. Ind. Appl. Microbiol.*, 1996, **17**, 295.
4. J. A. Barnett, R. W. Payne and D. Yarrow, *Yeasts: Characteristics and Identification*, Cambridge University Press, Cambridge, 3rd edn, 2000.
5. J. A. Barnett, *Yeast*, 2004, **21**, 1141.
6. C. P. Kurtzman and C. J. Robnett, *FEMS Yeast Res.*, 2003, **3**, 417.
7. S. A. James, J. P. Cai, I. N. Roberts and M. D. Collins, *Int. J. Syst. Bacteriol.*, 1997, **47**, 453.
8. S. A. James, M. D. Collins and I. N. Roberts, *Int. J. Syst. Bacteriol.*, 1996, **46**, 189.
9. C. P. Kurtzman and C. J. Robnett, *Antonie van Leeuwenhoek*, 1998, **73**, 331.
10. C. Belloch, A. Querol, M. D. Garcia and E. Barrio, *Int. J. Syst. Evol. Microbiol.*, 2000, **50**, 405.
11. A. Vaughan-Martini and A. Martini, in *The Yeasts: A Taxonomic Study*, ed. C. P. Kurtzmann and J. W. Fell, Elsevier, Amsterdam, 4th edn, 1998, p. 358.
12. J. P. van der Walt, in *The Yeasts: A Taxonomic Study*, ed. E. Lodder, Elsevier, Amsterdam, 2nd edn, 1970, p. 555.
13. G. I. Naumov, S. A. James, E. S. Naumov, E. J. Louis and I. N. Roberts, *Int. J. Syst. Evol. Microbiol.*, 2000, **50**, 1931.
14. D. R. Scannell, A. C. Frank, G. C. Conant, K. P. Byrne, M. Woolfit and K. H. Wolfe, *Proc. Nat. Acad. Sci. USA*, 2007, **104**, 8397.
15. T. Y. James, F. Kauff, C. L. Schoch, P. Brandon Matheny, V. Hofstetter, C. J. Cox, G. Celio, C. Gueidan, E. Fraker, J. Miadlikowska, H. Thorsten Lumbsch, A. Rauhut, V. Reeb, A. E. Arnold, A. Amtoft, J. E. Stajich, K. Hosaka, G.-H. Sung, D. Johnson, B. O'Rourke, M. Crockett, M. Binder, J. M.

Curtis, J. C. Slot, Z. Wang, A. W. Wilson, A. Schüßler, J. E. Longcore, K. O'Donnell, S. Mozley-Standridge, D. Porter, P. M. Letcher, M. J. Powell, J. W. Taylor, M. M. White, G. W. Griffith, D. R. Davies, R. A. Humber, J. B. Morton, J. Sugiyama, A. Y. Rossman, J. D. Rogers, D. H. Pfister, D. Hewitt, K. Hansen, S. Hambleton, R. A. Shoemaker, J. Kohlmeyer, B. Volkmann-Kohlmeyer, R. A. Spotts, M. Serdani, P. W. Crous, K. W. Hughes, K. Matsuura, E. Langer, G. Langer, W. A. Untereiner, R. Lücking, B. Büde, D. M. Geiser, A. Aptroot, P. Diederich, I. Schmitt, M. Schultz, R. Yahr, D. S. Hibbett, F. Lutzoni, D. J. McLaughlin, J. W. Spatafora and R. Vilgalys, *Nature*, 2006, **443**, 818.

16. S. B. Hedges, J. E. Blair, M. L. Venturi and J. L. Shoe, *BMC Evol. Biol.*, 2004, **4**, 2.

17. D. S. Heckman, D. M. Geiser, B. R. Eidell, R. L. Stauffer, N. L. Kardos and S. B. Hedges, *Science*, 2001, **293**, 1129.

18. S. E. Massey, G. Moura, P. Beltrão, R. Almeida, J. R. Garey, M. F. Tuite and M. A. S. Santos, *Genome Res.*, 2003, **13**, 544.

19. I. Miranda, R. Silva and M. Santos, *Yeast*, 2006, **23**, 203.

20. B. Dujon, D. Sherman, G. Fischer, P. Durrens, S. Casaregola, I. Lafontaine, J. de Montigny, C. Marck, C. Neuvéglise, E. Talla, N. Goffard, L. Frangeul, M. Aigle, V. Anthouard, A. Babour, V. Barbe, S. Barnay, S. Blanchin, J.-M. Beckerich, E. Beyne, C. Bleykasten, A. Boisramé, J. Boyer, L. Cattolico, F. Confanioleri, A. de Daruvar, L. Despons, E. Fabre, C. Fairhead, H. Ferry-Dumazet, A. Groppi, F. Hantraye, C. Hennequin, N. Jauniaux, P. Joyet, R. Kachouri, A. Kerrest, R. Koszu, M. Lemaire, I. Lesur, L. Ma, H. Muller, J.-M. Nicaud, M. Nikolski, S. Oztas, O. Ozier-Kalogeropoulos, S. Pellenz, S. Potier, G.-F. Richard, M.-L. Straub, A. Suleau, D. Swennen, F. Tekaia, M. Wésolowski-Louvel, E. Westhof, B. Wirth, M. Zeniou-Meyer, I. Zivanovic, M. Bolotin-Fukuhara, A. Thierry, C. Bouchier, B. Caudron, C. Scarpelli, C. Gaillardin, J. Weissenbach, P. Wincker and J.-L. Souciet, *Nature*, 2004, **430**, 35.

21. I. S. Hornsey, *The Chemistry and Biology of Winemaking*, Royal Society of Chemistry, Cambridge, 2007.

22. K. Rosen, in *Biotechnological Applications in Beverage Production*, ed. C. Cantarelli and G. Lanzarini, Elsevier Science, New York, 1989.

23. S. J. Singer and G. L. Nicholson, *Science*, 1972, **175**, 720.
24. B. Guérin, in *Yeasts*, ed. A. H. Rose and J. S. Harrison, Academic Press, London, 2nd edn, 1991, vol. 4.
25. F. Foury, T. Roganti, N. Lecrenier and B. Purnelle, *FEBS Lett.*, 1998, **440**, 325.
26. P. Mitchell, *Science*, 1979, **206**, 1148.
27. J. A. Barnett, *Yeast*, 2003, **20**, 1015.
28. J. A. Barnett and C. F. Robinow, *Yeast*, 2002, **19**, 151.
29. J. A. Barnett and C. F. Robinow, *Yeast*, 2002, **19**, 745.
30. A. C. Codon, T. Benitez and M. Korhola, *Appl. Microbiol. Biotechnol.*, 1998, **49**, 154.
31. A. Goffeau, B. G. Barrell and H. Bussey, R. W. Davis, B. Dujon, H. Feldmann, F. Galibert, J. D. Hoheisel, C. Jacq, M. Johnston, E. J. Louis, H. W. Mewes, Y. Murakami, P. Philippsen, H. Tettelin and S. G. Oliver, *Science*, 1996, **274**, 546.
32. G. F. Sprague, Jr., in *The Yeasts*, ed. A. H. Rose, A. E. Wheals and J. S. Harrison, Academic Press, London, 2nd edn, 1995, vol. 6.
33. A. E. Wheals, in *The Yeasts*, ed. A. H. Rose and J. S. Harrison, Academic Press, London, 2nd edn, 1987, vol. 1.
34. F. M. Harold, *Microbiology*, 1995, **141**, 2765.
35. G. Sherlock and J. Rosamond, *J. General Microbiology*, 1993, **139**, 2531.
36. A. Goffeau, *FEBS Lett.*, 2000, **480**, 37.
37. S. G. Oliver, Q. J. M. van de Aart, M. L. Agostoni-Carbone and 144 others, *Nature*, 1992, **357**, 38.
38. B. Dujon, *Trends Genet.*, 1996, **12**, 263.
39. T. Galitski, A. L. Saldanha, C. A. Styles, E. S. Lander and G. R. Fink, *Science*, 1999, **285**, 251.
40. A. J. P. Brown, in *Yeast Gene Analysis*, Methods in Microbiology, ed. A. J. P. Brown and M. F. Tuite, Academic Press, London, 1988, vol. 26, p. 79.
41. K. H. Wolfe and D. C. Shields, *Nature*, 1997, **387**, 708.
42. S. F. Altschul, W. Gish, W. Miller, E. W. Myers and D. J. Lipman, *J. Mol. Biol.*, 1990, **215**, 403.
43. C. Seoighe and K. H. Wolfe, *Gene*, 1999, **238**, 253.
44. T. F. Smith and M. S. Waterman, *J. Mol. Biol.*, 1981, **147**, 195.
45. R. B. Langkjær, P. F. Cliften, M. Johnston and J. Piškur, *Nature*, 2003, **421**, 848.

46. S. Ohno, *Evolution by Gene Duplication*, Allen & Unwin, London, 1970.
47. M. Lynch and J. S. Conery, *Science*, 2000, **290**, 1151.
48. D. B. Kaback, in *The Yeasts*, ed. A. E. Wheals, A. H. Rose and J. S. Harrison, Academic Press, London, 1995, vol. 6, p. 179.
49. G. Liti and E. J. Louis, *Annu. Rev. Microbiol.*, 2005, **59**, 135.
50. H. Jiang, W. Guan, D. Pinney, W. Wang and Z. Gu, *Genome Res.*, 2008, **18**, 1466.
51. M. J. A. van Hoek and P. Hogeweg, *Mol. Biol. Evol.*, 2009, **26**, 2441.
52. G. C. Conant and K. H. Wolfe, *Mol. Syst. Biol.*, 2007, **3**, 129.
53. F. S. Dietrich, S. Voegeli, S. Brachat, A. Lerch, K. Gates, S. Steiner, C. Mohr, R. Pöhlmann, P. Luedi, S. Choi, R. A. Wing, A. Flavier, T. D. Gaffney and P. Philippsen, *Science*, 2004, **304**, 304.
54. B. Dujon, D. Sherman, G. Fischer, P. Durrens, S. Casaregola, I. Lafontaine, J. De Montigny, C. Marck, C. Neuvéglise, E. Talla, N. Goffard, L. Frangeul, M. Aigle, V. Anthouard, A. Babour, V. Barbe, S. Barnay, S. Blanchin, J. M. Beckerich, E. Beyne, C. Bleykasten, A. Boisramé, J. Boyer, L. Cattolico, F. Confanioleri, A. De Daruvar, L. Despons, E. Fabre, C. Fairhead, H. Ferry-Dumazet, A. Groppi, F. Hantraye, C. Hennequin, N. Jauniaux, P. Joyet, R. Kachouri, A. Kerrest, R. Koszul, M. Lemaire, I. Lesur, L. Ma, H. Muller, J. M. Nicaud, M. Nikolski, S. Oztas, O. Ozier-Kalogeropoulos, S. Pellenz, S. Potier, G. F. Richard, M. L. Straub, A. Suleau, D. Swennen, F. Tekaia, M. Wésolowski-Louvel, E. Westhof, B. Wirth, M. Zeniou-Meyer, I. Zivanovic, M. Bolotin-Fukuhara, A. Thierry, C. Bouchier, B. Caudron, C. Scarpelli, C. Gaillardin, J. Weissenbach, P. Wincker and J. L. Souciet, *Nature*, 2004, **430**, 35.
55. M. Kellis, B. W. Birren and E. S. Lander, *Nature*, 2004, **428**, 617.
56. J. Piškur, E. Rozpedowska, S. Polakova, A. Merico and C. Compagno, *Trends Genet.*, 2006, **22**, 183.
57. A. Merico, P. Sulo, J. Piškur and C. Campagno, *FEBS J.*, 2007, **274**, 976.
58. H. Chen, L. Xu and Z. Gu, *Mol. Biol. Evol.*, 2008, **25**, 2513.

59. O. Winge and A. Hjort, *C. R. Trav. Lab. Carlsberg Ser. Physiol.*, 1935, **21**, 51.
60. O. Winge and C. Roberts, *C. R. Trav. Lab. Carlsberg Ser. Physiol.*, 1952, **25**, 141.
61. O. Winge and C. Roberts, *C. R. Trav. Lab. Carlsberg Ser. Physiol.*, 1949, **24**, 341.
62. C. C. Lindgren, *Bact. Rev.*, 1945, **9**, 111.
63. H. L. Roman, *Ann. Rev. Genet.*, 1986, **20**, 1.
64. J. A. Barnett, *Yeast*, 2007, **24**, 799.
65. M. N. Hall and P. Lindner (ed.), *The Early Days of Yeast Genetics*, Cold Spring Harbour Press, New York, 1992.
66. J. R. Johnston (ed.), *Molecular Genetics of Yeast: A Practical Approach*, IRL Press, Oxford, 1994.
67. A. E. Wheals, A. H. Rose and J. S. Harrison (ed.), *The Yeasts*, Academic Press, New York, 2nd edn, 1995, vol. 6.
68. A. J. P. Brown and M. F. Tuite (ed.), *Methods in Microbiology*, Academic Press, New York, 1998, vol. 26.
69. J. Weissman, C. Guthrie and G. R. Fink (ed.), *Guide to Yeast Genetics: Functional Genomics, Proteomics, and Other Systems Analysis*, Academic Press, New York, 2nd edn, 2010.
70. Z. Gu, L. David, D. Petrov, E. Jones, R. W. Davis and L. M. Steinmetz, *Proc. Nat. Acad. Sci. USA*, 2005, **102**, 1092.
71. J. Ronald, H. Tang and R. B. Brem, *Genetics*, 2006, **174**, 541.
72. R. K. Mortimer and J. R. Johnston, *Genetics*, 1986, **113**, 35.
73. O. W. Tawfik, C. J. Papasian, A. Y. Dixon and L. M. Potter, *J. Clin. Microbiol.*, 1989, **27**, 1689.
74. E. A. Shoubridge and P. W. Hochachka, *Science*, 1980, **209**, 307.
75. J. I. Harris, in *Structure and Activity of Enzymes*, ed. T. W. Goodwin, J. I. Harris and B. S. Hartley, Academic Press, New York, 1964, p. 97.
76. A. J. Ganzhorn, D. W. Green, A. D. Hershey, R. M. Gould and B. V. Plapp, *J. Biol. Chem.*, 1987, **262**, 3754.
77. M. Wiesenfeld, L. Schimpfessel and R. Crokaert, *Biochim. Biophys. Acta*, 1975, **405**, 500.
78. E. T. Young and D. Pilgrim, *Mol. Cell Biol.*, 1985, **5**, 3024.
79. B. M. Bakker, C. Bro, P. Cotter, M. A. Luttik, J. P. van Dijken and J. T. Pronk, *J. Bacteriol.*, 2000, **182**, 4730.
80. C. E. Paquin and V. M. Williamson, *Mol. Cell Biol.*, 1986, **6**, 70.

81. H. Feldmann, M. Aigle, G. Aljinovic, B. André, M. C. Baclet, C. Barthe, A. Baur, A. M. Bécam, N. Biteau, E. Boles, T. Brandt, M. Brendel, M. Brückner, F. Bussereau, C. Christiansen, R. Contreras, M. Crouzet, C. Cziepluch, N. Démolis, T. Delaveau, F. Doignon, H. Domdey, S. Düsterhus, E. Dubois, B. Dujon, M. El Bakkoury, K. D. Entian, M. Feurmann, W. Fiers, G. M. Fobo, C. Fritz, H. Gassenhuber, N. Glandsdorff, A. Goffeau, L. A. Grivell, M. de Haan, C. Hein, C. J. Herbert, C. P. Hollenberg, K. Holmstrøm, C. Jacq, M. Jacquet, J. C. Jauniaux, J. L. Jonniaux, T. Kallesøe, P. Kiesau, L. Kirchrath, P. Kötter, S. Korol, S. Liebl, M. Logghe, A. J. Lohan, E. J. Louis, Z. Y. Li, M. J. Maat, L. Mallet, G. Mannhaupt, F. Messenguy, T. Miosga, F. Molemans, S. Müller, F. Nasr, B. Obermaier, J. Perea, A. Piérard, E. Piravandi, F. M. Pohl, T. M. Pohl, S. Potier, M. Proft, B. Purnelle, M. Ramezani Rad, M. Rieger, M. Rose, I. Schaaff-Gerstenschläger, B. Scherens, C. Schwarzlose, J. Skala, P. P. Slonimski, P. H. Smits, J. L. Souciet, H. Y. Steensma, R. Stucka, A. Urrestarazu, Q. J. van der Aart, L. van Dyck, A. Vassarotti, I. Vetter, F. Vierendeels, S. Vissers, G. Wagner, P. de Wergifosse, K. H. Wolfe, M. Zagulski, F. K. Zimmermann, H. W. Mewes and K. Kleine, *EMBO J.*, 1994, **13**, 5795.

82. E. Valencia, C. Larroy, W. F. Ochoa, X. Parés, I. Fita and J. A. Biosca, *J. Mol. Biol.*, 2004, **341**, 1049.

83. C. Larroy, M. R. Fernandez, E. Gonzalez, X. Pares and J. A. Biosca, *Biochem. J.*, 2002a, **361**, 163.

84. C. Larroy, X. Pares and J. A. Biosca, *Eur. J. Biochem.*, 2002b, **269**, 5738.

85. H. M. C. Heick, J. Willemot and N. Begin-Heick, *Biochim. Biophys. Acta*, 1969,

86. C. Wills, *Nature*, 1976, **261**, 26.

87. A. D. Ellington and S. A. Benner, *J. Theoret. Biol.*, 1987, **127**, 491.

88. W. M. Fitch, in *Molecular Evolution*, ed. F. J. Ayala, Sinauer Associates, Sunderland, MA, 1976, p. 160.

89. J. W. Thornton, *Nat. Rev. Genet.*, 2004, **5**, 366.

90. L. Pauling and E. Zuckerkandl, *Acta Chem. Scand.*, 1963, **17**, S9.

91. W.-H. Li, C.-I. Wu and C.-C. Luo, *Mol. Biol. Evol.* 1985, **2**, 150.

92. S. A. Benner, *Adv. Enzym. Regul.*, 2003, **43**, 271.

93. J. Stackhouse, S. R. Presnell, G. M. McGeehan, K. P. Nambiar and S. A. Benner, *FEBS Lett.*, 1990, **262**, 104.

94. B. A. Malcolm, K. P. Wilson, B. W. Matthews, J. F. Kirsch and A. C. Wilson, *Nature*, 1990, **345**, 86.

95. S. A. Benner, M. D. Caraco, J. M. Thomson and E. A. Gaucher, *Science*, 2002, **296**, 864.

96. J. M. Thomson, E. A, Gaucher, M. F. Burgan, T. Li, J. P. Aris and S. A. Benner, *Nat. Genet.*, 2005, **37**, 630.

97. D. L. Swofford, *Phylogenetic Analysis Using Parsimony*, Sinauer Associates, Sunderland, MA, 1998.

98. Z. Yang, *Comput. Appl. Biosci.*, 1997, **15**, 555.

99. G. Sun, Q. Ji, D. L. Dilcher, S. Zheng, K. C. Nixon and X. Wang, *Science*, 2002, **296**, 899.

100. M. E. Collinson and J. J. Hooker, *Phil. Trans. Roy. Soc. London Ser. B*, 1991, **333**, 197.

101. D. Dilcher, *Proc. Nat. Acad. Sci. USA*, 2000, **97**, 7030.

102. R. T. Bakker, *Nature*, 1978, **274**, 661.

103. P. M. Barrett and K. J. Willis, *Biol. Rev.*, 2001, **76**, 411.

104. H. S. Pereira, D. E. Macdonald, A. J. Hilliker and M. B. Sokolowski, *Genetics*, 1995, **141**, 263.

105. M. Ashburner, *Bioessays*, 1998, **20**, 949.

106. M. L. Berbee and J. W. Taylor, *Can. J. Bot.*, 1993, **71**, 1114.

107. I. Schaaff, J. Heinisch and F. K. Zimmerman, *Yeast*, 1989, **5**, 285.

108. J. Ihmels, S. Bergmann, M. Gerami-Nejad, I. Yanai, M. McClellan, J. Berman and N. Barkai, *Science*, 2005, **309**, 938.

109. R. K. Mortimer and M. Polsinelli, *Res. Microbiol.*, 1999, **150**, 199.

110. U. E. Donalies, H. T. Nguyen, U. Stahl and E. Nevoigt, *Adv. Biochem. Eng. Biotech.*, 2008, **111**, 67.

111. J. P. van der Walt, *Antonie van Leeuwenhoek*, 1956, **22**, 265.

112. J. P. van der Walt, *Antonie van Leeuwenhoek*, 1965, **31**, 341.

113. J. P. van der Walt, in *The Yeasts, a Taxonomic Study*, ed. J. Lodder, North-Holland Publishing Co., Amsterdam, 1970, p. 316.

114. I. Campbell, *J. Gen. Micro.*, 1972, **73**, 279.

115. D. G. Sidenberg and M.-A. Lachance, *Int. J. Syst. Bact.*, 1986, **36**, 94.

116. A. Vaughan Martini and A. Martini, *Int. J. Syst. Bacteriol.*, 1987, **37**, 380.

117. G. B. Fuson, H. L. Presley and H. J. Phaff, *Int. J. Syst. Bacteriol.*, 1987, **37**, 371.

118. M.-A. Lachance, *Antonie van Leeuwenhoek*, 1993, **63**, 95.

119. J. Cai, I. N. Roberts and M. D. Collins, *Int. J. Syst. Bacteriol.*, 1996, **46**, 542.

120. A. Margaritis and P. Bajpai, *Appl. Env. Microbiol.*, 1982, **44**, 1039.

121. S. Grba, V. Stehlik-Tomas, D. Stanzer, N. Vahèiæ and A. Škrlin, *Chem. Biochem. Eng. Q.*, 2002, **16**, 13.

122. Y. D. Hang, E. E. Woodams and L. E. Hang, *Bioresource Technol.*, 2003, **86**, 305.

123. S. Limtong, C. Sringiew and W. Yongmanitchai, *Bioresource Technol.* 2007, **98**, 3367.

124. W. J. Yuan, X. Q. Zhao, X. M. Ge and F. W. Bai, *Appl. Microbiol.*, 2008, **105**, 2076.

125. G. G. Fonseca, E. Heinzle, C. Wittmann and A. K. Gombert, *Appl. Microbiol. Biotechnol.*, 2008, **79**, 339.

126. P. J. Anderson, K. McNeil and K. Watson, *Appl. Environ. Microbiol.*, 1986, **51**, 1314.

127. D. B. Hughes, N. J. Tudroszen and C. J. Moye, *Biotechnol. Lett.*, 1984, **6**, 1.

128. I. M. Banat, P. Nigam and R. Marchant, *World J. Microbiol. Biotechnol.*, 1992, **8**, 259.

129. M. R. Wilkins, M. Mueller, S. Eichling and I. M. Banat, *Process Biochem.*, 2008, **43**, 346.

130. E. N. Yablochkova, O. I. Bolotnikova, N. P. Mikhailova, N. Nemova and A. I. Ginak, *Appl. Biochem. Microbiol.*, 2004, **72**, 163.

131. M. M. Lane and J. P. Morrissey, *Fungal Biology Reviews*, 2010, **24**, 17.

132. J. Boidin and J. Adzet, *Bulletin de la Société Mycologique de France*, 1957, **73**, 331.

133. H. Schneider, P. Y. Wang, Y. K. Chan and R. Maleszka, *Biotechnol. Lett.*, 1981, **3**, 89.

134. C. P. Kurtzman, *Adv. Biochem. Eng. Biotechnol.*, 1983, **27**, 73.

135. R. Maleszka and H. Schneider, *Can. J. Micro.*, 1982, **28**, 360.
136. H. Schneider, R. Maleszka, L. Neirinck, I. A. Veliky, P. Wang and Y. K. Chan, *Adv. Biochem. Eng. Biotechnol.*, 1983, **27**, 57.
137. H. Schneider, G. Mahmourides, J. L. Mabelle, H. Lee, N. Maki and H. J. McNeil, *Biotechnol. Lett.*, 1985, **7**, 361.
138. H. Lee, A. P. James, D. M. Zahab, G. Mahmourides, R. Maleszka and H. Schneider, *Appl. Environ. Microbiol.*, 1986, **51**, 1252.
139. R. Maleszka, L. G. Neirinck, A. P. James, H. Rutten and H. Schneider, *FEMS Microbiol. Lett.*, 1983b, **17**, 227.
140. P. J. Slininger, P. Bolen and C. P. Kurtzman, *Enzym. Microb. Tech.*, 1987, **9**, 5.
141. G. Ditzelmuller, C. P. Kubicek, W. Wohrer and M. Rohr, *Can. J. Micro.*, 1984a, **30**, 1330.
142. G. Ditzelmuller, C. P. Kubicek, W. Wohrer and M. Rohr, *FEMS Microbiol. Lett.*, 1984b, **25**, 1958.
143. G. Ditzelmuller, E. M. Kubicek-Pranz, M. Rohr and C. P. Kubicek, *Appl. Microb. Biotechnol.*, 1985, **22**, 297.
144. J. E. van Cauwenberge, P. L. Bolen, D. A. McCracken and R. J. Bothast, *Enzym. Microb. Tech.*, 1989, **11**, 662.
145. J. Xu and K. B.Taylor, *Appl. Env. Micro.*, 1993, **59**, 231.
146. T. W. Jeffries, J. H. Fady and E. N. Lightfoot, *Biotechnol. Bioeng.*, 1985, **27**, 171.
147. W. A. Scheffers and T. O. Wikén, *Antonie van Leeuwenhoek*, 1969, **35**, A31.
148. V. Bravo, F. Camacho, S. Sánchez and E. Castro, *Bioprocess Eng.*, 1993, **9**, 159.
149. L. G. Neirinck, M. Bosch and B. A. Prior, *Arch. Biochem. Biophys.*, 1984, **228**, 13.
150. S. Sanchez, V. Bravo, A. J. Moya, E. Castro and F. Camacho, *Process Biochem.*, 2004, **39**, 673.
151. A. P. James and D. M. Zahab, *J. Gen. Micro.*, 1982, **128**, 2297.
152. R. Maleszka, A. P. James and H. Schneider, *J. Gen. Micro.* 1983a, **129**, 2495.
153. T. W. Jeffries, *Adv. Biochem. Eng. Biotechnol.*, 1983, **27**, 1.
154. A. Toivola, D. Yarrow, E. Van Den Bosch, J. P. Van Dijken and W. A. Scheffers, *Appl. Environ. Microbiol.*, 1984, **47**, 1221.

155. G. J. O. Martin, A. Knepper, A. B. Zhou and N. B. Pamment, *J. Ind. Micro. Biotechnol.*, 2006, **33**, 834.

156. P. A. Bicho, P. L. Runnals, J. D. Cunningham and H. Lee, *Appl. Env. Micro.*, 1988, **54**, 50.

157. D. N. Wedlock, A. P. James and R. J. Thornton, *J. Gen. Micro.*, 1989, **135**, 2019.

158. R. W. Detroy, R. L. Cunningham, R. J. Bothast, M. O. Bagby and A. Herman, *Biotechnol. Bioeng.*, 1982, **24**, 1105.

159. N. Fu and P. Peiris, *World J. Micro. Biotech.*, 2008, **24**, 1091.

160. J. Bader, E. Mast-Gerlach, M. K. Popović, R. Bajpai and U. Stahl, *J. Appl. Micro.*, 2010, **109**, 371.

161. L. O. Ingram, T. Conway, D. P. Clark, G. W. Sewell and J. F. Preston, *Appl. Environ. Micro.*, 1987, **53**, 2420.

162. B. S. Dien, M. A. Cotta and T. W. Jeffries, *Appl. Microbiol. Biotechnol.*, 2003, **63**, 258.

163. B. Bräu and H. Sahm, *Arch. Microbiol.*, 1986, **144**, 296.

164. F. Alterthum and L. O. Ingram, *Appl. Environ. Micro.*, 1989, **55**, 1943.

165. K. Ohta, D. S. Beall, J. P. Mejia, K. T. Shanmugam and L. O. Ingram, *Appl. Environ. Micro.*, 1991, **57**, 893.

166. D. P. Clark, *FEMS Microbiol. Lett.*, 1989, **63**, 223.

167. G. A. Sprenger and J. Swings, in *Bergey's Manual of Systematic Bacteriology, Vol 2: The Protobacteria (Part C)*, ed. J. T. Staley, D. R. Boone, D. J. Brenner, P. De Vos, M. Goodfellow, N. R. Krieg, A. Rainey, G. M. Garrity, K.-H. Schleifer, G. Garrity, Springer-Verlag, New York, 2005, p. 282.

168. J. L. Shimwell, *J. Inst. Brew.*, 1937, **43**, 507.

169. J. G. Carr and S. M. Passmore, *J. Inst. Brew.*, 1971, **77**, 462.

170. M. Coton, J. M. Laplace and E. Coton, *Lett. Appl. Microbiol.*, 2005, **40**, 152.

171. J. Swings and J. De Ley, *Bact. Rev.*, 1977, **41**, 1.

172. B. T. P. Barker and V. F. Hillier, *J. Agric. Science*, 1912, **5**, 67.

173. E. Coton and M. Coton, *J. Inst. Brew.*, 2003, **109**, 299.

174. G. A. Stanley, T. J. Hobley and N. B. Pamment, *Biotechnol. Bioeng.*, 1997, **53**, 71.

175. P. Lindner, *Atlas der mikroskopischen Grundlagen der Gärungskunde*, Tafel 68, Berlin, 3rd edn, 1928.

176. N. F. Millis, *J. Gen. Micro.*, 1956, **15**, 521.

177. A. J. Kluyver and C. B. van Niel, *Zbl. Bakt. (II Abt.)*, 1936, **94**, 369.
178. J. L. Shimwell, *J. Inst. Brew.*, 1950, **56**, 179.
179. J. De Ley and J. Swings, *Int. J. Syst. Bacteriol.*, 1976, **26**, 146.
180. J. Swings and J. De Ley, *Int. J. Syst. Bacteriol.* 1975, **25**, 324.
181. J. Swings, K. Kersters and J. De Ley, *J. Gen. Micro.*, 1976, **93**, 266.
182. M. Coton, J. M. Laplace, Y. Auffray and E. Coton, *Int. J. Syst. Evol. Micro.*, 2006, **56**, 121.
183. Y. A. Osman, T. Conway, S. J. Bonetti and L. O. Ingram, *J. Bacteriol.*, 1987, **169**, 3726.
184. L. O. Ingram, C. K. Eddy, K. F. Mackenzie, T. Conway and F. Alterthum, *Devel. Ind. Micro.*, 1989, **30**, 53.
185. J. P. Belaïch and J. C. Senez, *J. Bacteriol.*, 1965, **89**, 1195.
186. H. W. Doelle, L. Kirk, R. Crittenden, H. Toh and M. Doelle, *Crit. Rev. Biotechnol.*, 1993, **13**, 57.
187. P. Zikmanis, R. Kruce and L. Auzina, *Arch. Microbiol.*, 1997, **167**, 167.
188. J. P. Belaïch, P. Simonpietri and A. Belaïch, *J. Gen. Micro.*, 1969, **58**, vii.
189. J. P. Belaïch, A. Belaïch and P. Simonpietri, *J. Gen. Micro.*, 1972, **70**, 179.
190. P. L. Rogers, K. J. Lee, M. L. Skotnicki and D. E. Tribe, *Adv. Biochem. Eng. Biotechnol.*, 1982, **23**, 37.
191. L. Viikari, *CRC Crit. Rev. Biotech.*, 1988, **7**, 237.
192. S. Bringer-Meyer, K. L. Schimz and H. Sahm, *Arch. Microbiol.*, 1986, **146**, 105.
193. S. R. Elsden and J. L. Peel, *Ann. Rev. Micro.*, 1958, **12**, 145.
194. G. A. Sprenger, *FEMS Microbiol. Lett.*, 1996, **145**, 301.
195. T. Bauchop and S. R. Elsden, *J. Gen. Micro.*, 1960, **23**, 457.
196. T. Fuhrer, E. Fischer and U. Sauer, *J. Bact.*, 2005, **187**, 1581.
197. H. C. Aldrich, L. McDowell, M. F. Barbosa, L. P. Yomano, R. K. Scopes and L. O. Ingram, *J. Bacteriol.*, 1992, **174**, 4504.
198. J. P. Mejia, M. E. Burnett, H. An, W. O. Barnell, K. F. Keshav, T. Conway and L. O. Ingram, *J. Bacteriol.*, 1992, **174**, 6438.
199. E. A. Dawes and P. J. Large, *J. Gen. Micro.*, 1970, **60**, 31.
200. L. Viikari, *Appl. Microbiol. Biotechnol.*, 1984, **19**, 252.

201. H. W. Doelle and P. F. Greenfield, *Appl. Microbiol. Biotechnol.*, 1985, **22**, 411.
202. L. G. Preziosi, P. F. Michel and J. Baratti, *Arch. Microbiol.*, 1990, **153**, 181.
203. L. Viikari, *Appl. Microbiol. Biotechnol.*, 1984, **20**, 118.
204. K. D. Barrow, J. G. Collins, D. A. Leigh, P. L. Rogers and. R. G. Warr, *Appl. Microbiol. Biotechnol.*, 1984, **20**, 225.
205. D. Leigh, R. K. Scopes and P. L. Rogers, *Appl. Microbiol. Biotechnol.*, 1984, **20**, 413.
206. M. Zachariou and R. K. Scopes, *J. Bacteriol.*, 1986, **167**, 863.
207. H. Loos, R. Krämer, H. Sahm and G. A. Sprenger, *J. Bacteriol.*, 1994, **176**, 7688.
208. M. J. Hardman and R. K. Scopes, *Eur. J. Biochem.*, 1988, **173**, 203.
209. R. L. Kingston, R. K. Scopes and E. N. Baker, *Structure*, 1996, **4**, 1413.
210. P. Gunasekaran, T. Karunakaran, B. Cami, A. G. Mukundan, L. Preziosi and J. Baratti, *J. Bacteriol.*, 1990, **172**, 6727.
211. P. Gunasekaran, G. Mukundan, T. R. Kannan, S. Velmurugan, N. Ait-Abdelkader, E. AlvarezMacarie and J. Baratti, *Biotechnol. Lett.*, 1995, **17**, 635.
212. T. R. Kannan, A. G. Mukundan, N. Ait-Abdelkader, V. Augier-Magro, J. Baratti and P. Gunasekaran, *Arch. Microbiol.* 1995, **163**, 195.
213. G. Sangiliyandi and P. Gunasekaran, *J. Microbiol. Meth.*, 1998, **33**, 153.
214. G. Sangiliyandi and P. Gunasekaran, *Process Biochem.*, 2001, **36**, 543.
215. V. K. Ananthalakshmy and P. Gunasekaran, *Enzym. Microb. Tech.*, 1999, **25**, 109.
216. D. Goldman, N. Lavin, A. Schwartz, G. Shoham, D. Danino and Y. Shoham, *J. Biol. Chem.*, 2008, **283**, 32209.
217. A. A. DiMarco and A. H. Romano, *Appl. Environ. Micro.*, 1985, **49**, 151.
218. C. Parker, N. Peekhaus, X. Zhang and T. Conway, *Appl. Environ. Micro.*, 1997, **63**, 3519.
219. U. Kalnenieks, N. Galinina, M. M. Toma and R. K. Poole, *Microbiology*, 2000, **146**, 1259.
220. U. Kalnenieks, M. M. Toma, N. Galinina and R. K. Poole, *Microbiology*, 2003, **149**, 1739.

221. C. Wills, P. Kratofil, D. Londo and T. Martin, *Archi. Biochem. Biophys.*, 1981, **210**, 775.
222. S. Kinoshita, T. Kakizono, K. Kadota, K. Das and H. Taguchi, *Appl. Microbiol. Biotechnol.*, 1985, **22**, 249.
223. A. D. Neale, R. K. Scopes, J. M. Kelly and R. E. H. Wettenhall, *Eur. J. Biochem.*, 1986, **154**, 119.
224. P. J. O'Mullan, S. E. Buchholz, T. Chase, Jr. and D. E. Eveleigh, *Appl. Microbiol. Biotechnol.*, 1995, **43**, 675.
225. H. An, R. K. Scopes, M. Rodriguez, M. Rodriguez, K. F. Keshav and L. O. Ingram, *J. Bacteriol.*, 1991, **173**, 5975.
226. J. S. Seo, H. Chong, H. S. Park, K. O. Yoon, C. Jung, J. J. Kim, J. H. Hong, H. Kim, J. H. Kim, J. I. Kil, C. J. Park, H. M. Oh, J. S. Lee, S. J. Jin, H. W. Um, H. J. Lee, S. J. Oh, J. Y. Kim, H. L. Kang, S. Y. Lee, K. J. Lee and H. S. Kang, *Nat. Biotechnol.*, 2005, **23**, 63.
227. S. Yang, K. M. Pappas, L. J. Hauser, M. L. Land, G.-L. Chen, G. B. Hurst, C. Pan, V. N. Kouvelis, M. A. Typas, D. A. Pelletier, D. M. Klingeman, Y.-J. Chang, N. F. Samatova and S. D. Brown, *Nat. Biotechnol.*, 2009, **27**, 893.
228. V. N. Kouvelis, E. Saunders, T. S. Brettin, D. Bruce, C. Detter, C. Han, M. A. Typas and K. M. Pappas, *J. Bacteriol.*, 2009, **191**, 7140.
229. S. Bringer, T. Hartner, K. Poralla and H. Sahm, *Arch. Microbiol.*, 1985, **140**, 312.
230. M. Rohmer, P. Bouvier-Nave and G. Ourison, *J. Gen. Micro.*, 1984, **130**, 1135.
231. G. Flesch and M. Rohmer, *Biochem. J.*, 1989, **262**, 673.
232. M. Rohmer, P. Bouvier and G. Ourisson, *Proc. Nat. Acad. Sci. USA*, 1979, **76**, 847.
233. G. Ourisson, M. Rohmer and K. Poralla, *Ann. Rev. Microbiol.*, 1987, **41**, 301.
234. A. van Dorsselaer, A. Ensminger, C. Spyckerells, M. Dastillung, O. Sieskind, P. Arpino, P. Allbrecht and G. Ourisson, *Tetrahedron Lett.*, 1974, **15**, 1349.
235. L. O. Ingram and T. M. Buttke, *Adv. Microb. Physiol.*, 1984, **25**, 253.
236. L. O. Ingram, *Trends Biotechnol.*, 1986, **4**, 40.
237. S. Hayashida and K. Ohta, *Agric. Biol. Chem.*, 1978, **42**, 1139.

238. V. C. Carey and L. O. Ingram, *J. Bacteriol.*, 1983, **154**, 1291.
239. S. E. Buchholz and D. E. Eveleigh, *Biotechnol. Adv.*, 1990, **8**, 547.
240. M. Zhang, C. Eddy, K. Deanda, M. Finkelstein and S. Picataggio, *Science*, 1995, **267**, 240.
241. K. Deanda, M. Zhang, C. Eddy and S. Picataggio, *Appl. Environ. Micro.*, 1996, **62**, 4465.
242. A. Mohagheghi, K. Evans, Y. C. Chou and M. Zhang, *Appl. Biochem. Biotechnol.*, 2002, **98**, 885.
243. H. G. Lawford and J. D. Rousseau, *Appl. Biochem. Biotechnol.*, 2002, **98**, 429.
244. Y. Zhang, R. Ma, Z. Zhao, Z. Zhou, W. Lu, W, Zhang and M. Chen, *J. Microbiol. Biotechnol.*, 2010, **20**, 1156.
245. K. S. Makarova and M. J. Daly, in *Bacterial Stress Responses*, ed. G. Storz and R. Hengge, ASM Press, Washington, DC, 2nd edn, 2011, p. 445.
246. H. G. Lawford and J. D. Rousseau, *Appl. Biochem. Biotechnol.*, 1993, **39**, 687.
247. E. Joachimstahl, K. D. Haggett, J. H. Jang and P. L. Rogers, *Biotechnol. Lett.*, 1998, **20**, 137.
248. S. Yang, M. L. Land, D. M. Klingeman, D. A. Pelletier, T. Y. Lu, S. L. Martin, H. B. Guo, J. C. Smith and S. D. Brown, *Proc. Nat. Acad. Sci. USA*, 2010, **107**, 10395.
249. B. Bochner, V. Gomez, M. Ziman, S. Yang and S. D. Brown, *Appl. Biochem. Biotechnol.*, 2010, **161**, 116.
250. P. S. Panesar, S. S. Marwaha and J. F. Kennedy, *J. Chem. Technol. Biotechnol.*, 2006, **81**, 623.
251. Q. Xu, A. Singh and M. E. Himmel, *Curr. Opin. Biotechnol.*, 2009, **20**, 364.

CHAPTER 3

Alcoholic Fermentation

3.1 THE HISTORY OF SOME BASIC SCIENCE

When one considers that microbes were first observed some 350 years ago, it is difficult for us, in our modern world, to understand how it took around 200 years, until the time of Pasteur, to attribute unequivocally the phenomenon of alcoholic fermentation to the humble yeast. The religious atmosphere during those years was such that any doctrine other than spontaneous generation (abiogenesis) was unacceptable and prevented the promulgation of heretic notions that a minute organism, such as the yeast, might be responsible for fermentation. The story of how abiogenesis finally became impossible for most sentient people to believe in is one of the landmark scenarios in science. A few of the salient facts follow, and will hopefully indicate how tortuous the pathway was before man eventually understood how grape juice turned into wine, how honey and palm oil wines were produced and how milk soured. Having said that, since the mid-sixteenth century sake brewers have been using remarkably sophisticated technology, including heat-processing, which would suggest that they appreciated the dangers of "invisible", unwanted forms of life.

Thanks to the stimulus provided by the foundation of the Royal Society, there were a number of important scientific developments in the last couple of decades of the seventeenth century. From our perspective, two of the most significant involved the microscopical

Alcohol and its Role in the Evolution of Human Society
Ian S. Hornsey
© Ian S. Hornsey 2012
Published by the Royal Society of Chemistry, www.rsc.org

Figure 3.1   Artist's impression of Robert Hooke – no contemporary images
survive.

work of Hooke and his Dutch contemporary, van Leeuwenhoek.
Robert Hooke (1635–1703) (Figure 3.1) was curator of experiments at the Royal Society, making him, in effect, the first
professional scientist in Britain. He was certainly the first major
scientific figure to carry out scientific experiments and it is now
apparent that he developed many of the fundamentals of
gravitational theory before Newton.

It was while with the Royal Society that Hooke developed his
first compound microscope (Figure 3.2). Simple lenses (*e.g.* the
magnifying glass) had been in use since the end of the thirteenth
century, but their powers of magnification were limited. In 1590, a
Dutch spectacle-maker, Zacharius Janssen, used a second lens to
magnify the image produced by a primary lens – the basic principle
of the compound microscope. Hooke's work was an extension of
Janssen's and, encouraged by the Royal Society, led to the

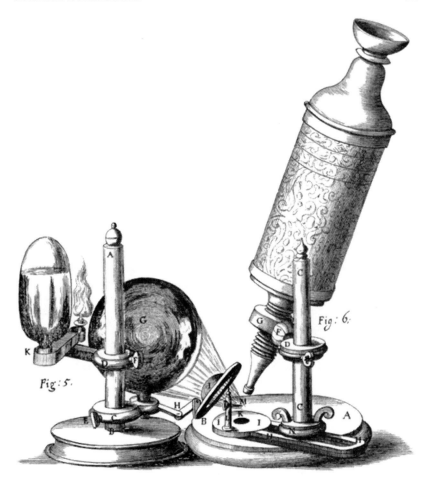

Figure 3.2 Hooke's compound microscope.

publication of *Micrographia* in 1665[1] (Figure 3.3), a work that he dedicated to Charles II. Hooke noticed the compartmentalized structure of most living organisms, *i.e.* that they formed tissues from individual cells, and was the first person to describe micro-fungi, which relate to much of this book.

Hooke did not observe organisms as small as bacteria, even though his microscope would have been powerful enough to have done so. This was probably because he viewed most of his objects, which were in the dried state, by reflected light, which would have made them somewhat opaque. Thus, the "discovery" of micro-

MICROGRAPHIA:

OR SOME

Phyſiological Deſcriptions

OF

MINUTE BODIES

MADE BY

MAGNIFYING GLASSES

WITH

OBSERVATIONS and INQUIRIES thereupon.

By *R. HOOKE*, Fellow of the ROYAL SOCIETY.

*Nampoſſis oculo quantum contendere Lincens,
Non tamen idcirco contemnas Lippus inungi.* Horac. Ep. lib. 1.

LONDON, Printed by *Jo. Martyn*, and *Ja. Alleſtry*, Printers to the
ROYAL SOCIETY, and are to be ſold at their Shop at the *Bell* in
S. *Paul's* Church-yard. M DC LX V.

Figure 3.3   "*Micrographia*", title page.

Figure 3.4 Antonj van Leeuwenhoek.

organisms was left to Antonj van Leeuwenhoek (1632–1723), from Delft, who was a self-educated man and not formally trained as a scientist. Leeuwenhoek (Figure 3.4), who was the son of a draper, was fortunate enough to hold a political sinecure in Delft, which allowed him plenty of time for pursuing his hobbies of metalwork and lens-making. His instruments were not compound microscopes, but single mounted lenses, with the object mounted in such a way that light was transmitted through it. With this apparatus he could achieve magnification of up to $\times 300$ and, by keeping his methods secret, he made sure that no other workers were able to use a single lens so effectively. Clifford Dobell[2] penned an excellent account of the life and work of van Leeuwenhoek, which has been reprinted.

From 1674 onwards, van Leeuwenhoek corresponded over 200 times with the Royal Society, one letter detailing his observations of fermenting beer in which he observed a variety of particulate matter:

"Some of these seemed to me to be quite round, others were irregular, and some exceeded the others in size, and seemed to consist of two, three or four of the aforesaid particles joined together. Others again consisted of six globules and these last formed a complete globule of yeast."

The letter, complete with appropriate diagrams, was written to Mr Thomas Gale, a Fellow of the Royal Society in 1680. The drawings (Figure 3.5) were taken from wax models made as a result of his observations, rather than straight from living material, and the letter commences: "I have made many observations concerning yeast and seen throughout that the aforesaid consisted of globules floating through a clear substance, which I judged to be beer ..." Some of the drawings made by van Leeuwenhoek were very precise, and it is possible to place taxonomic names on some of them. For whatever reason, systematic study of microbes was delayed for another 100 years or so, and it was not until 1786 that the Danish zoologist, O. F. Müller, studied bacteria and succeeded in discovering details of their structure. Müller also left accurate illustrative details, and it is possible to relate many of his drawings to extant taxa.

Yeasts from different environments were described, in some detail, but it was not until nearly 150 years later, in 1826, that Desmazières described the elongated cells from a film of growth on

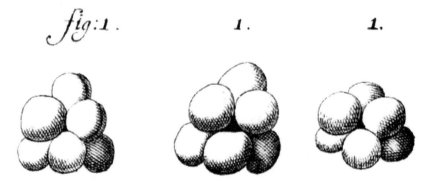

Figure 3.5   van Leewenhoek's yeast cells.

beer. He named the organism "Mycoderma Cerevisiæ" and regarded it as a member of the animal kingdom, rather than the plant kingdom. He did not record whether he thought that this organism had anything actively to do with fermentation. Leeuwenhoek, himself, made no mention of whether he recognized his "yeasts" as living organisms, nor did he allude to their *raison d'être* in beer.

One of the leading questions to be faced by seventeenth century natural scientists was that of the Aristotelian doctrine of "spontaneous generation", which, in essence, propounded that "life" was continually being created out of inanimate matter. The doctrine was not out of line with the biblical account of the Earth's creation, and for around 200 years there was a fierce debate about the topic. Spontaneous generation dominated some areas of scientific thought during the eighteenth century, and it was not until publications of Pasteur that it was laid to rest.

One of the first people to seriously question abiogenesis was the Italian physician Francesco Redi (1626–1679), who was interested in how maggots developed from unprotected meat. It had always been supposed that maggots in decaying meat were derived spontaneously from transformations occurring within the meat itself, so no searching questions were ever asked. In experiments performed around 1665, Redi placed clean linen cloths over jars containing fresh samples of meat. He observed that flies, attracted to the meat, landed on the cloth and laid eggs. Later on, maggots could be seen on the cloth, but not on the meat, showing conclusively that maggots grow from eggs and are not able to arise spontaneously from non-living meat. Unfortunately, the work was largely ignored; most of those who did read it regarded Redi as a heretic. In theory, Leeuwenhoek's discovery of microbial life should have signalled the "beginning of the end" for abiogenesis, but adherents of religious dogma simply accommodated these "newly found" forms of life in their explanations of how life was created. To them, non-living animal and vegetable matter contained a "vital" or "vegetative force" capable of converting such inanimate matter into new and different forms of life; Leeuwenhoek's "animalcules" were proof of this.

In 1710, Louis Joblot (1645–1723) observed that hay, when infused in water and allowed to stand for some days, gave rise to countless minute organisms ("infusoria"); this was an example of abiogenesis. Joblot then boiled a hay infusion and divided it into two portions, placing one in a carefully baked (sterilized) and closed vessel, which was heated thoroughly and kept closed. The other portion was not heated and was kept in an open vessel. The infusion in the open vessel teemed with microbial life after a couple of days, but no life appeared in the closed vessel – as long as it remained closed. Thus, Joblot proved that the infusion alone, once freed of "life" by heating, was incapable of generating new life spontaneously.

Two mighty eighteenth century protagonists, in this subject area, were two men of the cloth; the Englishman John Turberville Needham (1713–1781) and the Italian Lazzaro Spallanzani (1729–1799). Needham (Figure 3.6) was a devout believer in abiogenesis, and performed experiments along the same lines as Joblot, except that he used mutton broth, instead of hay infusion. Needham

Figure 3.6   John Needham.

found that "life" was created in heated, closed vessels as well as open ones, and maintained that he had confirmed abiogenesis, beyond all reasonable doubt, but it has since been shown that he had not heated his closed vessels sufficiently, and bacterial spores (which were unknown at that time – he conducted these experiments in 1748) had survived a sub-lethal heat dose. In addition, Needham's flasks were "sealed" by corks, which are, of course, porous.

Spallanzani set up two series of flasks; one was sealed at the top by melting the glass, the other was corked, *à la* Needham. The flasks had been filled with seeds and other vegetable matter, and then heated for one hour before being sealed. After a time he microscopically investigated their contents. The corked flasks contained innumerable small, swimming animalcules, just as Needham had found, but the properly sealed flasks contained none, or very few. Spallanzani concluded that the "life" in Needham's work had entered *via* the cork (*i.e.* it was in the air, and wouldn't have occurred if the necks had been sealed tight), expounding that: "animalcules do not exist that can survive boiling for one hour." In theory, this work should have settled the debate about abiogenesis, but it didn't. As we shall see, what it did do was to form the basis for the art of preserving food by canning, in the early nineteenth century. The main objection to Spallanzani's work was that by excluding air from some flasks, the "vital force" was unable to operate, and so no life was to be expected. Life in the absence of air was unheard of at this time.

Antoine-Laurent Lavoisier's (Figure 3.7) discovery of oxygen in 1775, and his work on the relationship between air and life, renewed interest in the controversy over the origin of life, and work on oxygen during the last quarter of the eighteenth century was to prove seminal to the advancement of chemistry and biology. Some of the most important work aimed at disproving spontaneous generation was carried out by the great German biologist (co-founder of the theory of cells) Theodor Schwann (1810–1882) (Figure 3.8), who allowed air to pass freely over previously heated organic substrates (meat and hay infusions), but only after it had passed through very hot glass tubes. Infusions receiving such air failed to yield "life". Franz Schulze (1815–1873), a contemporary of Schwann, but working independently, did much the same thing and obtained similar results, only he had passed the

Figure 3.7   Lavoisier's statue in the Louvre.

air supply in his experiments through solutions of sulfuric acid and potassium hydroxide, before allowing it to contact the infusions. Adherents to abiogenesis denigrated such work and said that it was

Figure 3.8 Theodor Schwann.

heat and the effect of harsh chemicals that was destroying the "vital force" necessary to create life.

Then, in the late 1850s, H. G. F.Schröder (1810–1885) and T. von Dusch (1824–1890) reported experiments in which they had studied the role of air in initiating decomposition of organic materials. It had already been established that decomposition could be prevented in many instances by heating the material and then excluding air from it (this technique was being used for the preservation of food). In their experiments, conducted between 1854 and 1861, they passed unheated air through cotton wool, and prevented their sterile organic broths from becoming contaminated. They suggested that microbes in the air were being excluded by filtration. In many ways, some of Pasteur's later experiments were an elaboration on Schwann's work.

From the earliest of times, natural philosophers have been fascinated about the seemingly spontaneous change that transformed grape juice into the physiologically interesting beverage called wine. Fermentation, although not understood, was an important entity to the alchemists and formed a core for many of their ideas (as did putrefaction, which they appreciated was a separate process). Ideas relating to the exact nature of fermentation were highly confused until the era of the phlogiston chemists, and the realization that, as Becher reported in 1682, only sweet liquids could give rise to true fermentation: "*Ubi flotandum, nihil fermentare quod non sif dulce*". The first positive ideas concerning the nature of fermentation were forwarded by the phlogistic chemists, Henry Willis in 1659, and Georg Ernst Stahl in 1697. Stahl's work is an extension of that of Willis. As Harden[3] says of these two workers:

"To explain the spontaneous origin of fermentation and its propagation from one liquid to another, they supposed that the process consisted in a violent internal motion of the particles of the fermenting substance, set up by an aqueous liquid, whereby the combination of the essential constituents of this material was loosened and new particles formed, some of which were thrust out of the liquid and others retained in it."

The newly formed particles in an alcoholic fermentation of a sugary substrate, at normal temperatures and pressures, are, of course, CO_2, which comes out of the liquid, and ethanol, which remains within it. Stahl also appreciated that a fermentation that is in such a state of internal turmoil, can readily transmit that "turmoil" to a hitherto static situation, thus causing similar changes in the *status quo*. He also observed that small amounts of acetic acid were formed during an alcoholic fermentation. The far-reaching consequences of this concept of fermentation are somewhat alien to Stahl's other main contribution to seventeenth century scientific thought, for it was he who propounded the phlogiston theory of combustion, which dominated chemical thinking for a century, or more. Phlogistic theory held sway until

Lavoisier's work heralded its downfall toward the end of the eighteenth century.

Until Lavoisier's work, the exact composition of organic compounds was unknown, and it was the Frenchman who established that they consisted basically of carbon, hydrogen and oxygen. He analysed the compounds relevant to alcoholic fermentation, applied the results of these analyses to a study of alcoholic fermentation, and was able to show, albeit somewhat fortuitously, that the products of fermentation equated to the whole matter of the original sugar. In order to do this, Lavoisier invoked his fundamental principle of experimental chemistry, which essentially states that there is the same quantity of matter before and after a chemical transformation. He constructed a balance sheet illustrating the quantities of carbon, hydrogen and oxygen in the pre-fermented sugar, and in the resulting CO_2, ethanol and acetic acid, confirming that the products contained the whole substance of the sugar. In hindsight, we find that there were substantial errors in the elemental analysis of sugar, but that these were balanced by errors in his other analyses, such that the overall stoichiometry balances.

Although Lavoisier was definite about the nature of the chemical change that he was studying, he was non-committal as to how fermentation was brought about. He suggested that the sugar was to be regarded as an oxide, and that when it was split into its two main fragments, one was oxidized at the expense of the other to form carbonic acid, whilst the other was deoxygenized in favour of the former to produce ethanol, which is combustible. In conclusion to this aspect of his work, Lavoisier proposed that "if it were possible to recombine carbonic acid and ethanol, then sugar would result". In hindsight, we can safely say that it is from Lavoisier's work that the modern studies on fermentation derive. One of the first theories regarding the cause of fermentation was forwarded by the Italian scientist Fabroni, at the very end of the eighteenth century. He maintained that it was attributable to the action of gluten [the definition of gluten in the OED says: "The nitrogenous part of the flour of wheat or other grain, which remains behind as a viscid substance when the starch is removed by kneading the flour in a current of water"] derived from the starch grain and sugar.

In 1803, the noted French chemist, Louis Jacques Thenard (1777–1857), refuted the gluten hypothesis, and noted that all fermenting liquids appeared to deposit a nitrogen-containing material, resembling brewer's yeast, during the course of the reaction. Results of some of his other experiments showed that when yeast is used to ferment pure sugar, it alters in its characteristics and is deposited as a white residue, which does not have the subsequent ability to ferment fresh sugar. This residue Thenard found to be devoid of nitrogen, and of much reduced weight. Thenard himself offered no alternative to explain the nature of fermentation, but he did feel that something akin to brewer's yeast was involved. He proposed that the causative agent was of animal origin and that it decomposed during the reaction to provide some of the observed CO_2 that is released; a view at variance with Lavoisier's findings.

The next outstanding contribution to the subject was published by the French chemist Joseph Louis Gay-Lussac (Figure 3.9) in

Figure 3.9   Joseph Louis Gay-Lussac.

1810. He proposed that fermentation was instigated by the action of oxygen on fermentable material, a conclusion that was reached by researching some of the methods used by the "father of canning", the Frenchman Nicholas Appert, the bulk of whose work was over the period 1795–1810. Appert's work was stimulated by a 12,000 franc prize offered by the French government for anyone who perfected a method for preserving food destined for the armed forces during the Napoleonic conflagrations. Basically, he submersed food contained in cork-stoppered, wide-mouthed glass bottles in boiling water for several hours. Such methods, simple as they may seem, were good enough to be used for many years by commercial canners, even though the relationship between micro-organisms and food spoilage was unknown. Coincidentally, Appert was the son of an innkeeper, and was able to gain experience in brewing and pickling before becoming a chef and a confectioner. Gay-Lussac found that food treated by Appert's method was quite stable but, as soon as it was exposed to air, fermentation and/or putrefaction set in. Further experiments enabled Gay-Lussac to prove that air was the causative agent, and that if liquid foods were actually boiled, and then exposed to air, then the onset of the two processes was delayed. He also found that if brewer's yeast was heated it was incapable of initiating fermentations.

The proposition that yeast was a living organism, not a chemical compound, was not made until the mid- to late-1830s, when the results from three totally independent pieces of work appeared. It should be emphasized that these treatises coincided with the development of much-improved microscopes and, in order of publication date, the first to appear was the French mechanical engineer Charles Cagniard-Latour, who, in 1835 (with additions in 1837), microscopically monitored the changes that yeast underwent over a period of fermentation. He noted the globular form of the yeast and observed reproduction by budding: "a small cell formed on the surface of a yeast globule; the two cells remained attached to each other for some time before becoming two separate globules." The lack of mobility of the organism led Cagniard-Latour to designate yeast to the vegetable (plant) kingdom, and he said that only living cells could cause alcoholic fermentation, which he deemed to be taking place in the liquid phase of the yeast suspension.

Also in 1837 was the publication of Theodor Schwann's work, which proved categorically that alcoholic fermentation was the result of a living organism, not an inanimate chemical mass. According to Schwann's experiments, solutions of cane sugar to which yeast had been added, and then boiled, fermented only when atmospheric air was passed through them but, contrary to Gay-Lussac's theory, not when heated air was used. Schwann also observed budding, and the formation of "several cells within one cell", what we now know as sporulation. The presence of a living organism during fermentation was confirmed by microscopical work, and Schwann was able to describe the morphology of yeast, which he named *Zuckerpilz*, or sugar fungus (from which the generic name *Saccharomyces* emanates). Schwann's concise explanation of alcoholic fermentation was as follows:

"The decomposition brought about by this sugar fungus removing from the sugar, and a nitrogenous substance, the materials necessary for its growth and nourishment, whilst the remaining elements of these compounds, which were not taken up by the plant, combined chiefly to form alcohol."

Unlike Cagniard-Latour, Schwann thought that the actual fermentation process was carried out inside the yeast cell. Schwann's report was very closely followed by a treatise by the German botanist Friedrich Traugott Kützing, which was also based on meticulous microscope work. There are accurate descriptions of what Kützing clearly thinks is a living vegetable organism responsible for fermentation. To sceptical chemists, of whom there were plenty (such as Berzelius), he issues the following statement:

"It is obvious that chemists must now strike yeast off the roll of chemical compounds, since it is not a compound but an organised body, an organism."

The summation of the work of these three great scientists was that "yeasts are plants consisting of individual cells which multiply in sugar solutions and as a result of their growth produce alcohol."

All three pieces of work were criticized, and even derided by the scientific establishment of the time. Probably the most influential

Figure 3.10   Jöns Jacob Berzelius.

chemist of the era was the Swede Count Jöns Jacob Berzelius (1779–1848), described by Harden[3] as "the arbiter and dictator of the chemical world". Berzelius (Figure 3.10) was totally unimpressed with the microscopic evidence presented to substantiate the above work and, although he accepted the role of yeast during fermentation, he regarded it as "being no more a living organism than was a precipitate of alumina". As far as Berzelius was concerned, he had published his views on fermentation, and similar chemical reactions, in 1836, and that was that. He had introduced the concept of "the catalytic force" (of which yeast was an example) that he held responsible for many chemical reactions, between substances both of mineral and of animal and vegetable origin. This force, he said:

> "Enabled bodies, by their mere presence, and not by their affinity to arouse affinities ordinarily quiescent at the temperature of the experiment, so that the elements of a compound body arrange themselves in some different way, by which a greater degree of electro-chemical neutralisation is attained."

If this was not bad enough for the credibility of the work of Cagniard-Latour, Schwann and Kützing, in the scientific world, in 1839 two more eminent chemists muddied the waters even more; they were the Germans Justus Liebig (1803–1873) and Friedrich Wöhler (1800–1882). Liebig (Figure 3.11), who at that time was

Figure 3.11   Justus Liebig.

regarded as being one of the first eminent "biochemists", was engaged in the debate on the cause of alcoholic fermentation and, although he did not believe that living organisms were involved, he did not agree with Berzelius's concept either. Liebig's reputation was immense, and his ideas concerning chemical processes were quite precise and generally accepted by the scientific community. His theory of fermentation, as Anderson[4] (1989) says, was merely an extension of Stahl's hypothesis, since it appears to be based upon very little original experimental work, and is seemingly based upon the findings of Thenard and Gay-Lussac.

Liebig felt that, as a result of alcoholic fermentation, all of the carbon in the sugar was converted into CO_2 and alcohol, the transformation being brought about by a body he called "the ferment", which was formed as the result of a charge set up by the access of air to the plant juices containing sugar, and which contains all the nitrogen of the nitrogenous constituents of the juice. The accumulation of the nitrogen caused instability, which was sufficient to trigger-off similar instability in the sugar, hence fermentation – which Liebig thought was a form of decomposition, rather than a reaction involving the generation of life. In a pure sugar solution, he maintained that the decomposition of the "ferment" finishes, and so fermentation ceases. This was not the case in the fermentation of plant juices and beer wort, because more "ferment" is continually being formed from nitrogenous constituents, and so the sugar in the juice/wort is completely fermented away. At the end of an alcoholic fermentation a yellow residue of yeast remained, which Liebig saw as being a non-crystalline, globular, solid material.

Wöhler's denigration of the trio's work was far more surreptitious, and consisted of an anonymous article in the journal *Annalen der Pharmacie*, of which Liebig was joint-editor. In 1839 the journal published (with Liebig's sanctioning) an article by the French chemist Turpin attesting to the accuracy of Cagniard-Latour's work. In order to counter the sentiments presented in Turpin's paper, a satirical attack was published in the same journal immediately afterwards, this being aimed at all proponents of the notion that yeast was a living organism. The article, one of the most amazing in the history of science, was reputedly written by Wöhler, with some minor adjustments by Liebig himself, and it portrays yeast as consisting of "eggs", which, when placed in a sugar solution, develop into minute animals. The authors claim to have observed fine details of the anatomy of these organisms:

"They have a stomach and an intestinal canal, and their urinary organs can be readily distinguished. The moment these animals are hatched they begin to devour the sugar in the solution, which can be readily seen entering their stomachs. It is then immediately digested, and the digested product can be recognised with certainty in the excreta from the intestinal canal. In a word, these infusoria eat sugar, excrete alcohol from their intestinal canals, and carbonic acid from their urinary organs. The bladder, when full, is the shape of a champagne bottle, when empty it resembles a little ball; with a little practice, an air-bladder can be detected in the interior of each of these animalculae; this swells up to ten times its size, and is emptied by a sort of screw-like action effected by the agency of a series of ring-shaped muscles situated in its outside."

This satirical account by Wöhler, which did the author no credit, seems a stark contradiction of behaviour in a man who is credited with perfecting the first laboratory synthesis of an organic compound (urea) from inorganic constituents (heat and ammonium thiocyanate). Liebig's sensible (albeit incorrect) account of fermentation was published immediately after the above skit. It is patently obvious that Berzelius and his cohorts were totally unwilling to have the process of alcoholic fermentation relegated from a cornerstone of theoretical organic chemistry to a nebulous reaction carried out by some minute living organism!

In spite of all the controversy surrounding the nature of fermentation, by the mid-nineteenth century there were enough influential adherents to the ideas of Cagniard-Latour to be able to convince the scientific world that yeast was a living organism; even Berzelius had admitted the fact by 1848. Liebig, however, remained adamant about his own theory. In 1841, Eilhard Mitscherlich (1794–1863), the discoverer of the chemical phenomenon of isomorphism, proved to be one of the first scientists to recognize yeast as a micro-organism, and showed that an aqueous suspension of it could invert sugar. He demonstrated that when yeast was placed in a glass tube closed by parchment, and plunged into sugar solution, the sugar entered the glass tube and was fermented there. There was no visible fermentation outside of the tube, and this was taken as proof that fermentation only occurred at the surface of yeast cells, and explained the process according to Berzelius's theory of catalytic activity, rather than Liebig's idea of the transference of molecular instability. In 1843, Helmholtz obtained results similar to those of Mitscherlich, only he used an animal membrane to close the tube instead of parchment. The aforementioned work of Schwann, Schulze, Schröder and von Dusch, and the later definitive work of Louis Pasteur, determined the true origin and function of yeast in alcoholic fermentation. In 1860, Pasteur concluded:

"Alcoholic fermentation is an act correlated with the life and organisation of the yeast cells, not with the death or putrefaction of the cells, any more than it is a phenomenon of contact, in which case the transformation of sugar would be accomplished in presence of the ferment without yielding up to it or taking from it anything."

He then proceeded to dismantle Liebig's theory *via* a series of carefully planned experiments. Pasteur's work is well documented in brewing literature (*e.g.* Redman;[5] Barnett[6]) and, for reasons of space, I shall not report the bulk of it here, save to quote the great man's conclusion as to the nature of fermentation:

"The chemical act of fermentation is essentially a phenomenon correlative with a vital act, commencing and ceasing with the latter. I am of opinion that alcoholic fermentation never occurs without simultaneous organisation, development, multiplication of cells, or the continued life of cells already formed. The results expressed in this memoir seem to me to be completely opposed to the opinions of Liebig and Berzelius. If I am asked in what consists the chemical act whereby the sugar is decomposed and what is its real cause, I reply that I am completely ignorant of it.
Ought we to say that the yeast feeds on sugar and excretes alcohol and carbonic acid? Or should we rather maintain that yeast in its development produces some substance of the nature of a pepsin, which acts upon the sugar and then disappears, for no such substance is found in fermented liquids? I have nothing to reply to these hypotheses. I neither admit them nor reject

them, and wish only to restrain myself from going beyond the facts. And the facts tell me simply that all true fermentations are correlative with physiological phenomena."

By 1872, Pasteur had answered all of Liebig's criticisms of his work, and most authorities accepted his tenet that "there is no fermentation without life". This applied not only to alcoholic fermentation, but to the myriad of other fermentations carried out by bacteria. By 1875, Pasteur had concluded that fermentation was the result of life without oxygen, whereby, in the absence of the free form of that element in the atmosphere, cells were able to obtain energy that was liberated by the decomposition of substances containing combined oxygen.

A synthesis of many of Pasteur's ideas was presented in two classic works: *Etudes sur le vin*, published in 1866,[7] and *Etudes sur la bière*,[8] published ten years later. Pasteur actually commenced his researches into the fermentation of sugar in the mid-1850s and, between 1855 and 1875, he established unequivocally: 1) the role of yeast in ethanolic fermentation, 2) fermentation as a physiological phenomenon and 3) differences between the aerobic and anaerobic utilization of sugar by yeast; indeed, Pasteur invented the terms "aerobic" and "anaerobic".

An assessment of the influence of Pasteur's work on the brewing industry has been given by Anderson,[9] and the same author has documented some of the highlights of the history of brewing science,[10] and the history of industrial brewing,[11] The significance of yeast and the role played by ethanolic fermentation during the formative years of the sciences of biochemistry and microbiology have been admirably reviewed by James Barnett.[12] Over the last decade or so, Barnett has written a series of articles on the history of yeast research, and his first three contributions dealt with the period from 1789 to 1850;[13] Louis Pasteur and his contemporaries[6] and the period from 1880–1900, which outlines the work of Fischer and Buchner and their contemporaries.[14] No account of alcoholic fermentation would be complete without reference to Arthur Harden's classic *Alcoholic Fermentation*, first published in 1911,[3] a book that has gone through several editions. Barnett's work on the history of yeast research has recently been synthesized in a work that deals with the contribution of men such as Emil Christian Hansen, as well as containing an excellent set of references.[15]

One conundrum that perplexed scientists for decades was exactly why wine and cider fermentations could commence spontaneously, without extraneous yeast addition, while brewery fermentations required yeast to be added first. It was argued that, during wine production, there was sufficient heat and air in must to "set particles in motion" (for the ferment), while in beer wort air levels had been reduced, through heat during the malt kilning process, and during wort-boiling. Yeast was, therefore, needed to "excite the separation and new arrangement on which the perfection of the products depends, and prevent the accidents to be apprehended from worts' disposition to ferment spontaneously through slow absorption of air from the atmosphere".

3.2 FERMENTATION BIOCHEMISTRY

Two major catabolic processes generate cellular energy in non-photosynthetic eukaryotic organisms: respiration, which requires oxygen, and fermentation, which occurs anaerobically. Respiration that is carried out in mitochondria produces around 80% of the energy by most eukaryotes under aerobic conditions, and mitochondria are, therefore, essential for cellular energy production in most higher organisms. When glucose is abundant, however, certain yeast species mostly undergo fermentation, even under aerobic conditions, and most can survive without a functional mitochondrial genome. The yeasts in question are those that underwent a process called whole-genome duplication (WGD) millions of years ago.

In overall terms, a major biochemical feature of a fermentation is that the NADH (or $FADH_2$) donates its electrons to a molecule produced by the same metabolic pathway that produced the electrons carried by the coenzyme. Mechanistically, fermentation is usually considered to be the "simplest" of the three main methods of energy-yielding metabolism, being a metabolic process in which carbohydrates (and related compounds) are oxidized, and thus release energy, in the absence of any external electron acceptors. During fermentation, electrons are not passed through an electron transport chain and oxidative phosphorylation does not occur. For example, many microbes will convert sugars to pyruvic acid with the concomitant production of NADH. If glycolysis and ATP production are to continue, NADH has to pass its electrons on to an acceptor, to regenerate NAD^+, and this is normally effected by using pyruvic acid (or one of its derivatives) as the final electron acceptor. Thus, organic compounds serve as both electron donors and electron acceptors. The waste products of these NAD^+ regeneration systems are known as "fermentation products". Microbes are especially diverse in the array of fermentation products that they produce, including organic acids such as acetic, lactic, succinic and butyric, as well as neutral compounds like ethanol, butanol and acetone.

Micro-organisms have evolved different methods of regenerating NAD^+ from NADH. In *S. cerevisiae*, acetaldehyde serves as the terminal electron acceptor, whence it is reduced to ethanol. Pyruvate is first decarboxylated, by pyruvate decarboxylase, to yield CO_2 and acetaldehyde (a step requiring thiamine pyrophosphate). Acetaldehyde is then reduced to ethanol by alcohol dehydrogenase (see page 53), thus regenerating NAD^+.

Some fermentations (in most breweries) employ relatively few species/strains of yeast, and the same can be said for the distillation industries, and for the yeasts used in a bakery. Other fermentations are much more complicated, those in wineries, for example, involving many more species. Over 200 species, representing some 28 genera, have been isolated from grapes, must and wine,[16] and over 100 strains of *S. cerevisiae* have been reported from spontaneous wine fermentations in Spain.[17] The diversity and evolution of non-*Saccharomyces* yeast populations during wine fermentations has also been investigated.[18]

In addition to being provided with a major source of carbon (*S. cerevisiae* can only grow on a limited range of carbohydrates, with glucose being the preferred carbon and energy source), and a nitrogen source, the yeast has day-to-day requirements for the macro-elements sulfur and phosphorus, and certain vitamins, such as biotin, pantothenic acid, inositol, thiamine, pyridoxine and nicotinic acid, if growth is to proceed normally. A range of metallic ions, such as Na^+, K^+, Fe^{++}, Zn^{++}, Cu^{++} and Mn^{++}, must also be provided, together with a range of trace elements. Most fermentable substrates, such as brewers' wort and grape must, will normally provide the yeast with the opportunities for maximal growth and metabolism.

There are two ways in which a fermentation can be initiated: either spontaneously (*i.e.* yeast cells arrive from production equipment, air or from raw materials [*e.g.* grapes]), or *via* an intentional yeast addition (seeding). With today's predilection for a consistent product, the latter is now more common, although some wines, ciders and lambic beers still require a "natural" fermentation. As is mentioned elsewhere, during the preparation of some beverages, such as palm wine, it is almost impossible to prevent spontaneous fermentation. Also, in our modern interest in all things "green", spontaneous fermentations are non-interventionalist, and cost nothing to initiate.

3.2.1 Glycolysis

Using vinification as an example, we can now consider the course of alcoholic fermentation. However vinification is actuated, the cells used to initiate the fermentation will usually have been subjected to near-starvation conditions, and are probably in the stationary phase of growth. Such cells are quiescent, unbudded and often referred to as being in phase G0. If seeding of a wine fermentation has arisen from cells naturally present in the atmosphere of the winery, or from wine-making equipment, then they will almost certainly be in the G0 phase. Cells that have come naturally from the grape surface will have come from a nutrient-rich environment, and so will probably not be in G0.

Introduction of yeast cells, by whatever means, into a fresh must sample results in a period of apparent inactivity, something that is an integral feature of the pattern of growth exhibited by every microbe when it is introduced into a fresh batch of growth medium. Figure 3.12 shows the pattern of a "normal" wine fermentation. The increase in absorbance is attributable to yeast growth, whilst the decrease in sugar concentration, measured here in °Brix, is the result of fermentation, and the concomitant release of ethanol and CO_2. From the practical winemaker's point of view, vinification may be divided into three distinct phases: an initial phase of yeast growth (multiplication), a middle phase of vigorous fermentative activity and a final phase in which fermentation declines and the inhibitory effects of ethanol, and other compounds, become apparent.

The initial stage of the growth cycle is called the "lag" phase, and represents a period when the introduced cells are adjusting to their new environment; in particular they are responding to the

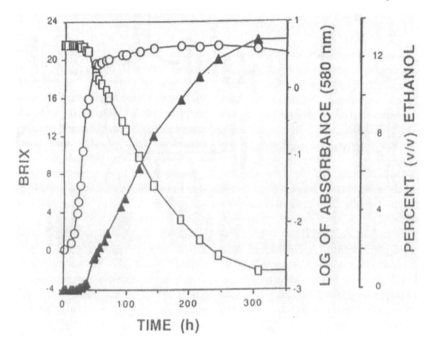

Figure 3.12   Fermentation profile of grape juice (after Boulton *et al.*, 1996[19]).
O = absorbance; ▲ = ethanol; □ = °Brix.

osmotic challenges presented by high sugar levels in must. Although there are no outward manifestations of metabolic activity, several important physiological and biochemical events are occurring within the cells; new enzymes and enzyme-carrier systems are being synthesized to enable the yeast to utilize the wide variety of must constituents. The duration of the lag phase is dependent to some extent upon the origin and condition of the inoculated cells, but is also governed by factors such as must composition and temperature. In wineries that employ "natural" methods of inoculating must, it is highly important that the lag phase of *S. cerevisiae*, and other wanted organisms, should be as short as possible, in order that the growth of less desirable microbes can be suppressed. As a general rule, the more metabolically active the inoculated cells (*i.e.* not in phase G0), the shorter the lag phase will be.

When cells are appropriately equipped, the lag period will give way to a short phase of the growth cycle in which the cells start to become metabolically active again, and embark upon cell division (budding). Thus, yeast multiplication commences as soon as the lag phase has ended and, although there is some fermentative activity, most of the carbon and energy source consumed is converted into essential cell components. Under normal conditions of vinification, this short period of accelerated growth inextricably leads to a period of exponential growth, in which cellular activity and cell division (generation time) are at their zenith. This is called the "logarithmic" ("log") phase of growth, and it is at this stage that the maximum production of ethanol is achieved. The observed high rates of cell multiplication are only possible if the original must contained sufficient dissolved oxygen to permit synthesis of the large quantity of cell membrane precursor material prerequisite for such an event. Normal grape handling during stemming and crushing invariably allows the must to absorb sufficient oxygen, and so, under most circumstances, a paucity of the element rarely becomes a limiting factor at the onset of fermentation.

According to lore, winemakers would regularly stir their crushed grapes before fermentation commenced, a practice that would unintentionally introduce sufficient oxygen for the completion of fermentation. In the case of yeast starter cultures, it is important that they are aerated during resuscitation/propagation.

After a short time, conditions in the fermenter become highly anaerobic, and the yeast converts fermentable sugars, principally glucose and fructose, to ethanol and CO_2 through the glycolytic, or Embden–Meyerhof–Parnas (EMP) pathway (Figure 3.13). This is the main energy-generating process for yeast, and can only proceed in the absence of oxygen (for an account of the evolution of carbohydrate pathways see Romano and Conway[86]) is not in the reference list.). The high level of glycolysis generates copious quantities of CO_2, which forms a blanket over the fermentation, and which prevents ingress of oxygen. This blanket persists until the ferment subsides, from which point there is a likelihood of some oxygenation occurring. Release of CO_2 bubbles at the base of the fermenting vat creates a turbulence that serves to disperse dissolved nutrients and, if sufficiently vigorous, will dislodge settled yeasts cells.

Glycolysis also precludes any significant increase in yeast cell biomass, since most substrate is converted to ethanol and CO_2, and it is also evident that the accumulating level of ethanol does not adversely affect the EMP pathway, as much as it does the rate of

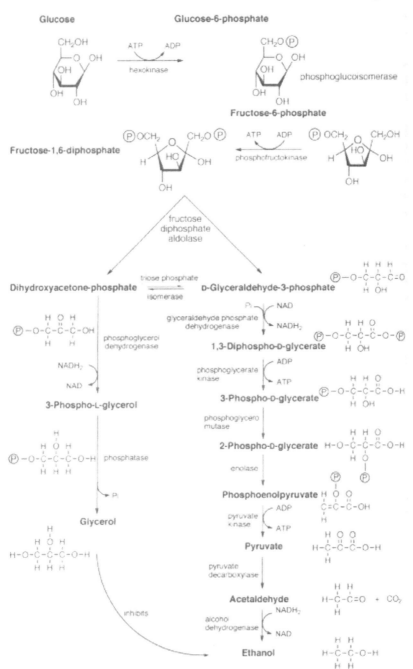

Figure 3.13 The Embden–Meyerhof–Parnas (glycolytic) pathway.

vegetative yeast growth. The latter effect is not too pronounced now, because, as the concentration of ethanol rises, so the cells are entering their non-growing phase, and no longer actively taking up nitrogen sources. One of the known effects of ethanol on the growth of yeast cells is that it interferes with the uptake of amino acids and ammonia (as NH_4^+).

In the continued presence of oxygen, the yeast will undergo aerobic respiration *via* the oxidation and decarboxylation of pyruvate and, through the Krebs cycle, the ultimate production of CO_2 and water. The events relating to the Krebs cycle also occur within the mitochondria. It was Pasteur who, in 1861, first described the deleterious effect of oxygen on wine fermentations. He noticed that when yeast was transferred from an anaerobic to an aerobic environment, growth was accelerated while uptake of sugar was diminished. This is known as the Pasteur effect, and is, in part, attributable to the difference in the Michaelis constant (K_M) for the accumulation of glucose under the two sets of conditions. It has been shown experimentally that, when yeast is grown anaerobically, the K_M value is 6.7 mM, whilst under aerobic conditions it is in the order of 17.4 mM.[20] These observations are linked to the affinity of glucose for its permease, which is decreased in the presence of oxygen. This reported enhancement of vegetative growth (in effect, cell material) is simply explained by the fact that *S. cerevisiae* can produce more ATP from each molecule of glucose under aerobic conditions. In practice, because of the vast excess of fermentable sugar, and the fact that the yeast favours the fermentative mode of metabolism, aerobic respiration rarely reaches significant levels. In theory, industrial fermentations should proceed for as long as anaerobic conditions prevail, and there is a source of hexose sugar available in the growth medium, but this rarely occurs during the fermentation of must. Part of the reason for this is that, under anaerobic conditions, sterol and fatty acid synthesis cannot keep pace with the level of cell membrane synthesis required for cell replication (budding).

It has been known for many years[21] that, at high glucose concentrations (>0.4%), even under aerobic conditions, *S. cerevisiae* metabolism is fermentative rather than oxidative. This is associated with morphological and biochemical alterations in the mitochondria, which resemble those in cells grown under anaerobic conditions. Biochemically, the most significant differences are the deficiency of certain Krebs cycle enzymes, and some respiratory chain components (notably cytochromes). This phenomenon is known as the Crabtree effect, the reverse-Pasteur effect or the glucose effect, and is an example of catabolite repression. In essence, all ethanolic fermentations are regulated by the Crabtree effect. It is the uptake of hexose sugars that is the rate-limiting step in glycolysis, and it has been shown that there is more than one type of glucose transport, depending primarily on substrate affinity. These mechanisms are referred to as "low-affinity" and "high-affinity" uptake. The regulation and expression of sugar uptake has been discussed.[22]

Once yeast growth gets under way, musts are rapidly depleted of ammonia and amino acids, especially the former. Amino acids present in only trace amounts are removed from must very quickly indeed. Arginine, the main form of storage nitrogen in the grapevine, is the most abundant amino acid in many musts, and whilst it is metabolized in conjunction with other nitrogen sources, when these are present they seem to be taken up in preference (*i.e.* arginine does not appear to be a preferred nitrogen source). When other sources are depleted, arginine metabolism increases, but this does not prevent substantial amounts invariably ending up in the final wine.

A lack of available nitrogen can lead to a "stuck" fermentation, and this sometimes occurs in musts that have been over-clarified, thus reducing the protein content. One of the known adverse effects of nitrogen limitation is the irreversible inactivation of hexose sugar uptake, which is caused by a decrease in turnover rate of hexose transporter proteins. Conversely, an excess of nitrogen can cause excessive cell multiplication, but a reduced conversion rate of sugars to ethanol in individual cells. If this occurs, there will appear to be a stimulation of fermentation, attributable purely to the high concentration of cells.

An active wine fermentation will support a yeast culture with cells in all four major stages of the cell cycle (one would not expect any cells to be in G0-phase). If any cells had been in phase G0 upon inoculation of the must, they will soon enter G1, a transition that is thought to be fuelled by a supply of endogenous glycogen within the metabolically inactive cell. The transition G0 → G1 is accompanied by renewed synthesis of membrane lipids and, within about two hours, cells start to enter S-phase, and the first signs of budding. If must has been inoculated with a starter culture, then around 90% of all cells will be showing signs of budding within 6–8 hours. Bud initiation will take much longer in fermentations that have been left to inoculate naturally, a lag of 3–4 days being unexceptional. Either way, progression through the cell cycle is critical for the formation of sufficient yeast biomass to support vigorous fermentation.

The time taken for a mother cell to prepare, produce and liberate a daughter cell is called the "generation time", and most of the strains of yeast that have found favour in wineries have a short period of replication under the conditions of their routine use. The Montrachet strain, for example, which is widely used in Californian wineries, has a generation time in grape juice of 1–3 hours at room temperature. Under conditions of industrial usage, most wine yeasts will multiply, through the log phase, until a population of around $1-2 \times 10^8$ cells/ml is attained. The overall yield of fermentable sugar to ethanol is around 50%, depending upon fermentation temperature. During active growth, cell division rate is limited by the growth of an individual cell. The initiation of a new round of division is regulated by signalling pathways comprised of networks of gene-mediated events that monitor cell size, age, nutritional status, the presence of mating pheromone and the completion of the previous round of cell division. These signalling pathways act together to form cell-cycle regulation points, which confirm that the previous steps have been completed without errors. Slow-growing cells have a relatively long generation time, and this additional time is spent in the G1 phase. In addition, daughter cells that have never produced a bud are smaller than their corresponding mother cells and so have longer G1 phases. This differential is known as mother-daughter asymmetry, and mother and daughter cells may be distinguished by both cell size and division rate. After the period of exponential growth, cells enter the phase of stationary growth, in which the number of viable cells in the fermentation appears to remain static. There is no stasis *per se*, but the number of cells being formed is being counteracted by the number that are becoming aged and non-viable. Most laboratory studies on yeast metabolism during the stationary phase of growth (which use defined media) have centred on its relationship to the gradual exhaustion of the carbon source. With yeast growing in grape must, entry into the stationary phase is not a simple matter of a response to carbon depletion. Indeed, it can occur when sugars are still present in high concentration (as much as 10% w/v). In a vinification fermentation, the onset of the stationary phase is rather more associated with the depletion of the nitrogen source, or some other macronutrient. Under such conditions, cells may still be metabolically very active, since there are plentiful reserves of ATP. The stationary phase of growth leads to the death, or decline, phase where the number of cells becoming moribund far exceeds the number of cells being liberated by budding, thus signalling a decline in overall viable cell number.

Cell ageing has been much studied in yeast and, from time to time, the popular press highlights work from this field that may well be relevant to ageing and death in humans! The subject is, of course, of huge consequence to the fermentation industries. Ageing is a predetermined progressive transition of an individual cell from youth to old age that finally culminates in death. Yeast replicative ageing is a function of the number of divisions undertaken by an individual cell, and may be measured by enumerating the number of bud scars on the cell surface. An aged yeast cell is morphologically and physiologically distinct from younger cells, and this greatly affects fermentation performance. As a consequence of fermentation, cell death may occur *via* one of two distinct pathways: necrosis and senescence. Necrosis may be defined as the accumulation of irreparable damage to intracellular components compromising cell integrity, leading to death and autolysis. Such damage occurs primarily as a result of exposure to excess stress, or repeated exposure to low-level stress. Necrosis may also occur following the deletion or disruption of specific essential genes due to a mutagenic event, or as a result of irreparable DNA damage. In contrast, senescence is the predetermined cessation of life as a result of the genetically controlled progression from youth to old age.

Even in the absence of lethal doses of stress or DNA damage during a wine fermentation, yeast will progress through a structured and defined lifespan, eventually reaching a senescent phase which will culminate in cell death. The lifespan of a fermenting yeast cell must not be thought of in purely chronological terms, but as the number of divisions it has undertaken. The number of daughters produced by a mother cell, termed the divisional age, indicates the relative age of the cell, whilst the maximum lifespan potential of a cell is referred to as the Hayflick limit. Yeast longevity is determined by genes and influenced by environmental factors. Each yeast cell is capable of dividing a number of times before reaching senescence, in which no cell division occurs, and death metabolism is initiated. This form of ageing is known as replicative senescence (and is a phenomenon shared by both yeasts and mammalian cells). Senescence is a consequence of termination of replication and is, therefore, intimately associated with cell division and the cell cycle.

In most industrial fermentations, cell division (budding) is not arrested primarily by nutrient deficiency, but by a lack of essential membrane lipids precursors. Should starvation conditions prevail, then yeast cells will stop budding, and enter the quiescent phase technically known as G0. Such shutting down of metabolism, and entry into this stationary phase of the cell cycle, is the cell's strategy for long-term survival (for months, or perhaps even years) and, because of it, businesses engaged in industrial fermentations are able to purchase freeze-dried cultures.

It is when the yeast population reaches its maximum, and cells just begin to embark upon the stationary phase of growth, that fermentation rates reach their zenith. The velocity of the fermentation during this middle phase of vinification is almost entirely dependent upon the preceding phase of yeast growth. Exactly how much sugar is fermented and when cells decide to enter the stationary phase appear to be determined by a combination of the fermentation temperature and the levels of nitrogenous matter present in the original must.

During the end-phase of fermentation, the inhibitory effects of the elevated level of ethanol begins to be felt by the yeast, and this, together with low levels of nutrients, results in a marked decrease in the rate of glycolysis, until, eventually, fermentation ceases. The final concentration of ethanol is dependent not only upon the initial sugar concentration, but also on the fermentation temperature (higher temperatures causing evaporative losses). In theory, there is a stoichiometric relationship between the amount of fermentable sugar initially present, and the amount of ethanol liberated by glycolysis, but such a relationship is rarely observed in practice. There are a number of reasons for this, one of which is the very way in which sugar levels are measured. Also, available must nitrogen plays a role, especially by determining how much carbon is converted into cellular components (*i.e.* yeast growth, rather than glycolysis). There is some evidence to show that the decline in fermentation rate is not necessarily associated with ethanol concentration, or loss of cell viability, but that it is simply the result of a permanent physiological alteration in the rate of glycolysis.

At the end of fermentation, amino acids are released into the wine by aged yeast cells. This release may be due to cell autolysis, or may result from ethanol toxicity. In respect of the latter, the phenomenon can be induced *in vitro* by adding alcohol to laboratory fermentations, thus enabling the role of ethanol in such leakage to be quantified.[23] The exact leakage mechanism, however, has yet to be elucidated. As indicated elsewhere, zymology has been a key area for the study of fundamental biochemistry, and historical aspects of fermentation in yeast have been presented by Barnett.[24] As part of an excellent series, the same author has performed similar feats with sugar metabolism,[25] enzymic adaptation and regulation,[26] solute transport[27] and active transport and uptake of various metabolites.[28] Walker wrote an erudite account of yeast physiology in 1998.[29]

3.2.2   The Entner–Doudoroff (E–D) Pathway

First reported by Nathan Entner and Michael Doudoroff in 1952[30] as being a novel pathway of glucose metabolism in *Pseudomonas saccharophila*, the E–D pathway (Figure 3.14) is now known to be widely distributed in nature, which suggests that it is more important in the biosphere than was first recognized. According to Conway,[31] since there are similarities in the overall schemes of the E–D and EMP pathways, the E–D pathway ought to be viewed as an alternative to the EMP pathway, although it can be argued that the former is more primitive than EMP glycolysis.

The overall pathways are similar because, in both of them, 6-carbon sugars are phosphorylated and then cleaved by aldolase enzymes into 3-carbon intermediates. The major difference between the two glycolytic pathways lies in the structure of the 6-carbon metabolic intermediates that serve as substrates for aldol cleavage. The E–D pathway involves cleavage of 2-keto-3-deoxy-6-phosphogluconate (KDPG) by KDPG aldolase to yield one molecule each of glyceraldehyde-3-phosphate and pyruvate,

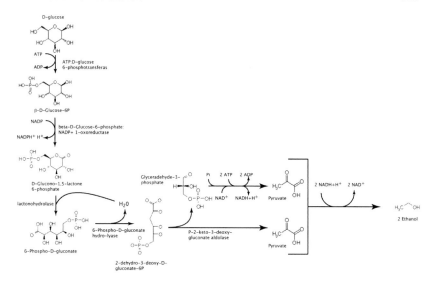

Figure 3.14   Entner–Doudoroff pathway.

whereas in EMP fructose-1,6-biphosphate is cleaved by fructose biphosphate aldolase to give glyceraldehyde-3-phosphate and dihydroxyacetone phosphate. Subsequently, the further metabolism of the triose phosphate intermediates is common to both pathways, and energy is generated *via* substrate-level phosphorylation.

The two key enzymes of the E–D pathway are 6-phosphogluconate dehydratase (6-phospho-D-gluconate hydroro-lyase, also known as the "Entner–Doudoroff dehydratase"; EDD) and KDGP aldolase, which are both highly conserved. EDD is unique to this pathway, whereas KDGP aldolase ("Entner–Doudoroff aldolase"; EDA) is multi-functional and has other metabolic roles, including hexuronic acid metabolism and hydroxyproline biosynthesis.

EDD catalyses dehydration of 6-phosphogluconate to form KDGP. Purified EDD has been obtained from *Z. mobilis*,[32] and the *edd* gene encoding the enzyme, which was partially purified from *Pseudomonas fluorescens*,[33] was cloned from *Z. mobilis* by Barnell *et al.*[34] who also determined the sequences of the gene.

KDPG aldolase is coded for by a gene labelled *eda*, again first purified from *P. fluorescens.*[35] The EDA enzyme from *Z. mobilis* has been purified and shown to act as a trimer of identical 23 kDa sub-units. Scopes[36] found the specific activity of this purified enzyme was 600 µmol min^{-1} mg protein^{-1}. The enzyme from *Pseudomonas putida* was purified and crystallized and its amino acid sequence determined,[37] and the entire nucleotide sequence of the *eda* gene from *Z. mobilis* has been determined by Conway's group.[38]

The E–D pathway is known to operate in a number of different modes, which may have taxonomic significance. As we have said elsewhere, in *Z. mobilis* (and a few other organisms), it is employed strictly for catabolism (being the only means of glucose catabolism); it is expressed constitutively and forms the core of central metabolism. The pathway operates in a linear fashion, with energy being provided by substrate-level phosphorylation. In coliforms (*E. coli* has been well studied[39]), the pathway is linear and has a peripheral role in central metabolism, only being employed for dealing with certain carbohydrates (such as gluconate). Most carbohydrate metabolism is *via* the EMP and pentose phosphate pathways but, when gluconate is available, enzymes of the E–D pathway are synthesized. This is described as the E–D in its inducible, linear mode. Pseudomonads (typically *P. aeruginosa*) operate E–D in a cyclical fashion, and the papers by Phibbs explain things beautifully,[40,41]

3.2.3 Fermentation Technology in Asia

Most truly ethnic fermented foods and beverages are a result of a "natural" fermentation, except for those in Asia, which are initiated *via* a range of microbes enmeshed into a cereal-based "starter" culture. Fermented foods are prevalent in many parts of Asia because of the need to address the considerable fluctuation in food availability and, historically, the *raison d'être* of food fermentation has been to extend the supply of food from times of plenty into times of dearth. Unlike Europe and the Middle East, where many alcoholic beverages are produced from fruit, most in the Asia-Pacific region are cereal based. Sake brewing is dealt with in Chapter 4, and one of the main characteristics of this process and many others is the use of a starter culture. The Asia-Pacific region, with its largely tropical and sub-tropical humid climate, is highly conducive to rice paddy cultivation and to mould growth and so the use of these fungi in food preparation should come as no surprise. Law *et al.*[42] have recently provided a review of the popular fermented foods and beverages of south-east Asia.

As we have seen, fruits and their juices, and some other substrates, such as milk and honey, need little encouragement to ferment and will do so spontaneously, but cereals, where "fermentable" sugar is locked away in a polymer, must have their food reserve (mainly starch) and protein content converted into fermentable sugars and amino acids, before fermentation can ensue. Hsing-Tsung Huang[43] summed things up beautifully: "There is no mystery as to how the process for making grape wine originated. It happened the first time grapes were pressed and the juice allowed to stand for some length of time ... The production of an alcoholic drink from a cereal is a much more sophisticated affair. It does not happen in a way that would catch the attention of an observer." As the *New Pharmacopoeia* (Hsin Hsiu Pên Tshao) of AD 650 put it: "Unlike wines from grain, grape wine and mead do not require chhü [ferment – see below], and to make grape wine, just allow the juice to stand and it will naturally ferment into wine."

Rice is the main staple throughout much of this huge region, but other crops are used as raw materials for fermentation. In its most ancient form, saccharification of cereal starches involved the use of salivary amylases, and this would be followed by "natural" fermentation by yeasts. Western cultures have practised malting (purposeful germination of grain) to achieve the same result, and the idea probably emanated from observing the reaction of grain that had dampened during storage. Asiatic peoples, however, have rarely used malting in the preparation of their traditional fermented drinks, and have tended to saccharify by means of mould starter cultures grown on raw or cooked grain. There are many local variations on a handful of themes.

As Huang[43] attests, since prehistoric times the Chinese have made an alcoholic beverage from cooked grains called *chiu*, a drink that has played an important part in their culture. In the absence of an exact English equivalent of *chiu*, the word has been traditionally translated as "wine" – although this is obviously unsatisfactory because, in terms of its manufacture, the drink bears more resemblance to ale or beer. In terms of ABV, however, which is greater than 10%, and its organoleptic properties, *chiu* is more akin to wine. Of interest is the fact that the great chemist Robert Boyle thought of this Chinese drink as a wine: "In that vast Region of China, which is enriched with so fertile a Soil, and comprizeth such variety of Geographical Parallels, they make not (as *Samedo* informs us) their Wine of Grapes, but of Barley, and, in the Northern parts, of Rice, where they make it also of Apples: but in the Southern parts, of Rice onely; yet not of ordinary Rice, but of a certain kind peculiar to them, which serves onely to make this Liquor, being used in diverse manners."

According to Haard *et al.*,[44] "the use of fermentation starters might very well have its origins in the process of Euchok, the daughter of the legendary king of Woo of 4000 BC, known as the Goddess of rice-wine in ancient China". In 1892 Calmette[45] was the first to scientifically investigate starter cultures when he reported the presence of several wild yeasts in ferments used in Indo-China to produce alcoholic beverages. They were accompanied by the moulds *Amylomyces*, *Rhizopus* and *Mucor*, and about 30 different bacterial species. The most abundant yeast was "similar to *Saccharomyces pastorianus*". The next milestone was when Went and Prinsen-Geerligs[46] identified *S. cerevisiae* and *Monilia javanica* (now *Pichia anomala*) as the main species in Indonesian *ragi* in 1896. *Saccharomyces* species were then shown to be widespread in Asian starters, and in 1913 Saito[47] isolated what he thought was an entirely new organism, which he designated *Endomyces linderi*, and which is now classified *Saccharomycopsis fibuligera*.

Tamang and Fleet[48] list three categories of inocula that are produced commercially for use as mixed starter cultures, the first of these being *koji*, which is discussed in Chapter 4. In the second type, wheat flour is moistened and fashioned into large compact cakes, which are then incubated to select for certain microbes. When deemed ready, the cakes are used to inoculate a starchy substrate, which is then hydrolysed and the products fermented. The microbes responsible for starch hydrolysis are yeasts and moulds, and this inoculum is used in the so-called *kao-ling* process.

The third category of starter contains a mixture of yeasts, moulds and bacteria, and usually takes the form of round or flattened, compact balls, which have been inoculated with material from a previous fermentation. This type of starter has a variety of names, depending on region: *chu*, *chiu-chu* or *chhü* (China); *bakhar ranu* or *murcha* (India); *nuruk* (Korea); *loogpang* (Thailand); *ragi*

(Indonesia); *bubod* (Philippines) and *banh men* (Vietnam). The fungi associated with some Asian fermented foods and beverages have been studied by Batra and Millner,[49] who reported that the significant fungi in *murcha* were *Hansenula anomala* var. *schneggii* (now considered a synonym of *Pichia anomala*), *Rhizopus arrhizus* and *Mucor rouxianus*. The functional fungi in these starter cultures have been reviewed by Aidoo and Nout,[50] and there is an extended treatment of some of them in Steinkraus.[51]

The first record of *chu* appears in *Shu-Ching* ("*Book of History*" or "*Book of Documents*"), written sometime in the Chou Dynasty (1121–256 BC), thought to be before 500 BC, in which it is stated that the starter is essential for making alcoholic beverages ("To make wine [*chiu*] or sweet liquor [*li*], you need to have *chhü nieh*"). The origin of "*nieh*" is controversial, and it is uncertain whether "*chhü nieh*" is a single entity or not. From Chou literature they seem to represent two separate entities, *chhü* being a preparation of ferments, and *nieh* being a sprouted cereal grain. During the late Chou (sixth–third century BC), it was found that *chuü* on its own was capable of saccharifying starch and fermenting sugar. Interpretations of *chhü nieh* have been extensively covered by Huang,[43] as have some methods of preparing *chiu*.

The use of *chuü* for the traditional manufacture of wine (*chiu*) from grains became *de rigueur* from the Han Dynasty (221 BC–AD 207) onwards. There is no exact English equivalent of *chuü*, and a number of translations exist, including "starter", "leaven", "barm" and "yeast". "Ferment" is used here, since *chuü* contains both live organisms and enzymes and, during incubation, the latter hydrolyse grain starch to form fermentable sugars, which provide sustenance for germinating fungal spores and yeast cells. Mycelium from fungal spores produces more amylases (and, thus, more starch breakdown), *etc. etc.* and yeasts ferment the sugar. For ages, this has been referred to as the "amylomyces" or "amylo" method. For Western tastes, contemporary use of *chuü*, or anything like it, for making wine, mead or beer would undoubtedly result in unacceptable "mouldy" notes.

In the *Chhi Min Yao Shu* ("*Main Techniques for the Welfare of the People*"), which was written by the Late Wei Dynasty official Jia Si-Xie in the sixth century AD, and which is the most completely preserved of the ancient Chinese agricultural texts, considerable attention is paid to *chu*. Ten different types are mentioned as being appropriate for alcoholic fermentations, and it is evident that *chu* could be prepared from wheat, barley and rice. The *Chhi Min Yao Shu* also tells us the activity of each type of ferment (expressed as the number of *tou* (*ca.* 1 bushel) of grain that can be digested by one *tou* of the ferment), and enabled four major types of ferment to be distinguished:

"Superior" (*shên chhü*); can convert 20–30 *tou* grain;
"Moderate" (*pai lao chhü*); 10 *tou* grain;
"Common" (*pen chhü*); 7 *tou* grain;
"Common" (*pai tuo chhü*); 15 *tou* grain.

The digestive and fermentative power exhibited during wine brewing is called *chhü shih* ("ferment" power), and this is illustrated by "the ability of the broth to dissolve the added cooked grain and keep itself bubbling with gas". Four chapters in the *Chhi Min Yao Shu* are concerned with the preparation of ferments and wines, there being nine varieties of the former and 38 kinds of the latter, which are dealt with in some detail.

Huang gives an excellent account of the origin of wine fermentation in China, and provides evidence for the fact that "the preparation of wine was already practised on a large scale by the time of the Shang". The Shang, China's first historical dynasty, flourished *ca.* the sixteenth to the eleventh centuries BC, and direct evidence for wine production was obtained in 1974 when winery artefacts from the mid-Shang period were recovered at Thai-hsi, near Kao-chhêng, in Henan. A variety of pottery vessels and a pottery funnel (Figure 3.15) plus pits of plums, peaches and jujubes were found as well as hemp and jasmine seeds. All this suggested that a flavoured fruit wine had been prepared at the site, and the recovery of a large amount of carbonized setaria millet a little later indicated the probable nature of the main ingredient. In 1980, an excavation at the late-Shang site at Thien-hu-tshun, Lo-shan County, Henan, yielded two tightly sealed bronze containers with their liquid contents *in situ*. The liquid was shown to contain alcohol.[43]

Huang[43] also tells us that the readings of oracle bone and turtle shell inscriptions from the late-Shang era (*ca.* 1200–1046 BC), the earliest writing in China, indicate that at least three types of alcoholic beverage were known at that time: a herbal wine called

Figure 3.15   Pottery vessels excavated from site 14 at Thai-hsi, Kao-chhêng, Hopei (from Huang[43] - with permission).

chang, li, which is interpreted as a sweet, low-alcohol beverage made from rice or millet, and *jiu*, a fully fermented, relatively clear, rice or millet drink with a strength seemingly similar to our present-day wine (probably 10–15% ABV). Another two beverages are described from the Zhou period (*ca.* 1046–221 BC): *luo*, probably fruit-based, and *lao*, an opaque rice or millet drink.

The ferment *chhü* was almost certainly "discovered" when someone inadvertently left a basket of steamed rice exposed to the air for several days. The mouldy mass that was produced was dried (intentionally?) and provided the basis for initiating subsequent fermentations. It was later shown that flour from other grains, such as wheat and millet, could support mould growth if it was mixed with water to form a cake. The original rice ferment became known as *chhü* in north China and *chiu yao* ("wine medicament") in southern China.

3.2.4   The Mycoflora of Starter Cultures

Aidoo *et al.*[52] have given an account of the occurrence and function of yeasts in Asian indigenous fermented foods, and paid particular attention to amylolytic fermentation starters, of which they state "are used in the form of starchy tablets containing mixed cultures of starch-degrading moulds and fermenting yeasts. They are used for the manufacture of beers, wines and pasty snacks from various kinds of cereals and starchy crops, such as rice, sorghum, millet and cassava". They are prepared from uncooked dough of rice or wheat flour (sometimes mixed with cassava meal), water and a variety of herbs and spices. The dough is inoculated with a "back-slop" (obtained from a previous fermentation) before being fashioned into balls or (flat) cakes, and these are then incubated at *ca.* 30 °C for a week. Incubation allows the relevant microbial floras to develop and also encourages dehydration. In such a condition, they will remain viable for up to six months at ambient temperature.

The functional microfloras of most amylolytic starter cultures have been studied to some extent and many microbes have been elucidated, some of which are documented here. In starters such as *chuii* and *ragi*, the main amylolytic moulds are *Amylomyces rouxii*, *Mucor* spp., *Rhizopus* spp. and *Aspergillus* spp., whilst frequently encountered yeasts are *S. cerevisiae*, *Saccharomycopsis fibuligera*, *Hansenula* spp. and *Candida* spp.[51] The exact microflora of widely used starters like these varies with location. A study of around 100 yeast strains from *ragi*, *murcha*, *bubod* and "Chinese yeast" was conducted by Hesseltine and Kurtzman[53] and it transpired that the predominant yeast was *Saccharomycopsis fibuligera*. *Sm. malanga* ("a phenotypically similar species that also produces amylase") was the next in abundance, but was not found in *murcha* or *bubod*.

Murcha (*marcha*), widely used as a starter in Nepal, Bhutan and the Himalayan regions of India, was the subject of an intensive study by Tamang and Sarkar,[54] and the mould species encountered were *Mucor circinelloides*, *Mucor* sp. (close to *hiemalis*), *M. praini*, *Rhizopus chinensis*, *R. oryzae* and *R. stolonifer*, with the *Mucor* spp. being more prevalent than *Rhizopus* spp. Other genera encountered were *Aspergillus*, *Penicillium*, *Amylomyces* and *Actinomucor*. Over 20 yeast strains have been reported, with *Saccharomycopsis fibuligera* often being dominant. Other prominent species include *S. cerevisiae* and *Pichia burtonii*. In a study of the characteristics of 20 yeast strains recovered from *murcha* samples in Sikkim, India, Tsuyoshi *et al.*[55] found it possible to place them into four groups according to physiological and morphological traits. *Saccharomyces bayanus* constituted Group I, *Candida glabrata* comprised Group II and *Pichia anomala* Group III. Group IV was represented by *Sm. fibuligera*, *Sm. capsularis* and *Pichia burtonii*. Groups I–III were ethanologenic, while Group IV were highly amylolytic yeasts.

The properties of the Vietnamese starter *banh men* were studied by Lee and Fujio,[56] who found that the mean pH of their samples was 5.76, and the moisture content 13.6%. Total mould and yeast counts amounted to 1.3×10^6 and 4.3×10^6 cfu g^{-1} fresh sample, respectively. Some 53 fungal isolates were reported with the moulds being identified as *Mucor indicus, M. circinelloides, Rhizopus oryzae* and *Amylomyces rouxii*. Yeast isolates were identified as *S. cerevisiae, Pichia anomala, P. burtonii, Saccharomycopsis fibuligera* and *Candida* sp. The characteristics of *banh men* suggest that it is closely related to some forms of *ragi*. In a later study involving PCR-mediated DGGE,[57] 53 starter cultures were examined and 13 fungal species and 23 species of bacteria were identified. The mycoflora was remarkably consistent with the amylase producers being *Rhizopus oryzae, R. microsporus, Absidia corymbifera, Amylomyces* sp. and *Sm. fibuligera*. The ethanologens were *S. cerevisiae, Issatchenkia* sp., *Pichia anomala, P. ranongensis, Candida tropicalis* and *Clavispora lusitaniae*. There were two regular opportunistic contaminants: *Xeromyces bisporus* and *Botryobasidium subcoronatum*.

Limtong *et al.*[58] carried out an extensive study of yeast diversity in Thai traditional alcohol starters, including those for rice wine, *loog-pang-lao*. *Loog-pangkao-mag*, starters for alcoholic sweetened rice, were also examined, and in both types of *loog-pang Sm. fibuligera* predominated. All strains of this organism showed high amylolytic activity but low ethanologenic properties. Other species identified included *Pichia anomala, P. burtonii, P. fabianii, P. mexicana, Issatchenkia orientalis, Candida glabrata, C. rhagii, Torulaspora globosa* and *S. cerevisiae*.

Of the 20-odd yeast strains isolated from a Nepalese *murcha* culture, one, identified as *Pichia burtonii*, exhibited evidence of extracellular amylolytic activity.[59] The enzyme was purified and its N-terminal amino acid sequence determined. Hydrolytic activity toward various substrates showed that it was an α-amylase (named *Pichia burtonii* α-amylase; PBA). It is a glycoprotein with an apparent molecular mass of 51 kDa, and optimum activity at pH 5.0 at 40 °C. The amylase is capable of randomly hydrolysing the α-1, 4-glycosidic linkages in soluble starch, and producing malto-oligosaccharides ranging from maltose (G2) to maltohexaose (G6). During the early phase of hydrolysis the dominant products are maltotetraose (G4) and maltopentaose (G5). The regulation of this enzyme is of interest because it was shown that *P. burtonii* produces PBA *in vitro* even when cultured in the presence of soluble starch. With *Sm. fibuligera* being the major amylolytic yeast in *murcha*, the precise role of *P. burtonii* in saccharification is unclear.

Hesseltine and Ray[60] isolated 69 strains of lactic acid bacteria from *murcha* and *ragi* starters used in Java, Bali and Nepal, and found that most belonged to the genus *Pediococcus* (probably *P. pentosaceus*), and to *Streptococcus faecalis*. None of the isolates were able to break down starch, and when inoculated (as pure cultures) onto rice most of them showed no signs of growth – until yeasts and moulds were added as "contaminants". It was suggested that the role(s) of these microbes may well be to produce secondary products from glucose formed by the amylolytic mycoflora.

Sm. fibuligera is known to have high amylolytic activity[61] and this property has been used in the food industry to produce high-sugar syrups.[62] The capability of *Sm. fibuligera* to degrade starch is due to the production of two types of amylases: endo-acting α-amylase and exo-acting glucoamylase, some strains producing synthesizing both enzymes while others produce just one. A few *Sm. fibuligera* strains produce enzymes with specificity for α-D-glucosidic linkages (mainly α-glucosidases and transglucosylases), and the ability of yeast amylases to digest raw starch is technologically very useful. The *Sm. fibuligera* IFO 0111 strain produces only one amylolytic enzyme, glucoamylase, which can degrade raw starch, and its molecular and genetic properties were studied by Hostinová,[63] who found that it produces glucoamylase Glm (as opposed to Gla). The enzymatic properties of Glm and Gla are slightly different (they have differing amino acid compositions) and the former seems to be secreted in only one glycosylated form. According to Ueda and Saha,[64] the ability of glucoamylase Glm to hydrolyse granular starch is very high and on a par with the *Aspergillus* and *Rhizopus* glucoamylases. An evaluation of *Sm. fibuligera* IFO 111 glucoamylase Glm in the hydrolysis of corn starch has been made.[65] Results showed that thermal modification of the starch substrate prior to enzyme application is unnecessary, and that the optimum concentration of the enzyme was 33–75 U g starch^{-1}. The extracellular α-amylase from strain ST2, which is copiously produced during the stationary phase of growth, was purified and characterized by Gogoi *et al.*[66] The enzyme had a molecular weight of 40.9 kDa, a temperature optimum of 60 °C, and was active in the pH range 4.0–9.4. Starch hydrolysis products are glucose, maltose, maltotriose and maltotetraose.

Because not all *Sm. fibuligera* strains have the ability to produce quality products, rapid diagnostic methods are desirable for screening purposes. One method, based on the application of inter simple sequence repeat (ISSR) markers, and used on Thai ferment organisms, including *loog-pang*, was described by Dansakul *et al.*[67] Results showed that ISSR markers can be successfully applied to study the intra-specific variation in this species and that there are genetic variations in *Sm. fibuligera* specimens taken from different fermented food and beverage starters.

The commercial applications of amylolytic yeasts have been discussed,[68] and over 150 starch-degrading yeasts have been described.[69] During the last couple of decades, numerous attempts have been made to widen the spectrum of yeast strains possessing amylolytic activity, and increased attention has been paid to somewhat neglected amylolytic genera such as *Lipomyces* – particularly *L. kononenkoae* and *L. starkeyi*, which utilize starch as a sole source of carbon. Both of these species have highly efficient extracellular amylolytic systems, which allow growth on starch with very high biomass yields.[70] Sequence analyses and the substrate specificity of *L. kononenkoae* α-amylases have been studied in detail.[71] *Schwanniomyces alluvius* (now *Debaryomyces occidentalis*) was shown to be able to effect a direct and quantitative conversion of starch to ethanol[72] and, at one time, it was described as a "potential super-yeast".[73] The extracellular amylolytic system of *Filobasidium capsuligenum*, the basidial stage of *Candida japonica*, has been investigated and shown to be comprised of an α-amylase and two forms of glucoamylase.[74]

Amylolytic strains among yeasts used in the brewing industry have been reported,[75] and new amylolytic yeast strains have been searched for in a number of unusual, albeit promising, habitats such as cassava flour factories.[76] Starch-degrading strains have been constructed; Favaro *et al.*[77] attempted to develop an efficient amylolytic *S. cerevisiae* strain, when they took a wild-type with promising industrial fermentative traits and engineered it to secrete a fungal glucoamylase. Just to illustrate the enormous possibilities, in the same year, Wang et al.[78] constructed an amylolytic industrial brewing yeast strain with high glutathione content for making beer with improved anti-staling properties. Sivaramakrishnan et al.[79] penned an overview of α-amylases from microbial sources, surveying this family of enzymes and outlining their engineering capabilities and recent research.

3.2.5 Why Ferment?

As has been known since the time of Pasteur, fermented food and beverages harbour a wide range of microbes, many of which have arrived surreptitiously, which transform the chemical composition of the raw materials. Authorities have tended to define "fermented foods" according to their own field of study and so opinions vary. van Veen,[80] for example, defined fermented foods as "foods that are fermented until they, or at least one of the constituents, have been subjected to the action of micro-organisms for a period, so that the final products have undergone considerable changes in chemical composition and other respects. Sometimes these changes are due not only to microbial action, but also to autolytic processes brought about by the enzymes of the product itself". In an early work on food microbiology, F. W. Tanner[81] was of the opinion that preservation and improvement of flavour were the two main reasons for treating food in this way, although he was somewhat sceptical about the latter and preferred to call it "change of flavour".

Steinkraus[51] attributes five possible roles played by food fermentations:

1. Enrichment of the diet through the development of a greater range of flavours, aromas and textures.
2. Preservation of food via lactic acid, acetic acid, ethanolic and alkaline fermentations.
3. Biological enrichment of food substrates through proteins, essential amino acids, essential fatty acids and vitamins.
4. Detoxification of some foodstuffs and degradation of anti-nutritive matter.
5. Expedition of cooking times (reduced fuel use). Related to this last point is the fact that fermentation often improves the digestibility of food.

Diet enrichment, amply discussed by Platt many years ago,[82] is usually directly attributable to the functional microbes associated with a fermented food, and Platt called this "biological ennoblement". Platt was particularly interested in the B-group vitamins in relation to kaffir beer brewing, but nowadays we like to consider compounds such as antioxidants, omega-3 polyunsaturated fatty acids and the bioavailability of minerals when dealing with fermented food bio-transformations. The field can be expanded to encompass the therapeutic and immunological value of foods. Some fermented foods are now intensively marketed as "health foods", "functional foods" or "neutraceutics".

From his experiences in Africa, Platt illustrated how a simple diet may be "ennobled" by replacing the staple grain (maize) by beer made from that grain. One diet consisted of 700 g maize, supplied as either whole grain or flour, and this was consumed as a porridge. Diet two comprised 275 g maize plus five pints of maize beer (the amount that could be made from 425 g maize). Analysis showed that the levels of riboflavin and nicotinic acid were almost twice as great in the beer diet as in diet one. It was also significant that the incidence of the disease pellagra (classically associated with a maize diet) was seldom encountered in drinkers of kaffir beer. It was found that, with the exception of thiamine, levels of all of the B vitamins increased during grain germination (malting), with some (folic acid, nicotinic acid and riboflavin) as much as three- or four-fold.[83,84] Ascorbic acid is also produced during grain malting. Riboflavin levels increased noticeably as a result of yeast activity, with other B vitamins increasing less so, and it was suggested that these were likely to be stabilized in the acidic beer environment produced by LAB activity at the end of fermentation.

Apart from B vitamins, other important contributions made by yeast include trace elements (chromium, selenium), enzymes, glycans and carotenoid pigments.

Writing in 1994, Campbell-Platt[85] wrote: "All over the world, fermented foods continue to provide an important part of our diet. Fermented foods and beverages everywhere provide some 20–40% of our food supply; typically around one-third of our food intake is of fermented foods." The situation has changed somewhat since then because of the "Westernization" of diets in some parts of Asia, particularly China.

REFERENCES

1. J. Hooke, *Micrographia: or Some Physiological Descriptions of Minute Bodies Made by Magnifying Glasses*, Royal Society, London, 1665.
2. C. Dobell, *Anthony van Leewenhoek and his "Little Animals"*, John Bale, Sons and Danielsson Ltd, London, 1932.
3. A. Harden, *Alcoholic Fermentation*, Longmans, Green & Co., London, 1911.
4. R. G. Anderson, *J. Inst. Brewing*, 1989, **95**, 337.
5. N. B. Redman, *The Brewer*, 1995, **81**, 371.
6. J. A. Barnett, *Yeast*, 2000, **16**, 755.
7. L. Pasteur, *Etudes Sur Le Vin*, Victor Masson et Fils, Paris, 1866.
8. L. Pasteur, *Etudes Sur La Bière*, Gauthier Villars, Paris, 1876.
9. R. G. Anderson, in *Proc. 25th European Brewery Convention Congress*, Brussels, 1995, p. 13.
10. R. G. Anderson, *Ferment*, 1993, **6**, 191.
11. R. G. Anderson, in *Handbook of Brewing*, ed. G. G. Stewart and F. G. Priest, Marcel Dekker, New York, 2006.
12. J. A. Barnett, *Microbiology*, 2003, **149**, 557.
13. J. A. Barnett, *Yeast*, 1998, **14**, 1439.
14. J. A. Barnett and F. W. Lichtenthaler, *Yeast*, 2001, **18**, 363.
15. J. A. Barnett and L. Barnett, *Yeast Research: An Historical Overview*, ASM Press, Washington, DC, 2011.
16. R. E. Kunkee and R. W. Goswell, in *Economic Microbiology Volume 1, Alcoholic Beverages*. ed. A. H. Rose, Academic Press, London, 1977.
17. M. J. Torija, N. Rozès, M. Poblet, J. M. Guillamón and A. Mas, *Antonie van Leewenhoek*, 2001, **79**, 345.
18. N. Hierro, Á. González, A. Mas and J. M. Guillamón, *FEMS Yeast Res.*, 2006, **6**, 102.
19. R. B. Boulton, V. L. Singleton, L. F. Bisson and R. E. Kunkee, *Principles and Practices of Winemaking*, Chapman and Hall, New York, 1996.
20. A. Kotyk and A. Kleinzeller, *Biochim. Biophys. Acta*, 1967, **135**, 106.
21. H. G. Crabtree, *Biochem. J.*, 1929, **23**, 536.

22. R. E. Kunkee and L. F. Bisson, in *The Yeasts, ed. A. H. Rose and J. S. Harrison, Academic Press, London, 2nd edn*, 1993, vol. 5.
23. S. P. Salgueiro, I. Sá-Correia and J. M. Novais, *Appl. Environ. Microbiol.*, 1988, **54**, 903.
24. J. A. Barnett, *Yeast*, 2003, **20**, 509.
25. J. A. Barnett and K. D. Entian, *Yeast*, 2005, **22**, 835.
26. J. A. Barnett, *Yeast*, 2004, **21**, 703.
27. A. A. Eddy and J. A. Barnett, *Yeast*, 2007, **24**, 1023.
28. J. A. Barnett, *Yeast*, 2008, **25**, 689.
29. G. M. Walker, *Yeast Physiology and Biotechnology*, John Wiley & Sons, Chichester, 1998.
30. N. Entner and M. Doudoroff, *J. Biol. Chem.*, 1952, **196**, 853.
31. T. Conway, *FEMS Microbiol. Rev.*, 1992, **103**, 1.
32. R. K. Scopes and K. Griffiths-Smith, *Anal. Biochem.*, 1984, **136**, 530.
33. R. Kovachevich and W. A. Wood, *J. Biol. Chem.*, 1955, **213**, 745.
34. W. O. Barnell, K. C. Yi and T. Conway, *J. Bacteriol.*, 1990, **172**, 7227.
35. R. Kovachevich and W. A. Wood, *J. Biol. Chem.*, 1955, **213**, 757.
36. R. Scopes, *Anal. Biochem.*, 1984, **136**, 525.
37. N. Suzuki and W. Wood, *J. Biol. Chem.*, 1980, **255**, 3427.
38. T. Conway, R. Fliege, R. Jones-Fitzpatrick, J. Liu, W. Barnell and S. Egan, *Mol. Microbiol.*, 1991, **5**, 2901.
39. N. Peekhaus and T. Conway, *J. Bacteriol.*, 1998, **180**, 3495.
40. T. G. Lessie and P. V. Phibbs, *Ann. Rev. Micro.*, 1984, **38**, 359.
41. P. V. Phibbs, in *Microbial Metabolism and the Carbon Cycle*, ed. S. R. Hagedorn, R. S. Hanson and D. A. Kunz, Harwood Academic, New York, 1988.
42. S. V. Law, F. Abu Bakar, D. Mat Hashim and A. Abdul Hamid, *Int. Food Res. J.*, 2011, **18**, 474.
43. H. T. Huang, in *Science and Civilisation in China, vol. 6, Biology and Biological Technology*, Cambridge University Press, Cambridge, 2000.
44. N. F. Haard, S. A. Odunfa, C.-H. Lee, R. Quintero-Ramirez, A. Lorence-Quiñones and C. Wacher-Radarte, *Fermented Cereals: A Global Perspective*, Cultural Services Bulletin

No. 138, Food and Agriculture Organization of the United Nations, Rome, 1999.

45. A. Calmette, *Ann. Institut. Pasteur*, 1892, **6**, 604.
46. F. A. C. Went and H. C. Prinsen-Geerligs, *Verh. Kon. Ned. Akad. Wetensch., Afd. Natuurh., Tweede Sect. II*, 1896, **4**, 3.
47. K. Saito, *Zeitschr. f. Gärungsphysiol. 2*, 1913, **3**, 151.
48. J. P. Tamang and G. H. Fleet, in *Yeast Biotechnology: Diversity and Applications*, ed. T. Satyanarayana and G. Kunze, Springer, Berlin, 2009, p. 169.
49. L. R. Bahtra and P. D. Millner, *Mycologia*, 1974, **66**, 942.
50. K. E. Aidoo and M. J. R. Nout, in *Fermented Foods and Beverages of the World*, ed. J. P. Tamang and K. Kailasapathy, CRC Press, Boca Raton, 2010, p. 127.
51. K. H. Steinkraus (ed.), *Handbook of Indigenous Fermented Foods*, Marcel Dekker Inc., New York, 2nd edn, 1996.
52. K. E. Aidoo, M. J. R. Nout and P. K. Sarkar, *FEMS Yeast Res.*, 2006, **6**, 30.
53. C. W. Hesseltine and C. P. Kurtzman, *Anales Inst. Biol. Nac. Autón. México, Ser. Bot.*, 1990, **60**, 1.
54. J. P. Tamang and P. K. Sarkar, *Microbios*, 1995, **81**, 115.
55. N. Tsuyoshi, R. Fudou, S. Yamanaka, M. Kozaki, N. Tamang, S. Thapa and J.P. Tamang, *Int. J. Food Micro.*, 2005, **99**, 135.
56. A. C. Lee and Y. Fujio, *World J. Microbiol. Biotechnol.*, 1999, **15**, 51.
57. V. N. Thanh, le T. Mai and D. A. Tuan, *Int. J. Food Micro.*, 2008, **128**, 268.
58. S. Limtong, S. Sintara, P. Suwanarit and N. Lotong, *Kasetsart J. (Nat. Sci.)*, 2002, **36**, 149.
59. A. Takeuchi, A. Shimizu-Ibuka, Y. Nishiyama, K. Mura, S. Okada, C. Tokue and S. Arai, *Biosci. Biotechnol. Biochem.*, 2006, **70**, 3019.
60. C. W. Hesseltine and M. L. Ray, *J. Appl. Bact.*, 1988, **64**, 395.
61. A. M. Knox, J. C. Preez and S. G. Kilian, *Enzym. Microb. Tech.*, 2004, **34**, 453.
62. D. K. Sandhu, K. S. Vilkhu and S. K. Soni, *J. Ferm. Technol.*, 1987, **65**, 387.
63. E. Hostinová, *Biologia, Bratislava*, 2002, **57**(11), 247.
64. S. Ueda and B. C. Saha, *Biotechnol. Bioeng.*, 1983, **15**, 1181.

65. V. Horváthová, K. Šlajsová and E. Šturdík, *Biologia, Bratislava*, 2004, **59**, 361.
66. B. K. Gogoi, R. L. Bezbaruah, K. R. Pillai and J. N. Baruah, *J. Appl. Bact.*, 1987, **63**, 373.
67. S. Dansakul, V. Leelawatcharamas, C. Charoenchai and H. Urairong, *Kasetsart (Nat. Sci.)*, 2009, **43**, 339.
68. R. S. Tubb, *Trends Biotechnol.*, 1986, **4**, 98.
69. A. K. McCann and J. A. Barnett, *Yeast*, 1986, **2**, 109.
70. R. De Mot, K. Adries and H. Verachtert, *Appl. Microbiol.*, 1984, **5**, 106.
71. N. Ramachandran, I. S. Pretorius and R. R. Cordero Otero, *Biologia, Bratislava*, 2005, **60**(16), 103.
72. G. B. Calleja, S. Levy-Rick, C. V. Lusena, A. Nasim and F. Moranelli, *Biotechnol. Lett.*, 1982, **4**, 543.
73. W. M. Ingledew, *Crit. Rev. Biotechnol.*, 1987, **5**, 159.
74. R. De Mot and H. Verachtert, *Appl. Environ. Micro.*, 1985, **50**, 1474.
75. A. M. Sills and G. G. Stewart, *J. Inst. Brew.*, 1982, **88**, 313.
76. C. Laluce, M. C. Bertolini, J. R. Ernandes, A.V. Martini and A. Martini, *Appl. Environ. Microbiol.*, 1988, **54**, 2447.
77. L. Favaro, M. Basaglia, R. Saayman, S. H. Rose, W. H. Van Zyl and S. Casella, *Chem. Eng. Trans.*, 2010, **20**, 97.
78. J.-J. Wang, Z.-Y. Wang, X.-P. He and B.-R. Zhang, *J. Microbiol. Biotechnol.*, 2010, **20**, 1539.
79. S. Sivaramakrishnan, D. Gangadharan, K. M. Nampoothiri, C. R. Soccol and A. Pandey, *Food Technol. Biotechnol.*, 2006, **44**, 173.
80. A. G. van Veen, *Fermented protein-rich foods* FAO Report No. FAO/57/3/1966, 1957.
81. F. W. Tanner, *The Microbiology of Foods*, Garrard Press, Champaign, IL, 1932.
82. B. S. Platt, *Food Tech.*, 1964, **18**, 68.
83. B. S. Platt and R. A. Webb, *Proc. Nutr. Soc.*, 1946, **4**, 132.
84. L. Goldberg and J. M. Thorp, *S. Afr. J. Med. Sci.*, 1946, **11**, 177.
85. G. Campbell-Platt, *Food Res. Int.*, 1994, **27**, 253.
86. A. H. Romano and T. Conway, *Res. Micro.*, 1996, **147**, 448.

CHAPTER 4

Grasses

4.1 INTRODUCTION

Grasses (Poaceae or Gramineae) are angiosperms and, within that primary division of the plant kingdom, they belong to the large group known as monocotyledons, which relates them to onions, orchids, lilies, palms and thousands of other species. Monocotyledons possess only one seed leaf (cotyledon), whereas plants in the other major group of flowering plants, the eudicotyledons, have two. Other differences include a primary root which is unbranched, short-lived and replaced early in plant development by a profusion of stem-borne (adventitious) roots. In contrast, eudicotyledons generally have a long-lived and branched primary root. Monocotyledons have a characteristic vascular patterning, the vascular bundles being scattered within the stems (they are arranged in a ring in the eudicots). The major leaf veins are arranged parallel to each other (reticulate in eudicots), and inter-mediate leaf veins always connect with other veins, rather than ending blindly in the mesophyll as in eudicots.

The grass family is of particular interest to humans, and the economic and ecological significance of the grasses has given rise to widespread interest in their evolution and classification. The cereals, bamboos, sugar-cane and forage and weedy grasses are of pre-eminent importance in the economy of mankind. Grasses, which occur in virtually every terrestrial habitat, cover as much as one-fifth of the Earth's land surface. Members of the family are

Alcohol and its Role in the Evolution of Human Society
Ian S. Hornsey
© Ian S. Hornsey 2012
Published by the Royal Society of Chemistry, www.rsc.org

also ecological dominants, covering approximately 20% of the planet's land surface.[1]

Grasses have long been recognized as a "natural" group, and the family includes approximately 10,000 species, which have been categorized into over 700 genera on the basis of architectural, anatomical and embryological characteristics.[2,3] Genera have been unified into sub-tribes, tribes, super-tribes and sub-families, making the grasses one of the largest families of flowering plants. Attempts to produce a comprehensive, formal taxonomic structure of the family began over two centuries ago, while serious studies of grass evolution began in the late-nineteenth century.

Over the past couple of decades, cladistic methods have been applied to grass phylogeny and evolution, the first attempt being by Kellogg and Campbell,[4] who analysed 33 characters scored for virtually all grass genera. Plant molecular systematists rely upon a relatively small number of molecular markers for phylogeny reconstruction, including, for example, the chloroplast DNA (cpDNA) genes *mat*K, *ndh*F and *rbc*L and, less commonly, cpDNA intergenic spacers. The nuclear genome is usually represented by sequences from the highly repetitive rDNA arrays – either the genes themselves, for studies of ancient divergence events, or spacer regions, for analyses at lower taxonomic levels. The justifications for choosing these markers are strong. First, because the chloroplast genome and the rDNA arrays occur in high copy numbers, they are technically relatively easy to work with, thus allowing systematists to amass data for large numbers of taxa. Second, individual systematists working with these markers are able to interpret their data within enormous databases of sequences that have been gathered by the systematic community. An additional advantage, specific to chloroplast markers, is that because the chloroplast genome is clonally inherited, problems associated with recombinant gene copies will not be encountered. The problem plant molecular systematists are now addressing is the limited number of markers available.

The Grass Phylogeny Working Group (GPWG) was established in 1996 explicitly to combine available data on the phylogeny of the grass family and to use the data to propose a new classification. Most contributors had already published papers on grass phylogeny and were invited to contribute their data. All sequences available by the end of 1997 were assembled by Elizabeth Kellogg into a single large data set in NEXUS format, and this was distributed to all participants, who had the chance to comment. After this, a final version was distributed, and the results from these first analyses were published.[5] Analysis of eight data sets (four plastid, three nuclear and one structural) produced a single most parsimonious tree, and a revised classification recognizing twelve sub-families was proposed based explicitly on the phylogeny. This phylogenetic hypothesis and available fossil evidence will enable the evolutionary history of the grasses to be explored with greater clarity.

The origin of the grasses can be dated by the appearance of grass pollen in the fossil record. The grasses and their relatives have distinctive pollen that is nearly spherical and with a single pore. Grass pollen itself can be distinguished by minute channels or holes that penetrate the outer, but not the inner, pollen wall.[6] The earliest firm records of grass pollen are from the Palaeocene of South America and Africa, between 60 and 55 Mya.[7] This date is after the major extinction events that ended the age of dinosaurs and the Cretaceous period.

Additional fossil pollen grains that may be grasses or may be grass relatives have been found in Maastrichtian deposits of Africa and South America (approximately 70 Mya); these were fossilized just before the end of the Cretaceous. Because of the way the pollen was preserved, however, it is impossible to tell whether it had the channels in the outer wall that are characteristic of the grasses.[7,8] For an excellent account of the evolutionary history of the grasses see Kellogg,[9] and for detailed information on grasses generally see the excellent *Flora of North America*.[387,388]

4.2 BARLEY

The origins of barley agriculture are closely allied to the domestication of wheat, and there is an overlap of several archaeological sites considered to be important in the early development of the two crops. Cultivated barley (*Hordeum vulgare* L., Figure 4.1) is a founder crop of Old World Neolithic food production, and one of the main cereals of early Mediterranean agriculture. Over these areas, wheat seems to be a universal companion of barley and is generally considered to be a superior staple to barley. That said, we have to question the reason for barley domestication. What was it used for? Was its domestication a unique event, and how long did

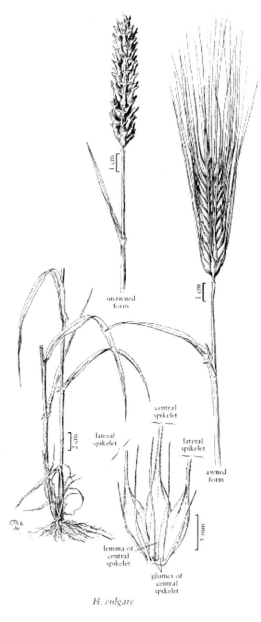

Figure 4.1 *Hordeum vulgare* L. (from *Flora of North America*[387] - by kind permission).

the process take? Being "the poor person's bread" is not a satisfactory answer, and there are some who maintain that barley was not domesticated for the same reasons that led to the domestication of other cereals (such as the dwindling of other resources, seasonal shortages, the desire for a sedentary food base or the need for a surplus for trade purposes).

Today, barley is widely cultivated in all temperate regions from the Arctic Circle to the tropics and represents a central component of the majority of agro-ecosystems in temperate climes. In addition, wild barley species are adapted to some of the most extreme ecosystems around the globe, and their germplasm offers us an, as yet, unexploited potential for the future development of improved barley strains. Barley ranks number five in world crop production (as measured by harvested area), with the European Community producing over 60% of the world harvest.

Figure 4.2   *Hordeum vulgare* subsp. *spontaneum* (C. Koch) Thell. (By kind permission of Mark Nesbitt.).

Barley is a diploid ($2n = 14$) and mostly self-pollinated crop that shows close affinities to a group of wild and weedy forms that have been traditionally described as *Hordeum spontaneum* C. Koch. (Figure 4.2). These are, in fact, the wild races or sub-species of the cultivated crop, and so the correct designation for the wild type is therefore *H. vulgare* L. subp. *spontaneum* (C. Koch) Thell. As such, they are cross-compatible and fully interfertile with cultivated barley. Morphologically, it is rather difficult to distinguish between wild *H. spontaneum* and cultivated two-rowed varieties, since they differ primarily in their methods of seed dispersal.

According to Weiss *et al.*,[10] barley was among the earliest crops to be exploited during the Mesolithic/Neolithic transition and the first to become domesticated, somewhere around 9500 yBP. The primary role of barley in the origins of agriculture can be attributed to its resilient qualities, such as rapid development to maturation, xerophytic adaptations (such as long awns[11]), as well as tolerance of a wide range of edaphic factors such as pH and salinity. As Mark Blumler[12] related, these attributes made barley well suited to the early Holocene climate in which the spring growing season was shorter than it is now, and the summer longer, hotter and drier.

The differences between wild and domesticated barley are relatively small, and this fact, together with the apparent ease with which wild forms become domesticated, suggest that is unlikely that barley domestication only occurred once. Agricultural historians and geneticists alike think that the crop was domesticated in dispersed small areas and spread by human migrations and trade. There are some, myself included, who feel that barley may have been brought into cultivation purely because of the relative ease with which beer may be obtained from it. Whatever, once in the hands of man barley offered a versatile, hardy crop with a tolerance for a wide range of climatic and ecological conditions.

In the ear of barley three single-flowered spikelets are borne at each node of the rachis. These alternate on the rachis so that the triplet of spikelets immediately above or below any one triplet is on the opposite side of the rachis or axis of the ear. This arrangement gives six rows of spikelets. The middle spikelet of each triplet is usually termed the median spikelet and the two flanking it, one on either side, the lateral spikelets. In all barley varieties the median spikelet is fertile and develops a grain which invariably has its ventral or grooved side next to the broad face of the rachis. The exact way in which the spikelets develop in domesticated forms gives us two main types:

1) Two-rowed, traditionally called *Hordeum distichum* L., in which only one median spikelet in each triplet is fertile and usually bears a prominent awn. The two lateral spikelets are reduced and borne on longer stalks. They contain no grains and have no awns, and so each ear contains just two rows of fertile spikelets.
2) Six-rowed forms, traditionally referred to as *Hordeum hexastichum* L., in which all three spikelets in each triplet bear seed and usually all are awned. Ears in these forms have six rows of fertile spikelets.

When the rachis internodes are short, the grains are closely packed and are compressed together, giving a dense ear. When the internodes are longer, the grains have more space and the ears are lax. In two-rowed barleys, dense ears tend to stand upright (sometimes referred to as *erectum* forms) and in some cases the grains and awns may spread out like a fan (*zeocrithon* type), while lax ears have the tendency to bend over and "nod" (*nutans* forms). When six-rowed barleys have lax ears, the lateral rows of grains tend to overlap, giving rise to what are known as "four-rowed" barleys. This is a misnomer, although a separate species designation (*H. tetrastichum*) was assigned to such forms at one time. The two-rowed condition is regarded as being primitive and is found in the wild progenitor of the crop as well as in all other Old World *Hordeum* species. The difference in spikelet type is controlled primarily by a single gene *vrs1* with the two-row type being dominant. The *vrs1* gene also has considerable effects on many other characteristics in barley. For example, two-row lines show higher seed weight and resistance to lodging, and larger and more circular seeds than six-row lines. Chemical analyses show that two-row forms have higher protein content than six-rowed types. There is also evidence that β-glucan content is lower in two-rowed barleys, and that they also contain more starch, but less fibre.

In an early barley classification, Harlan[13] recognized four species: *H. vulgare*; *H. intermedium*; *H. distichon* and *H. deficiens*. The first two were six-rowed types, the latter two were two-rowed, and division within those groups depended upon floret details. Since there are no biological barriers to hybridization between wild and cultivated barley, all forms are now deemed to belong to a single biological species[14] and, as a result of gene-flow, any rigorous taxonomic distinction between wild and cultivated forms is problematical.

The cytology and cytogenetics of barley are relatively well known[15] and the plant forms seven closed bivalents with distal chiasmata, one in each arm, at meiosis. The chromosomes are relatively large, but reports of their lengths vary somewhat due to differences in contraction. Work by Kakeda and Fukuki,[16] for example, gives mean chromosome lengths between 6.0 and 13.7 μm. The 2C-value is *ca.* 11 pg.[17]

The designation of barley chromosomes has become equivocal of late because of the use of two differing numbering systems. The "standard" or "Burnham and Hagberg" system based on chromosome size and morphology[18] gave designations of barley chromosomes as 1 2 3 4 5 6 and 7, while the more recently introduced "Triticeae" system is based on the homoeologous relationships of the barley and wheat chromosomes. In the latter the designations, when aligned with the above, are 7H 2H 3H 4H 1H 6H and 5H, respectively. Literature on the cytology, cytogenetics, molecular cytogenetics and gene structure of barley, and on the relationship of the barley and wheat genomes, has been reviewed.[19]

Barley chromosome terminology was tidied up at the 7th International Barley Genetics Symposium held at the University of Saskatchewan, Canada, in 1996. The following points were recommended for the designation of barley chromosomes and their arms:

1. Each of the seven barley chromosomes to be designated by a figure from 1 to 7 according to its homoeologous relationships with chromosomes of other Triticeae species. The figure is followed by the letter H, *e.g.* 2H.
2. The genomes of *Hordeum vulgare* and *H. bulbosum* are symbolized by the letter H.
3. The chromosome arms are designated by the letters S and L.
4. The barley genome present in the variety "Betzes" becomes the reference genome in the Triticeae to which definitions of translocations, short arm/long arm reversals, *etc.* are standardized in all species.

4.2.1 Barley Genome

Barley combines the least complicated crop genome (diploid) in the Triticeae with a wealth of genetic and genomics resources developed over the past couple of decades (*i.e.* 400,000 ESTs, Affymetrix chip, transcript map with over 1000 genes, isolation of 30,000 gene-containing BAC clones). In addition, its genome exhibits a high colinearity with other species of Triticeae (*e.g.* wheat, rye, raygrass) with which it also shares numerous agronomic traits. Thus, barley holds a central position, not only as a crop plant, but also as a model species within the Triticeae.

With 5.3 billion letters of genetic code, or 5300 million base pairs (Mbp), the barley genome is one of the largest found in cereal crops and is almost twice the size of the human genome. It is slightly smaller than the three individual genomes in hexaploid wheat, and some 40% smaller than that of rye. As with other plant species with medium to large genomes, the vast bulk of the DNA in barley is represented by repetitive DNA sequence motifs, typically 2 to 10,000 bp long, that are repeated hundreds or even thousands of times in the genome.

Besides its obvious role as a crop, barley is an established model species for genetic and physiological studies.[20] It is a convenient experimental organism because: it is an annual with a short life-cycle; it is diploid with only seven pairs of chromosomes; it is true breeding allowing multiple testing; it exhibits wide diversity in physiology, morphology and genetics; a wide range of stocks is available; and it has well-defined genetic maps.

The cultivated and wild species share a common genome (HH, $n = 7$), are inter-fertile and can be recombined. Until the late-nineteenth century, all cultivated barleys existed as landraces (mixtures of inbred lines and hybrid segregants). Some landraces persist to the present day, especially in developing countries, though selection and breeding has largely replaced landraces with pure cultivars. Barley is an inbreeding species and single plant selection, which promotes uniformity, has been common since the 1800s.

The widespread presence of wild forms is a bonus for the barley breeder and, once good genetic maps for wild and cultivated types have been constructed, and physiological models of crop responses to the environment have been developed, wild barley offers a veritable "goldmine" of untapped genetic reserves. The improvement of abiotic stress tolerance in the barley crop depends on understanding the range of genetic variation possessed by both cultivated and wild barley. In turn, the rate of progress depends upon the occurrence of genetic variation desirable for crop development and the availability of precise methods of gene transfer and selection.

Surveys of the genetic variation in wild barley and studies on the process of domestication show that cultivars represent only a fraction of the range of variation shown in wild populations, and it would seem that only around 40% of the alleles found in wild barley are present in cultivars. It has been shown that genetic diversity is greatest in wild populations from the most highly stressed environments, such as are found in some parts of the Near East.[21] Wild barley and Middle Eastern landraces have already proved to be very fruitful sources of genes for modern crop improvement, a noteworthy example being the development of the *mlo* resistance to powdery mildew.[22,23]

The story of *Mlo* resistance to barley powdery mildew started in 1942, when a resistant mutant was first described. The condition has been mutagen-induced repeatedly since then, and around 1970 it was also recognized in barley landraces that had been collected from Ethiopia in the 1930s. *Mlo* resistance is unique in a number of ways, including the fact that it is effective against all isolates of the pathogen (*Erysiphe graminis hordei*), and resistance is conferred by rapid formation of large cell wall appositions at the encounter sites preventing penetration by the fungus.

There are fewer examples of useful genes for grain quality than for disease resistance, and this has been put down to the fact that quantitative variation in endosperm composition offers very few evolutionary advantages.[24] Despite this fact, it has been shown that wild (desert) populations consist of slender plant types with thinner husk and lower grain milling energy.[25] In addition, allele-dependent barley grain β-amylase activity has been reported.[26]

The development of dense chromosome maps, based on molecular markers, has facilitated the detection of quantitative trait loci (QTL) for a range of characters in barley,[23,27] and such information gives the potential for studying genetic mechanisms underlying field performance and reveals the need for new genetic variation. Of particular importance is the improvement of abiotic stress tolerance of cultivars in order to enhance crop reliability.

The state of molecular barley breeding at the onset of the twenty-first century was assessed by Thomas,[28] who also looked at the prospects for this branch of science. As he related: "molecular breeding can be applied to either single gene or polygenic characters but is not widely used in commercial barley breeding, other than as a marker for resistance to the Barley Yellow Mosaic Virus Complex. There are many reports of potential targets for use in molecular breeding but the few validation studies that have been carried out to date are disappointing."

Characters controlled by single genes of large effect (so-called "major genes", such as dwarfing genes and disease resistances) are relatively easily recognized and can be placed directly on genetic maps. Conversely, some important traits, such as yield and quality, are the result of the influence of a number of genes and their individual effects cannot be distinguished phenotypically. It is, however, possible to identify regions of the genome that affect such characters, and these are the quantitative trait loci (QTL). QTL can be tagged with molecular markers and manipulated in the same way as major genes can, and the use of molecular markers as genetic tags is called marker assisted selection (MAS).

Knapp[29] showed that MAS could be of importance in selecting superior genotypes during a breeding programme and reduce the amount of multi-site trialling, and MAS has been used to determine the barley varieties most suitable for malt whisky distilling.[30] This latter is an interesting case, because a major gene is implicated, and the scenario is worth documenting. Epiheterodendrin (EPH) is a cyanogenic glucoside produced during the germination of barley, and a breakdown product released during mashing and fermentation can, under certain conditions, react with ethanol in copper stills to produce ethyl carbamate, a carcinogen, during whisky manufacture.[31] Certain barley cultivars (such as "Derkado" and "Decanter") have inherited an allele at the *eph* locus that appears to block formation of EPH. The Scottish Crop Research Institute (SCRI) identified simple sequence repeats (SSRs) that are linked to *eph* and can be used in MAS schemes to select non-producers. These markers are not tightly enough linked to *eph* to be diagnostic across all germplasm, but they can be used effectively in MAS schemes.

In a subsequent paper on molecular barley breeding,[32] the intricacies of "conventional selection" *versus* modern molecular breeding methods were discussed in the light of the competitive nature of barley breeding, especially in the UK. The authors found that, whilst the amount of variation in the landrace and wild barley gene pools is undoubtedly far greater than that of cultivated, there appears to be more than sufficient variation to enable barley breeders to sustain breeding progress for the near future. It was pointed out that marker assisted selection (MAS) protocols can be substituted for conventional selection for a number of major-gene targets but, in the majority of cases, conventional selection is more resource efficient. Results from (then) current QTL mapping studies had not identified sufficiently robust and validated targets for UK barley breeders to adopt MAS to assist in the selection of more "complex" traits such as yield and malting quality.

In the development of barley cultivars for malting purposes, it is usually not possible to select for malt quality characteristics in early generations because of the high cost associated with micro-malting large numbers of grain samples and assessing quality traits on the resulting malt samples and because of the sensitivity of quantitative malt quality traits to environmental variation. Ayoub *et al.*[33] used marker-based selection to manipulate α-amylase activity in a barley breeding population. They crossed a malting barley cultivar ("Morex") with an unrelated but adapted feed barley cultivar ("Labelle") and selected among the progeny based on genotypes at marker loci near a QTL region that had been mapped in a population derived from a cross between Morex and another feed barley cultivar ("Steptoe"). Their objective was to assess whether marker-based selection could be effective in manipulating a QTL region in a barley breeding population that shared only one parent with the population in which the QTL region had originally been detected and mapped. Marker-based selection was applied among $F_{2:3}$ lines from the cross Morex/Labelle, targeting Morex alleles in a region of chromosome 5H (7) that had previously been found to affect α-amylase activity in the cross Steptoe/Morex. The target region was represented by two polymerase chain reaction (PCR) markers. Selected lines were grown in field plots for two years, and agronomic and grain quality data were collected. Grain samples were micro-malted and assessed for malt quality traits. Selection for the Morex allele at two PCR markers on chromosome 5H was effective in increasing α-amylase activity, and it was concluded that marker-based selection for a QTL could be effective even when applied in a population other than the mapping population.

4.2.2   Wild Ancestry

H. vulgare L. subp. *spontaneum* (C. Koch) Thell. (Figure 4.2), the wild ancestor of the cultivated crop, is well known,[34,35] and has brittle ears (spikes), which, at maturity, disarticulate into individual, wedge-shaped, spikelets. These fragments are highly specialized devices that ensure the survival of the plant in the wild. Wild barley has complementary genes, *Btr1* and *Btr2*, which control the formation of the brittle rachis, and cultivated barley carries mutant recessive alleles at either of the loci resulting in a non-brittle rachis.[36] The *btr1* and *btr2* loci have been mapped to chromosome 3HS, and the two loci are tightly linked.[37] Takahashi[38] has shown that most occidental cultivars carry the *btr1* allele, and most oriental cultivars carry the *btr2* allele, indicating a clear differentiation pattern of the two groups.

Most occidental cultivars are of genotype *btr1Btr2* and are referred to as W-type, while most oriental ones are *Btr1btr2* (E-type). Using markers derived from a high-density AFLP-based genetic map based on an E-type × W-type cross, a phylogenetic analysis showed a clear separation between the E- and W- clades (Komatsuda *et al.*,[389]) and that barley consists of at least two lineages. The same group confirmed that non-brittle rachis of W-type barley lines is controlled by a single gene (*btr1*) on chromosome 3H, but that non-brittle rachis of E-type lines is controlled by a major gene (*btr2*) on chromosome 3H and two QTLs on chromosomes 5HL and 7H.

The brittle rachis is one of the most critical traits in the evolution and domestication of barley, and the mutation mentioned above will rarely occur in totally wild populations. It may, however, quickly be selected for in a population that is subjected to the man-made environment of reaping, threshing and sowing.[39] Experimental field trials and computer simulations indicate that, under certain conditions, the homozygous recessive (non-brittle) genotype may become predominant in a remarkably short time (around 20 years[39]). Thus, what we are saying is that the domestication of an important crop like barley is dependent on a single mutation!

A single recessive mutation is also responsible for the fertility of lateral florets in the spike, and the conversion from the two-row to the six-row condition.[40] The fertility of lateral spikelets is controlled predominantly by the alleles at the *vrs*1 locus on chromosome 2HL and is modified by the alleles at the *int-c* locus at chromosome 4HS. Komatsuda and Mano[41] integrated the *int-c* and *btr1* loci in molecular linkage maps of barley, and obtained a high-resolution map from which the markers will be of use to expedite further high-density mapping.

The close affinity between wild and cultivated barley can be judged by the sporadic spontaneous hybridization that occurs when the two forms grow side-by-side. In the past, some of the products of hybridization have been incorrectly classified as separate wild types, and been given species status (*e.g. H. agriocrithon* Åberg). The many brittle and semi-brittle rachis variants of barley, some of them six-rowed,[42,43] provide grounds for objection to the above-mentioned domestication scenario, and some geneticists have for years tried to incorporate six-row brittle forms (*e.g. H. agriocrithon*) into an evolutionary taxonomy of the barleys. Most now agree that the minor genetic differences, wide genetic diversity and ease of hybridization and introgression with domesticated forms account for the large number of varieties found in the wild.

The relationship between traditional and modern barley classifications is given in Table 4.1. The modern classification relies on isozyme, seed storage protein and DNA analyses (reviewed in Nevo[44]).

As Figure 4.3 shows, *H. vulgare* subsp. *spontaneum* is spread over the east Mediterranean basin and the countries in western Asia, and penetrates as far east as Turkmenia, Ladakh and Tibet.

Table 4.1   Classification of barleys (after Zohary and Hopf[45]).

Traditional Classification	*Modern Grouping*
A section (sect. *Cerealia* Åberg) within the genus *Hordeum* L. containing the following species:	A single species containing both wild and cultivated forms. Collective name: *H. vulgare* L.
1. Wild two-rowed barley *H. spontaneum* C. Koch Brittle, hulled.	1. *H. vulgare* subsp. *spontaneum*
2. Cultivated, two-rowed barley *H. distichum* L. Non-brittle, mostly hulled.	2. *H. vulgare* subsp. *distichum* [= *H. vulgare* convar. *distichon*]
3. Cultivated six-rowed barley *H. vulgare* L. [= *H. hexastichum* L.] Non-brittle, both hulled and naked forms.	3. *H. vulgare* subsp. *vulgare* [= *H. vulgare* convar. *vulgare*]
4. Brittle six-rowed barley *H. agriocrithon* Åberg.	4. *Agriocrithon* forms are now known to be secondary hybrid derivatives between 1 and 3.

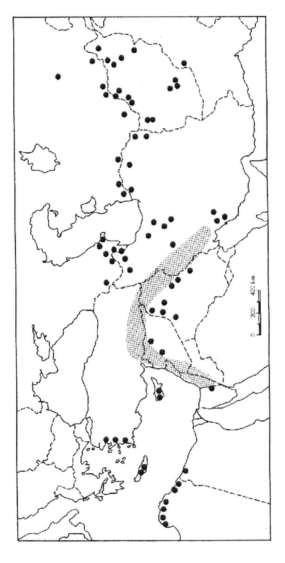

Figure 4.3 Distribution of wild barley. The area in which wild barley is massively spread is shaded. Dots represent additional sites, mainly of weedy forms (from Zohary and Hopf[45] - by kind permission)

It occurs both in native habitats and as weed in man-made habitats. Its centre of distribution lies in the "Fertile Crescent", where it is continuously and hugely distributed. As Zohary and Hopf[45] explain, in this region: "it constitutes an important annual component of open herbaceous formations, and it is particularly common in the summer-dry deciduous oak park-forest belt, east, north, and west of the Syrian desert and the Euphrates basin, and on the slopes facing the Jordan Rift Valley. From here, it spills over the drier steppes and semi-deserts."

4.2.3 Archaeological Evidence for the Domestication of Barley

Most archaeological evidence suggests that barley domestication went hand in hand with the evolution of Neolithic villages in the Levantine arc of the "Fertile Crescent". Pre-Neolithic peoples in this area, most notably Natufian foragers, became increasingly dependent on plant foods for their existence, and may have cultivated plants. Their known cultural remains include grinding stones, sickle blades and storage pits, but such objects only indicate general plant utilization, and there has been debate as to which plants were actually processed, Joy McCorriston,[46] for example, having highlighted the use of the acorn. Thus, there is no evidence that the Natufians domesticated barley, wheat or any other plant.

The earliest archaeological specimens of barley seeds were recovered from pre-agricultural sites in the Near East, dating from 19,000 to 9000 yBP.[45] The kernel specimens were essentially identical to present-day wild barley, being two-rowed with a brittle rachis. The earliest domesticated barley to be recovered (9500–8400 yBP) also had two-rowed spikes. Cultivation of six-rowed barley was apparently a little later, with estimated dates being in the 8800–8000 yBP range. Then, around 7000–6000 yBP, when barley became widely cultivated in the alluvial soils of Meso-potamia (and, later, in Lower Egypt), the six-rowed form soon became dominant and ousted its two-rowed sister to become the most important crop in these early civilizations.[47]

4.2.4 Molecular Evidence for Barley Domestication

There are still some unresolved matters regarding the origin of cultivated barley, the principal one being whether it was mono-phyletic or not. We have Takahashi's evidence,[38] which suggests that the brittle to non-brittle rachis mutation must have occurred on at least two independent occasions, and then Zohary[48] made the case for a polyphyletic origin (*i.e.* domestication was a multiple event). Subsequently, Badr *et al.*[49] found from genetic studies that out of 317 wild and 57 cultivated lines, wild populations from Israel/Jordan were, in molecular terms, more similar than any others to the cultivated gene pool. It was proposed, therefore, that this area was the region where barley was brought into culture, and that it was a monophyletic event.

Several studies have attempted to define genetic variability within wild barley populations. Nevo[44] reviewed studies of genetic diversity in wild barley utilizing enzyme systems, DNA markers and other sources of variation. RFLP has also served as a tool for the same purpose[50,51] and, more recently, reports of a number of studies using RAPD analysis of wild barleys have appeared – the first such documentation being by González and Ferrer.[52] RAPD analysis of some wild barley accessions from Israel followed,[53] and a larger number of plants from Israel, Turkey and Iran were similarly analysed a little later,[54] before Baum *et al.*[55] extended this particular area of study later in the decade. The primary objective of the latter group was to define the diversity existing in wild populations and to assess whether entities or groupings of similar genotypes could be identified. Such information is important in the context of wild barley germplasm conservation, and their analyses revealed extensive polymorphism between the different genotypes of wild barley in the "Fertile Crescent".

Characterizing the origins of domesticated barley and its spread in the Old World is crucial to our understanding of the expansion of agriculture. Together with pastoral animals such as goats, sheep and cattle, the movement of cultivated barley across Eurasia and Africa may have been one of the key events in a global adoption of agriculture among Neolithic societies. Unfortunately, resolution of the debate on the origins and spread of barley was, for some time, hampered by the dearth of East and South Asian barleys used in molecular studies. The focus on cultivated samples mainly from Europe and North Africa has undoubtedly influenced the way we have thought about barley's passage into agriculture.

Michael Purugganan's New York group changed all this when they undertook a phytogeographic analysis of five genes in a large, worldwide sample of cultivated barley landraces as well as several wild barley accessions.[56] The study drew heavily from cultivars in East and South Asia, with 177 landraces from China, India, the Himalayas, Korea and Japan. Analysis showed that an area of genetic discontinuity exists in the region between the Near East and Southwest Asia and that different genetic lineages appear to predominate westward into Europe and Africa, eastward to Asia, the Indian sub-continent and the Himalayas, and southward into Ethiopia.

The data were consistent with a contemporary series of experiments[57] that had suggested two domestication sites for barley; one within the "Fertile Crescent", the second 1500–3000 km further east. The "Fertile Crescent" domestication contributed the majority of diversity in European, North African and American cultivars, whereas the second domestication (possibly at the eastern edge of the Iranian Plateau) contributes most of the barley diversity from Central Asia to the Far East. The suggestion was that cultural diffusion or independent innovation was responsible for the expansion of agriculture to areas of South and East Asia during the Neolithic revolution.

Although the proposed second site of barley domestication had only a limited impact on the overall genetic diversity of European barley, work by Jones *et al.*[58] revealed that the impact on the form of the European landraces, and the modern cultivars derived from them, has, in fact, been significant. They used an association-based study to relate variation in flowering time to sequence-based polymorphisms in the photoperiod response gene, *Ppd-H1*, and identified a causative polymorphism (SNP48) that accounts for the observed variation in the flowering time of barley. As Turner *et al.*[59] confirmed, the gene has responsive (*Ppd-H1*) and non-responsive (*ppd-H1*) alleles present. After the cloning of *Ppd-H1*, the molecular diversity of responsive and non-responsive alleles within an extended collection of diverse barley germplasm was investigated. Mapping landrace *Ppd-H1* alleles according to accession origin shows that the observed diversity has a definite geographic structure. Genotypes predicting a non-responsive phenotype were prevalent in landraces from Central and Northern Europe, where long growing seasons with moisture available for grain filling over the summer favour the late flowering that such alleles impart. Photoperiod-responsive alleles dominated in landraces from Southwest Asia, Southern Europe and the Mediterranean Basin, and were found in all *H. spontaneum* lines studied. The non-responsive allele was not found in *H. spontaneum*, suggesting that the mutation occurred post-domestication during the human-led radiating spread of cereals throughout Europe.

Matters concerning the control of flowering time in temperate cereals have been reviewed[60] and, in relation to the above work, observe, "the latitudinal cline in photoperiod response found in barley landrace material indicates that the non-responsive *ppd-H1* mutation created a phenotype that early farmers selected and maintained because it conferred an ecological advantage in the environmental conditions of Northern Europe. This, along with the selection of crops with no vernalization requirement, represents important milestones in post-domestication adaptation".

Further evidence for the diphyletic origin of barley was presented by Komatsuda's group,[61] who looked at a collection of cultivated barleys, wild barleys (ssp. *spontaneum*) and weedy brittle rachis varieties (ssp. *vulgare* var. *agriocrithon*). Using a DNA sequence closely linked to the brittle rachis complex, which was amplified and re-sequenced, a phylogenetic tree was constructed.

The situation regarding the status of *Hordeum vulgare* ssp. *vulgare* var. *agriocrithon* is an interesting one, and a topic that has caused some controversy over the years. The finding of wild forms with six-rowed ears and brittle rachises in wild stands in Tibet, as documented by Åberg,[62,63] led to the proposal that this region was a centre of barley domestication. Such a notion was later rejected when these wild six-rowed types were deemed to be feral contaminants of other crops.[64,65] Then it was found that true wild *spontaneum* barley is present in Tibet, Nepal, India, Pakistan and Afghanistan, and this encouraged Xu[66] to conclude that the six-row naked forms of cultivated Chinese barleys may originate from genotypes domesticated in Tibet. Thus, the possibility of the Himalayas being a possible centre of barley domestication was seriously reconsidered.

The development of a six-rowed spike is controlled by a single allele, *vrs1*, that is recessive to the dominant allele responsible for the two-rowed spike (*Vrs1*). The presence of the recessive gene *vrs1* is, by itself, sufficient to cause two-row barley to become six-row barley, even though other genes may be associated with the change. In the latter category we may cite *Intermedium spike-c.h*

(*Int-c.h*), which occurs in six-rowed plants and is involved in the enlargement of the lateral spikelets.

In addition, some other genes (responsible for reduced size and reduced fertility of lateral spikelets) that had been introduced to different chromosomes *via* artificially induced mutations have been shown to be conspicuously absent from known six-rowed cultivars. In view of such information, it became apparent that *Vrs1* had been the primary target of spontaneous mutation during the evolution of the six-rowed condition. Precise information on the actual origin of six-rowed barley was obtained by Takao Komatsuda and colleagues,[67] who isolated the *six-rowed spike 1* (*vrs1*) gene by means of positional cloning. Their study showed that the wild-type *Vrs1* allele (for two-rowed barley) encodes a transcription factor that includes a homeodomain with a closely linked leucine zipper motif. Analysis of induced mutants clearly confirmed the identity of *Vrs1* and demonstrated its biological function.

Expression of *Vrs1* was localized in the lateral-spikelet primordia of immature spikes, suggesting that the VRS1 protein suppresses development of the lateral rows of spikelets. Loss of function of *Vrs1* results in complete conversion of the rudimentary lateral spikelets in two-rowed barley into fully developed fertile spikelets (*i.e.* the six-rowed phenotype). Over the years, humans have tended to deliberately select individuals from wild populations such that seed recovery and seed yield are increased. Through this, it is evident that during the domestication of barley (and wheat) rudimentary spikelets or flowers of wild species have had their functionality restored in order to increase seed number. Phylogenetic analysis demonstrated that the six-rowed phenotype originated repeatedly, at different times and in different regions, *via* independent mutations of *Vrs1*.

Thus, to summarize, Komatsuda's team showed that six-rowed barley originated from a mutation in a homeodomain-leucine zipper I-class (HD-ZIP I) homeobox gene (put another way, HD-ZIP I determines the number of rows of spikelets in barley).

The Japanese-led group suggested that the DNA sequence of *Vrs1* may trace the origin and migration of two-rowed barley. As they note, both two- and six-rowed barleys emerged in Greece between 8000 and 6000 yBP, but the six-rowed form dominated in the Balkans and Central Europe (5000–2000 yBP) and in Southern Europe and North Africa (7000–4000 yBP). As Helbaek[47] reported, after its disappearance from ancient Mesopotamia and ancient Egypt, two-rowed barley does not reappear in the archaeological record of these two regions until around 1100 yBP. Cultivation of two-rowed barley is virtually unrecorded in Central and Northern Europe until 1000 yBP, and some authorities have assumed that it was introduced into Europe around 900–800 yBP by crusaders returning from the Near East.[68] Such a notion was partially supported by a segment of Komatsuda's work, which reported that an allele (*Vrs1.b3*) carried by 14 two-rowed cultivars did not show any polymorphism in its DNA sequence. This indicated that the origin of the *Vrs1.b3* allele was a fairly recent event and that the allele had spread rapidly throughout the world and become dominant. Because the DNA of archaeological samples remains unravelled, such a hypothesis does not necessarily conflict with Helbaek's[47] suggestion that two-row barley was cultivated earlier than its six-rowed counterpart. The authors noted the lack of detailed genetic analysis of archaeological barley specimens, and made a plea for this situation to be rectified.

4.2.5 Qasr Ibrim

A couple of years later, Palmer *et al.*[69] reported on work in this very field when they examined archaeobotanical remains of barley from the site at Qasr Ibrim. This fortress site was built in Pharaonic times (*ca.* 680 BCE by command of King Taharqa, 25th Dynasty), although there is evidence of fortification around 300 years before. It was built overlooking the Nile in Lower Nubia, between the first and second cataracts, and mass human activity lasted there through to Ottoman times. Qasr Ibrim was an important boundary settlement between the Nubian and Roman Empires, and in all was occupied by five successive cultures: Napatan, Roman, Meroitic, Christian and Islamic. The settlement, which also contains remains of religious buildings, was manned by Ottoman soldiers until the early nineteenth century. Throughout all of these periods, the inhabitants successfully grew barley by using the natural water cycle provided by the River Nile ("basin irrigation"), not *via* artificial irrigation.

Surprisingly, the archaeobotanical remains of barley from all of the cultural stages at Qasr Ibrim (and from Nauri, another important site further upriver between the second and third cataracts) have all been two-rowed. This contrasts with contemporaneous barley remains recovered from sites downstream of the first cataract, such as Tell el-Amarna, which are all six-rowed. The latter are "typical" of ancient Egypt and would almost certainly have been available to the inhabitants of Qasr Ibrim. Why did they use the two-rowed form, and why was it so popular that it persisted as a favoured variety over several successive generations? It is thought that the main use for barley at Qasr Ibrim was as animal feed, in which case the six-rowed form would normally be preferable. Was two-rowed barley imported, and, if so, why?

The six-row condition is a derived state in which a loss of function mutation occurs in the *Vrs1* gene. Three different loss of function mutations have been identified in the *Vrs1* gene, all of which result in the loss of lateral bud suppression leading to the six-rowed cultivated varieties. Two of these mutations are geographically restricted to the Western Mediterranean and East Asia, respectively. The third, defined by *vrs1.a1* clade, occurs worldwide, and is considered the most ancient of the three mutations. It is responsible for most six-rowed barley varieties.

In order to investigate the reason for the two-rowed plants at Qasr Ibrim, Palmer *et al.* amplified the *Vrs1* gene from ancient DNA retrieved from archaeobotanical remains of barley covering the whole range of occupancy of the site. Observing standard procedures for preventing cross-contamination of DNA from other sources, they managed to retrieve relatively large amounts of the nucleic acid from even the most ancient of barley samples (almost 3000 years old). To their surprise, the team found that the barley at Qasr Ibrim carried the non-functional *Vrs1* allele, and was, therefore, derived from an ancestor that was six-rowed! This was the first report of such an event.

Thus, the Qasr Ibrim two-row phenotype was caused by a different mechanism than that responsible for modern two-rowed barley. A switch from six-row to two-row would have necessitated extreme selective pressure in favour of the latter condition, and the consistency of the two-row phenotype throughout all of the various strata over the three millennia indicates that the mechanism of lateral floret inhibition is more likely to have been genetic, not environmental. As a result, it was proposed that the two-row condition probably resulted from a gain of function mutation at another locus that reasserted the two-row condition (*i.e.* these grains carry a gene that negates/cancels the effect of the mutation that led to the growth of six-rowed spikelets.

Robin Allaby was of the opinion that it was unlikely that the ancient farmers around Qasr Ibrim wilfully decided to grow two-rowed barley, because the six-rowed phenotype produces higher grain yields, so the reasons behind such selection must be related to local environmental conditions – such as water stress. As he explained: "Qasr Ibrim is located in the upper Nile which is very arid in relation to the lower Nile where six-row remains are found, and we know from previous studies that two-row barley can tolerate water stress better than the six-row variety." Such a finding may well have enormous implications for the development of drought-resistant crops.

4.2.6 Beverages from Barley

The main beverage type from barley is, of course, beer in its various forms. I have dealt at length with the history of this drink,[70] and documented its manufacture,[71] and space precludes any repetition here. A number of major texts on the subject have appeared over the years, and [72–87] may be recommended. Before barley becomes useful for brewing, it must first be germinated (malted), and Dennis Briggs[88] has written an exemplary tome on this subject.

4.3 WHEAT

Wheat was the first domesticated crop and is the youngest polyploid species among the agricultural crops. Together with rice and maize, wheat provides greater than 60% of the calories and proteins for our daily life. Wheat is best adapted to temperate regions, unlike rice and maize, which prefer tropical environments. The trade value of wheat exceeds that of any other cereal crop. Having first originated as a domesticate roughly 10,000 years ago, wheat now feeds around 35% of the world's population. It spread from the Near East, where it first emerged, to flourish in a wide variety of environments – from the short summers of far northern latitudes, to cool uplands, to irrigated regions of the tropics. Humans domesticated wheat and, to a certain extent, our dependence on the crop has helped to domesticate *Homo sapiens*.

The Neolithic Revolution fundamentally altered human development, and both wheat and barley were destined to feed the great
civilizations of Mesopotamia, Egypt, Greece and Rome.

In essence, the history of wheat is the history of agriculture. By
1915, botanists had described three classes of wheat; the one-
seeded diploid "monococcum", the two-seeded tetraploid
"emmer" and the hexaploid "dinkel". The one-seeded wild relative
of monococcum was reported in Greece and Anatolia between
1834 and 1884. The two-seeded wild relative of emmer was
discovered by Aaronsohn in 1910 in Lebanon, Syria, Jordan and
Israel. As a result, it became accepted, as de Candolle[89] had
suggested in 1886, that since wild wheats grow in the Euphrates
basin, wheat cultivation must have originated there. Between 1918
and 1925, T. Sakamura and his student H. Kihara at Hokkaido
University in Japan and K. Sax at Harvard reported their classic
studies on the genetic architecture of the three groups of wheats.
They analysed meiosis in wheat species and hybrids and were the
first to establish the basic chromosome number of seven and
document polyploidy in the wheat group. This was a crucial
observation and established polyploidy as a major macro-
speciation process, and wheat as a polyploidy genetic model.

It was the Russian botanist Nikolai Vavilov[90] who found that
the greatest diversity in the gene pool of wild wheats (and barleys)
is in Southwest Asia. Thus, with the premise that "where diversity
is greatest plants have been growing, fixing mutations, and
interbreeding longest", Southwest Asia is the ancestral home of
wheat. Then, Robert Braidwood narrowed things down a little
when he identified the modern ecological range of wild wheats as
the semi-arid Mediterranean woodland belt known as the "hilly
flanks" of the Fertile Crescent.[91] Braidwood reasoned that pre-
farming peoples had adapted culturally and ecologically to specific
environments over long periods of time (what he called "settling
in") and that the first wheat farmers had long been living among
natural swards of wild wheat.

The genus *Triticum* is, of course, a member of the grass family
(Poaceae), and within that group of plants falls into the tribe
Triticeae (which, in turn, is a division of the sub-family Pooideae).
Rye and barley are also classified in this tribe, as well as members
of the genus *Aegilops* (goatgrasses). The major grass sub-families,
including Pooideae, radiated 50–80 Mya. The tribes Triticeae and

Poeae (*e.g. Lolium rigidum*) diverged around 35 Mya, and barley and rye diverged from the *Triticum/Aegilops* lineage around 11 Mya and 7 Mya, respectively. Subsequent events, approximately 8000 years ago, increased wheat ploidy to four and later to six.

Like many other important plant groupings, the wheat genus, *Triticum* L., has been subjected to a confusing array of taxonomic treatments. Modern classifications of cultivated wheats and their closely related wild types is based on cytogenetic criteria, and Table 4.2 is from the scheme used by MacKey.[92] Traditionally, wheat classification has suffered from excessive "splitting" and, until fairly recently, workers would regard every main morphological type of cultivated wheat as a separate species. In the genomic age, it is now evident that such species delimitation is not warranted, and most wheat scientists today split *Triticum* into five biological species (see below).

Several distinct species of *Triticum* were brought into cultivation, but nearly all modern wheat cultivars belong to two species: bread wheat, *T. aestivum*, ideal for high-rising bread, and hard, or durum-type, wheat (*T. turgidum*), used for the production of macaroni and low-rising bread. Other species, plus more primitive forms of the above two, such as einkorn, have been important in the past but survive today only as relicts.

Domestic wheats fall into four cytogenetic groups: one diploid, two tetraploid and one hexaploid, and forms within each group are interfertile and share the same chromosome constitution. Hybrids between groups are highly sterile. The cultivated forms with the diploid chromosome number ($2n = 14$) contain two sets of a single genome (designated AA). Tetraploid forms ($2n = 28$ chromosomes) combine two distinct genomes, either AABB or AAGG. Hexaploids ($2n = 42$ chromosomes) contain three different genomes, AABBDD.

As Table 4.2 shows, five biological species of *Triticum* are recognized today. Three of them, one diploid and two tetraploids, contain both cultivated varieties and their wild counterparts. Another diploid wheat species contains only wild forms, and the fifth species is hexaploid and contains only cultivated forms.

1. Einkorn (*T. monococcum* L.) is diploid with a genomic designation AA or AbAb; has both wild and cultivated forms. The latter, with its characteristic hulled grains, was

Table 4.2 Wheat classification (after Zohary and Hopf[45]).

Modern grouping (biological species)	*Traditional classification*
(i) Diploid ($2n = 14$) einkorn wheat Genomic constitution: AA (AbAb) Both wild and cultivated forms Collective name: *T. monococcum* L.	1. Wild einkorn *T boeoticum* Boiss. emend. Schiem. (brittle, hulled). Including single-grain forms (subsp. *aegilopoides*) and two-grain forms (subsp. *thaudar*) 2. Cultivated einkorn *T. monococcum* L. (non-brittle, hulled)
(ii) Diploid ($2n = 14$) *T. urartu* Tuman. Genomic constitution: AA (AuAu) Only wild forms	1. Wild *T. urartu* Tuman. (brittle, hulled)
(iii) Tetraploid ($2n = 28$) wild and tamed emmer wheats, durum-type wheats, *etc.* Genomic constitution: AABB Both wild and cultivated forms Collective name: T. *turgidum* L.	1. Wild emmer *T. dicoccoides* (Körn.) Aschers. & Graebner. (brittle, hulled) 2. Cultivated emmer *T. dicoccum* Schübl. (non-brittle, hulled) 3. Macaroni or hard wheat *T. durum* Desf. (cultivated, free-threshing) 4. Rivet wheat *T. turgidum* L. (cultivated, free-threshing) 5. Polish wheat *T polonicum* L. (cultivated, free-threshing) 6. *T. carthlicum* Nevski [= *T. persicum* Vav.] (cultivated, free-threshing) 7. *T. parvicoccum* Kislev small grained archaeobotanical forms (cultivated, free-threshing)
(iv) Tetraploid ($2n = 28$) Timopheev wheat Genomic constitution: AAGG Both wild and cultivated forms Collective name: *T. timopheevi* Zhuk	1. Wild Timopheev's wheat *T. araraticum* Jakubz. (brittle, hulled) 2. Cultivated Timopheev's wheat *T. timopheevi* Zhuk. (non-brittle, hulled)
(v) Hexaploid ($2n = 42$) bread wheat Genomic constitution: AABBDD Only cultivated forms Collective name: *T. aestivum* L.	1. Spelta *T. spelta* L. (non-brittle, hulled) 2. *T. macha* Dekr. & Men. (non-brittle, hulled) 3. *T. vavilovii* Tuman. (non-brittle, hulled) 4. Bread wheat *T. aestivum* L. [= *T. vulgare* Host; *T. sativum* Lam.] (free-threshing) 5. Club wheat *T. compactum* Host. [= *T. aestivo- compactum* Schiern.] (free-threshing) 6. Indian dwarf wheat *T. sphaerococcum* Perc. (free-threshing)

an important grain crop for early farmers, but is now only a relic. Nesbitt and Samuel[93] report wild einkorn as being "widespread, growing from the Balkans to Iran in weedy habitats such as roadsides and field edges".

2. Diploid *T. urartu* Tuman. ex Gandil. has a genomic designation of AA (or AuAu), and comprises only wild forms. Morphologically, *T. urartu* closely resembles wild forms of *T. monococcum*; crossing experiments indicate that it is reproductively isolated from both wild and cultivated einkorns. Isozyme tests indicate a similar scenario.[94]

3. Tetraploid *T. turgidum* L. (genomic designation AABB) comprises wild emmer wheat, cultivated emmer, durum wheat and several other cultivated tetraploid forms. Molecular analysis has shown that genome A of *T. turgidum* is closely related to the diploid genome of *T. urartu*, whilst that of *T. monococcum* is more distant.[94] It is now evident that a *T. urartu*-like diploid wheat was involved in the polyploid origin (in the wild) of wild emmer (*T. turgidum* subsp. *dicoccoides*, or *T. dicoccoides*). From the start of agriculture, emmer emerged as the principal stock of wheat, and it gave rise to the wide range of modern, free-threshing, tetraploid, durum-type wheats. By hybridization with a wild *Aegilops* species, cultivated tetraploid *T. turgidum* gave rise to another species; the hexaploid bread wheat.

4. Tetraploid *T. timopheevi* Zhuk. (genomic designation AAGG) includes both wild and cultivated hulled forms. This species is a rare domesticated glume wheat endemic to a small area in Georgia, and is thought to have played a minor (localized) part in overall wheat-crop evolution. The A genome of *T. timopheevi* is most similar to that of the diploid *T. monococcum*, indicating that a *monococcum*-like diploid ancestor was involved in the setting up of polyploidy in wild Timopheev's wheat.

5. The hexaploid *T. aestivum* L., or bread wheat, has a genomic designation AABBDD and originated under cultivation by the addition of the DD chromosome complement from the wild grass *Aegilops squarrosa* L. (syn. *A. tauschii* Coss.) to the cultivated tetraploid AABB *turgidum* wheats. Put another way, bread wheat originated from the hybridization of three related diploid species: *T. urartu* (A genome), an extinct or undiscovered population closely related to *T. speltoides* (B genome) and

A. squarrosa (D genome). Each of these has seven chromosomes and these three genomes are descendents of extant diploid species that have widely been used as germplasm in genetic studies and crop improvement. The *T. aestivum* group of wheats is very variable and represents the most important wheat crop of today. The group comprises several primitive, hulled, spelta-type wheats as well as numerous free-threshing forms, including modern bread wheat.

4.3.1 Einkorn (*T. monococcum*)

T. monococcum (Figure 4.4) is a relatively uniform diploid wheat with hulled grains and delicate ears and spikelets. As would be expected from the specific name, most cultivated varieties produce one grain per spikelet, but cultivars with two grains do exist. Archaeobotany tells us that this plant was one of the founder grains crops of Neolithic agriculture in the Near East and a major component of the early crop assortment of Europe. Einkorn is a small (rarely >70 cm high), low-yielding plant that can survive on soils that are too impoverished for other wheats. It has a nutritious, yellowish flour of poor rising capability. Since the Bronze Age, the importance of einkorn has diminished, mainly due to huge competition from free-threshing ("naked") wheats.

The wild ancestry of einkorn is well documented, and *T. monococcum* is closely related to a group of wild and weedy wheats spread over the Near East and adjacent areas, and usually referred to as "wild einkorn" (*T. boeoticum* Boiss. emend Schiem.). According to Nesbitt,[95] both archaeological and botanical evidence suggests that current-day distribution of wild wheats is a reasonable guide to their distribution 10,000 years ago, "with the proviso that the distribution of wild einkorn has shrunk". *T. boeoticum* and *T. monococcum* are morphologically and genetically very close, and hybrids between them are fully fertile.

Since most domesticated einkorn has one-grained spikelets, it is most likely that it was derived from *T. boeoticum* subsp. *aegilopoides*, a plant that has exclusively one-grained spikelets and grows at the western end of the wild form's range. Spikes of wild einkorn from the eastern edge of its range (*T. boeoticum* subsp. *thoudar*) contain a mixture of one- and two-grained spikelets. The distribution of wild einkorn is shown in Figure 4.5.

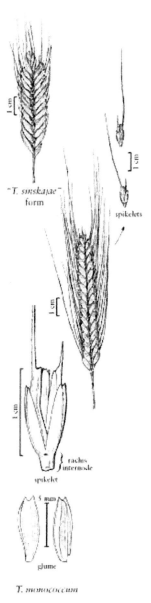

"T. sinskajae" form

spikelets

rachis internode

spikelet

5 mm

glume

T. monococcum

Figure 4.4   *Triticum monococcum* L. (from *Flora of North America*[387] - by kind permission).

As Zohary and Hopf[45] state, the main distinguishing trait between wild einkorn and cultivated einkorn is the mode of seed dispersal. Wild einkorn has brittle ears and the individual spikelets

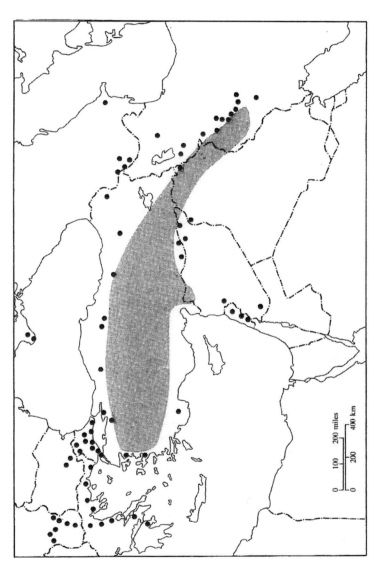

Figure 4.5 Distribution of wild einkorn. The area in which wild einkorn is massively spread is shaded. Dots represent additional sites (after Zohary and Hopf[45] - by kind permission).

disarticulate at maturity to disperse the seeds. In cultivated forms the mature ear remains intact and only breaks up into individual spikelets as a response to external pressure (*i.e.* threshing). Survival depends on sowing and reaping by humans.

Einkorn, in both its wild and cultivated forms, is noticeably absent from many of the classical archaeological sites in the Near and Middle East. This is because it has a preference for much cooler climates, and is, consequentially, almost totally absent from hot regions, such as Lower Mesopotamia and Egypt. Wild einkorn is widely distributed over western Asia, and its distribution centre lies in the Near East arc, *i.e.* northern Syria, southern Turkey, northern Iraq and adjacent Iran, as well as parts of western Anatolia (see Figure 4.5). It was used by the ancients primarily for making gruels and porridges, but its specific use in brewing is somewhat unclear. The rather sparse carbonized remains of both forms at Jarmo is about as near as we get to the alluvial plains of Mesopotamia, where attempts to grow the cultivated plant seem to have met with failure. The botanical remains tell us that wild einkorn must have grown at Jarmo with wild emmer – just as it does in the Kurdish mountains today.

Einkorn was probably extensively collected from the wild as a food source before it passed into cultivation. Charred, narrow kernels (kernels from cultivated forms are wider than those from wild forms) have been retrieved from tenth and ninth millennia uncal. BC Tell Abu Hureyra in northern Syria[96,97] and from pre-agricultural layers (ninth and eighth millennia uncal. BC.) in Tell Mureybit,[98] also in northern Syria. Then, we find that Neolithic sites at Abu Hureyra, Cafer Höyük and Çayönü (the latter two both in southeastern Turkey) confirm the farming of einkorn by 7800–7500 uncal. BC. This accords with the widely used figure of 8000 BC for the origins of agriculture. The concentration of the earliest agricultural sites in and near the Fertile Crescent confirms that einkorn (and emmer) was first domesticated in this area. It should be pointed out, however, that very few sites have yielded plant remains prior to 7000 uncal. BC, and there is no evidence that any one of the hulled wheats, or barley, was domesticated before the others, or that domestication first took place in one area rather than another.

We find that einkorn tended to prefer areas with relatively cool climates, and was totally absent from hot regions such as Egypt and lower Mesopotamia. Remains of the crop continue to be frequent in Chalcolithic, Bronze Age and Iron Age sites in the Near East, but they seem to have become replaced by free-threshing wheats. Einkorn played an important role in the early spread of Neolithic agriculture, and occurs regularly in the Neolithic farming settlements that appear first in Greece and Cyprus, then in other areas. Einkorn was one of the main cereal crops of the first farming settlements of Central Europe (the Linearbandkeramik (LBK) culture; 4500–400 uncal. BC), and was often grown in association with emmer. *T. monococcum* remains are scarce from early farming cultures of the western Mediterranean, and seem to have disappeared from such areas soon after the end of the fifth millennium uncal. BC. It persisted in many parts of Europe until medieval times, and had all but disappeared as a cultivar by the beginning of the tenth century.

4.3.2   Emmer and Durum-type Wheats (*T. turgidum* L.)

These wheats are regarded as a single biological species because they all share AABB chromosomes and are all fully interfertile with each other. Domesticated turgidum wheats fall into two categories according to their response to threshing:

i)   Hulled, non-shattering emmer wheat, *T. turgidum* L. subsp. *dicoccum* (Schrank) Thell. (traditionally called *T. dicoccum* Schübl.) in which the products of threshing are individual spikelets; the grains remain invested by the glumes and pales. As in einkorn, threshing results in breaking the rachis of the ear at its weakest points below each spikelet, and this is considered to be a primitive trait.

ii) More advanced, free-threshing tetraploids, which evolved from hulled emmer as a result of domestication. Best known of the group, perhaps, is durum wheat, *T. turgidum* Desf. var. *durum* MacKey (also known as *T. durum* Desf.). Less well known are rivet wheat, Polish wheat and two other forms (see Table 4.2). Durum wheat is an allotetraploid that originated through intergeneric hybridization and polyploidization of two diploid grasses; *T. urartu* (AA genome) and a B-genome diploid related to *Ae. speltoides* (SS genome).

Hulled emmer was the principal wheat of Old World agriculture in Neolithic and early Bronze Ages, and was certainly used for brewing beer. Later on, it was gradually replaced by free-threshing tetraploid and hexaploid forms and, today, it is a relic crop occasionally grown in some parts of Europe and Southwest Asia. Wild emmer (*T. dicoccoides*) (Figure 4.6) was formed by the hybridization of two diploid wild grasses, *T. urartu* (closely related to wild einkorn) and an as yet unidentified species of *Aegilops* (related to *Ae. searsii* or *Ae. speltoides*).

The wild ancestor of emmer was first identified as long ago as 1873, when part of a spike was found in a collection of wild barley from Mount Hermon in southern Syria. The plant was later found in abundance in Israel. Two morphologically distinct forms of *T. dicoccoides* have been recognized; one narrow-eared and native to the whole range of wild emmer, and a wide-eared, more robust form with a more restricted distribution. Unlike wild einkorn (and wild barley), wild emmer has failed to spread outside of the Fertile Crescent and, of all cereal progenitors, extant wild emmer is most highly representative of its early distribution range.

Emmer was a much more successful crop than einkorn and became the chief wheat in many parts of the Near East. The importance of this crop in the brewing of beer in the Near East cannot be overestimated.[99] The earliest signs of the pre-agricultural gathering of wild emmer wheat (and wild barley) is from Ohalo II, an early Epi-Palaeolithic, now submerged, site on the southern shore of the Sea of Galilee, dating from *ca.* 17,000 BC.[100]

Emmer was also the chief wheat in ancient Egypt, and was used for both baking and brewing. Its eventual demise began with the introduction of free-threshing wheats (first *T. durum*, then *T. aestivum*) during the first Ptolemaic period.[101] Contrary to some reports, the ancient Egyptians did not grow spelt, and speculation that einkorn was one of their crops is just that. The latter has been recorded, but it was almost certainly a weed species, and not a crop in its own right.

As far as we know, the earliest form of wheat agriculture *per se* was practised at the Syrian site at Tell Aswad, some 25 km southeast of Damascus, with identifiable remains of cultivated emmer being found in a layer dated to 7800–7600 uncal. BC. Since no wild, *dicoccoides*-type emmer has ever been found from this site (the region is too arid to support emmer), it is reasonable to assume that cultivated stock was brought into this area from elsewhere. Zohary and Hopf[45] suggest that this is not later than

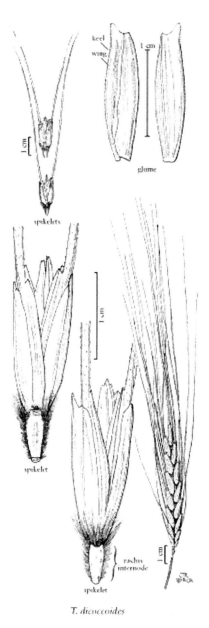

Figure 4.6 *Triticum dicoccoides* (Körn. Ex Aschers. and Graebn.) Schweinf. (from *Flora of North America*[387] - by kind permission).

7800 BC, thus implying that wheat cultivation is an even more ancient practice. All the evidence suggests that emmer cultivation must have been well under way in the Near East by the middle of the ninth millennium BC. From 7600–7500 uncal. BC onward, for example, charred grains corresponding to *T. dicoccum* appear at Abu Hureyra,[96,97] and in Pre-Pottery Neolithic B (PPNB) Jericho.[102]

From the very onset of agriculture in the Near East emmer is the principal cereal of the newly established PPNB settlements, certainly in quantitative terms. It continues to be the main grain crop during the later stages of the Neolithic, and was widely grown there up to the Bronze Age, whence free-threshing wheat started to prevail. Emmer was also the main crop in the spread of Neolithic agriculture from its Near Eastern heartland. It was: the most prolific constituent of the assemblage that started agriculture in the Aegean, the principal cereal of the LBK farmers, present at places belonging to the Impressed Ware culture that introduced agriculture into the western Mediterranean basin and was THE wheat of Neolithic and Bronze Age Egypt. Emmer and barley were the principal grains that started Neolithic agriculture on the Indian sub-continent.[103] Of the three main Neolithic grain crops (einkorn, emmer and barley) the wild progenitor of emmer (tetraploid AABB *T. dicoccoides*), being confined to the Near Eastern "arc", has the most limited distribution. This strongly suggests that early finds of wild emmer outside of this range can be taken as a sign of agricultural activity.

Free-threshing wheat forms appeared in the Near East soon after cultivated emmer made its debut. These forms are certainly present among the plant remains of seventh millennium uncal. BC. Can Hasan III in Turkey[104] and Tell Aswad in Syria. They also occur at a number of other sites.[105] For some time, these PPNB finds in the Near East were referred to as *T. turgidum-T. aestivum*, or "*aestivo-compactum*" wheats, because of some difficulties apportioning "tetraploid" or "hexaploid" status to recovered remains.

Many years ago, it was realized that wheats comprised a polyploidy series of plants, and by the 1920s it was clear that the hexaploid spelt wheats must be an allopolyploid of a tetraploid *Triticum* and one, or more, diploid *Aegilops* species. This notion was confirmed by the classical experimentation of McFadden and Sears,[106] who showed that *Aegilops squarrosa* was the species involved in the hybrid. Subsequent work, using other techniques, has confirmed the role of *A. squarrosa* as an ancestor of both the hulled and free-threshing wheats.[107] When *T. dicoccum* and *A. squarrosa* were crossed, and the chromosome number doubled in the first generation, a fertile, hybrid wheat with great morphological similarity to *T. spelta* was formed. The hybrid possessed certain features of domestication that would select against its survival in the wild. Further experiments have shown that all crosses of tetraploid wheats, whether or not free threshing, with *A. squarrosa* result in hulled, spelt-type, hexaploid wheats.[108] The first hexaploid wheat, therefore, would have been a hulled wheat.

Durum, or macaroni, wheat is the hardest of all wheats (*durum* in Latin means "hard"), and its density, combined with its high protein content and gluten strength, make it the wheat of choice for making premium pasta products. It is also used for making couscous and burghul. Durum kernels are amber in colour and larger than other types of wheat, and (uniquely) have a yellow endosperm, which gives pasta its colour. Milled endosperm is called semolina. The crop was developed by selection of domesticated emmer strains grown in Central Europe and the Near East around 7000 BC, which developed a free-threshing form.

It is reckoned that the dispersal of durum wheat was associated with the rise of Islam, whose devotees greatly admired the crop. Through them durum spread throughout the Middle East, north Africa and some parts of Southern Europe (*e.g.* southern Spain). This is in accord with the fact that durum gives greater yields than other wheats in areas of low rainfall. Today, durum is an important wheat crop in western Asia, and is extensively cultivated in the Mediterranean basin, Argentina, Australia, Canada, India, the USA and several other countries. Durum reached the USA during the nineteenth century and, presently, production there is concentrated in North Dakota and surrounding area, which is responsible for around 75% of the crop. North Dakota durum is noted for its colour and strong gluten characteristics.

Durum was involved in the events leading to the production of hexaploid bread wheat (see below) and the A and B genome chromosomes in these two cultivated species are greater than 99% identical.

Genomic analyses have shown beyond doubt that *T. aestivum* is a hybrid of a tetraploid *turgidum* wheat (genomic composition

AABB) and *A. squarrosa* (genomic constitution DD), and it was later confirmed[109] that the tetraploid parent of hexaploid wheat was the cultivated form of emmer rather than the wild form. It has since been appreciated that hexaploid bread wheats must have originated outside of the "Fertile Crescent", and then only after tetraploid Neolithic wheat agriculture had met the distribution area of *A. squarrosa*. The latter extends from northern China westwards to west and southwest of the Caspian Sea (well northeast of the Fertile Crescent). The contact between the two probably happened in the Caspian belt between 6000 and 5000 uncal. BC. As van Zeist[110] noted, agriculture did not reach the Caspian Sea until after 6000 uncal. BC, and so hybridization could not have occurred until then. Thus, the early naked wheats in the Near East cannot possibly have been hexaploid *T. aestivum*, and so they should be classified as tetraploid *T. turgidum*.[45]

4.3.3   Bread Wheat (*Triticum aestivum* L.)

The first grains of the even more advanced, cultivated, naked-seeded, free-threshing wheats (of the *T. aestivum* or "bread wheat" group) appear in the record shortly after 6000 BC at Tell es-Sawwn, near Samarra (Iraq). Because they do not have to be de-hulled, such varieties are ideal for making bread and other forms of food. This being the case, why did Mesopotamian farmers continue to grow hulled emmer and, to a far lesser extent, einkorn? Were these hulled forms of wheat better for storing, or were they less likely to succumb to damage in the field? We don't know. Maybe they were preferred for brewing purposes!

T. aestivum (Figure 4.7) is now the most important wheat species and accounts for around 90% of the world's wheat production. The name encompasses a number of contrasting types, but all are hexaploid and interfertile when crossed with one another. *T. aestivum* evolved under cultivation from *T. turgidum* stock that was already domesticated. It does not have a wild hexaploid counterpart (unlike the diploid and tetraploid wheats), and represents a classic example of evolution by polyploidy. Thus, the hexaploid bread wheats originated *via* the addition of a third genome to the two genomes already present in *T. turgidum*. As intimated, no AABBDD hexaploid is known from the wild and so the genome addition could only have taken place under

Figure 4.7   *Triticum aestivum* L.

cultivation. Bread wheat has been synthesized in the laboratory by crossing the two aforementioned parents and doubling the chromosomes in the hybrids. Hexaploid wheats can be assigned to two groups according to their response to threshing: hulled and free-threshing.

In addition to naked forms (see below), there are several hulled hexaploid types, which survive today mostly as relic crops. Most important, certainly historically, is the above-mentioned spelta wheat (spelt), *T. aestivum* subsp. *spelta* (L.) Thell. Modern taxonomists regard it as a sub-species of the bread wheat, rather than a species in its own right, which means that the original designation, *T. spelta* L., is now largely defunct. As would be expected from its likely origin (see above), spelt has no wild hexaploid ancestor.

The reports of the presence of spelt at some of the early "digs" appear to be mis-identifications, because as Hans Helbaek[47] pointed out:

"Spelt has never been found in prehistoric deposits outside Europe, and present cultivation is restricted to certain Central European mountainous districts and a few other places where it is known to have been introduced in historical times by people coming from Central Europe."

Since Helbaek's statement was made, there appears to have been one positive identification of spelt, at the northern Iraqi site of Yarym Tepe, dating from *ca.* 5000 BC.[111] Nesbitt and Samuel,[93] however, regard such early archaeobotanical records of spelt from outside Europe as doubtful, especially as fragmentary remains of spelt spikelets can be difficult to distinguish from those of some *Aegilops* species.

The most abundant and best documented archaeological evidence for spelt comes from Europe. Remains of the crop occur at later Neolithic sites (2500–1700 BC) in eastern Germany and Poland, and Jutland,[112] and during the Bronze Age it spread widely in Northern Europe. Records of spelt from elsewhere are sparse and poorly documented. One of the problems is that spelt is usually present as a small proportion of other wheats.

As intimated above, a clear model for the history of spelt had been accepted by many workers. From its origin, the migration of spelt to Europe can then be traced by way of archaeobotanical remains in Transcaucasia, Moldavia and Bulgaria. The problem is that archaeobotanical facts are not fully consistent with the model. One simple question, for example, is "why is spelt absent from the Near East?" There are other anomalies, as well. As Nesbitt[95] says, "neither biological nor archaeological evidence allow any definite statement on where and when current day populations of spelt had their origin." The experimental evidence of genetics demonstrates that a hybridisation of spelt must have occurred before the evolution of hexaploid, free-threshing wheat. Evidence from DNA[113,114] and from isoenzymes[115] points to a monophyletic origin of hexaploid wheat.

Isoenzyme evidence[116] shows relatively consistent differences in ADH enzymes between Asian and European forms of spelt, suggesting that European forms did not simply originate by migration from Asia. This is supported by RFLP DNA analyses that find European spelts to be most closely related to European bread wheats, and only distantly related to Asian spelt.

Archaeobotanical evidence for the presence of hexaploid free-threshing wheat and emmer in Neolithic Europe shows that spelt *could* therefore have originated from a hybridization event of a free-threshing hexaploid and a hulled tetraploid wheat in this region. Such a notion has, however, been shaded by more recent evidence from Transcaucasia and north of the Black Sea, which suggests an alternative route of spelt travel to Europe that avoids the Near East (explaining why there are no Near Eastern finds of spelt). The most widespread interpretation of events is that spelt originated once, in Transcaucasia or northwestern Iran, and travelled westwards to Europe as a second-wave crop several millennia after the arrival of agriculture.

Spelt wheat remains from the Bronze Age and the Iron Age are more numerous and come from all over east, central and Northern Europe, as well as from Greece. The crop was well known to the Romans, and in Europe it persisted until the start of the twentieth century. Small quantities of spelt are still grown today, particularly in southern Germany and northern Switzerland[117] and southern Spain.

Free-threshing bread wheats are predominant in this day and age, and most important (and the most universal) is bread wheat, or common wheat (*T. aestivum* subsp. *vulgare* (Vill.) MacKey; also known as *T. aestivum* L., *T. sativum* Lam. or *T. vulgare* Host.). Two other free-threshing hexaploid types are: club wheat (*T. aestivum* subsp. *compactum* (Host.) MacKey (syn. *T. compactum* Host.), with compact ears. It is grown in the northwestern USA and Afghanistan, and has been documented in early European cultures. Finally, there is Indian dwarf wheat, *T. aestivum* subsp. *sphaerococcum* (Perc.) Mackey (syn. *T. sphaerococcum* Perc.), with small grains, which is native to India and Pakistan.

Interestingly, it has been shown in a field survey that the *T. turgidum*- *A. squarrosa* association hypothesized in the theory of bread wheat evolution still exists in parts of the Alborz Mountain region, northern Iran, part of the area where bread wheat probably evolved.[118]

Unlike rice and barley, where domestication occurred at the diploid level, the evolution of polyploid genomes played an important role in bread wheat domestication. In addition, somewhere along the way to the emergence of this important crop, two other *Triticum* species, diploid einkorn and tetraploid emmer, were domesticated independently and also served as important crops. The events that increased wheat ploidy to four, and later to six, occurred approximately 8000 years ago.

Before molecular techniques can be used to speed the development of improved wheat varieties and enable the creation of novel germplasm, important genes need to be isolated. Gene mapping and isolation in hexaploid wheat is hampered by the complex genome. Complexity is due to both the allohexaploid nature (*i.e.* the presence of three homoeologous genomes, A, B and D) and the enormous size of the genome. Fractionation into individual chromosomes has long been found to be an attractive route to simplify the analysis of the complex wheat genome, and this has been carried out by micro-dissection.[119] Flow cytometric sorting, however, seemed more interesting because large numbers of chromosomes can be purified in a short time.[120] Since the early 1990s, Jaroslav Doležel and co-workers[121,122] have been using flow cytogenetics for targeted analysis of plant genomes. The strategy relies on purification of specific chromosomes using flow cytometry and is especially attractive for species with large genomes. Flow cytometric analysis and sorting of mitotic chromosomes (flow cytogenetics) were originally developed for humans,[123] and since then have been modified for other animals and for plants. Flow cytometric classification of chromosomes (flow karyotyping) is now a useful tool for detecting chromosome aberrations, for gene mapping and for constructing chromosome-specific gene libraries. Flow sorting of mitotic chromosomes in *T. aestivum* has been reported.[124]

Recently, chromosomes sorted from hexaploid wheat were used for the preparation of the first sub-genomic BAC library specific for chromosomes 1D, 4D and 6D.[125] BAC libraries specific for chromosome 3B and the short arm of chromosome 1B have also been established. These genomic resources should greatly facilitate physical mapping and gene cloning in hexaploid wheat. The potential of flow cytometry for chromosome sorting in durum wheat has been reported by Kubaláková *et al.*[126]

Rice, maize and wheat, which co-evolved from a common ancestor around 55–75 Mya,[9] differ greatly in genome size, and among agricultural crops the allohexaploid *T. aestivum* has the largest genome at 16,000 Mb. This is eight-fold larger than maize, and 40 times larger than that of domesticated rice.

Duplication of chromosome segments, coupled with amplification of transposable elements (TEs), has been a major driving force for cereal genome expansion, although polyploidization also contributed to the large genome size of wheat. About 90% of the wheat genome consists of repeated sequences and 70% of known transposable elements (TEs), the remaining fraction belonging to single or low copy sequences (*i.e.* genes). Amplification of TEs, coupled with duplication of chromosome segments, was a major driving force for cereal genome expansion, although polyploidization also contributed to the large genome size of wheat.

In small genomes, such as that of rice, gene distribution along chromosomes is relatively homogeneous, but in the large wheat genome gene clusters (gene-rich regions) are separated by long stretches of TEs (gene-poor, or gene-free regions). Recent analyses of wheat genome organization indicate that where genes are clustered in these gene-rich islands, gene density is similar to that found in small genome species.[127] The presence of such islands should facilitate gene cloning based on recombination mapping – something that had always been considered difficult in polyploidy species. Isolation of important genes in wheat is a major challenge, and a prerequisite for the exploitation of molecular techniques in the development of new varieties.

All TEs have two basic properties, the first being the ability to move from place to place in the genome (hence their designation). The second is their ability to amplify their copy number within the genome. The contributions of TEs to plant gene and genome evolution have been discussed by Bennetzen.[128]

Readers interested in a review of wheat taxonomy, and in modern wheat science, are directed to the Wheat Genetics Resource Centre site (Kansas University): www.ksu.edu/wgrc.

4.3.4 Wheat and Beer

In our modern world, "wheat" is regarded as a style of beer, rather than what it is (and always has been), one of the ingredients of beer. It was used by ancient Near Eastern and Egyptian brewers, and was earmarked by classical Greek scholars as a brewing ingredient of the Celts and other "inferiors". As Max Nelson admits,[79] it is to the third century AD Greek author Athenaeus that we owe our knowledge of the attitude of the Greeks toward beer.

In the fourth book of his *Deipnosophists* ("dining scholars") there is long list of the various eating and drinking habits of different peoples. Of the lower Celtiberian classes he says: "What is drunk is a beer made of wheat prepared with honey, and oftener still without any honey, and they call it *korma.*"

In Roman times, the historian Tacitus observed in his ethnographical treatise *Germania* that the drink of the Germans, consumed apparently from ceramic containers, was "a liquid from barley or wheat, which, once rotted, has a certain resemblance to wine".

Dr Heinrich Knaust, the German beer writer, and a "sixteenth-century Michael Jackson", knew all about the beer styles of his day. In what was one of the first books written about German beer (the first edition was published in Erfurt in 1575), he catalogued around 120 different styles and made it known that there was a clear distinction between beers based on wheat and those based on barley. He says, for example: "The noble Hamburg beer is the queen of all other wheat, or white, beers, just as the Dantzic beer has the precedence and is queen of all the other barley, or red, beers." The statement encapsulates the major characteristic of most beers brewed with a fair proportion of wheat; the light colour ("whiteness"). It seems as though the terms "white" (for wheat) and "red" (for barley) were first coined by Johann Brettschneider (alias Placotomus), professor of medicine at the University of Königsberg, who penned *Über Natur und Kräfte der Biere* in 1549.

The oldest archaeological evidence of wheat-beer brewing in Europe dates from around 800 BC. It comes from a deposit in an earthenware amphora found in a Celtic grave near the north Bavarian village of Kasendorf, around seven miles west of Kulmbach. Analysis of the residue pointed to it being from a dark, wheat ale flavoured with oak leaves. The artefact can be seen at the Bavarian Beer Museum in Kulmbach.

In his excellent exposé on brewing in Holland over the years,[86] Richard Unger says that, from the earliest of records, it seemed that beer could be made from any grain – with oats, wheat, rye and barley being the main components. Wheat was invariably the most expensive and barley was usually the cheapest, but prices would vary with availability.

The extremely intimate relationship between grain and beer, and the importance of the former to everyday life, can be gleaned from the fact that in periods of grain shortage measures were taken. In

1408 and 1416, for example, the authorities in Amsterdam only allowed the export of beer if an "equivalent" quantity of rye or wheat was imported (*i.e.* sufficient to make the volume of the beer being exported).

By the mid-seventeenth century, barley became the principal source of fermentable material for Dutch brewers, although spelt wheat was still used. Bakers and brewers often competed with each other for grain supplies, and it was generally appreciated that, because of losses during production (malting, for example), bread gave much higher levels of nutrition from a set amount of grain. With wheat, nutritional loss during brewing compared to bread-making was over 75%. The only way to adjust such a loss was to make weaker beer!

As agricultural production rose in Europe, it seems as though Dutch brewers used less wheat and more oats and barley in their grists. Records show that during the fifteenth to seventeenth centuries, Dutch brewers obtained much wheat and barley from the Baltic as well as barley and oats from France.

As Unger reports, a sixteenth-century Portuguese captain, Monzo Vazques, said Dutch brewers used wheat, barley or oats and rye. He maintained that beer made with wheat was light and clear and served directly from fermentation vats, without it having to be put into barrels first. Beer made with barley, the captain claimed, was "good but cheaper" and has "less of a head". The old sea dog also maintained that beer made with oats and rye "had a different colour as well as body and so cost less and was weaker". It is interesting to note that a commentator in the second half of the sixteenth century observed that buckwheat was used for brewing in Delft, as well as the four grains mentioned above. It is also evident that Haarlem brewers used buckwheat in the 1570s and 1580s, as did Rotterdam brewers in the early seventeenth century.

During the fifteenth and sixteenth centuries, town governments in Holland laid down standard recipes for different categories of beer. The standard, or *pegel*, specified the quantities of each type of grain that brewers should use in a beer. There was room for manoeuvre, depending upon the availability or cost of grains. Unger outlines some of the complicated situations that could arise from the *pegel.* In 1544, Haarlem *dubbelbier*, for example, used "20 sacks of barley malt and 16 sacks of oat malt with the possibility of substituting wheat for barley, but at the rate of two of wheat for every four of barley, and replacing oats at the rate of three of oats for two of barley. Even spelt could be used; two sacks of this grain replacing one sack of oats and one sack of barley"! To add to all this, the measures for the various grains might differ and, in addition, there was no strict relationship between the quantity of beer produced by a brewery and the amount of grain used.

In Europe, beer was made from oats, rye, wheat and barley, and the combination actually used by brewers would be adjusted according to availability (and price). Records tell us that, by the late thirteenth century, the proportion of grains supplied to the canons of St Paul's Cathedral in London by their estates comprised 46% wheat, 46% oats and only 9% barley. Most of the wheat went to make bread, and any surplus, together with nearly all the oats and barley, was used for brewing. A common brewing ingredient in England around this time was "dredge", a mixture of barley and oats. Wheat, and other grains used for bread-making, continued to be used for brewing until late medieval times, when periodic shortages left bakers without raw materials. It was then that rulers started to preclude the use of all but barley for brewing purposes.

Early edicts were issued by officials in Nuremberg in 1290 and 1305, which forced brewers to use solely barley as a source of fermentable material. A similar ruling was made in Munich in 1447, and permitted only barley, hops and water to be used in brewing. The much more famous *Reinheitsgebot* ("German Purity Law") was a ducal decree issued on 23rd April, 1516, by the then co-rulers of Bavaria, Duke Wilhelm IV and Duke Ludwig X. Initially confined to Bavaria, the Law later spread throughout Germany, and applied until 1987, when it became a victim of EC law! The very wording of the *Reinheitsgebot*, which, in effect, prohibits the use of other grains in brewing, intimates that wheat, oats and rye were very important food crops. The term "Purity Law" is really a misnomer, and has often been interpreted incorrectly. Its real aim was to prevent the use of grains useful for baked (and other) foods being "wasted" on beer production. Intriguingly, the term "reinheit" (purity) did not appear in the original document, and did not manifest itself until around 1918. In addition, the law was only meant to apply to bottom-fermented, lagered beers; the rules governing top-fermented (ale-style) beers were much more generous. So the ruling classes could still brew their wheat beers!

Similar restrictions, but with different objectives, were to apply in other countries at various times. In Britain, for example, the 1697 Malt Tax proscribed that barley was the only grain to be used by brewers for mashing, and duty was levied at an appropriate rate. The Tax remained in force until 1880, when William Gladstone repealed it (the "Free Mash-tun Act"). This, of course, had nothing to do with food economics/logistics, it was purely a tax.

The situation regarding wheat use in the USA in the nineteenth century is evocatively summarized in Thausing's textbook,[129] where the publishers' preface openly admits: "In this country the beer brewing business has been established exclusively by Germans, who still prefer reading books printed in the German language." Of wheat usage, Thausing says: "In some countries wheat is used for the production of local beers, either unmalted as raw fruit or as wheat malt, for instance, in Belgium and in North Germany for the production of weiss beer. Wheat malt gives to beer the so-often-desired light color much better than barley malt. But the malting of wheat is a very difficult matter, partly because it is a naked fruit, and partly because in germinating the cotyledon appears on the same point of the grain shortly after the radicle has made its appearance. The gaining of wort from wheat is difficult on account of the tenacious layer of grains. The wort is frequently cloudy, the beer from wheat is difficult to clarify, and is said not to keep well. The causes of the last-mentioned phenomena may be the larger quantities of proteine substance which are dissolved in mashing; vegetable glue, mucedine, and vegetable albumen ... Should the brewer use wheat for fabrication of beer, he should prefer the mealy to the glassy wheat." Thausing then goes on to engage us with the types of wheat prevalent at that time: "Three kinds of grain must be mentioned here as varieties of wheat: spelt or German wheat, Emmer [called here *Triticum amyleum*], one-grained wheat or horse spelt [called *T. monococcum*; *i.e.* einkorn]. These kinds of wheat are largely cultivated in the southern part of Germany and in Switzerland, and are also known under the name of 'Swabian wheat'."

Within today's world-wide brewing industry, small amounts of wheat, not necessarily malted, are incorporated into many beers, usually at a level of 3–10% of the grist. Such additions are reckoned to enhance the diastatic activity of the grist and to increase the supply of yeast nutrients to wort. Wheat is also said to improve "mouthfeel" and/or head-retention of a beer. Wheat malt is used for brewing certain beer styles, such as German-style weissbier/weizenbier, where it may comprise 75–80% of the grist.

It would appear that only hexaploid bread wheats are used for malting, and the lack of a husk, with the concomitant lack of rigidity, presents the maltster with handling problems, especially when the grain is wet. Wheat grains tend to pack down tightly in the steep, in the germination compartment and on the kiln. Because the grain is readily crushed, the steep, germination compartments and kiln should be loaded less deeply than would be the case for barley. Water uptake is more rapid than for barley (steeping time usually around 38–55 hours) because there is no husk to resist grain swelling. During germination, the acrospires grow free (unlike barley) and are easily damaged by turning. Later on, during wort separation in the brew-house, the lack of a husk contributes to slow run-off.

During the last couple of decades there has been a revival of interest in the white beers of Belgium and Germany. Belgian *witbier* is usually brewed with 20–40% unmalted wheat and has a typical spicy flavour. German *weissbier* is made with malted wheat (up to 80%) and a special strain of yeast that can convert ferulic acid into 4-vinyl guaiacol, resulting in a characteristic phenolic flavour. While turbidity is unacceptable in ales and pilsener-style beers, haze intensity and stability are important quality character-istics in white beers. Haze can, of course, result from the presence of yeast, but the major non-yeast components of haze in white beers are proteins and polyphenols. When 40% wheat is used in a grist, the permanent haze stability of a beer is reduced to a level lower than that obtained from an all-malt beer. Experiments indicated that it was wheat gluten, most likely wheat gliadins, causing this negative effect.[130] Wheat gluten proteins were found to be haze active in that they interact with polyphenols and protein-polyphenol complexes. At low gluten levels, a haze is formed, although at higher gluten levels, these insoluble complexes are too large to stay in suspension and precipitate out. In subsequent work, Delvaux *et al.*[131] investigated the influence of wheat malting on beer haze, and found that it had a positive effect. They suggested that, by increasing the level of protein degradation, less precipitation occurred, and a more stable haze was formed.

Belgian *witbiers* are often flavoured with plant additives, such as coriander seed and orange peel, and are usually lightly hopped. Both categories of white beer normally employ top-fermenting "ale" yeasts, not bottom-fermenting strains. With the rekindling of interest in brewing with wheat, there are now various offerings on the market and an appropriate array of terminology. For an account of brewing with wheat see Hieronymous.[132]

4.4 MAIZE

Maize (*Zea mays* ssp. *mays*) (Figure 4.8), a fairly recent domesticate, is the planet's third most important cereal crop, the pre-eminent

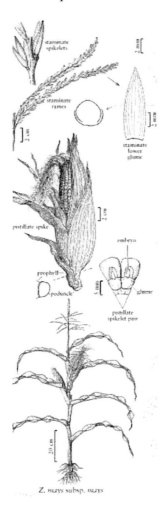

Figure 4.8 *Zea mays* subsp. *mays* L. (from *Flora of North America*[388] - by kind permission).

grain in the New World (in the USA it is twice as important as any other crop) and was the first plant to be developed into a commercial platform for molecular "pharming" in the field. Genetically modified (GM) maize has been grown in Canada and the USA since 1997, and some 80% of the latter's crop is now genetically modified. Unlike most crops, maize, a member of the grass tribe Andropogoneae, has no morphologically equivalent wild form.

The term "maize" emanates from the indigenous Taino word *maiz*, but once the plant spread to North America it became known as "Indian corn", a term that was shortened to "corn" during the nineteenth century. Indian corn is now usually a term reserved for the multi-coloured flint corn (*Z. mays* var. *indurata*). The British tend to call the plant "sweet corn", although this should strictly apply to particular varieties of maize (*e.g. Z. mays* var. *saccharata*).

Maize is a tall, warm weather annual and, although it is deep-rooted, it requires relatively high moisture for abundant growth. The botany of the maize plant is unique in that the male inflorescence is a staminate tassel and the separate female inflorescence eventually turns into a highly specialized structure, the cob, which bears large grains. Superficially, the plant resembles bamboo, the stems being erect and typically 2–3 m in height. Stems bear many nodes from which the flag-like leaves arise – those nearer the base being the broadest (50–100 cm long × 5–10 cm wide). The ears (female inflorescences) develop under leaves, close to the stem, and they are so tightly covered by leaves that they do not become apparent until the emergence of the pale-yellow silks from the leaf whorl at the end of the ear. Looking like tufts of hair, the silks are elongated stigmas, and initially appear green, changing to yellow or red. At the apex of the stem is the tassel, essentially an inflorescence of male flowers. When the tassel is mature, and conditions are sufficiently warm and dry, the anthers dehisce and shed their pollen, which is spread by wind. Each silk may be pollinated and will eventually produce one maize kernel. Young ears can be eaten raw with both cob and silk, but this is not advisable when the plant matures.

Typical of grasses, the maize kernel is surrounded by a pericarp, and whilst the cob can almost be considered as a multiple fruit, the kernels never actually fuse into a single mass. Grains are about the size of peas, and are attached, almost in vertical rows, around a central pith. Modern ears contain anything from 200 to 400 kernels, can be 10–25 cm in length and are of variable colour. Ground maize kernels produce more flour than an equivalent weight of wheat seeds, and this will contain less bran than wheat flour. On the other hand, maize flour lacks gluten, making it unsuitable for baked goods requiring good rising capability.

As is documented below, maize was malted by the earliest settlers in North America and is still malted in African villages today. Preparations made from the raw grain are major sources of extract for brewers and distillers. Thousand corn dry weight (TCW) values vary from 100–600 g, and dimensions are length 8–17 mm and width 5–15 mm with a thickness of 4 mm. The flattened, naked grain has a narrow base and a broad apex; there is no furrow. Pericarp and testa are fused as a tough hull, which may have various colours, but is often yellow/orange. The embryo is comparatively large, often 10–12% of the grain, and is placed to one side at the base.

Columbus came across maize in the New World and took it back to Spain, where it spread throughout Europe, to North Africa, the Middle East, India and China. The increasing use of maize as a staple food reflected the large yields per hectare that could be achieved, as compared to wheat and barley. Basically, because maize was "cheap", it became a dominant food for poorer and underprivileged parts of society. Unfortunately, wherever maize went, pellagra almost certainly followed,[133] and there was a suspicion that maize contained some kind of toxin. Since pellagra was rare in Mexico, despite widespread maize consumption in that country, this could not be the case. When preparative methods were examined, it was shown that the ancient Aztec and Maya peoples used to soak maize in an alkaline solution (water + lime) prior to eating. We now know that maize is niacin-deficient, and that pellagra is caused by a lack of that vitamin (not actually proven until the 1930s). Soaking maize meal overnight liberated bound (unavailable) niacin and the amino acid tryptophan, from which niacin can be formed. These compounds then became "available" for human nutrition. Immature maize shoots produce measurable amounts of the hydroxamic acid, 2,4-dihydroxy-7-methoxy-1,4-benzoxazin-3-one (DIMBOA), which confers a natural defence capability against a wide range of pests, including arthropods, fungi and bacteria. DIMBOA levels decrease as a plant matures.

4.4.1 Maize Progenitor

The molecular revolution of the last 20 years has provided compelling evidence that teosinte (*Zea mays* ssp. *parviglumis*), a small annual plant growing naturally on Mexican hillsides, is the progenitor of modern maize. By using microsatellite data of diverse maize landraces and the three wild taxa of *Z. mays*,[134] it was revealed that *Zea mays* ssp. *parviglumis* is the ancestor of maize and supported a single domestication event in southern Mexico. By these molecular data and archaeological records, the maize domestication event is estimated to be between 6000 and 9000 years ago.[134,135] Teosinte is the trivial name for annual and perennial species of the genus *Zea* native to Mexico and Central America: *Z. diploperennis* Iltis, Doebley & Guzman; *Z. perennis* (Hitchc.) Reeves & Mangelsdorf; *Z. luxurians* (Durieu & Ascherson) Bird; *Z. nicaraguensis* Iltis & Benz; *Z. mays* ssp. *mexicana* (Schrader) Iltis; *Z. mays* ssp. *parviglumis* Iltis & Doebley; and *Z. mays* ssp. *huehuetenangensis* (Iltis & Doebley) Doebley (reviews in [136,137]). Teosinte plants are taller and broader-leaved than most grasses. The name "teosinte" is derived from "teocintl", which is of Nahuátl Indian origin, and has been interpreted to mean "grain of the gods". Some species of this grass are distinct from maize, both genetically and taxonomically, and they appear not to have played any role in the origin of maize. *Z. mays* ssp. *parviglumis*, on the other hand, shares a particularly close genetic relationship with maize, and all available evidence indicates that it is the direct ancestor of maize. The plant grows in the valleys of southwestern Mexico and, in these regions, it grows commonly as a wild plant along streams and on hillsides, although it can also invade cultivated fields as a weed. It is most common in the Balsas River Valley on the Pacific slopes of the states of Michoacán and Guerrero in southwest Mexico and hence it is also known as "Balsas teosinte". Much of the Balsas region receives an annual precipitation of between 1200 and 1600 mm, and has mean annual temperatures between 20 and 28 °C and so, if this is the key area, it would make the potential vegetation and ecological context of maize's origins a tropical broadleaf deciduous forest. With its meagre ear containing just two entwined rows of well-armoured kernels, this grass might well have been overlooked had it not been for its tremendous variation. This trait was not overlooked by the early agriculturalists and, within the past 10,000-odd years, native North Americans have transformed teosinte into a major global food plant. This transformation was so striking and so complex that, for over one hundred years, many botanists did not believe that the transformation was possible. This led to a series of competing theories and much lively debate. For many workers, one of the main problems has been the lack of intermediate forms (excluding hybrids), either in nature or from the archaeological record. Another worry is the fact that teosinte is a most unpromising source of edible grain. Add to this the fact that teosinte is so unlike maize in the structure of its ear that nineteenth-century botanists failed to recognize the close relationship, and placed teosinte in its own genus (*Euchlaena*).

Maize and teosinte differ in many aspects of plant morphology and productivity. For example, long lateral branches are observed in teosinte. These lateral branches are tipped with tassels (male inflorescence) in teosinte, whereas maize reveals short ear (female inflorescence)-tipped lateral branches. Although the plants share a similarly robust growth form, their female inflorescences are strikingly different. The teosinte ear has a small number of kernels (about 5–12) and each kernel is enveloped tightly by a stony casing. Collectively, the kernel and its stony casing are known as the cupulate fruitcase. At maturity, the teosinte ear disarticulates such that the individual fruitcases become the dispersal units. Protected within its casing, the teosinte kernel can survive the digestive tracts of birds and grazing animals, enabling the seed to be dispersed with ease. By comparison, the massive maize ear can produce 500 or more kernels, each of which is attached to a central axis of the ear (or cob). The kernels are naked without adequate protection from predation and are easily digested by animals. Since the kernels are firmly attached to the cob and the ear does not articulate (*i.e.* is non-shattering), a maize ear left on the plant will eventually fall to the ground with its full complement of kernels. It is likely, therefore, that when hundreds of maize kernels germinate the following season in such close proximity, the emerging plants are unable to obtain sufficient light and soil to grow and reproduce – meaning that maize is totally dependent on humans for its survival. From an anthropomorphic point of view, of course, retention of kernels on the head makes maize cobs easy to harvest and consume. A summary of some of the major theories

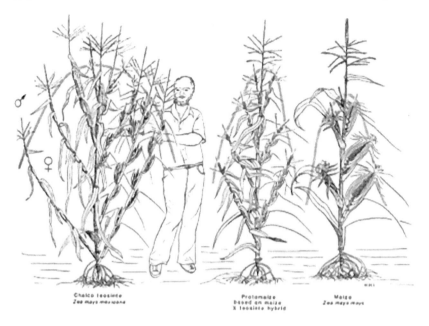

Chalco teosinte
Zea mays mexicana

Protomaize
based on maize
X teosinte hybrid

Maize
Zea mays mays

Figure 4.9   From teosinte to maize (after Iltis, 1983).

propounded for the origin of the maize ear appears in the paper by Hugh Iltis,[138] who also forwards one of his own: the evocatively named "Catastrophic Sexual Transmutation" theory (Figure 4.9). Other accounts of matters concerned with the long-running controversy have been given by Beadle[139–141] and Manglesdorf.[142,143] The maize domestication event resulted in the original maize landrace varieties, which were spread throughout the Americas by Native Americans and adapted to a wide range of environmental conditions.[144] Starting with landraces, twentieth century plant breeders selected maize inbred lines for use in hybrid maize production.[145] They improved yield, resistance to biotic and abiotic stresses and seed nutritional quality. Despite its selection history, however, most maize genes retain high levels of nucleotide diversity.[146–148] It was George Beadle[139] who forwarded the first compelling argument that teosinte was the sole progenitor of maize. It was not a new idea, but Beadle went further into the matter and used experimentation to prove his point. Decades later, he also followed up by leading a movement dedicated to settling the question of the origin of maize.[149] The proposal that teosinte was the sole progenitor of maize is known as *the teosinte hypothesis.* As outlined by Beadle, the hypothesis states that: (a) teosinte provided a useful food source and ancient peoples cultivated it for this purpose; (b) during the cultivation of teosinte, mutations that improved teosinte's usefulness to humans arose and were selected by ancient peoples; (c) as few as five major mutations would be sufficient to convert teosinte into a primitive form of maize; (d) different mutations controlled different traits, *e.g.* one mutation would have converted the disarticulating ear-type of teosinte into the solid ear-type of maize; and (e) over the course of time, humans selected additional major mutations plus many minor ones. Molecular evidence has supported many of Beadle's ideas. A study to locate quantitative trait loci (QTL) involved in morphological differences between maize and teosinte found five QTL of strong effect.[150] Two or three of these QTLs have been resolved to probable individual genes, all of which encode transcriptional regulators. Moreover, each maize gene bears a strong signature of selection. The first such gene to be identified was *tb1* (*teosinte branched1*), so named because *tb1* mutants have a plethora of tillers tipped with male inflorescences, just like teosinte.[151] *tb1* controls the fate of axillary meristems by repressing organ growth in tissues where it is expressed; maize alleles express *tb1* mRNA at higher levels than teosinte alleles.[152] A selected gene that maps to a second of the five regions, *ba1* (*barren stalk1*), also affects plant architecture, and is required for the initiation of all aerial axillary meristems, leading to the speculation that *ba1* may interact with *tb1* to regulate vegetative shoot architecture.[153] According to Wang *et al.*,[154] the third QTL corresponds to the gene *tga1* (*teosinte glume architecture1*). Maize forms of *tga1* confer soft glumes, whereas teosinte forms confer a hardened, stony fruitcase. *tga1* has been cloned and the selected maize *tga1* alleles appear to be mutant relative to extant teosinte. This situation is unlike *tb1*, *ba1* and most selected genes that have been identified where the selected alleles are also found in extant teosinte at moderate frequency. Thus, while many elements of the teosinte hypothesis hold up, it is notable that the remarkable morphological difference between maize and teosinte may be largely attributable to new combinations within the tremendous genetic diversity exhibited by the latter, not to mutation as Beadle had hypothesized.[155] As Buckler's team[155] emphasized, maize molecular diversity is roughly 2- to 5-fold higher than that of other domesticated grass crops, and Tenaillon *et al.*[146] reported that in 25 maize samples, one nucleotide every 28 base pairs is polymorphic, and overall nucleotide diversity is almost 1.3%. The latter study found almost no evidence of selection in 21 genes from chromosome 1. The wild *Z. mays* ssp. *parviglumis* often has levels of nucleotide diversity in excess of 2%, and these two observations amplify the notion that the tremendous diversity of maize and teosinte has been the genetic raw material for the radical transformation of a wild grass into a high-yielding grain crop. Several other maize genes affected by artificial selection have been identified and studied, and these include: *c1* (an anthocyanin regulator),[156] *su1* (sugary1)[157] and *Y1* (phytoene synthase – for carotenoid synthesis).[158] The history of that American favourite "sweetcorn" is interesting because five independent mutation events have been involved (in different geographical areas), four of which have been identified to the nucleotide level.[159] These mutations occurred at the *Sugary1* (*Su1*) locus, where a recessive *sugary1* (*su1*) allele is responsible for "sweetness". The gene is known to code for an isoamylase functioning in starch synthesis in maize endosperm.[160] Recessive *su1* results in the accumulation of phytoglycogen, a highly branched, water-soluble polysaccharide, rather than starch. Phytoglycogen confers the creamy texture to sweetcorn. Alleles at *Su1* were the first to be genetically characterized following the "rediscovery" of Gregor Mendel's work at the start of the twentieth century.[161] In 1890 Willet Hays[162] had anticipated Mendel's laws by reporting the phenomena of dominance, recessiveness, segregation and independent assortment by clearly describing the 3 : 1 segregation ratio in controlled crosses of flint maize and *su1* maize; landmark work in the study of genetics. As John Doebley observes, maize and its wild relatives, the teosintes, present a paradox. Maize and the teosintes exhibit such extreme differences in their adult morphologies that taxonomists initially considered the teosintes more closely related to rice than to maize, whereas the maize and teosinte genomes are so similar that they share the same chromosome number ($n = 10$), have similar or identical chromosome morphologies and they can be easily cross-hybridized. Moreover, F_1 hybrids between maize and some forms of teosinte exhibit completely normal meiosis and full fertility. Doebley stresses that it was because of these conflicting observations on morphology and genetics that interpreting the relationship between the two plants generated a contentious debate over much of the twentieth century.

It is now accepted that the striking differences in the morphologies of the two plants were the result of human selection during the domestication process and that, since maize was domesticated fairly recently, there has been insufficient time for their genomes to become differentiated at the cytogenetic level.

In a fascinating paper entitled "The origin of the naked grains of maize", Doebley's team[154] maintain that the most critical step in maize domestication was the liberation of the kernel from the hardened protective casing that envelops the kernel in teosinte. This evolutionary step exposed the kernel on the surface of the ear, so that it could be readily used as a food source by humans. The team demonstrated that the key event in maize domestication is controlled by a single gene, *teosinte glume architecture* (*tga1*), belonging to the SBP-domain family of transcriptional regulators. The factor controlling the phenotypic difference between teosinte and maize maps to a 1-kilobase region, within which the two plants, remarkably, exhibited only seven fixed differences in their DNA sequences. One of these differences encodes a non-conservative amino acid substitution and may affect protein function, and the other six potentially affect gene regulation. Molecular evolution analyses indicate that this region was the target of selection during maize domestication. Their results showed that modest genetic changes in single genes can induce dramatic changes in phenotype during domestication and evolution.

A brief description of what must have happened to the teosinte seed head indicates the enormous transformation that occurred. Each of the five to twelve cupulate fruitcases in a teosinte ear is formed from an invaginated internode (cupule) within which the kernel sits, and a glume that covers the opening of the cupule such that the kernel is completely hidden from view. When mature, the teosinte ear disarticulates into the individual fruitcases, each of which contains one kernel. The fruitcase functions to protect the kernel from predation, and passes unscathed through the digestive tracks of animals, providing a means of biotic seed dispersal. At maturity, teosinte fruitcases are heavily lignified and the epidermal cells are filled with silica, giving the fruitcase a stony appearance. Cupules and glumes are present in maize, but reduced in size relative to the kernel such that they do not surround the kernel. In maize, these organs form the central cob of the ear to which the kernels are attached. Maize glumes are less lignified and contain less silica than their teosinte counterparts. Thus, maize domestication involved a change in ear development such that the cupules and glumes form the internal axis of the ear, rather than casings around the kernels. In a sense, maize domestication involved turning the teosinte ear inside out.

As Edward Buckler's group forcefully announced,[163] maize domestication is one of the greatest feats of artificial selection and evolution, whereby a weedy plant from Central Mexico was converted through human-mediated selection into the most productive crop in the world. The changes were so astounding, in fact, that it took much of the last century to identify the true ancestor of modern maize but, thanks to modern genetic studies, the molecular basis of this evolution is now being unravelled.

4.4.2 Maize Genomics

Few, if any, plants rivalled the contribution of maize as a Mendelian genetic model during the twentieth century, and today the plant continues to be a leading botanical model system for analysis in several areas. Although the maize genome behaves genetically as a simple diploid with ten pairs of chromosomes, its organization is quite complex. Early genetic analyses of duplicated genes suggested that maize had homoeologous regions.[164–166] This early work was later supported by comparative restricted fragment length polymorphism (RFLP) mapping across the cereals, which showed that two maize chromosome sets aligned with one chromosome each of rice and sorghum, thereby demonstrating a WGD event.[167,168] Evolutionary analysis of duplicated genes indicated that maize may have arisen by allotetraploidy, suggesting that maize was formed by the hybridization of two slightly diverged progenitors rather than the duplication of a single progenitor.[169] Maize and sorghum (which is clearly diploid) shared a common ancestor about 12 Mya, but the maize genome is much larger (that of B73 being 2300 Mb) because of recent polyploidization (tetraploidy) and repetitive DNA propagation. According to Gaut and Doebley,[169] the most recent WGD in maize happened about 12 Mya, just after the "split" from sorghum and, in the time since that tetraploidy, the maize genome "diploidized" by deleting most of the duplicated centromere regions and also deleting or tolerating degeneration of one member of most of its paired gene sets, sometimes fragmenting ancestral gene orders across multiple chromosomes and obscuring similarities in gene order that existed among its ancestors. As implied, the results of fractionating a tetraploid back toward a diploid can be somewhat complicated. For many years it has been generally believed that maize arose as a tetraploid, even though the two progenitor genomes could not be unequivocally traced within the genome of modern maize. The notion that maize is a tetraploid first arose from the fact that the plant has a haploid chromosome number of 10 ($2n = 20$), whereas many closely related grasses (*e.g. Saccharum* spp.) have only five chromosomes in the haploid nucleus. In 1951, Rhoades[164] demonstrated by genetic linkage mapping that non-tandem gene duplicates are common in the maize genome and, later on, mapping results of isozymic variants were also consistent with the hypothesis that the maize genome contains large duplicated regions. More recently, molecular mapping studies using RFLP markers have shown that most of the ten maize chromosomes contain duplicated segments. Comparative genomic studies based on the colinearity of grass genomes[168] indicated that the maize genome aligns with the diploid rice and sorghum genomes in two chromosome sets, implying WGD.

The tetraploid origin of the maize genome is thought to have arisen from hybridization and genomic fusion of diploid parents, and this is consistent with a haploid chromosome number of 10 (as opposed to the more typical number of 5 for other members of the grass sub-family Andropogoneae) and the presence of two unlinked copies of many genes. In their seminal work on the origin of the maize genome, Gaut and Doebley[169] suggested a segmental allotetraploid mode of evolution, based on analysis of synonymous substitution distances (dS) of 14 duplicated maize loci that appeared to show a bimodal dS distribution. Segmental allotetraploidy arises from the hybridization of species whose genomes are partially distinct, in contrast to allotetraploidy involving the fusion of genomes of two distinct species and autotetraploidy involving a doubling of chromosome number within a species. Both genomic allotetraploidy and autotetraploidy are predicted to show unimodal dS distribution patterns around a mean that corresponds to the divergence time of the two distinct diploid ancestors (allotetraploidy) or the time of the switch from tetrasomic to disomic inheritance (autotetraploidy). Segmental allotetraploidy essentially represents a combination of allotetraploidy and autotetraploidy; thus, the dS distribution pattern is predicted to be bimodal, with one set of sequences grouped around a mean representing the time of divergence of the diploid ancestors and the other representing the time of the switch from tetrasomic to disomic inheritance. The model of Gaut and Doebley[169] also suggested that one of the two ancestral diploids was related more closely to modern sorghum than it was to the other ancestral diploid and postulated three distinct divergence/duplication times: 20.5 Mya, corresponding to the divergence of the maize and sorghum ancestral species; 16.5 Mya, corresponding to the divergence of modern sorghum and the sorghum-like maize subgenome; and 11.4 Mya, corresponding to the switch from tetrasomic to disomic inheritance after genomic hybridization within the maize genome. Volker Brendel's laboratory in Iowa revisited Gaut and Doebley's work (see Eckardt, 2003[170]) and made use of improved distance estimation methods to re-examine the same data set that they had analysed, and further analysed an enlarged set of putative duplicated maize gene pairs. Their study found dS to be distributed normally around a single mean, interpretable as a single duplication event that occurred 16 Mya, which is close to the time of speciation between maize and sorghum. More recent still, results obtained by Swigonova *et al.*[171] supported the tetraploid origin of maize and indicated that the contemporaneous divergence of the ancestral sorghum genome and the two maize progenitor genomes at about 11.9 Mya. They also maintained that diploidization must have occurred before 4.8 Mya, and thus preceded the major maize genome expansion by gene amplification and retrotransposition.

Wei *et al.*[172] reported on what was essentially an exercise in maize genome palaeoethnobotany when they constructed a sequence-ready fingerprinted contig-based physical map that covered 93.5% of the genome. The physical map contained 25,908 genetic markers and it enabled the team to align nearly 73% of the anchored maize genome to the rice genome. Comparative genomics confirmed that the euchromatic regions between rice and maize are highly conserved and, when they physically delimited these conserved regions, they detected many genome rearrangements. Extensive definition of the duplication blocks within the maize genome permitted the reconstruction of chromosomes of the maize progenitor. The work confirmed that the maize genome had experienced two rounds of genome duplications, an ancient one prior to the maize-rice divergence and a recent one after tetraploidization.

In 2005, the US National Science Foundation, Department of Agriculture (USDA) and Department of Energy formed a consortium to sequence the B73 maize genome. Primary sequencing was completed in 2008. The following year, the results of this sequencing work were published. Writing in *Science*,[173] researchers at the Genome Sequencing Center (GSC) at Washington University in St. Louis, Missouri, along with teams from the University of Arizona, Iowa State University and Cold Spring Harbor Laboratory in New York, uncovered the complete genome sequence in B73 maize. The map produced shows the order of genes along each of maize's ten chromosomes and the physical distance between those genes. The sequence spans 2.3 billion DNA base-pairs and contains some 32,500 genes, or about one-third more than the human genome. The maize genome contains 2.3 billion nucleotide sequences, about 85% of which are repetitive.

It took 150 researchers four years to complete the reference genome, and the project cost $29.5 million. Funding came from the National Science Foundation and the US departments of agriculture and energy. At the time of the release of the report, maize had annual value of $47 billion in the USA economy, and the country was the source of 44% of the world's corn.

4.4.3 Early Records of Maize Use in Alcoholic Beverages

Although maize beer was not necessarily the key ingredient that enabled the Inca to establish the largest empire of the New World (and one of the largest in history), it was certainly one of a number of interrelated factors that enabled the Cuzco rulers to advance and maintain their power over a vast region of the Andes. In this region, maize has always had a profound religious and even a magical significance, as well as being of great economic importance. The elevated status of *Zea mays* remains with us today in some parts of the world.

Although the early colonists to North America took their own supplies of malt, hops and beer with them, they were anxious to learn how to brew from the indigenous "Indian corn". The earliest report that we have of making beer from maize was given by cartographer, mathematician and natural philosopher Thomas Hariot (*c.* 1560–1621),[174] who said: "Wee made of the same in the countrey some mault, whereof was brued as good ale as was to be desired." Although maize was easy to grow in the new colonies, it did not instantly become a popular raw material for brewing, for records show that malt was constantly being imported from England, and increasing quantities of barley were being grown in the colonies. Contemporary accounts tell us that the colonist's beer in the early days was notoriously variable, and could be made from other raw materials, such as molasses. The problem of malting maize engaged the mind of John Winthorp, Jr.,[175] son of the first governor of Massachusetts, and himself governor of Connecticut. In 1662, Winthorp presented a paper on the subject to the Royal Society of London, to which he had recently been admitted as a fellow. After vivid descriptions of the plant and how to cultivate it, Winthorp explains how a decent loaf of bread can be made from corn. He then touches on the subject of beer: "The English have also found out a way to make a very good Beer of Grain: that is, either of bread made thereof, or else by Malting it. The way of making Beer of Bread, is by breaking or cutting it into great lumps about as big as a mans fist, to be mash'd, and so proceeded with as Malt, and the impregnated Liquor as Woort, either adding or omitting Hopps, as is desired.

To make good Malt of the Corn, a particular way must be taken. The Barly Malt-Masters have used all their skill to make good Malt hereof the ordinary way; but cannot effect it; that is, that the whole Grain be Malted, and tender and flowry, as in other Malt. For it is found by experience, that this Corn, before it be fully Malted, must sprout out both ways (*i.e.* both Root and Blade), to a great length; of a finger at least; if more, the better. For which it must be laid upon an heap a convenient time. Wherein on one hand, if it lyeth of a sufficient thickness for coming, it will quickly heat and mould, and the tender Sprouts be so entangled, that the least opening of the Heap breaks them off; and so hinders the further maturation of the Grain into Malt. On the other, if it be stirred and opened to prevent too much heating, these sprouts which have begun to shoot, cease growing, and consequently the Corn again ceaseth to be promoted to the mellowness of Malt.

To avoid all these difficulties, this way was try'd and found effectual. Take away the top of the Earth in a Garden or Field two or three inches, throwing it up half one way, and half the other. Then lay the Corn, for Malt, all over the Ground so as to cover it. Then cover the Corn with the Earth that was pared off; and there is no more to do, till you see all the Plot of Ground like a green Field covered over with the sprouts of the Corn, which will be within ten days to a fortnight, according to the time of the year. Then take it up, and shake the earth from it and dry it. For the roots will be so entangled together, that it may be raised up, in great pieces. To make it very clean, it may be washed, and then presently dry'd on a Kiln, or in the Sun, or spread thin on a Chamber floor. This way, every Grain that is good will grow, and be mellow, flowry and very sweet; and the Beer made of it, be wholsom, pleasant, and of a good brown colour.

Yet Beer made of the Bread, as aforesaid, being as well coloured, as wholsom and pleasant, and more durable; this therefore is most in use. And the rather, because the way of Malting this Corn, last described, is as yet but little known amongst them."

4.4.4 Double-mashing

The double-mashing system was developed in North America, to deal with grists containing large proportions of rice or maize grits (as much as 60% of the mash bill) and to utilize the nitrogen- and enzyme-rich malts that were available. An account of the process is given in Briggs.[88] Double-mashing is used in areas such as North America, where large proportions of the brewer's grist (25–60%) consist of starchy adjuncts such as rice, maize or sorghum grits: materials that require cooking because of the high gelatinization temperatures of their starches. Two mashes are prepared. In the first, or adjunct, mash, cereal grits together with a proportion of highly diastatic malt (diastatic power (DP) 80–200 L) or microbial α-amylase are mixed ("doughed in") at about 35 °C. After a 30–60 minutes stand, the cooker contents are heated and, after a hold of *ca.* 20 minutes at about 70 °C to allow any available starch to be liquefied, the mash is heated further and is held at 100 °C for 45 minutes to disrupt the grits and complete starch gelatinization. In past times, the adjunct mash was boiled to liquefy starch, but to economize on heat when bacterial α-amylase is being used, it may be held at a lower temperature. Once the adjunct mashing sequence is underway, the second mash, with a mainly, or all, malt grist is prepared. This is mashed in (doughed in) at about 35 °C and is allowed to stand for around one hour, when the adjunct mash is pumped into it and the two are mixed. The temperature of the combined mash will be around 68 °C, and this will be held for around 15 minutes, during which time the starch of the adjuncts is rapidly saccharified. The temperature of the whole mash is then raised to 73 °C and any residual starch will be converted. Wort separation is then achieved using a lauter tun or a mash filter. Grits are uncooked, nearly pure fragments of starchy endosperm derived from cereal grains. The preparations most employed are from maize, sorghum and rice. During their preparation, the surface layers and embryos of the grains are removed, reducing the lipid, ash and fibre contents of the material – thus, there is a higher starch content than is found in the original grains. Maize grits, prepared from yellow dent corn, are widely used as adjuncts in the USA, and they may be prepared by several different milling processes. One such, an integrated process, permits the extraction of other valuable products, such as oil. Maize corns are screened and washed, then conditioned with steam to soften the hull and the germ. The grain is then decorticated and de-germed by abrasive milling, and is successively passed through break rolls and screens to yield the pure grits fraction (*ca.* 55% yield), which are starch-rich with very little oil content. Small amounts of malt are made from wheat, rye, triticale, sorghum, millet, oats and maize. Attempts to malt rice have resulted in only limited success. With the high gelatinization temperatures of its starches, maize malts less well than sorghum but, generally, its malting characteristics have been relatively little studied. Maize grains are large in relation to other cereals, and robust seedling shoots soon emerge during malting. There is evidence that the nutritive value of maize grains is enhanced as a result of malting. As is shown above, John Winthorp reported that European settlers in North America brewed beers from maize, but that the grain could not be malted successfully if barley malting methodology was employed. This was because the shoots and rootlets had to be well developed before adequate modification ensued. By the end of the nineteenth century, when brewing was becoming more of a science, it began to be appreciated just why maize was difficult to malt, and even more difficult to protect from microbial infection. Accordingly, special malting methodology was devised. Steeping was prolonged; at 25–30 °C it would last for around 50 hours, and if the temperature fell to 20 °C, it could last for around 120 hours. Moist grain was then consigned to the floor, and covered with wet sacks. After the first 50-odd hours, during which time the temperature would be held at 20–24 °C, temperature would be allowed to rise until, after 60 hours, it reached 24–26 °C and finished at around 30 °C. Grain would be regularly sprinkled with water (warm in winter) and germination proceeded for 7–8 days, by which time the rootlets were about three times the length of the grain and the shoots were about 5 mm in length. More recent methods employ chemical additions (*e.g.* salt, sulfuric acid, salicylic acid and calcium bisulfite) to the steep water in order to reduce microbial growth, since maize is extremely prone to mould infestations. Problems with microbes can be minimized by changing the steep water at regular intervals or, better still, by steeping in running water.[176] Maize varieties vary greatly in their malting qualities but, like sorghum, they should be malted "wet and warm". There appear to have been very few attempts to select for improved malting quality in maize. Modern clear-beer brewers find little favour with maize, because of its low enzyme content and low extract levels (as determined by conventional techniques). To illustrate the point, Aderinola[177] reports that maize malted for six days at 28 °C would not saccharify adequately and, for conversion to occur, needed to be mixed with 20% barley malt when mashed. Aniche[178] studied the effect of germination time on maize malt quality. Two maize varieties were germinated for different periods of time (up to seven days) at a temperature of 28 °C. It was found that the germination time had a significant effect on the diastatic power, α-amylase activity, colour, fat, protein, cold and hot water extracts and malting loss of maize malt. Notably, the protein and fat contents of the maize malts decreased as germination time increased. During malting, maize (and sorghum) seedling growth is quite extensive, causing substantial malting losses. In contrast to barley, wheat, rye and triticale, it seems that most, or all, of the enzymes involved in modification of maize and sorghum appear to originate exclusively from the scutellum. There are conflicting reports, but the enzyme levels in these grains usually change little, or not at all, in response to applications of gibberellic acid.

4.5   RICE

4.5.1   General

Rice (*Oryza sativa* L.) (Figure 4.10) is a plant of south and east Asia, and has long been the principal cereal in this part of the

Figure 4.10   *Oryza sativa* L. (from *Flora of North America*[387] - by kind permission).

Figure 4.11   Paddy field.

world. It is a diploid ($2n = 24$), predominantly self-pollinated grass, that can be highly productive when grown in paddies (Figure 4.11), where fixed nitrogen is naturally provided by symbiotic blue-green algae. During its long history of cultivation, Asian rice has undergone considerable differentiation, and thousands of cultivars have evolved as a response to the wide range of environmental conditions into which it has been introduced. They fall into three groups:

a)   short grained "japonica" or sinica" forms, adapted to a relatively cool climate;
b)   long grained "indica"; and
c)   broad grained "javanica" forms, which thrive under more tropical conditions.

The crop is closely related to, and fully interfertile with, an assemblage of wild and weedy rice forms that are widely distributed over south and east Asia including the Indian sub-continent, south China and Indonesia.[179] They include both perennial and annual wild forms, often referred to as *O. rufipogon* Griff. and *O. nivara* Sharma & Shastry, respectively, as well as shattering weedy rices (known as *spontanea* forms of *O. sativa*), which often infest rice cultivation. In many places, cultivated varieties, weedy forms and wild races grow side by side, hybridize with one another and form huge complexes.

The distribution range of the wild relatives of cultivated rice is huge, and the present day distribution of *rufipogon*, *nivara* and *spontanea* rices give only the merest hint of where the progenitor of cultivated rice might have originated and been taken into cultivation. Archaeological finds in China and India, and some other countries, give certain clues as to where rice domestication could have originated, but the distribution of some 20 wild species across four continents, and the inherent difficulties in characterizing wild specimens, has not helped. To this has to be added the fact that there has been a long running feud concerning the relative antiquity of rice in India and China!

The earliest archaeobotanical indications of rice cultivation emanate from the middle and lower parts of the Yangtze River valley in China, where several early sites have yielded rice remains. The earliest come from Peng-tou-shan in the Hupei basin,[144,180] and they include charred grains embedded in pottery, as well as husks retrieved from fired earth (dated by AMS radiocarbon testing to 6400–5800 BC). Somewhat later, in the fifth millennium BC, remains of cultivated rice appear in the Lower Yangtze Valley.[144,181]

The other major rice cultivar is "African rice" (*O. glaberrima* Staud.), which, it has been suggested, was domesticated from the wild annual *O. barthii*, a plant which grew in the flood basin of the central Niger delta. It is thought that the domestication of this crop was under way as early as 3500 years ago (although the earliest evidence of domestication, dating to *ca.* AD 200, comes from the site of Jenne-Jeno near the bend of the Niger River) and, from the Niger delta, it was introduced into other parts of West Africa, where farmers further developed the grain into diverse cultivars that could thrive in deepwater basins, water holes in savannahs, forest zones, swampy areas or dry highlands. The wild ancestor was a savannah plant that grew in water holes that filled up during the rains and then dried out in the dry season.[144]

Like its better-known relative, the African form is an excellent source of energy and some essential vitamins and minerals. In terms of protein, it surpasses whole wheat and maize, and is devoid of cholesterol. Sodium and fat levels are low. The fact that Asian types have largely supplanted *O. glaberrima*, even in regions where it was an important part of everyday life, is largely due to the fact that the former are easier to grow, harvest and mill. In addition, most varieties of African rice have grain with a reddish colour rather than the more polished, white, Asian types. For culinary purposes, African rice is prepared in the same manner as its Asian sister, and it is also used to make regionally popular beers.

The "wild rice" of North America is *Zizania palutris* (ex-*Z. aquatica* L.) [$2n = 30$], which belongs to one of the 11 related genera in the same tribe. Traditionally, this species was self-propagating and harvested only by Native Americans in the Great Lakes area. It is now commercially grown in limited areas. The naming of species of *Zizania* as wild rice in North America is rather unhelpful, because in most other regions of the world the name "wild rice" generally refers to *O. rufipogon*.

When compared to wheat, rice has a relatively low protein content (around 8% in brown rice and 7% in milled rice *versus* 10% in wheat), but brown rice ranks higher than wheat in available carbohydrate, digestible energy (kilojoules [kJ] 100 g^{-1}) and net protein utilization. Rice protein has a greater lysine content than wheat, maize and sorghum, and milled rice has a lower crude fibre content than any other cereal. Based on the criterion of mean grain yield, rice crops produce more energy and protein per hectare that wheat and maize, and can thus support more people per unit of land than either of those two.[182] As a human food, rice has, in recent years, gained in popularity in parts of the world where other staples have traditionally dominated, and the greatest rise in rice consumption during the last few decades has been in Africa.

The basic nomenclature of *Oryza* species has changed little since the 1960s and, according to Vaughan and Morishima[183] the genus *Oryza* (coined by Linnaeus in 1753) includes 21 wild species and 2 cultivated species, *O. sativa* and *O. glaberrima*. The latter, "African rice", is cultivated only in restricted areas of western Africa, but the former, generally known as "Asian rice", is cultivated globally. Since the time of its original domestication, Asian cultivated rice has moved around the planet with migrating human populations, and rice cultivation now takes place on all continents except Antarctica and feeds more than half of the world's population (supplying *ca.* 20% of its calorific intake).

Despite rice's central role in world agriculture, the evolutionary history of *O. sativa* is incompletely understood, but excellent reviews have been provided by Vaughan *et al.*[184,185] Within *O. sativa*, extensive morphological, ecological and physiological variation exists, the result of selection for adaptations to different habitats and growing conditions across the world. *O. sativa* includes an estimated 120,000 different, named cultivars ranging from traditional rice varieties preserved by local farmers to the commercially bred "elite" cultivars developed during the green revolution.

The majority of rice cultivars can be placed within two sub-species or races of rice, sometimes described as *O. sativa japonica* and *O. sativa indica*.[186] This distinction is made on the basis of a number of physiological and morphological traits such as drought tolerance, potassium chlorate resistance, phenol reaction, plant height and leaf colour. These two sub-species are commonly associated with differences in growth habit as well, with *indica* rice usually found in the lowlands of tropical Asia and *japonica* rice typically found in the upland hills of southern China, Southeast Asia and Indonesia, as well as outside of Asia (Africa, South America, North America). *O. sativa japonica* can be further differentiated into tropical (*javanica*) and temperate (*japonica*) forms with, according to Garris *et al.*,[187] the temperate form appearing to be a derivative of tropical *japonica*. In addition to the two major races, several other minor rice types have been identified by using genetic markers, and these include the upland, drought-tolerant *aus* cultivars of India and Bangladesh, the deep-water *ashinas* varieties of Bangladesh and the aromatic *basmati* rice of India. Whether all these types have arisen from a single domestication event is not known, but for fragrant rice the situation is a little easier to assess because it has been shown that this variety has resulted from a mutation at the *Badh2* locus. As has been shown,[188] this encodes the gene for betaine aldehyde dehydrogenase. In most rice varieties the gene product prevents formation of 2-acetyl-1 pyrroline. In contrast, fragrant varieties contain a loss of function mutation in the gene, thus permitting synthesis of the volatile compound, which provided the aroma for which the rice is known. Because all fragrant varieties seem to contain the same mutant haplotype, fragrant rice is inferred to have arisen once.

Rice may not have been the first cereal domesticated by man but it has definitely become the most cultivated. Today, it is grown in more than 100 countries, and it is probably the Earth's most geographically diverse crop, being grown as far north as Manchuria in China and as far south as Uruguay in South America and New South Wales in Australia. It grows at more than 300 metres elevation in Nepal and Bhutan and 3 metres below sea level in Kerala in India. This spectacular diversity exists because of its long history of cultivation and selection by humans under diverse environments. Rice growing environments are classified into four major categories: irrigated, rain-fed, upland and flood-prone.

4.5.2 Domestication

Rice domestication is thought to have begun ≈9000 years ago within a broad geographic range spanning eastern India, Indochina and parts of southern China. *O. rufipogon* Griff., the generally recognized progenitor (called "common wild rice"; CWR) of cultivated rice, grows across this entire range. In contrast to the better-known domestication and evolution of modern maize, the origins of *O. sativa* from its wild ancestor *O. rufipogon* have been debated for decades, and the main, unresolved, question seems to be whether it originated monophyletically or polyphyletically.

Most neutrals would agree that the conundrum has not yet been unequivocally resolved. Two major hypotheses, suggesting either a single origin of cultivated rice or multiple, geographically independent domestications from its wild ancestor, have been proposed. By the 1980s, prominent rice botanists, particularly Oka[189] and Chang,[190] favoured a single origin for rice followed by differentiation into *indica* and *japonica* sub-species under cultivation. Both archaeological evidence (based on rice grains[191]) and some genetic markers[192] support a single domestication of *O. sativa indica* from *O. rufipogon* populations growing in the lowland regions of southern China. In this scenario, *O. sativa japonica* developed later in upland growing regions, selected from *indica* rice.[193]

Many molecular studies, on the other hand, indicate a clear genetic differentiation between the two main rice types, and this has led some authors to propose that Asian rice originated from two geographically distinct gene pools of the wild ancestor *O. rufipogon*. This has been referred to as the "double domestication". Genetic distance data, in particular, reveal *indica* and *japonica* are genetically distinct from each other, promoting the possibility that the two major rice types may have arisen from different ancestral gene pools. As long-established reports by Kato *et al.*[194] and Second[195] illustrate, however, the polyphyletic hypothesis is not merely a child of the genomic age.

Using genomic palaeontology, convincing experimental evidence has been provided for double domestication.[196] The workers compared the transpositional history of 110 LTR retrotransposons in the genomes of two rice varieties, "Nipponbare" (*japonica*) and "93–11" (*indica*), whose complete sequences had just been elucidated. The group concluded that these two genomes diverged from one another at least 200,000 years ago, *i.e.* at a time that is patently older than the late-Neolithic date given for the domestication of the crop. Unfortunately, such studies do not permit the unambiguous identification of the geographical region associated with the domestication.

The domestication history of cultivated rice has been complicated by large-scale movements associated with the trading of domesticated rice varieties and with the migration of humans. By examining the spatial variation of wild rice haplotypes, however, it is possible to identify the geographic regions from which cultivated rice arose[197] and, using this approach, Londo *et al.*[186] revealed multiple independent domestications of cultivated rice.

A little later, Gao and Innan[198] indicated the non-independent domestication of the two major rice sub-species, as demonstrated by multi-locus microsatellites. They genotyped a total of 92 individual plants from the two *O. sativa* sub-species and *O. rufipogon* for 60 microsatellites. An approximate Bayesian method was applied to estimate demographic parameters for *O. rufipogon vs. O. sativa* ssp. *indica* and *O. rufipogon vs. O. sativa* ssp. *japonica*. The team showed that the *japonica* sub-species suffered a more severe bottleneck than the *indica* sub-species and thus a greater loss of genetic variation during its domestication. Across microsatellite loci there was a significant positive correlation in the reduction of genetic diversity between the two sub-species. The results suggested that completely independent domestication of *indica* and *japonica* sub-species may not explain the data obtained and that there is at least partial sharing of their ancestral populations and/or recent gene flow between them.

Duncan Vaughan and co-workers gave everyone food for thought when they made a good case for a single domestication event.[184,185] They noted that, despite the evidence of changes to the genome that long predate domestication, one can find *indica* types and *japonica* types growing in the same region as stands of wild perennial rice. They also emphasized that the domesticated types are not restricted to specific areas either, even in places more or less untouched by modern varieties. Also worthy of note was the fact that in China, in the early part of the twentieth century, *japonica* grew mainly in the colder north and *indica* in the warmer, moister south, but there were places where the two grew side by side. Vaughan asked, somewhat rhetorically, "Is it coincidental that this is close to where the first evidence of rice domestication has been reported?"

But the crucial piece of evidence, as far as Vaughan was concerned, surrounded a mutation to the *sh4* gene, the gene that controls shattering. It was shown that the non-shattering mutation to *sh4* is the same in all *indica* and *japonica* varieties tested. After admitting that a dual or multiple origin of domesticated rice has been the prevailing opinion among rice scientists because rice is clearly differentiated into two major groups and several minor groups, Vaughan[185] then offered the following:

"Based on information currently available, neither geography nor genetics of rice and wild rice exclude the possibility that rice was only domesticated once. The fact that the key domestication allele for non-shattering is the same in varieties of *indica* and *japonica* rice tested to date means that a single domestication of rice, at least for the major varietal groups, was not just possible but probable. It is not difficult to believe that during the period after the emergence of the first domesticated rice there would have been an abundance of cultivated (semi-cultivated) and wild rice occurring in close vicinity. Gene flow among different forms of rice (cultivated, semi-cultivated and wild) would have made the genetic structure of these populations "messy". Subsequent gene flow among domesticated, semi-domesticated and wild rice would have started the process of introgression. A feature of this environment must have been populations with high genetic variation from which early farmers selected different types of rice. Therefore, hybridisation–introgression between cultivated and wild rice has occurred since the onset of rice domestication. This process, together with farmers' selection, has resulted in the great diversification of cultivated rice genepool. The indica- or japonica-associated alleles found in wild rice are most likely the result of hybridisation–introgression of *indica* or *japonica* cultivars with wild rice."

Thus, Vaughan is strongly of the opinion that both Asian and African rice becoming domesticated was a single event and that it was the subsequent evolution that led to their genetic complexity.

4.5.3 Taxonomy

Introductory grass taxonomy is presented elsewhere in this book (see p. 141). Within the aforementioned BEP lineage, rice belongs to the sub-family Ehrhartoideae, which contains three tribes, one of which is the Oryzeae, to which rice belongs. The other major tribe is the Ehrharteae, and it is estimated that these two diverged some 34.5 ± 6.8 Mya,[199] in the late Eocene or early Oligocene when the planet as a whole was becoming cooler and global CO_2 concentration was dropping to pre-industrial levels.

The ancestral spikelet structure for Ehrhartoideae is three-flowered, with the lower two flowers being sterile. This arrangement is best seen in the tribe Ehrharteae, which have two glumes, two proximal sterile lemmas (remnants of the lower two flowers) and a single hermaphrodite floret. Some members of the tribe Oryzeae, including *O. sativa* itself, retain this ancestral structure, except that the glumes are minute. In other members of Oryzeae, the glumes are reduced to a cup-like rim, and the sterile lemmas may be very small or missing. The ancestral number of stamens, six, is still retained in the genus *Oryza*.

The leaf anatomy of the sub-family was studied in detail by Tateoka,[200] and later by Watson and Dallwitz.[201] Both works confirm that all species of each of the two major tribes have a double bundle sheath around the veins, as in all C_3 grasses, with the outer sheath parenchymatous and the inner sheath with thick walls. All investigated species possess micro-hairs somewhere on the leaf surface. Midribs in most Oryzeae species are complex, with two or more vascular bundles placed at the top and bottom of the leaf.

Most Ehrhartoideae are perennials, something that appears to be the ancestral life history for the group, but the annual habit has arisen independently in the A genome species of *Oryza*. The switch from perennial to annual seems to occur frequently throughout evolutionary time. Like rice, most Ehrhartoideae have a chromosome base number of $n = 12$, the only major exceptions being in the genus *Zizania*, where $2n = 30$.

Although members of the Ehrhartoideae are widely distributed, most species occur in the Old World. The common ancestor of the family may have originated in Australia/New Zealand or in tropical Africa. The common ancestor of *Oryza* (and *Leersia*) occurred in tropical Africa, and species in other parts of the world arrived there by independent dispersals.[202] For example, the ancestor of *O. granulata* arrived dispersed to India and China from Africa, and *O. rufipogon* was dispersed to India and eastern Asia from the same source. Phylogeny does not permit the assessment of the direction of dispersal, but west to east is certainly plausible. Phylogeny suggests at least three independent colonizations of North America: one each in *Leersia*, *Zizania* and *Luziola*. It is likely that each occurred *via* different route.

All available evidence suggests that the ancestor of Ehrhartoideae was a plant of moist habitats, and reconstructions reveal that likely ancestors were plants of high rainfall (680–1400 mm per annum). It is significant that all extant Oryzeae grow naturally in damp areas, many occupying flooded sites with some (*e.g. Hygroryza*) even being truly aquatic.

The tribe Oryzeae, as conventionally delimited, includes approximately 12 genera and more than 70 species distributed throughout the tropical and temperate regions of the world (Clayton and Renvoize[2]). As the largest tribe in the sub-family, Oryzeae contain more than half of both genera and species of the sub-family.[3,203] In this tribe, *Oryza* L. with approximately 23 species and *Leersia* Sw. with approximately 18 species are the two largest genera and are distributed worldwide.[3] Except for *Zizania* L., which is disjunctly distributed in eastern Asia and eastern North America, the remaining nine genera are distributed regionally or confined to a specific continent.[3]

Guo and Ge[204] presented a molecular phylogeny of the tribe Oryzeae, based on DNA sequences from chloroplast, mitochondrial and nuclear genomes. They concluded that the tribe Oryzeae is monophyletic and falls into two main clades corresponding to the traditionally recognized sub-tribes (Oryzinae and Zizaniinae). Oryzinae comprise the genera *Oryza* and *Leersia*, and it was proposed that they branched off from the remaining genera of Oryzeae ~20 Mya, and separated from each other ~14 Mya. A divergence time of ~9 Mya was obtained for the most basal split within the genus *Oryza*. Their estimates confirmed what Second had suggested,[205] *i.e.* that Oryzeae diverged during the Miocene epoch, and not by the end of the Eocene (36–40 Mya), which fossil evidence had suggested.

The genus Oryza appears to have a common ancestor dating back between 10.2 and 8.8 Mya.[204] Molecular phylogenies show that each of the genomic groups in *Oryza* is monophyletic, as expected from cytogenic studies. The diploid species with the A genome (which include *O. sativa* and its wild progenitor *O. rufipogon*) all form a clade and, likewise, the C genome diploids are sisters. The B, E, F and G genomes are each represented by a single species. All species of *Oryza* have distinctive epidermal outgrowths known as "tubercles" on the upper (fertile) lemma and palea. These structures are otherwise unknown in the Poaceae, and provide good evidence for monophyly of the genus.

4.5.4 Aspects of Rice Genomics

The first higher plant genome to be sequenced was that of *Arabidopsis thaliana* (L.) Heynh., an annual crucifer trivially named "thale cress". When two rice genomes were first published in draft format,[206,207] the comparative analyses that could be performed were hindered by a lack of long-range contiguity. Now these three plant genomes have multi-megabase contiguity. Comparisons of *indica* and *japonica* reveal strikingly little difference in the gene content, but there are massive intergenic differences. Rice and *Arabidopsis* diverged from a common ancestor around 200 Mya and, although the existence of *Arabidopsis*-rice synteny has been controversial, evolutionary models based on estimated mutation rate have predicted some syntenic relationships between such distantly related species. Various workers have detected a low, but detectable, synteny between these monocot and eudicot plants – even after 200 million years of divergence!

The genome of the *japonica* sub-species was sequenced and assembled by whole-genome shotgun sequencing. The assembled sequence covered 93% of the 420-megabase genome. Gene predictions on the assembled sequence suggested that the genome contains 32,000 to 50,000 genes. Homologues of 98% of the known maize, wheat and barley proteins are found in rice. Synteny and gene homology between rice and the other cereal genomes were shown to be extensive, and it was suggested that the rice genome sequence could provide a foundation for the improvement of cereals, our most important crops.[207]

With a compact genome spanning approximately 430 megabase (Mb) pairs, an extensive genetic map[208] and established synteny with other cereal crops,[167] the cultivated rice species *Oryza sativa* represents a model for cereals as well as other monocot plants.[209] The completion of the genome sequence of rice, in December 2004, opens a new and exciting chapter in our quest functionally to characterize all of the approximately 50,000 annotated genes in rice.[210–212] An account of rice functional genetics research in China has been written.[213]

An excellent web-based resource on the species of *Oryza* has been provided by the International Rice research Institute (IRRI), at http://www.knowledgebank.irri.org/wildRiceTaxonomy/default.htm. At this site there is a key to the species, along with individual species descriptions, photographs and maps.

It was at the Gordon Conference in Plant Genetics in 1997 that the notion of an international effort to sequence the rice (*Oryza sativa*) genome was first forwarded. It took about 8 years and 14 major laboratories from 9 countries to complete the project and analyse the results.[214]

As outlined previously, rice is considered a model cereal crop because it has a relatively small genome size as compared with other cereals, a vast germplasm collection, an enormous repertoire of molecular genetic resources and an efficient transformation system. The scientific value of rice is further enhanced with the elucidation of the genome sequence of the two major sub-species of cultivated rice, *Oryza sativa* ssp. *japonica* and ssp. *indica*. The sequence of the *japonica* cultivar "Nipponbare" was recently completed by a consortium of 10 countries, which comprised the International Rice Genome Sequencing Project (IRGSP), and represents a map-based finished sequence of the entire genome obtained using the hierarchical clone-by-clone sequencing strategy.[215] The sequence of the *indica* cultivar was derived by a whole-genome shotgun sequencing approach.[206,216] These genome sequences are invaluable resources not only in understanding the structure and function of the rice plant itself but also in deciphering the organization of other cereal genomes, which share an appreciable degree of synteny with rice (see, for example, [217,218]).

As a result of work carried out over a number of years, the species of *Oryza* have now been classified into ten distinct genome types. By assessing the degree of meiotic pairing in interspecific hybrids, traditional genome analyses grouped the majority of Oryza species into five diploid and two allotetraploid genome types (A-, B-, C-, E-, F-, BC- and CD-genomes), as documented in Nayar's milestone effort of 1973.[219] Because of the difficulties in obtaining hybrids with presumably more distantly related species, three additional genomes, G-, HJ- and HK-, were later recognized based on total genomic DNA hybridization[220] and molecular phylogenetics.[221]

In *Oryza*, one-third of extant species are allotetraploids that originated through hybridization between diploid genomes and, in particular, four (B-, E-, F- and G-genomes) out of the six diploid genomes each have a single species. Consequently, elucidating the phylogenetic relationships of the diploid rice genomes is critically important for understanding the evolutionary history of the entire genus.

Despite extensive studies on evolutionary relationships among rice genomes and species, the phylogenetic relationships among genomes remained elusive until a study that sampled all recognized *Oryza* species and utilized sequences of two nuclear genes and one chloroplast gene.[221] This study supported the monophyly of each of the previously recognized genome types and reconstructed the origins of the tetraploid species.

Zou *et al.*[222] sampled and sequenced 142 single-copy genes in order to clarify the relationships among all diploid genome types of the rice genus. Their analysis identified two short internal branches around which most previous phylogenetic inconsistency emerged. These represent two episodes of rapid speciation that occurred approximately 5 and 10 Mya and gave rise to almost the entire diversity of the genus. Their phylogenetic analysis successfully resolved the phylogeny of rice genome types, and laid a solid foundation for comparative and functional genomic studies of rice and its relatives. The study also highlighted the fact that organismal genomes might be mosaics of conflicting genealogies because of rapid speciation, and amply demonstrated the power of phylogenomics in the reconstruction of rapid diversification.

Although comparative analysis of plant genomes has provided important insights into genome organization, shared ancestral gene order (synteny) and mechanisms underlying their conservation and disruption (see reviews by Bennetzen[223] and Tang *et al.*[224]), such studies have lacked the phylogenetic breadth to elucidate thoroughly the mechanisms, rates or directionality of genome evolution. Ammiraju *et al.*[225] looked at the evolution of the collective *Oryza* genome by sequencing and comparing nine orthologous genomic regions encompassing the *Adh1-Adh2* genes (from six diploid genome types) with the rice reference sequence. They reported the generation and analysis of a unique genus-wide vertical comparative sequence data set, encompassing all diploid genome types of *Oryza*, using a single biologically important genomic region. The work constituted the largest comparative genomic sequence layout then available for any plant genus.

They chose the *Adh1-Adh2* region for their comparative analysis for several reasons. Firstly, this region has long been the subject of intense genetic, evolutionary and functional investigation across many plant lineages. Secondly, it was one of the first local chromosomal segments developed as a comparative exploratory model for microsynteny among the cereals. The rice *Adh1-Adh2* region on chromosome 11 is microsyntenic to that of maize (*Zea mays*) chromosome 4.[226,227] The study of the evolutionary history of this region was extended to a broad array of closely related *Oryza* species and, overall, the analyses unveiled several significant insights into the history and tempo of *Oryza* evolution. First results indicated that the *Adh1-Adh2* region has undergone a number of physical changes in a relatively short evolutionary timeframe; second, a large number of these changes are very recent (and thus narrowly lineage specific), and at least some are frequent (*e.g.* duplications).

4.5.5 Importance of Rice

Ever since the Neolithic revolution, cereals have provided a staple diet in the nutrition of mankind and, for thousands of years, humans have shuffled genes by breeding and selection in order to create the familiar domestic varieties of these grasses. Much progress has been made in taste, nutritional value and productivity, most notably during the "Green Revolution", many important events of which took place during the 1960s. There have been some notable successes resulting from developments in the science of agronomy, as exemplified by the fact that cereal production roughly doubled in developing countries over the period 1961–1985. Unfortunately, because of the world population explosion, the loss of arable land and the effects of climate change, we can no longer depend on the universal distribution of a few "high yield" crop varieties. This is where genomics has a role to play, for we can no longer rely on past methodologies for feeding the world. Accordingly, it was decided that a cereal genome should be sequenced, with a view to:

- Creating an exhaustive inventory of genes of that cereal, and trying to assign a function to them;
- Identifying "candidate genes" for agronomic characteristics that have been mapped in a genetic interval;
- Accessing a large quantity of new chromosomal markers to assist in selection and help in the creation of new varieties;
- Creating "DNA chips" for the global analysis of gene expression in cereals under different conditions.

Rice was the cereal selected to be sequenced as a matter of urgency. For many years, workers in China and Japan had been studying the rice genome, but there was little, or no, commercial interest or interest from Europe and the USA. This all changed in the early 1990s, when the rice genome was beautifully described by Chris Somerville, molecular biologist at Stanford University, as "the Rosetta stone" of the cereal world.

Rice has the smallest genome of all the cereals: 430 million nucleotides. The maize genome is five times larger, and that of wheat 40 times larger! Preliminary comparisons between different cereal genomes had revealed large blocks of homologous genes whose order is relatively conserved. This phenomenon, known as synteny, makes rice a good entry point for characterizing the genes of other cereals, and associating them with various agronomic traits. Furthermore, the rice genome can serve as a model for the monocotyledons, one of the two major flowering plant groups. The sobriquet of "model plant" for rice is supported by the existence of numerous resources for a genomic approach, such as excellent genetic maps and efficient techniques for genetic transformation, which make rice the easiest cereal to transform.

In fact, rice was already a "model" in several respects, because it has for some time been the subject of studies on yield, hybrid vigour, genetic resistance to disease and adaptive responses. Scientists have taken advantage of the existence of plentiful varieties adapted to a very wide range of environmental conditions, from dry soil in temperate regions to flooded cultures in tropical regions.

The significance of the rice genome becomes more apparent when one realizes that worldwide rice production has doubled in the last 30 years, in part due to the introduction of new varieties. Ominously, its present growth barely follows consumption and in 2025 there will be 4.6 billion people who depend on rice for their daily nourishment (compared with around 3 billion today). A further increase in production is therefore necessary, but many small producers will have to use land that is less favourable for cultivation, such as brackish or briny soils. To complicate matters, it is envisaged that the availability of water resources will become more and more problematic in the future.

4.5.6 Sake

Rice has long been the main staple of the traditional Japanese diet, and is not only consumed daily as a staple food, but is also used to brew sake. Japanese cuisine has developed the art of providing side dishes to complement consumption of this staple, and table manners have been evolved with a view to establishing the most refined ways of eating rice and drinking sake. In one way or another, Japanese diet is inseparable from rice cultivation. Agriculture did not reach the Japanese archipelago until the very end of the Neolithic period (the Jōmon period; 14,000–400 BCE), and it is assumed that rice cultivation arrived with migrants from China (either *via* the Korean peninsula or directly by sea). From the beginning of cultivation, only short-grain rice was known in Japan (the long grain type is regarded as inferior and unpalatable).

Throughout the ages, sake has had profound social and religious significance in Japan, and its production has largely survived the "Westernization" of Japanese culture. Thus, the basic techniques for making the drink have not changed very much since the sixteenth century. There are now over 2000 sake breweries (*kuramoto*) in Japan, each priding itself on the distinct local character of its products. Even so, over 50% of liquor taxes in

Japan are now accounted for by beer, and only around 6% are attributable to sake. As Japanese food gains in popularity abroad, so sake is being increasingly imported (around 1% of production – and increasing). Around half of exported sake is bound for the USA and Taiwan.

One thing that has not changed much is the connection between sake brewing and Japan's indigenous religion, Shinto. Every brewery in the country has a small Shinto shrine on the grounds, and often a larger one near the brewery. At the beginning of the brewing season, the brewers, owner and other employees will gather with a priest for a ceremony to pray for a successful and safe brewing season. This takes place at even the largest breweries, amidst gleaming, modern equipment.

Sake is made in every prefecture from Hokkaido to Okinawa, with the exception of Kagoshima. Traditionally sake breweries have been identifiable by their characteristic white walls and tiled roofs, although modern premises tend to be more contemporary in their architecture. In ancient times, breweries would hang out large balls of tightly bound cedar leaves, called *sakabayashi*, in order to inform the neighbourhood that a new batch of sake was ready for consumption. The *sakabayashi* has now been adopted as the symbol of the Japanese sake brewery. Some extant sake breweries can trace their heritage back several hundred years, and are true family businesses.

Sake brewing (the process being called *tsukuri*) – see Figures 4.12 and 4.13 – is a proud art that has been practised in Japan for over two millennia, and has traditionally been a seasonal affair, with brewing commencing in autumn, peaking in winter and petering out with the onset of warmer late spring. Brewing would normally last for 100 days, starting at the end of October and continuing through to the beginning of April (give or take). For taxation and accounting purposes the sake-brewing year commences on October 1st (in 1798, the shogun had specified that no sake brewing was permitted before the autumn equinox).

Nowadays, some breweries produce all year round although, generally speaking, sake brewing retains much ritual and tradition, and methods have been passed down from generation to generation. The *toji*, or head brewer, is ultimately responsible for the brewery (the *kura*) and all production, and is in charge of a number

Figure 4.12 Sake brewing.

Figure 4.13 Sake brewing.

of assistants, known as *kurabito*. *Toji* were/are master brewers who have served extensive apprenticeships, including carrying out the most menial of tasks. During the brewing season, brewers would live in their *kura*, and would not be allowed to leave until after the final mash had been completed! In those far-off times, most *toji* and *kurabito* were farmers, who would work on their land in summer and in the brewery during the winter months. Sadly, as the number of farmers in Japan declines, there are fewer aspiring brewers around.

As is the case with many of our ancient beverages, numerous words have been written about sake over the centuries, and I would like to use a couple of paragraphs from Samuel Morewood's 1824 (pre-Pasteur!) epic[228] to lay down some of the facts about the drink – as our forebears saw them. He writes:

"Of rice wine there are several sorts, but none of them have any resemblance to the wines of Europe, either as to taste or quality, being variously compounded, and never allowed in the manufacture to preserve the mere flavour of the original material. That called *mandarin*, being considered of a superior class, is drawn from rice of a particular description, different from that which is eaten. The grain is steeped for twenty or thirty days in water, and then gently boiled. When it is quite soft and pulpy, and completely diluted and dissolved by the heat, it is allowed a considerable time to ferment, in proper vats prepared for the purpose, generally of glazed earthenware. Several wholesome ingredients are added during the process, mostly simples, and consisting of such fruits and flowers as impart an agreeable flavour and pleasing colour [given as yellow, red, white or pale colour].

At the end of several days, when the motion or agitation occasioned by the fermenting process has subsided, and when the liquor has thrown up all the scum or dross, it is drawn off into glazed vessels, where, by a second species of fermentation, it clears itself, and develops, by the taste and smell, its good or bad qualities. When sufficiently fined, so as to shew by standing for some time, its body and colour, it is put into small jars, in which way it is commonly sold … This wine is so strong that it will keep for a great many years … Within the empire it is principally consumed within the higher orders, who can afford to buy it; and when exported, it sells very dear."

4.5.7 Sake Taxonomy

The Liquor Tax Law in Japan defines and regulates sake as alcoholic beverages with an alcohol content of less that 22% made according to the following processes:

a. Fermented and filtered beverages made from rice, moulded rice (*komekoji*) and water.
b. Fermented and filtered beverages made from rice, moulded rice, water and sake lees or any other substance authorized by Ministerial ordinance (which stipulates that the total amount of "other substance" for use as a primary ingredient be limited to no more than half the amount of rice [including *komekoji*] used in the product).
c. Alcoholic beverages made by adding sake lees to sake and then filtering.

There are two basic types of sake: *futsū-shu* and *tokutei meishō-shu*. The former is "ordinary" sake (*c.f.* a table wine), whilst the latter is "special designation sake", or premium sake. As would be expected from the outline production methods above, there are numerous types of sake, and the Japanese government recognizes the following special denominations (products that satisfy the various requirements for special denominations are labelled as such):

Ginjō-shu, made artisanally (no machinery) from rice, water and *komekoji* (as are all types), with the addition of small amounts of pure, distilled ethanol to enhance fragrance and flavour. It undergoes a protracted fermentation at low temperature, to give a light and delicate flavour with a certain degree of fruitiness and floweriness.

Daiginjō-shu is considered to represent the pinnacle of the sake brewer's art. Production is as above, but even more painstaking and labour-intensive, and the result is light, fragrant and very complex.

Junmai-shu is brewed only with rice, water and *koji* mould, and has a fuller flavour than other forms of sake, with a perceptible acidity. Before 2004, the Japanese government stipulated that *Junmai-shu* must be made with rice polished down to 70%, or less, of original weight, but this restriction has now been removed.

Junmai-Ginjō-shu, essentially a sub-class of *Ginjō-shu*, but brewed solely with rice, water and koji.

Junmai-Daiginjō-shu, a sub-class of *Daiginjō-shu* but, again, brewed only with rice, water and koji.

Honjōzō-shu, brewed using a small amount of pure distilled alcohol, which is added before pressing. This is added to extract extra flavours and aromas from the mash, and the result is a sake that is lighter and usually more fragrant than *Junmai-shu*. The designation was created in the late 1960s to distinguish what is a premium product from cheaply made varieties to which large volumes of alcohol had been added purely to increase volume. To conform to this designation, a sake must be made with no more than 116 litres of alcohol added to every 1000 kg rice.

Tokubetsu-Junmai-shu, the name translates as "Special" *Junmai*, and it is made in the same manner as this drink, but with either special rice or more highly milled rice.

Tokubetsu-Honjozo-shu, as above, but is "Special" *Honjōzō*.

These grades are collectively known as "*Tokutei Meishoushu*" (special designation sake), and are regarded as premium products. They constitute around 25% of Japan's sake consumption. By far the largest quantity of sake consumed in Japan comes under the heading *Futsu-shu* (normal or "Table" sake), and substantial amounts of distilled alcohol are added to cheaper types to increase yield. The rice milling rate will vary for the above sake categories. *Daiginjō-shu*, for example, requires milling to retain only 50% of the grain, *Ginjō-shu* and *Junmai-Ginjō-shu* 60% and, until recently, *Honjōzō-shu* 70% retention. *Ginjō-shu*, *Daiginjō-shu*, *Junmai-Ginjō-shu* and *Junmai-Daiginjō-shu* are brewed to the *Ginjō-tsukuri* method, which specifies the use of highly polished rice and fermenting the sake at cold temperature to create a characteristic fragrance.

Other significant sake categories arise from treatment post-fermentation. *Nigori-zake* is a cloudy (milky) product, because the mash is passed through a coarse cloth to separate off the liquid, and some of the rice lees is allowed through to the final product. Technically, it fails the Japanese legal definition of a sake, which is *Seishu* (literally, "clear/clean sake"). *Nama-zake* refers to sake that is unpasteurized. It requires refrigerated storage and, although it tastes "fresher", it is not very robust, and does not lend itself to transportation. *Mukora* means "unfiltered" but, more specifically, relates to a drink that has not been carbon-filtered. It will have been pressed and separated from the lees, and is thus clear. Sake is generally not aged, but there are some exceptions, one being *Koshu*, which has been aged years, often turning yellow and acquiring a honey flavour. Some types have a sherry-like bouquet. Sake that has been aged in casks, and thus takes on flavour and aroma from the wood, is called *Taruzake* (cask sake). Because wood can impart a strong flavour, premium products are rarely used for this style. *Fukurozuri* is a method of separating sake from the lees without external pressure, by hanging the mash in bags and allowing the liquid to seep out under its own weight. Sake made in this way is called "drip sake" (*shizukazake*). *Genshu* is "undiluted sake".

Generally, sake is pasteurized (a process called *hi-ire*) twice before being sold, but *Nama-chozo shu* is a bottled product that has been pasteurized only once after reaching maturation. Conversely, *Nama-zume chu* is bottled sake that has been pasteurized once before reaching maturity.

Until April of 1992, there was another system classification in place. At that time, sake was designated as Second Class, First Class or Special Class. By default, all sake was classified as Second Class unless proved otherwise. If brewers wanted their sake to be known as superior to most, they would have to submit a sample to the National Tax Office for assessment. The National Tax Office maintains a staff of professional sake tasters for precisely this kind of thing (the higher the category, the more expensive the product, and the more tax payable).

4.5.8 Brewing Sake

Sake is classified as a beer rather than a wine, since it emanates from grain rather than fruit. The mode of brewing, however, is vastly different from that used for "normal" ales or lagers. The earliest account of the sake brewing processes aimed at a Western patronage was Isaac Titsingh's *Bereiding van Sacki* of 1781. For a start it is produced by multiple parallel fermentation (known as *heiko fukuhakkoshiki*), whereby rice starch is saccharified and fermented simultaneously. The micro-organisms involved are *S. cerevisiae* (sometimes as *S. cerevisiae* var. *sake*) and the filamentous fungus *Aspergillus oryzae*. The latter organism is important commercially because of its ability to produce high levels of hydrolase enzymes, such as amylases, proteases, phytase and phosphatase, and is used to manufacture various other food products, including soy sauce and miso.

As in beer-brewing, water is all-important in making sake, and is used in a ten-to-one ratio to polished rice. It should be colourless, tasteless and odourless, and is subjected to much more stringent standards than is tap water. Many established sake breweries have been built in areas renowned for their pure spring water, and the two most prominent are Nada (Hyogo prefecture) and Fushimi (Kyoto prefecture). Potassium, magnesium and phosphate can be tolerated, whilst manganese and iron are undesirable (especially the latter). The story is told of an old sake brewer called Yamamura Tazaemon, who, in 1837, discovered a water supply from Nishinomiya, a port in Nada, which proved to be perfect for brewing sake. He called it *miyamizu*, "heavenly water". Nowadays, filtration and other techniques permit a wider range of raw waters to be used for brewing, and it quite possible to create *miyamizu*.

Most of the rice used for brewing sake is of the "japonica" variety (*O. sativa* subsp. *japonica*), which becomes characteristically very sticky when cooked. Some two-thirds of this will comprise the cultivar "*Yamada Nishika*" – the so-called "King of sake rice", which originated in the Hyōgo prefecture. Other varieties are used (nine in all), including "*Omachi*" (one of the oldest), "*Gohyakumangoku*" and "*Miyama Nishiki*", and each will bestow its own character to the final product. Many breweries grow their own rice, but much is supplied by sub-contractors.

Rice must be highly polished before use (to a greater degree than required for food), and around 15% of the surface material is removed. This reduces levels of proteins, lipids and minerals that could jeopardize clarity. By definition, such rice will be rich in starch, which will be exposed to enzymes from *A. oryzae*. Many sake types have strict milling requirements (called *seimaibuai*), and most specify that at least 30% of the kernel must be milled away. Generally, the more the rice has been milled, the cleaner, lighter and more delicate the sake flavour. Some polished rices are shown in Figure 4.14.

$$\text{Milling rate (\textit{Seimaibuai}) (\%)} = \frac{\text{Polished rice weight (kg.)}}{\text{Brown rice weight (kg.)}} \times 100$$

Figure 4.14   Polished rices.

Most taxonomists would agree that the yeast involved in sake fermentation is a strain of *S. cerevisiae*. Over a century ago, Yabe claimed it as *S. sake*, but that description was not universally accepted. Bamforth[229] has it as *S. cerevisiae* var. *sake*. One thing is indisputable: it certainly has an above-average ethanol tolerance.

To commence the process, polished rice (*shinpaku-mai*) is washed, soaked in water (a normal rate would be 25 hL water per ton of rice), steamed and cooled. Prior to being steeped, newly polished rice is left out to "rest" until it has absorbed enough moisture from the air to prevent shock (splitting) when immersed in water. Brewers time the soaking time very precisely, because over-hydration of the rice can render it useless for brewing, and immersion times can vary from overnight for an ordinary milling, to a few minutes for highly polished rice. Likewise, steaming is an art, and rice kernels should be steamed in a way that results in firm outer layers and a soft inner core (normally around 50 minutes in a continuous steamer). The traditional manner of steaming was to place soaked rice in a large wooden tub (*koshiki*) with holes in the base. This would be positioned over a metal vessel containing boiling water. Steam would pass through the apertures in the bottom of the *koshiki*, and so sterilize and gelatinize the rice, thus rendering it susceptible to the action of microbes (particularly *koji* mould).

Elsewhere in the brewery, in a warm, damp room called a *koji-muro* (traditionally lined with "Japanese cedar" panelling), *A. oryzae* (koji mould, or *koji-kin*) spores are mixed with steamed rice and left for a few days until fungal mycelium forms a covering. At this stage, the rice is known as *komekoji* (moulded rice), and will contain essential starch-degrading, and other enzymes. Seeded rice is tightly wrapped in fine cloth, and placed in a long, shallow tub (a *toko*). At night, the cloth bag is manipulated such that all of the developing mould has a chance to benefit from the ambient warmth and moisture in the *koji-muro*. Traditionally, the manipulation of the *koji* bag was a convivial ritual for the brewery staff, and was a job for delicate hands! On the morning of the second day, the mixture is moved from the *toko* and placed in cedarwood trays, the *tana*, which are stacked and covered with cloth. It is at this stage that *koji* growth is at its peak, and the temperature rises accordingly. It is usual to mix at regular intervals to help disseminate heat. In the afternoon, the *koji* is spread out thinly on a heated table to dry carefully and, on the morning of the third day, the steaming *koji* is removed from the *koji-muro*, whence it is spread in long lines on trays to reduce temperature and minimize further fungal growth. The *koji* will soon now be ready for brewing the next day. Around 20% of each day's steamed rice production is used for making *koji*, and the process is all-important in determining the flavour-profile of the resultant sake.

The next step is to prepare the starter mash, known as *shubo* (colloquially as *moto*). The Japanese characters used in writing the word *shubo* are the characters for "mother" and "sake", and the *moto* is one of the aspects of sake brewing that make it a unique process in the world of alcoholic beverages. It basically consists of a mixture of *koji* rice, a yeast starter culture and water. Modern methodology sees the addition of lactic acid, but artisanal practice would encourage the growth of lactic acid bacteria, which naturally sour the mixture. One of the oldest ways of making *moto* is *bodai moto*, which has great cultural significance because it relates to the time when sake was largely brewed in temples and shrines by priests and monks. The method is reputedly based on techniques developed at a temple in Nara some six centuries ago, whereby *moto* was made using raw, uncooked rice. *Bodai moto*, which literally translates as "monk's sake", is still employed by a few extant breweries. Steamed rice, casually seeded by airborne yeast cells, is sealed in a cloth bag and submerged in a batch of uncooked, polished rice which is immersed in water. This is left to its own devices and after 4–5 days the water becomes cloudy and frothy as result of yeast activity. This yeast-laden water is then filtered off and the uncooked, polished rice is steamed. This rice is then used for a second mash, together with the "yeasty" water and a further batch of *koji* rice. This constitutes the *moto* and will be ready for brewing after 5 days. The main disadvantage of this ancient procedure is that large numbers of lactic acid bacteria develop, which confer sour notes to the sake.

For three centuries or so, right until the first half of the twentieth century, another traditional method of making *moto* was widely used. Termed *kimoto*, it employs steamed rice, *koji* rice and water in the initial mix, which, in the early evening, is distributed into small, shallow wooden tubs (of around 70 cm diameter). The mixture is then stirred every four hours by the *toji*, and this promotes cooling in the chilly night air. There are various ways of achieving a mix, and nearly every brewery seems to have its own little quirks. Ultimately, the *toji* are aiming for a thorough mix and a stable and uniform temperature throughout the tub. The following morning, the developing *moto* is ground by long bamboo poles with wooden slats attached. This abrades the rice against the bottom of the tub, and this process is continued until the grains are reduced to about one-third of their original size. By now, the mash will be in the form of a thick paste, and the agitation will have stimulated the activity of the *koji* mould. The pole-ramming activity is termed *yama-oroshi*, and is extremely exhausting work. It is one of the classic "typical" scenes of the sake brewery, and is the subject of numerous ancient paintings. During the pounding, brewers would traditionally sing old brewing songs to keep themselves awake, and to count the strokes! Making *moto* by this method is rather rare today.

The paste produced by *yama-oroshi* is transferred to a large wooden vat and left for 2–3 days at 8 °C. Buckets of hot water are then introduced into the middle of the mix, in order to raise the temperature, an act that stimulates fermentative activity from "infecting" airborne yeasts. The mixture is then maintained at 25 °C and, between 20 and 25 days later, the *moto* is fit to be used as a started culture. *Moto* made by the *kimoto* method contains a relatively high concentration of amino acids, which yields a dry, rich-tasting sake.

The *yamahai* method for making *moto* was developed during the Meiji era (1868–1912) as a labour-saving development of the *kimoto* method, for it was during the very early twentieth century that brewers realized that all the hard work was not absolutely necessary for getting all the *koji* enzymes to work on rice starch properly. An employee of the National Institute for Brewing Studies discovered that *koji* enzymes would work quite well without all that pounding, as long as a little more water was added, and the temperature of the mash was raised a little. In addition, there is an initial mixing of pure koji rice with hot water, which hastens rice saccharification before bulk steamed rice is added. The development of this new methodology signalled the end of *yama-oroshi* (if necessary), and the new technique was named *yama-oroshi hai-shi*, which literally means "*yama-oroshi* ceased", or "discontinuation of *yama-oroshi*". This mouthful was later shortened to *yamahai*. While the *yamahai* method was developed to shorten production time, it is slower than the modern method of making *moto*, and is now only used in speciality brews for the "earthy" flavours it produces.

A couple of years after the *yamahai* process was developed, it was found that, if lactic acid was added to the *moto* at the beginning, the whole process could be completed in around half the "normal" time. This was called the *sokujō* process and, today, most sake is made with *sokujō moto* (meaning "fast-developing"), although it took about a decade before the method became widely accepted. A pure yeast culture is used to seed a *sokujō* mash. *Yamahai moto* takes around one month to develop, whilst *sokujō moto* can be ready in a couple of weeks. Flavour-wise, *yamahai* has a higher sweetness and acidity, with richer, deeper, more pronounced flavours than *sokujō*.

As the *moto* develops, by whichever method, all kinds of pleasant aromas arise, and the brewer can tell a great deal about how the *moto* is developing from the smells alone. These aromas will vary greatly depending on what kind of sake is being brewed. *Ginjō-shu moto*, for example, will often be heavily suffused with fruity essences, with lots of apple, banana and strawberry notes.

When ready, *moto* is transferred to a larger tank where more freshly steamed rice, water and *koji* rice (and yeast, if necessary) are added three times over the next four days. At each addition (on days one, three and four), the volume of what is already in the tank is just about doubled, and the temperature of the mass is allowed to rise to 15 °C over the first two days, during which time it will be left uncovered in the vat. Potential spoilage organisms from the atmosphere are prevented from proliferating by the endogenous acidity of the mash. By day three, mash temperature is lowered to 9–10 °C, and this helps to suppress contaminants, and give the yeast the best chance.

By making sequential additions, the brewer allows the yeast to keep pace with the increased volume. After the incremental additions have been made, the mixture is now known as the main mash, or *moromi*, and it is this that will undergo alcoholic fermentation, which can last for anything from two to six weeks. Whilst this is going on, more fermentable sugars will be slowly released through continued saccharification of steamed rice starch by *A. oryzae*.

From now on, the *toji* have to fight to keep the temperature within bounds – especially if there is no modern cooling apparatus. The temperature of the *moromi* is maintained at around 15 °C. Most sake fermentations will be complete within 18 days, but some high-grade products have their fermentation slowed by reducing the temperature to 10 °C or less.

When fermentation is deemed to be complete, the *moromi* will have attained an alcohol content of 18–20% ABV, and its flavour and fragrance will have developed. The raw sake is now filtered, and in olden times this was effected by holding the mash in a weighted, long, narrow cotton sack held over a wooden "sake boat" (*sakafune*). The raw sake would trickle through a spigot in the base of the boat and would be collected. Nowadays, pressing and filtration technologies are generally used to separate liquid from lees. Young sake is then held for ten days at low temperature, which encourages glucose and acidity levels to be enzymatically reduced. It is then pasteurized at 60 °C, and transferred to sealed wooden vats where it will be held for 6–12 months. Storage allows the product to mellow, and to take up some "wood" characters. After storage, blending may take place, and then most sakes are diluted to lower their strength slightly (usually around 15–17% ABV).

Some breweries carbon-filter their beers and, while this process can remove unwanted odours and flavours, it can remove desirable ones as well. The solids left after pressing and filtering are further pressed and become known as *kasu*. These are used to make *shōchū* and pickles, and for livestock feed.

High levels of ethanol can be achieved in sake, maybe as high as 20% ABV (the sake yeast can tolerate up to 30% ethanol) – much higher than for any other directly fermented (non-distilled) beverage. This is because both sugar production and fermentation are going on at the same time, and the yeast is not subjected to a high initial sugar concentration (which might prove inhibitory).

During World War II, when there was a general shortage of rice, many Japanese sake manufacturers started to incorporate industrial alcohol into their products, in order to overcome the shortfall in raw materials. This was a *volte-face* for sake brewers who had hitherto used nothing but rice to procure their alcohol, and represented the end of 2000 years of tradition. Japanese sake was immediate polarized into two types; one with additives and one without. This distinction continues today.

4.5.9 Organoleptic Considerations

As devotees of the drink will tell you, there are some 700 identifiable constituents in sake, which is a greater number than can be claimed for wines or beers. Add to this the fact that there are over 5000 brands of the drink, and one can see that the job of the taster is an onerous one. Accordingly, over the years, distinctive paraphernalia has evolved with the aim of standardizing tasting as much as possible. One such item is the tasting cup, or *kikichoko*, which is designed specifically to make it easier to observe the different shades and colour clarity of a sample. To aid this, the interior of the cup is lined with concentric blue circles, which helps the gauger to differentiate between samples.

The key points for a sake taster are colour, aroma and taste, and the latter category is divisible into body, sweetness/dryness and purity. Colour is assessed for depth and clarity, whilst there are a number of identifiable fragrances by which aroma is described. Body can be generally categorized as "full-bodied" or "clean", whilst one can also distinguish between "sweet" and "dry". Sakes with higher alcohol content, sugar content, acidity and amino acid content will err toward the full-bodied end of the scale, while those with less of these determinants are more likely to taste cleaner. Most sakes will have a sugar content between 4 and 7% by volume, those at the higher end being sweet, and those at the opposite end dry. Sakes also contain a certain amount of acidity (0.05–0.15% by vol.), mainly attributable to lactic, malic and succinic acids, which offsets the sweetness of sugars. The amino acid (particularly glutamic acid) content of a sake is also an important criterion, for a sample that contains higher levels of these compounds is said to be more full-bodied and rich-tasting (*i.e.* have a greater *umami*[†]). If taken to the extreme, very high levels of these compounds can give an off-flavour (*zatsumi*), and the defect will become more apparent as the sample matures.

Sake has the greatest range of "ideal drinking" temperatures of any alcoholic beverage, although it is most commonly consumed when warm. Recommended temperatures vary from 5 °C to 55 °C, although most Japanese find that sake tastes best when consumed close to body temperature, or very slightly higher. The custom of drinking warm sake (*kan-zake*) became commonplace in Japan at the end of the seventeenth century, and continues to be in vogue today. Conversely, sake probably offers less variety of aroma and taste than most alcohol-based drinks (the basic elements of taste being sweetness and *umami*).

One of the unique parameters in the world of sake is the "sake meter value" (SMV), which is an almost obligatory feature of a bottle of the drink (together with ABV, capacity and serving temperature advice). The SMV (*nihonshu-do*) is essentially an indication of sweetness or dryness of a sake. The basic measurement requires a saccharometer, which will indicate the amount of residual sugar in the sample. Alcohol will, of course, be lighter than water and sugar (glucose) will read heavier than water (the more unfermented sugar, the denser it is). The scale, as used by brewers, has numbers assigned in a way such that lower numbers indicate increasing sweetness, and higher numbers indicate a drier product. Thus, samples are interpreted as being dry (*karakuchi*), with positive values (*e.g.* +5), or "sweet" (*amakuchi*), with negative values (*e.g.* −5). Originally, a value of 0 was considered to be neutral but, with perceptions and preferences having changed over the past few decades, +2 or so is now considered neutral. The *nihonshu-do* is not the sole factor contributing to sweetness or dryness, since acidity plays an important role. Sake with higher

---

[†]*Umami* is one of the five generally recognized basic tastes sensed by receptor cells of the human tongue. It is a loan word from Japanese, roughly meaning "tasty". The *umami* taste is due to the detection of the decarboxylate anion of glutamic acid.

acidity will generally taste drier than it actually may be based on numbers alone (and a sake with lower than usual levels of acidity can taste a little sweeter than the *nihonshu-do* would indicate.

Tasting sake properly includes several phases of employing one's sense of smell. Initially, there is a sniff, which will determine the general fragrance. Next there is the *fukumi-ka*, which is a fragrance that arises as one holds the sake in one's mouth and breathes. Finally, there is the *modori-ka*, a third fragrance that becomes discernible immediately after one swallows (or expels) a sip.

4.5.10 Ceremonial Use

Sake is often used as part of Shinto purification rituals. In a ceremony called *kagami biraki*, wooden casks of sake are opened with mallets during Shinto festivals, weddings and the like. This drink, called *iwai-zake* ("celebration sake"), is served freely to all present to spread good fortune.

The first alcoholic drink produced in Japan may well have been *kuchi-kami* ("mouth-chewed sake"), which was made by chewing grain, or other plant material (such as nuts), and expectorating the mass into a container. Amylases in saliva would break down starches and "wild" yeasts would convert sugars to ethanol. For certain religious purposes, it was necessary to have the plant material chewed by virgins! Such a method is not dissimilar to that used for various Amerindian products, such as *chicha*, pulque and cauim. Inscriptions from the fourteenth century BC indicate that Chinese millet wine was made in a similar fashion. How the custom of chewing grains was incorporated into fermentation processes is obscure, but it may have arisen when mothers pre-chewed grain for their infants, and inadvertently left some material lying around. In the Andean regions of Bolivia and Peru, Indians still make *chicha* by salivary amylolysis (see Figure 1.2).

Over the past couple of decades there have been numerous advances in the development of sake yeast strains, and dozens of new organisms have been isolated, characterized and introduced. Of particular interest, to some, are those strains that can introduce enhanced levels of alcohols, esters, acids and other compounds that can affect the nuances of flavour and fragrance.

The Japanese sake yeast story really starts in the early twentieth century when the Central Brewers' Union first began collecting and distributing known commercial strains (mostly from large breweries), and making them available to smaller concerns throughout the empire. The CBU assigned numbers to these isolates, which still apply today, although not all are still used. Yeasts #1 to #6, #8 and #11 to #13, for example, are now considered to be producers of too much acid for most brewers, and are in the "not now used" category. Yeasts #7, #9 and #10 are the most important these days, with #7, with its mellow fragrance and robustness during fermentation, being the single most commonly used strain in Japan. Yeast #9, owing to its excellent fragrance-creating ability, and stout performance during fermentation, is the most commonly used strain for brewing *ginjō-shu*. Yeast #10 gives a lower-acid, fine-grained flavour in sake, but can be problematic at higher fermentation temperatures. In recent years, #14, which is low in acidity and gives fruity fragrance (mainly "apple" and "pear"), and #15 (fragrant, but not quite so robust during fermentation) are favourites for very fine sakes.

According to Casal,[230] it is likely that wines made from millet predate rice wines in both China and Japan, and that rice wine was probably produced in China a long time before it was developed in Japan. Steinkraus[231] says: "Sake is a clear, pale yellow rice wine with an alcoholic content of 15–16% or higher, a characteristic aroma, little acid, and a slight sweetness. Japanese sake is closely related to Chinese rice wine, *shaosing chu*, although the latter has a deeper colour and a more oxidised 'sherrylike' flavour (and often contains some wheat, as well)."

4.5.11 Zutho

While sake is becoming a product that is available world-wide, rice beer is a relatively rare product in the world and is indigenous to, and only really found in, Nagaland, India, where it is known as *zutho*.[232] Results of the first scientific study on *zutho* were reported by Teramoto's team in 2002.[233] They found that a product from the Kohima district in Nagaland appeared as a whitish, porridge-like slurry containing 5.0% (v/v) ethanol, with a pH of 3.6. It had a fruity aroma and sour taste. According to the team, its unique aroma had similar character to Japanese sake and sprouted rice sake. A brewing strain of *S. cerevisiae* was found to be the fermentative organism. Results of an analysis of Kohima *zutho* are shown in Table 4.3.

Table 4.3 Characteristics of a Nagaland *zutho* sample (after Teramoto et al.[233]).

Ethanol (%, v/v)	5.0
2-Methylpropanol (mg/l)	36
3-Methylbutanol (mg/l)	144
Propan-1-ol (mg/l)	10
Ethyl acetate (mg/l)	285
Isoamyl acetate (mg/l)	3
Acetaldehyde (mg/l)	31
pH	3.6
Acidity (ml)	5.1
Reducing sugar (mg/ml)	6.3
Total sugar (mg/ml)	39.7

An extremely interesting account[234] of developments in rice culture has recently been provided, which puts into context the historical significance of the plant in South Asia.

4.6 SORGHUM

4.6.1 General Biology

Sorghum (*Sorghum bicolor* (L.) Moench) (Figure 4.15), which is now described as the "fifth most important cereal crop in the world" (in terms of both crop production and acreage planted), was mentioned by the Greek philosopher Theophrastus (372–c. 287 BC), sometimes known as "The Father of Botany", who, relying on a report from India, thought that it was a "kind of wild barley". The plant was first accurately described by Pliny the Elder (AD 23–79), who classified it as a millet, and it is then hardly mentioned again until the sixteenth century. The plant was named *Holcus bicolor* by Linnaeus in 1753, and then transferred to its own genus by Conrad Moench in 1794. Harlan and de Wet[235] developed a nomenclature by which cultivated sorghum contained five principal genetic races (which older taxonomies refer to as "sub-species"): "Bicolor" from North Africa, "Guinea" from West Africa, "Durra" and "Caudatum" from Northeast Africa (and also found in India) and "Kafir" from the Nile region. There are then ten hybrid races under *S. bicolor* subsp. *bicolor* (synonym *Sorghum vulgare* Pers. – which is used by some to describe the cultivated plant). The genus *Sorghum* has 25 species distributed in Africa, Asia and Australia, and is closely related to maize (*Zea mays* L.) and sugar cane (*Saccharum officinarum* L.).

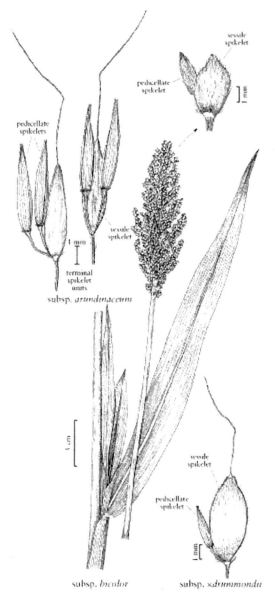

Figure 4.15 *Sorghum bicolor* (L.) Moench (from *Flora of North America*[388] - by kind permission).

According to many authorities, the progenitor of the cultivated sorghums is *Sorghum arundinaceum* (Desv.) Stapf. Ex-Prain, an annual wild species from the humid tropics of West and Central

Africa. Nikolai Vavilov considered that Abyssinia was the centre of origin of the crop. Others maintain that sorghum may well have been domesticated several times (in different places) and the existence of distinct cultivated races in different parts of its range suggests that human selection in each of these places has been going on for a very long time. Sorghum is a cereal of remarkable genetic variability – more than 30,000 varieties have been described from worldwide collections – and the domesticated sorghums consist of a complex of intergrading forms of wild, weedy and strictly cultivated forms. The variation in morphology is immense and species infertility, coupled with much hybridization, means that sorting out the origins and relationships of cultivated sorghums is far from easy. Sorghum is a diploid, self-pollinated crop, and during domestication underwent considerable differentiation; for example, *S. bicolor* is typically an annual grass, but some cultivars are perennial.

Wild forms of sorghum, closely related to the crop, are confined to Africa, south of the Sahara, and (probably) westwards to Yemen, indicating that the cereal must have been domesticated in this area, probably in the Savannah belt south of the Sahara.[45] Certainly, the Sudan-Chad-Ethiopia area of northeast Africa is a region of high sorghum genetic diversity today, and it is thought that the first uses of wild plants for domestic purposes occurred in this area around 7000 BC. The East African Cushites were traditional grass collectors who lived in Ethiopia prior to 5000 BC, and who cultivated wheat and barley and probably domesticated sorghum from wild plants. We know that sorghum of the "Durra" type was under cultivation in the Nile valley around 4000 years ago, and there is evidence for the domestic use of wild sorghum at Napth Playa, an early Neolithic site in the Western Desert, southern Egypt.

The spread of sorghum from its native lands is quite interesting, and it would appear that it was taken by early migrants from (say) Ethiopia to the Sudan, and then to the Niger River where it reached West Africa and became an important Stone Age crop there. The tropical forests of the Congo stopped its spread southwards from West Africa, and sorghum cultivars were actually spread south from East, not West, Africa. This assessment is based on the fact that the modern sorghums from Central and South Africa are genetically nearer the plants of Tanzania than to those of West Africa. Carbonized grains have been found in parts of East Africa dating to the twelfth century; it is likely that sorghum reached the region at least 1000 years earlier. The Bantu people adopted sorghum from East Africa and this helped to facilitate their explosive expansion into the savannah countries of eastern and southern Africa over the past 1000 years.

Sorghum probably was taken from Africa to India during the first millennium BC by Arab traders travelling by dhow across the Indian Ocean. The sorghums of India are related to those of East Africa. There is no Sanskrit name for sorghum, so it is presumed to have reached India after the Sanskrit people had arrived there in around 1500 BC. Sorghum then moved along the Silk Route from India to China in the early Christian era. Archaeological remains of cultivated sorghum have been found in Arabia, and later in India, dating to the third and second millennia BC, confirming the early spread of the plant out of Africa into India.

During the seventeenth century, West African cultivars were taken by slave-trading ships to the New World. Whether this was a deliberate attempt to introduce the crop to the Americas, or whether the plant was simply a ship victual (with the "surplus" being landed with the slaves), is open to speculation, but it was often the practice to feed landed slaves with a food with which they were familiar, and which they had eaten aboard ship. Whatever, sorghum was certainly introduced as a grain crop to the USA in the mid-nineteenth century. It was originally grown along the Atlantic coast, and then spread westwards to drier regions that were too arid and hot for maize. Before 1900, sorghum was well established in the southern Great Plains and California. It became the fourth most important crop for sugar by the 1930s, but never rivalled the importance of sugar cane or maize (although it is an important forage crop for hay or silage).

Today, the world sorghum economy consists of two distinct sectors: a traditional, subsistence, smallholder farming sector where most of the crop is consumed directly as food (mainly in Africa and Asia), and a modern, mechanized, high-input, large-scale sector where output is largely used as animal feed (mainly in developed countries and in Latin America). In the former, the crop is grown under age-old farming systems with low rates of fertilizer use, and the adoption of improved moisture conservation technologies is limited. Yields are generally less than 1 tonne/

hectare and variable year-on-year. Conversely, the use of modern agricultural practices in developed countries (hybrid seed, fertilizer and water management) permits yields of 3–5 tonnes/hectare.

Sorghum is still a very important crop on its native continent, with 21.76 million metric tonnes being produced in 2008/2009, which represented around 18.7% of total African cereal production for that year. *Per capita* consumption of the crop is highest in Africa, with some 90–100 kg/yr being consumed in Burkina Faso and Sudan, for example, and this represents over one-third of the total calorie intake in these two countries. In Nigeria, sorghum is the most important staple crop, and is produced mainly in the north (the "Nigerian Savannah"). This zone also supports the cultivation of other cereals, such as millet, maize and wheat. Millet, in particular, may be intercropped with sorghum, since the two crops have similar agronomic characters. Nigeria fulfils a substantial part of its nutrient energy demand from sorghum (*ca.* 95% of the total harvested crop is consumed as human food of some kind).

The most common and simplest food prepared from sorghum is porridge but, as a sign of its overall importance, sorghum is used to produce a wide range of traditional food and beverage products in various parts of Africa, a practice that apparently goes back to prehistoric times. In particular, a wide variety of fermented beverages can be produced from sorghum, and evidence suggests that it was at an early stage that sorghum grains were used to make a "small beer", using a souring lactic fermentation and resulting in a very nutritious fermented gruel, rather than a beer as we know it.

Sorghums are very variable in form and colour, and have an even wider range of synonyms, including Kaffir corn, Milo, Milo maize, Great Millet, Durra, Feterias, Shallu, Jowar, Indian millet, Cholam, Guinea corn, Intama and Koaliang. Such a list (albeit incomplete) suggests how widely the crop is grown in warmer climes.

Unfortunately, sorghum has long suffered from being considered as "poor people's food", owing to the fact that it was generally consumed in deprived rural areas, where it might comprise over 70% of the food intake. This stigma has been compounded by the fact that sorghum proteins are poorly digested relative to the proteins of other cereals. In one study[236] carried out in Peru, for instance, sorghum gruels from different cultivars had an apparent protein digestibility mean value of 46%, while rice, maize and wheat gruels gave values of 66%, 73% and 81%, respectively. As we shall see, sorghum proteins do not make brewing with the grain any easier either.

Nevertheless, sorghum is a staple food for about 300 million people worldwide, and Durra is the basic foodstuff of much of the population of the Horn of Africa.

Total world production of sorghum in 2007/2008 was 63.26 million metric tons; the leading producers are listed below:

United States	12.64 million metric tons
India	7.93
Nigeria	10.00
Mexico	6.20
Sudan	4.50

4.6.2 Sorghum Beverages

Fermented beverages can be essentially non-alcoholic or alcoholic, according to processing, and the latter, which are essentially beers, fall into two categories: clear and opaque. African beers go under many names, including kaffir beer (Bantu beer), sorghum beer (South Africa), opaque beer (Zimbabwe) and *pito* (West Africa). In South Africa, the kaffir or Bantu beers have numerous tribal names, including *utywala* (Xhosa), *jwala* (Tswana) and *utshwela* (Zulu).

Opaque beers (such as "sorghum beer") are low in alcohol and contain semi-suspended particles of starch, yeast and cereal debris. Such beer is generally pinkish-brown in colour, due to anthocyanidin pigments in the grains, and has an enhanced viscosity attributable to yeast. It also has a sour taste (caused by the activity of lactic acid bacteria), is un-hopped and is consumed whilst actively fermenting. *Pito*, on the other hand, is less cloudy (even clear) and less sour, but is still un-hopped and consumed whilst fermenting. These traditional African drinks are normally less alcoholic than European beers, principally because sorghum grains have considerably lower amylolytic activity than barley (or wheat). The very nature of sorghum grains means that these traditional products are very different from "European-style" beers. To supply the ever-increasing urban population of some parts of

Africa, industrialization of traditional methods has been encompassed, and many of the breweries that have followed this route
employ high proportions of maize (*Zea mays* L.) in their grists.
Probably the best known non-alcoholic beverage is *mageu*
(*mahewu*, or *magou*) from southern Africa, which is a lactic acid
bacterium-fermented gruel.

The art of kaffir beer brewing predates written history and, over
the years, the drink has assumed great significance in the nutrition
of Bantu tribesmen. In the villages, kaffir beer brewing is the
domain of women, and every girl learns to brew before marriage.
Because of its importance, South Africa has spawned several
modern, commercial kaffir beer breweries, to service the needs of
Bantu tribesmen who have left their homelands for employment
purposes. As Steinkraus[231] noted: "Kaffir beer production is the
only large modern industry founded on tribal art of African origin
… thus, it is of unusual interest to those studying indigenous
fermented foods and the effects modern processing methods can
have on them." Haggblade and Holzapfel[237] have described the
industrialization of Bantu beer brewing.

As Schwartz[238] reports, the essentials of kaffir beer brewing
involve malting, mashing, souring, boiling, conversion, straining
and alcoholic fermentation. During artisanal processing, kaffir
beers are made in drum-like (115–180 L) pots, and each litre of
beer can utilize anything from 180–360 g of grain, depending on
type. Sorghum malt is produced by soaking grain in water for 24–
48 hours, and then allowing it to germinate for a few days.
Sprouted grain is sun-dried, and then left to mature for several
months, before being pulverized and made into a thin gruel. This
slurry is then boiled, and then cooled before a small amount of
fresh malt is added (both for amylolytic activity and for
introduction of microbes). This mixture is left for 24 hours to
accommodate the growth of lactic acid bacteria. On day two it is
boiled in a pot, and then returned to the brewing vessel to allow the
alcoholic fermentation to commence. On the third and fourth days
more pulverized, uncooked malt is added, and on the fifth day the
brew is strained through a basket to remove the husks. The beer is
then deemed ready to drink.

About equal quantities of malted and unmalted grains are
mashed in cold and boiled water, respectively, at the time that they
are combined so that the net temperature of the mash mixture

turns out to be 37 °C. As is the case with brewing with barley, the ability of the grain to yield sufficient diastatic power is critical, but Platt and Webb[239] concluded that most saccharification was not attributable to sorghum amylases, but to filamentous fungi growing on the grain. In traditional fermentations, there is no purposeful addition of yeast; the organism is either introduced with the malt, or persists within the fermentation pots. The absence of a sterilizing, "cooking", stage following the mashing of kaffir beer is deleterious to the keeping qualities of the final product. There is a boil after lactic souring, which does reduce microbiological loading, but there is re-infection *via* subsequent malt additions. The final pH of kaffir beers is usually in the 3.3–3.5 range, and even this degree of acidity cannot prevent rapid microbial activity in the product.

Industrial sorghum (kaffir) beer brewing was initiated by municipal governments in South Africa and Zimbabwe in the early years of the twentieth century, but was perfected many years later. The modification of such a traditional craft proved a great challenge, and there were many technical difficulties to be overcome. In 1954, the Council for Scientific and Industrial Research (CSIR) in South Africa was commissioned to carry out applied research on behalf of municipal sorghum beer brewers. This assistance helped to foster a spectacular increase in scale from home brewing (*ca.* 0.5–2 hL brews) to an industrial scale (*ca.* 150–270 hL brews), and this was made possible by the development and adoption of carefully specified raw materials, and the use of specialized equipment. With the new protocol, it became possible to partition individual phases of kaffir beer brewing, which could be carried out at optimum temperatures. During artisanal brewing regimes, it is not uncommon for several important reactions to occur simultaneously. Industrial brewing of kaffir beer also employs two distinct fermentations (lactic and alcoholic) and, besides being carried out on a larger, more scientific, scale, differs mainly from the artisanal process inasmuch as a yeast culture is added to initiate alcoholic fermentation. Souring (lactic acid production) is achieved by holding a mixture of sorghum malt and water at 48–50 °C for 8–16 hours until the desired degree of acidity has been attained (pH 3.0–3.3, with lactic acid representing 0.3–1.6% (average 0.8%) of total acidity). This souring step is critical, because the nature of the end-product depends upon it. The "sour", as it is called, has an influence on mashing, fermentation and the body of the beer. Soured malt (sufficient to give one-third of the brew length) is pumped to a cooker, where it is mixed (diluted) with two volumes of water. An adjunct (usually maize or sorghum grits) is added and the mixture is boiled for 2 hours, before being cooled to 60 °C, whence unsoured sorghum malt is added, and the temperature is held for 1.5–2 hours to allow starch conversion. This thinner, sweeter mixture is then cooled to 30 °C and inoculated with a top-fermenting strain of *S. cerevisiae* (the yeast, which is kept locally in dried form, is slurried prior to pitching). The pitched mixture passes through coarse strainers to remove gross debris (husks, *etc.*), and the wort is then fermented for 8–24 hours at 30 °C, prior to distribution into large containers. If the beer is destined for small containers (cartons), fermentation may only last around six hours.

Lactic acid bacteria for the souring step are introduced *via* a previously soured malt/water slurry, not as a pure inoculum, and it has been shown that, if the mixture is kept at 50 °C prior to addition, essential bacteria are preferentially selected. As with barley, malting results in the formation of amylases and proteases, as well as contributing to flavour improvement. In respect of the latter, polyphenol content of the grain is reduced during malting (some sorghum varieties are considered "birdproof" because of their high polyphenol content). This is also important because high polyphenol levels inhibit souring organisms, and hence increase pH.[240]

Barley malt contains more β-amylase than α-amylase, but this situation is reversed in sorghum malt, where the latter can exceed the former in a ratio of 2–3 : 1. In addition, the solubility of sorghum amylases is dependent upon the polyphenol content of the grain. Insoluble amylases are, of course, of little value to the brewer. Maltase (α-glucosidase) is also present in ungerminated sorghum, and its level does not increase during malting. Both sorghum maltase and amylases function at the rather acidic pH (3.7–4.0) of a kaffir beer mash and, although most amylolysis occurs during mashing, some starch will still be broken down during fermentation. The optimum temperature for saccharification is 60 °C, and because souring (lactic acid production) takes place simultaneously, starch breakdown in the mash is incomplete.

Besides lowering the pH, lactic acid is important because it gives a sharp, refreshing feel to the beer, and this is much admired by the Bantu people. Lactic acid also "softens" the protein sheath surrounding each starch granule, and thus makes starch more susceptible to enzyme attack. Starch is also an important component of Bantu beer, providing much-needed calories for imbibers. A considerable quantity of both gelatinized and ungelatinized starch is present in the drink, the gelatinized form helping to keep ungelatinized material in suspension, and thus adding body and "creaminess" to the final product.

An artisanal kaffir beer would be expected to contain: 2–4% w/v ethanol, 0.3–0.6% acid (as lactic) and 4–10% total solids. The pH would be between 3.3 and 3.5. Data for an industrially brewed beer are given in Table 4.4.[241] As Steinkraus[231] avers: "Kaffir beer is probably one of the least expensive alcoholic drinks in the world. Because of its low cost and the traditional role it plays in Bantu culture, demand for it is likely to remain strong for the foreseeable future."

The inability to cultivate barley in tropical countries and the restriction on barley and malt imports in some African countries (such as Nigeria) has stimulated research into the use of sorghum for brewing clear beers, both raw and malted grains being used for this purpose.

Sorghum is not high in nutritional status when compared to other cereals and, while it is an important starchy staple, it does

Table 4.4 Composition of an industrially brewed kaffir beer (after Aucamp *et al.*[241]).

	Range	*Mean*
pH	3.2–3.7	3.4
Alcohol (% w/v)	1.8–3.9	3.0
Solids (% w/v)		
Total	3.0–8.0	5.4
Insoluble	2.3–6.1	3.7
Nitrogen (% w/v)		
Total	0.059–0.137	0.093
Soluble	0.010–0.017	0.014
Thiamine (µg/100 ml)	20–230	93.0
Riboflavin (µg/100 ml)	27–170	56.0
Nicotinic acid (µg/100 ml)	130–660	315.0
Ascorbic acid (mg/100 ml)	0.01–0.15	0.04

not provide a complete diet. Having said that, it has some very useful health-promoting components that have yet to be fully exploited. Sorghum is generally rich in phenolics (with their antioxidant activity), and the wax of sorghum, which is on the pericarp surface of the grain, may also have some useful medical potential. Of specific interest here are the primary, long-chained alcohols in the wax, which are called policosanols and are reportedly effective in lowering the amount of low-density lipoprotein (LDL) cholesterol and raising the amount of high-density lipoprotein (HDL) cholesterol, thus improving the LDL/HDL ratio.

Over the last few years, the use of sorghum in the brewing of European-style lager has proved highly successful, and has been reported in Nigeria, East Africa, southern Africa and the USA. Most work on sorghum brewing has focused on the use of malted grain, but it has been argued that using raw (unmalted) grain is a more logical and cost-effective approach to brewing with sorghum because of the intrinsic problems associated with malting sorghum, and the need for additional exogenous enzymes when using sorghum malt.

One of the major problems when brewing with sorghum is that of very low free amino nitrogen (FAN) levels, and there is considerable evidence to show that the levels of FAN in sorghum grain wort mashed with commercial enzymes are considerably lower than those obtained with sorghum malt. In order to achieve adequate levels of FAN (130–150 mg/L) to support optimal yeast growth and fermentation efficiency, unacceptably high levels of proteolytic enzymes must be used. The low levels of FAN obtained when using exogenous proteolytic enzymes can be attributed primarily to the factors that contribute to the low digestibility of the kafirin fraction of sorghum protein, which accounts for around 70% of total grain protein, and which is rendered less soluble by wet cooking during brewing. A number of methods have been used to increase the digestibility of sorghum protein, including the use of reducing agents such as sodium bisulfite, 2-mercaptoethanol and ascorbic acid. Recently,[242] it has been shown that potassium metabisulfite (KMS) significantly increases the rate of sorghum protein hydrolysis because of the reduction of intermolecular disulfide bonds in the kafirin polymers and oligomers, which presumably allows better access of protease to kafirin.

At 65 °C, the temperature generally used when mashing barley malt, sorghum starch is inadequately gelatinized, and saccharification is sub-optimal for brewing purposes. Accordingly, the use of exogenous enzymes is essential. If mashing is carried out at 85 °C, sorghum starch is adequately gelatinized and sugar release is greater, although the ratio of glucose to maltose does not change (probably owing to the structure of sorghum starch and the pattern of development of amylases). It is generally recognized, however, that, for most purposes, the gelatinization of sorghum starch occurs at too high a temperature in order for it to be used without prior gelatinization in a preliminary boil. This puts restraints on the type of brewing equipment that can be used, and the proportion of adjunct that can be used in the grist. To circumvent such problems, grain extrusion has been used as a means of pre-gelatinizing brewing adjuncts such as sorghum. Extruded preparations of cereals are widely used as snack foods, breakfast cereals and animal feeds. Extrusion cookers are screw-operated, and ground ingredients are introduced from a feed at the base, whence they are mixed with steam, water or other substances before passing to a mixing cylinder. From here the mixture passes to the extruder, which is driven by a worm screw. During its passage through the extruder, the mixture is worked into a dough that is continually heated, and subjected to pressure, shearing and forced mixing. Temperatures can exceed 200 °C. At the end of a barrel, the plastic material emerges from a die, and it expands and is cut into sections by a rotating knife before being cooled and dried.

Early work on extruded malt has been carried out by Dennis Briggs and co-workers in Birmingham,[243] and Fors and Nordlöv[244] instigated sensory analyses. As the latter reported, a number of *N*-heterocyclics, especially the pyrazines, methylpyrazine and 2,5 (and/or 2,6)-dimethylpyrazine result from the extrusion process, and these can contribute undesirable flavours to beer. Sorghum malt differs from barley malt inasmuch as it has a higher starch gelatinization temperature, and lower β-amylase content. Sorghum bran can contain high levels of undesirable phenolics, and it may be necessary to remove the outer layers by roller-milling prior to brewery use. De-branned sorghum can be used for extrusion purposes. A review concerning brewing beer with sorghum has been provided by Owuama.[245]

Sorghum was first recognized as an important brewing adjunct during World War II, when malting barleys were at a premium. Technically, adjuncts are essentially starchy materials containing little or no protein. They are a potential source of "additional" alcohol and may add to the colour, taste, aroma and vitamin content of beer, as well as contributing to its head retention capability. Unmalted materials, such as sorghum, bajra, tapioca, soy beans, wheat and barley flours have all been added to grists as adjuncts. In the USA and Mexico, the use of sorghum as an adjunct in brewing lager beer is well established.

Prolamin proteins in the cereal grains wheat (gluten), rye (secalin) and barley (hordein) are known to bring about an allergic response or a detrimental autoimmune reaction in some individuals. The latter is a condition known as coeliac disease (CD), and is a chronic, genetic disease characterized by the formation of autoantibodies and the destruction of the mucosal lining of the small intestine, which results in malabsorption. Typical symptoms suffered by coeliacs are abdominal pain, diarrhoea and constipation, and there are a number of long-term complications (including anaemia and osteoporosis). Coeliacs are condemned to life-long avoidance of products containing proteins such as gluten.

Increased awareness and diagnosis of CD and gluten sensitivity have spurred the demand for gluten-free products. There is now a burgeoning need to develop nutritious, palatable and affordable foods and drinks suitable for coeliacs. Sorghum grain and its proteins are safe for coeliac patients and individuals with varying levels of gluten intolerance. The main sorghum proteins, the kafirins, however, are resistant to digestion, and are difficult to extract, and this limits their use in foods. Attempts are underway to increase the nutritional and functional value of sorghum proteins.[246] In terms of brewing and ethanol production, the impact of mashing on sorghum proteins has been studied,[247] as has the effect of fermentation on sorghum proteins.[248]

Pito and *dolo* are two important West African drinks, the former being common in Ghana, Togo and Nigeria, with *dolo* (and similar products, like *tchpalo* and *tchoukoutou*) being favoured in Burkina Faso, Ivory Coast, Mali and Benin. These beers, together with other similar products, are an essentialm part of the diet in these countries.[249] Because of their importance, a number of scientific studies have been made, including the fermentation flora.[250–253] In Burkina Faso, *dolo* is widely produced and is the most important fermented beverage consumed, with 40% of the sorghum grown in the country used for its production. In Ghana, *pito* has important socio-economic implications.[254] The production of *pito* has been described by Ekundayo,[255] and Onaghise and Izuagbie.[256]

An initial analysis of the *S. bicolor* genome has been carried out.[257] Using whole-genome shotgun sequencing, ratified by genetic, physical and syntenic information, around 98% of the plant's genes were placed in their chromosomal context. Genetic recombination was found to be largely confined to about one-third of the sorghum genome with gene order and density similar to those of rice. As the team said: "Retrotransposon accumulation in recombinationally recalcitrant heterochromatin explains the roughly 75% larger genome size of sorghum compared to rice. Although gene and repetitive DNA distributions have been preserved since palaeopolyploidization – 70 Mya – most duplicated gene sets lost one member before the sorghum-rice divergence. Concerted evolution makes one duplicated chromosomal segment appear to be only a few million years old. Around 24% of genes are grass-specific and 7% are sorghum-specific. Recent gene and microRNA duplications may contribute to sorghum's drought tolerance."

4.7 MILLETS

The millets are often referred to as "minor cereals", but they are important components of African and Asian agriculture. According to de Wet,[258] cereals that do not belong to the wheat, barley, oats, maize or rice genera can conveniently be grouped as "millets". Roughly 95% of the world's millet area lies in the developing countries, mainly Africa and Asia, and these plants are primarily grown in poor areas that are subject to low rainfall and drought, where other grain crops would struggle. Whereas sorghum is grown for both human food and animal feed, the major millets are produced almost entirely for human food. Most millets are notably rich in phytochemicals, particularly phenolics, which may have health-promoting properties.

4.7.1 Some Millet Botany

Botanically, millets are an agronomic group of highly variable grasses that produce numerous small seeds, rather than being a discrete taxonomic grouping. At some stage, *Sorghum* spp. were included in the millets. The most important of the group are: pearl

Figure 4.16   *Pennisetum glaucum* (L.) R. Br. (from *Flora of North America*[388] - by kind permission).

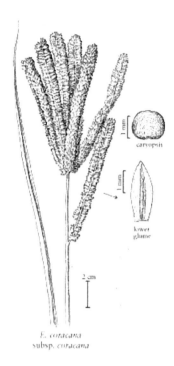

Figure 4.17 *Eleusine coracana* (L.) Gaertn. (from *Flora of North America*[388] - by kind permission).

millet (*Pennisetum glaucum* (L.) R.Br. ex-*P. typhoides* Stapf. and C.E.Hubbard) (Figure 4.16), finger millet (*Eleusine coracana* (L.) Gaertn.) (Figure 4.17), broomcorn (common or proso) millet (*Panicum miliaceum* L.) and foxtail millet (*Setaria italica* (L.) P. Beauv.). Other millets of lesser and/or regional importance include: kodo (*Paspalum scrobiculatum* L.), Indian barnyard (or sawa) (*Echinochloa frumentacea* Link), Japanese barnyard (*E. esculenta* (A.Braun) H. Scholz), Guinea (*Brachiaria deflexa* (Schumach) C.E.Hubbard), browntop (*B. ramosa* (L.) Stapf.) and little (*Panicum sumatrense* Roth. ex Roem. & Schult.). Tef (*Eragrostis tef* (Zucc. Trotter); ex-*E. abyssinica*), fonio (*Digitaria exilis* (Kippist) Stapf.) and Job's tears (*Coix lacryma-jobi* L.), the latter a minor food and fodder crop of northeastern India,[259] are also frequently called millets. An excellent on-line identification atlas for the millets of archaeobotanical interest has been prepared by Dorian Fuller,[260] and Zhang *et al.*,[261] using phytolith evidence, have just devised a means of differentiating between *S. italica* and *S. viridis* in archaeobotanical material.

Of the four millets that are widely distributed throughout temperate Europe and Asia, two of them, crab grass (*Digitaria sanguinalis* (L.) Scopoli) and Japanese barnyard millet (which has also been accorded the specific name *E. crus-galli* (L.) P. Beauv.), are not generally well known. The former is a cosmopolitan weed, which became semi-domesticated in southeastern Europe after many centuries of harvesting from the wild. Even now, it seems as though it has still retained the ability to disperse its seeds "naturally". It was a popular cereal in Southern Europe during Roman times and persisted as such in southeastern Europe until the early nineteenth century.[262] Currently, it is grown as a minor crop in the Caucasus and Kashmir. Japanese millet is a plant of central Eurasia, and is grown as a cereal in China, Japan and Korea, but is no longer used as such anywhere in Europe.

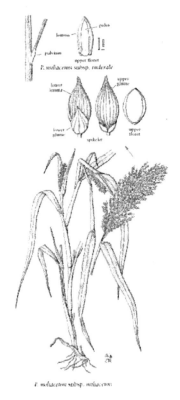

Figure 4.18   *Panicum miliaceum* L. (from *Flora of North America*[388] - by kind
          permission).

The two other widespread Eurasian millets, broomcorn and foxtail, are the oldest domesticated dry farming crops in Eurasia. They were staple foods in the Far East and even in the entire Eurasian continent prior to the popularity of rice and wheat, and are still important foods in the semi-arid regions today. The genus *Panicum* is widely distributed throughout the warmer parts of the world, and *P. miliaceum* (Figure 4.18) is one of the hardiest of all cereal crops. This plant is the "true millet" of Classical times, "milium" of the Romans and "dokhan" of the Hebrews and, today, is cultivated mainly in eastern and central Asia, in India and, to a lesser extent, in the Middle East. Being a warm season plant, it tolerates intense heat, poor soils and severe droughts, and is tolerant of brief rainy seasons. It is a summer crop, being sown in late spring and harvested in late summer, thus completing its life-cycle in 60–90 days.

4.7.2 Broomcorn Millet

Although best known for the domesticated broomcorn millet, there are around 500 species of *Panicum* (the "panic grasses"), some of which are important wild cereals. Broomcorn millet has the lowest water requirement of all grain crops. It is also a relatively short-season crop and can perform well in poor soils. Both the location of domestication and the wild ancestor of cultivated broomcorn millet have not been unequivocally identified, but weedy forms of broomcorn millet are widespread in central Asia and are generally referred to as *P. miliaceum* subsp. *ruderale* (Kitag.) Tzvelev. According to Kitagawa,[263] these weeds are native to central China, and have shattering panicles, which Scholz[264] maintained have developed from cultivated forms that regained the ability to reproduce naturally (*via* a mutation). He is also convinced that these weeds recently spread to Central Europe and North America. Broomcorn has long been grown in central China,[265] where it is still widely grown, and is an important crop in Mongolia, Korea and northern India. A cultivar with glutinous endosperm is favoured in China for bread-making, but in India and Mongolia non-glutinous forms are favoured because the grains are cooked in the same way as rice. Broomcorn millet has been cultivated in Southern Europe for at least 3000 years, and was widely distributed throughout the continent during the Bronze Age – its popularity declining during the last century. Some of the earliest European finds are from

Linearbandkeramik sites in the Ukraine, Moldavia and Poland.[266] Despite fragmentary archaeobotanical evidence, Zohary and Hopf are satisfied that proso millet does not belong to the Near East crop assemblage, and that it is a central Asian plant that was picked up and added to barley and wheat agriculture soon after its arrival and establishment in that area.

4.7.3 Oldest Evidence for Millet

Around 30 years ago, the world's oldest millet remains, dated to *ca.* 8200 cal. yBP, were discovered at the Early Neolithic site of Cishan, northern China. The site covered an area of around 80,000 m^2, and contained over 500 grain storage pits, the largest of which were 5 m deep and capable of storing up to 1000 kg of grain.

The Cishan site (36°34.511′N, 114°06.720′E; Figure 4.19) spreads across the mesa on the northern banks of the Luo River, southwest of Wu'an County in Hebei Province, northern China. It is situated near the junction between the Loess Plateau and the North China Plain at an elevation of 260–270 metres above sea level. The site was discovered in 1973, and excavated from 1976–1978, and is the earliest Neolithic site in North China, being generally dated to around 6000 BC. Because of its unique features (*e.g.* stone grinders, stone sickles and tripod pottery), it defined a specific type of culture, the "Cishan Culture", and several more similarly defined sites have been unearthed in China.

The site contained >50,000 kg of millet grain stored in pits but for some while it had been impossible to determine whether the remains were from foxtail millet or from broomcorn millet (all evidence for millet domestication in China is based on the morphology of carbonized grain). The doubt arose from a lack of agreement regarding the validity of some of the earlier (1980s) *S. italica* finds. Whatever the precise nature of the grain, the large quantity stored at Cishan is a strong indication that millet was being regularly cultivated there. Using new chemical differentiation methods, and husk phytolith identification,[267,268] convincing evidence has emerged to show that the crop in a newly excavated pit at Cishan contained solely broomcorn millet. Moreover, the remains were dated between 10,300 and 8,700 cal. yBP – which pushes millet cultivation back about 2000 years, and gives that crop an even bigger head-start over rice than was previously

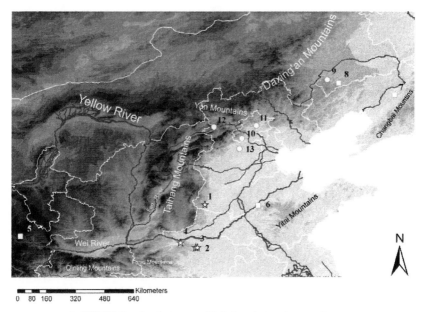

Kilometers
0 80 160   320   480   640

☐ Pre-5000 BC sites where broomcorn millet is the major cereal recovered

☆ Pre-5000 BC sites where foxtail millet is the major cereal recovered

○ Pre-7000 BC site with Neolithic pottery, lacking published archaeobotanical evidence

Figure 4.19   Cishan, and other important sites in China (from Liu *et al.*,[280] - with kind permission). Key to sites: 1. Cishan; 2. Peiligang; 3. Shawoli; 4. Wuluoxipo; 5. Dadiwan; 6. Yuezhuang; 7. Xinle; 8. Xinglonggou; 9. Xiaohexi; 10. Donhulin; 11. Zhuannian; 12. Hutoliang; 13. Nanzhuangtou.

thought! Millet remains at Cishan can be from as late as 5500 BC, but after 8700 cal. yBP, grain crops begin to contain small quantities of foxtail millet (estimated at 0.4–2.8% of finds).

By examining the anatomy and silicon structure patterns in the glumes, lemmas and paleas from the inflorescence bracts in a variety of modern broomcorn millet, foxtail millet and a number of closely related grasses, professor Lu's team[269] were able to establish a means of distinguishing between these two important crops. They showed that five key diagnostic characters in phytolith morphology can be used to differentiate broomcorn and foxtail millets in archaeobotanical remains.

Work by You[270] and Li and Wu[271] indicates that the geographical distribution of both common millet and foxtail millet in China show different patterns, and that the former is more adapted to the drier interior areas; its optimal growth occurs at a mean annual temperature (MAT) from 6 to 8 °C and mean annual precipitation (MAP) from 350 to 450 mm. Foxtail millet is more common in semi-wet eastern areas, and its optimal growing conditions are 8 to 10 °C (MAT) and 450 to 550 mm (MAP).

Grains of broomcorn have also been found during the initial stages of various other Early Neolithic sites, including Xinglonggou (*ca.* 8000–7500 cal. yBP); Yuezhuang (*ca.* 7800 cal. yBP) and Dadiwan (*ca.* 7800–7350 cal. yBP), but foxtail millet was barely present. At the latter, a site in northwest China (Figure 4.19), there is good evidence of early millet cultivation. This is the type site of the Dadiwan Culture, and was built on a mountain slope south of the Qingshui River near the Wei River, in Qin'an County, Ganshu Province. The site was originally excavated in the 1970s, and revisited in 2006.

Stable isotope biochemistry ($\delta^{13}C$ and $\delta^{15}N$) and radiocarbon dating of people, dogs and pigs from the Dadiwan site[272] have provided new data about the origins of millet agriculture in this area. Analysis of ancient human and animal bones indicated two distinct phases of plant and animal domestication at Dadiwan. The first (at various times between 7900 and 7200 cal. yBP) was brief and non-intensive, whereby people harvested and stored sufficient broomcorn millet to victual themselves and their hunting dogs throughout the year (referred to as "pre-Yangshao"). The second, much more intensive, phase was manifest by 5900 cal. yBP and, during this time, both broomcorn and foxtail millets were being cultivated (termed "Yangshao"). The systems represented in both phases developed elsewhere: the first emerged with hunter-gatherers in the arid north, while the more intensive, later phase evolved further east and arrived at Dadiwan with the Yangshao Neolithic package (the first evidence of Neolithic culture in China was discovered at Yangshao, Henan Province, in 1921 – see Kwang-chih Chang's book[273]).

The archaeological records of Dadiwan over the past 60,000 years and the origin of millet agriculture in the light of recent finds have been reported by Zhang *et al.*[274] Artefact distributions there revealed patterns of human behavioural variation and adaptation over this time period from primitive hunting and gathering to advanced hunting and gathering, to primitive Neolithic agriculture, and finally to advanced Neolithic agriculture. Archaeological remains at the new excavation (2006) of a site at Dadiwan (one of the important early millet agriculture sites in North China) are continuous from Late Pleistocene to Holocene. Dadiwan is the westernmost expression of early agriculture in northern China, and has produced China's earliest painted pottery.

It is widely believed that East Asian agriculture evolved in isolation from early agricultural developments elsewhere on the planet, and produced a distinct set of domesticates including rice, broomcorn millet, foxtail millet, pigs, dogs and chickens.[275–277] Existing evidence suggests that two historically independent evolutionary phenomena operated in two ecologically distinct parts of mainland China: a rice-based system in the warmer and more humid south, and a millet-based system in the colder, more arid north. In addition to this geographic divide, many authors have noted apparent east-west distinctions of broomcorn and foxtails millets in prehistory. Broomcorn is more typical of the western part of the Yellow River basin, while foxtail predominates further east. Shelach[278] postulated that broomcorn millet may have been domesticated several times or spread rapidly across North China.

Humans in the Yangzi Drainage may have harvested wild rice as early as 12,000 yBP,[279] but the earliest evidence for its actual domestication varies in age from 10,000 to 6000 yBP. The precise timing is still unresolved, but we still know more about rice in southern China than we do about millet in the north.

In an erudite contribution to the history of millet cultivation in China, Professor Martin Jones's Cambridge team echoed new thoughts on the locale of early agriculture in Southwest Asia.[280] With the rapidly expanding volume of work being carried out in northern China, they remark that a number of reports have drawn attention to the fact that some of the early sites in China are located in foothills rather than in valley bottoms. Accordingly, they drew on recent evidence to question the valley-based model for agricultural origins, which had shaped discussion thus far, and suggested that they should be considered in an alternative geographical framework.

As some previous workers[276,278] have argued, the traditional view of the Yellow River as the "core area" is inadequate to explain the development of agriculture in North China, and one should look further to the northeast. Here, the key geographic features would be the hilly chains at the plateau-floodplain boundary, rather than the river valley itself. As the Cambridge workers point out: "The change in this model for the beginnings of agriculture in North China has parallels with changes in the equivalent model in southwest Asia. In both regions, the focus was once on great river valleys, the Nile, Tigris and Euphrates in the West, the Huanghe in the East (and also the Changjiang in South China). Following improved archaeological survey and excavation, the emphasis in the West moved from the valleys themselves to the 'hilly flanks' above them.[91] The emergence of an improved data set for North China seems to be pointing to an equivalent shift in the East."

4.7.4 Foxtail Millet

Like broomcorn millet, foxtail millet (or Italian millet) is a warm season cereal, surviving well under dry conditions and completing its growth cycle in a short time (sometimes as little as 40 days). Unlike broomcorn, however, its wild progenitor is well established. It is widely cultivated in India, Japan and China, and is still grown in southeast Europe, albeit as a minor crop. The cultivated plant is a diploid ($2n = 18$), typically self-pollinated, and is interfertile with its progenitor, *Setaria viridis* (L.) P.Beauv., which is a common summer weed across Eurasia (indeed, Harlan[281] maintains that green millet is one of the world's worst weeds of arable land!). As with many other crops, the cultivated and wild/weedy forms differ mainly in the way that their seeds are dispersed, cultivated plants retaining their seeds, rather than losing them through shattering. Domestication of foxtail millet led to the production of much larger inflorescences with a much more complex branching pattern, as well as a reduction in basal (tillering) and axillary vegetative branching.[282,283] Foxtail has smaller grains than those of common millet and they appear flattened on one side.

Vavilov[284] maintained that the principal centre of diversity for foxtail millet is East Asia, including Japan and China, although, based on archaeological evidence, Harlan[281] suggested independent domestication in China and Europe. The notion of multiple origins is supported by archaeological, isoenzyme and morphological evidence but, in 1987, Sakamoto[285] suggested that foxtail millet originated somewhere in Central Asia, Afghanistan, Pakistan and India, because strains with less restricted compatibility and with primitive morphological traits are found there.

Sakamoto's hypothesis excluded China as a centre of origin of foxtail millet, and relegated it to a secondary centre of origin! The power of RFLP in phylogenetic studies has permitted the analysis of the origin and genetic differentiation of crop species and their wild relatives, and some of the "mysteries" surrounding the history of foxtail millet are being resolved (see reference to Fukunaga's work below). Yellow foxtail millet (*S. pumila*) is an Indian domesticate, and has been cultivated as a minor crop there since 2000 BC.

As Doust *et al.*[286] point out, some of the changes during foxtail millet domestication can also be seen during the domestication of maize from its wild progenitor. *S. italica* and *S. viridis* are in the same grass sub-family as maize, sugar cane and sorghum, but in a different tribe, having diverged from the maize lineage some 28 Mya.[287] Using common markers among foxtail millet, maize and rice, it was possible to suggest candidate genes for controlling inflorescence branching. Such comparison is possible because the millet genome is broadly collinear with those of rice and maize.[288,289] The work showed that the inflorescence characteristics that differentiate the two millet species are each controlled by a small number of loci. Under cultivation, foxtail millet underwent a reduction in number of flowering tillers, and flowering panicle enlargement. The shift from shattering to retention is governed by two complementary recessive mutations.[290]

Some interesting experiments with *S. viridis* have been reported by Tracey Lu,[291] who looked at the botanical characteristics of the plant and conducted some harvesting experiments. The latter were aimed at emulating prehistoric foragers, and included spike-shaking, spike-picking, uprooting and cutting by lithic flakes. The most selective method was spike-shaking, whereby only ripe seeds were collected; green or flowering florets were left on the plant to continue their growth.

As Zohary and Hopf[45] aver, "*Setaria italica* is the principal crop of Neolithic agriculture in north China", citing Ho,[292] An[293] and Crawford[275] as sources, and remarking that "numerous remains of this cereal appear in the early farming villages that developed in the loess soil belt in the upper Yellow River basin".

A cluster analysis of almost 3000 accessions (from 16 Chinese provinces and 22 other countries) was carried out[294] using nine morphological and agronomic descriptors. The populations from different geographical origins were grouped into 12 clusters. Results showed the existence of high levels of variation for all characters observed among regions and a multiple domestication hypothesis with three centres (China, Europe and Afghanistan/ Lebanon) was suggested. The group also proposed a new system of classification in which four races ("maxima", "moharia", "indica" and "nana") were recognized.

4.7.5 Some Genomics

The small genome size and diploid nature of foxtail millet should have made it a suitable plant for genetic and molecular studies but, because it has been regarded as a minor cereal of regional importance, work on it has lagged behind that of other staple cereals (foxtail millet research commenced in the 1930s).

Genetic maps of foxtail millet based on RFLP markers have been constructed[288,289] and many markers became available for research purposes. A little later, Kenji Fukunaga and a series of co-workers, using some of these markers, sought to elucidate the genetic relationships of foxtail millet landraces.[295] They found that a dendrogram constructed from RFLP information was divided into five major clusters. Clusters I and II contained strains mainly from East Asia, while cluster III consisted of strains from tropical and sub-tropical regions in Asia (*e.g.* Taiwan, India and the Philippines). Cluster IV was represented by some strains from East Asia, a strain from Nepal and a strain from Myanmar, and cluster V contained strains from central and western regions of Eurasia (*e.g.* Afghanistan, Central Asia and Europe). The results indicated that foxtail millet landraces have differentiated genetically during their history. Chinese landraces could be classified into four clusters, thus again illustrating their variability. The extensive gene diversity of Chinese strains was confirmed when the mitochondrial DNA (mtDNA) variation in foxtail millet was studied.[296] Here, three types were discernible, type I being predominant in India, Central Asia and Europe. When a phylogentic analysis of the ribosomal DNA (rDNA) intergenic spacer sub-repeats was carried out,[297] the situation became more complicated, with "type I" just mentioned being divisible into seven sub-types (Ia–Ig – see Figure 4.20). Some questions were answered by this work, but the authors emphasized the need for more data.

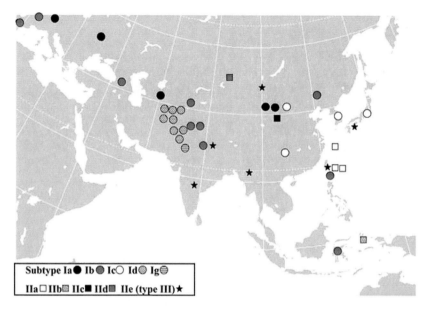

Figure 4.20 Geographical distribution of the rDNA types of *Setaria italica* (from Fulunaga *et al.*,[297] by kind permission).

More extensive study of neglected areas is always liable to throw up interesting material, as was exemplified by the recent discovery of *S. italica* convar. Moharia in northern Iran.[298] Historically, this is a crop that was widely grown from Europe through to Southwest Asia, and was very important in Persia, being a traditional cereal in Mazandaran for many decades.

In a paper entitled "Millets across Eurasia", Hunt *et al.*[299] collated and reviewed the published records of the genera *Panicum* and *Setaria* in pre-5000 cal BC sites across the Old World. The sites, from an area covering China through to central-eastern Europe, taking in the Caucasus, Iran, Syria and Egypt, were included with accompanying calibrated radiocarbon dates. As the authors explain, the period of time chosen for this study encompasses the early phases of food production in the areas studied. Admitting that this might be considered arbitrary, they feel that this is better than using the term "Neolithic", which is defined differently in different parts of the world.

Although the review focused on the two principal Asian millet domesticates, archaeobotanical literature contains records of a spectrum of within-genus plants. In the case of *Panicum*, the continuum ranges from "domesticated *P. miliaceum*" through "*Panicum miliaceum*", "*Panicum* sp.", "*Panicum capillare*", through to entirely distinct species (*e.g. P. turgidum*, not currently known to have any close relationship to broomcorn millet). A similar sort of continuum exists for *Setaria*. The survey showed that 41 sites had records of *Panicum* while 33 reported *Setaria*, and the data emphasized contrasting patterns of origin for the two principal Eurasian millets. Both occur prior to 5000 cal BC in North China, but only *P. miliaceum* occurs prior to that date in Western Asia and Europe.

Several hypotheses have been proposed to explain the separate distributions of these two millets: a single domestication in either northern China or Eastern Europe followed by rapid spread across the Central Asian steppe, or multiple domestications that could represent either two discrete events at either end of the region, or diffuse domestication across the whole steppe zone.[300]

Broomcorn and foxtail millets were the oldest cultivars in the Korean peninsula, where dry-field crops preceded wet-field production. Available data are consistent with a time-transgressive model with these crops first appearing in China, then in Korea and then in Japan.[301]

4.7.6   Other Millets

Pearl millet, also known as bulrush millet, is, after sorghum, the most important native African cereal. Because of its tolerance of low rainfall and high soil temperature, it is cultivated across the semi-arid tropics of that continent, and is also widely grown under similar conditions in India and Pakistan. *Pennisetum glaucum*, which accounts for about 50% of all millet grown and is the world's sixth most important cereal crop, is a morphologically complex species with several recognized cultivated, weed and wild forms. The grains are ovoid and, with a length of 3–4 mm have the largest seeds of the millets. Wild pearl millets occur extensively from coastal Senegal and Mauretania to northeastern Ethiopia, and also extend into the Sudan, and can be found in the foothills of the mountains in the central Sahara. According to Jack Harlan,[302] pearl millet could have been domesticated anywhere along the southern borders of the Sahara. Some authors have proposed multiple domestications,[281] others a single domestication.[303]

Several authors agree that domestication has taken place in Africa, but geographical origins vary along the Sahelian zone from Mauretania to Sudan.

As Poncet *et al.* related,[304] the wild forms of the crop (the *P. mollissimum* and *P. violaceum* ecotypes of *P. glaucum* ssp. *monodii*) are only found in Africa, where they have been involved in the domestication of the crop for several thousand years.[305] As is to be expected, domestication has caused genetic modifications of some of the original traits of the wild plants, the most important being the suppression of spikelet shedding, the size reduction of bristles and bracts leading to uncoated seed, and increase in seed size and spikelet pedicel length, and the loss of dormancy. The group undertook a genetic analysis of the domestication syndrome in pearl millet, particularly in terms of the inheritance of the major characters.

Being an outcropping species, pearl millet cultivations invariably contain weedy populations, and such hybrids hinder phylogenetic analyses (although statistical programmes are now available to identify hybrid genotypes – see reference 306. Oumar *et al.*[307] used microsatellite loci to analyse the diversity and evolutionary history of this crop. Their aims were to: 1) compare the diversity between wild and cultivated samples; 2) identify hybrid genotypes between wild and cultivated samples; and 3) document the evolutionary history and domestication of pearl millet. They found significantly higher diversity in the wild pearl millet group. The cultivated millet sample possessed 81% of the alleles and 83% of the genetic diversity of the wild millet sample. Using Bayesian approaches, they were able to identify intermediate genotypes between the cultivated and wild groups, and then analyse the phylogenetic relationship among those accessions that did not show introgression. From the results, the group were able to conclude that a monophyletic origin of pearl millet in West Africa is the most likely scenario.

In a later piece of work,[308] the genetic control of the above domestication traits was investigated. Many morphological differences could be attributed to the effect of a small number of loci with relatively large effects. A comparison of the domestication events of pearl millet with the closely related foxtail millet show that seed production per primary branch in pearl millet increases only slightly above that found in *P. mollissimum*, its presumed wild progenitor.[282] Inflorescences, however, are much longer, with a resultant increase in primary branch number. Inflorescence differences in domestication of foxtail millet are the result of changes in primary branch number and density, spikelet number and bristle (sterile branchlet) number. Genetic aspects of the domestication syndrome in pearl millet have been studied by Poncet *et al.*[304,308]

Pearl millet ($2n = 2x = 14$) and the increasingly important elephant grass ($2n = 4x = 48$) are sexually compatible and can undergo hybridization, which favours the amplification of their genetic background and the introgression of favourable alleles into breeding programmes. The greatest problem relating to the practical use of these hybrids is infertility caused by triploidy; a problem that is now being addressed.[309]

Pearl millet is used widely as a brewing raw material in Nigeria, and some work has been carried out on its suitability for such a role. The nutrient composition of both grains and malt has been presented,[310] and it was shown that the vitamin content increased during malting, as did, to a lesser extent, protein and total phenol content. Lipid, phytate and oxalate levels decreased during malting. In a study of the *in vitro* digestibility and amino acid composition of pearl millet protein,[311] it was shown that its digestibility values were higher than that of sorghum and comparable to that of maize. The amino acid profile of pearl millet is more favourable than that of "normal" sorghum and "normal" maize and was deemed comparable to wheat, barley and rice, making it a promising brewing material.

Panicum sonorum Beal and *Setaria macrostachya* Kunth. are both New World millets, the former emanating from the American Southwest, and the latter, which was important prior to the rise of maize as a crop, from Mesoamerica. *P. turgidum* ("merkba"; "afezu") is used by the Tuareg peoples of southern Algeria, the grains being used to make a porridge. *P. sumatrense* ("sama") is still cultivated in South Asia, and was once widely cultivated in the Indian sub-continent where remains can be identified from sites dating from 3000 BC.[312]

Finger millet is an important tropical cereal in eastern and southern Africa and in India. It is not as drought tolerant as pearl millet, and was domesticated from *Eleusine africana*, which grows in Uganda and Ethiopia. The earliest *bona fide* record for the crop from Africa, *ca.* AD 600, comes from Axum, Ethiopia. Acreage of this crop is decreasing in Africa, even though it is regarded as a healthy food, but in India, where it is known as "*ragi*", it has increased in importance.

Browntop millet, which is now only cultivated in a few dry areas in southern India, is a good example of crop evolution in progress. Three forms are known, which illustrate the transition from the wild plant to a fully domesticated crop. The wild form retains all of its "wild" characteristics, the (intermediate) weedy form is used for fodder and for grain in drought years and the domesticate is taller, has non-shattering ears and is cultivated in pure stands. According to Nesbitt,[312] this crop is threatened with extinction as farmers adopt new commercial crops, and food habits change.

Although somewhat indigestible and regarded as a low-quality food, kodo millet is still cultivated in central India, and is known to have been cultivated in the sub-continent for around 4000 years. Grains are either consumed whole, like rice, or are roasted and ground into flour.

Sawa millet has as its ancestor the tropical grass *Echinochloa colona*, although the time of domestication is imprecisely known. The crop is widely grown in India, Pakistan and Nepal mainly on lands that do not support other crops very well. The grains are usually cooked in water in a similar way to rice, and can be used as a grain for brewing. Japanese barnyard millet is closely related to sawa and is grown in Japan, China and Korea, albeit on a small scale. The earliest records of domestication date to 2000 BC from the Japanese Jomon period. Again, it is cultivated in areas where rice is likely to struggle. Its wild ancestor is the widespread temperate grass, *E. crus-galli*.

Tef (also known as "annual bunch grass" and "lovegrass") is native to Ethiopia and is an ancient grain that was cultivated there long before recorded history. It is the basic ingredient of *injera*, a popular unleavened bread that is a major component of the Ethiopian diet. Tef is also an important ingredient in the nutritious gruel, *muk*, and in a number of indigenous beverages. The grain is a good source of protein and is rich in iron, calcium, phosphorus, potassium and lysine and, since the grain is not processed, none of the goodness is lost. Tef is gluten-free, and this has been partly responsible for an increase in its popularity in other parts of the world. It has the smallest of all millet grains.

4.7.7   Some Millet Beverages

One beer made from pearl millet that has been the subject of laboratory studies, and has been brewed under controlled conditions, is "oyokpo", a traditional opaque product from Nigeria. Traditional methods of brewing oyokpo invariably result in beverage that becomes undrinkable after a couple of days. A vastly improved beer was produced by controlled malting and employing precise temperatures for mashing and fermentation (which employed a pure culture of *S. cerevisiae*). For example, an artisanally brewed beer with a commencing sugar content of 7.37 g/100 ml fermented out to 2.4% ABV (pH 3.8), while the "controlled" version, with improved malt, started at 19.73 g/100 ml sugar and fermented out to an ABV of 5.56 (pH 4.2).[313] No attempts were made to extend the shelf-life of oyokpo until the use of sodium benzoate was proposed.[314] Careful malting of foxtail millet results in reasonable amylolytic activity (maximum within 30 hours of the onset of germination), and millet α-amylase has been used to saccharify cassava.[315]

Relatively little commercial pearl millet malting is carried out, and this has been partly due to a paucity of information about its technology. The situation changed during the early years of this millennium when Pelembe *et al.* looked at the effect of malting conditions[316] and then investigated the effect of germination moisture and time on the capacity of the malt for brewing opaque beers and lager-style beers.[317] In context of the latter, it was suggested that pearl millet might have better potential than sorghum malt in lager beer brewing, at least as a barley malt extender in areas where barley cannot be grown. It was also mooted that pearl millet might be able to replace sorghum in brewing opaque beers (such as "uphutsu" in Mozambique) in low-income communities.

Bilbil, a millet beer produced in Cameroon, has been the subject of a thesis by André Ganava,[318] who studied the relevance of brewing this beer among the Mafa women living in the northern province of that country. The work shows that brewing is the only profitable economic activity available for these females, and that it occupies an important position in their lives. This is in spite of the efforts of various bodies who would like to see an end to female domination of brewing. The central role of the "cabaret" (basically a pub) in Mafa society was emphasized as it serves as a social centre and a platform for political and social discussion. On the plus side, brewing gives economic autonomy to women, but there is cognizance of the fact that beer can have negative effects on society. The most obvious problem is alcoholism, but there is always the fact that grain used for brewing might be better used for food production.

Mafa populations mostly inhabit mountain environments, where they have developed what has been called "terrace farming",[319] and the fieldwork for this study was carried out in the outskirts of the town of Maroua. While millet production is their main preoccupation, they practise annual crop rotation, with millet one year and sorghum the next.

As has been shown by many other anthropological studies relating to drink, *bilbil* has a religious and socio-cultural relevance. Consumption is controlled by the elders and largely meant for sacred occasions. The beer is usually inaccessible to women. It has been argued that millet beer is much closer to the hearts of the Mafa people than wine used to be in the Christian mass,[320] a point taken up by Muller-Kosack,[319] who said: "Ceremonial beer is the 'Eucharist' of Mafa religion. It represents the link between God, the ancestors and the living within a patrilocal community. The 'way of the beer' begins with those who have a deceased father and who are therefore responsible for the ceremony of the sacred pot called *baba*, and ends with the mountain priest (*biy dza*) who libates beer at the sacred site of his local kin."

While it is the man who controls everything in the household, brewing beer for money has to be seen as one of the few opportunities that women have for achieving some economic independence. Compared to their traditional position in the household, where they depend on men, women are able, through brewing beer, to finance certain expenditure relating to both daily domestic needs and their individual needs.

Ganava notes that during the dry season almost all women are brewing beer because farming, the main activity, is over and they have nothing else to do – certainly nothing else that will earn them money. They are waiting for the next rainy season and are basically "saving money". The practice is, apparently, spreading: in 1970, there were around 700 women brewing *bilbil* in Maroua, and this number had increased to 1200 by 2001.

The importance of millet in societies like this one cannot be overestimated and, large trees excepted, local belief is that "millet is the only plant that has a soul, like humans". Millet is food and drink, and is considered a symbol of life itself, and the importance that the Mafa attach to this grain can be illustrated by the way that a variety of rituals accompany its production – from sowing through until harvesting.

The ban on the importation of barley malt in Nigeria in 1988 (as a result of the high cost of importation) brought about a major change in brewing technology in that country, and in other parts of Africa. Attention turned to the use of local raw materials, such as maize, sorghum and millets. Sorghum has long been used for brewing (see p. 241), but the amount available for brewing, as opposed to food, use is limited, and there has long been a need to seek another substitute for barley to supplement sorghum.

Some millets, notably finger millet, have been used as barley substitutes in East Africa and India for ages, and some maltsters and brewers prefer these grains to sorghum on account of their higher enzymic activity and flavour. Late last century, work was carried out to evaluate the suitability of other millets for brewing and in one extensive study in Nigeria[321] the brewing potential of a *Digitaria exilis* ("acha") malt was compared with three pearl millet (*Pennisetum typhoides*) malts and two sorghum (*S. bicolor*) malts. As would be expected, diastase, α-amylase, amyloglucosidase and proteases increased with malting time and were associated with grain modification. Development of hydrolytic enzymes was significantly higher in pearl millet and acha than in sorghum. The major starch-degrading enzyme in the pearl millets was α-amylase, whereas β-amylase was prevalent in the acha malt (as in barley). Gibberellic acid (GA) stimulated the diastatic activity of the pearl millets, acha and one of the sorghum varieties, but inhibited the proteolytic activity of all pearl millets, acha and the sorghums. Potassium bromate had little, or no, effect on malting losses. The promising high β-amylase content of acha malt was counteracted by its high malting loss, thus making it an uneconomical grain to brew with. The work concludes with the notion that a blend of acha malt with either a pearl millet or sorghum malt will produce a malt of the same profile as barley malt!

In a comparative study of lager beers brewed from barley, millet and sorghum malts, Agu[322] found that the non-barley beers were darker in colour and had a different flavour from "the norm". It was also found that the millet beer showed better head retention than the sorghum sample, but that the ethanol content of the latter was higher than in millet beer. The general opinion was that barley beer outscored the beers brewed from sorghum and millet.

With the cost of malting barley being around double the cost of millet, it makes sense in a developing country such as Nigeria to employ as much indigenous raw material as possible. With this in mind, the viability of locally available millet as an adjunct to barley in brewing was assessed.[323] Results indicated that a combination of 65% malted barley and 35% of millet would improve brewing capacity, reduce labour costs and shorten the brewing cycle, thus providing substantial financial savings. Reduction in costs was deemed to be superior to those obtained when either rice or maize were employed as adjuncts. From a brewing point of view, malted millets are covered in Dennis Briggs's excellent tome,[88] and have received detailed attention in a recent review of the uses and potential uses of sorghum and millets.[324]

4.7.8 An Archaeological Perspective

For several years, the cemetery at the Dawenkou site in Shandong has provided important evidence for social ranking and other aspects of social organization during the Neolithic period in China. Dawenkou (5000–3000 BC) was a millet-farming culture located on the low hills and lakeshores of the Shandong highland.[325] Foxtail millet was the dominant crop, but some broomcorn was grown. A distinctive feature of Dawenkou culture is the rich grave furnishings, and a large variety of pottery cooking, serving and storage vessels is found at Dawenkou sites, especially in burials. It appears that wheel-throwing technology was introduced during the late phase of the Dawenkou culture. There are many forms of tripod vessels and several kinds of decorative techniques. It seems as though tall-stemmed drinking cups, possibly for alcohol, had an important role in mortuary ritual and, if these cups were filled with an alcoholic drink, then it is reasonable to assume that millets were the source of the fermentable material.

In a paper that examined chronological changes in a number of aspects of mortuary ritual at the Dawenkou site, Anne Underhill[326] documented the presence/absence of drink-related burial artefacts – most notably those tall-stemmed (expensive) pottery vessels. Some scholars (Maisels,[327] quoting Gina Barnes) maintain that drinking rituals lay the basis for the ancestor rites of historical China. Underhill noted that over time burials were accompanied by an increasing number of pots, saying: "It is clear that by the Late Period, families using the cemetery at Dawenkou invested more pottery vessels for funerals. I suggest that during each period, families with greater resources (status as defined by wealth, at least) were able to feed more mourners at funerals and to provide more vessels for the deceased to use in the afterlife. At least some of these vessels contained food."

It was also evident that alcohol use in mortuary ritual was not restricted to one age or sex. Keightley[328] suggested that the location of tall-stemmed cups and other kinds of vessels for alcohol in graves can identify those vessels intended for the deceased *versus* those used by mourners during funerals. Chang[329] proposed that during the Shang Dynasty, shamans drank alcohol from bronze vessels to help transport them to the world of the gods and ancestors, and Underhill proposed that during the Neolithic Period, alcohol was favoured for its transformative properties (*i.e.* by making mourners feel that they could communicate with their deceased loved ones). She also notes that alcohol would have been produced by many families, and many kinds of containers could have been used to drink it, although the tall-stemmed cups were the most prestigious.

It seems as though the use of alcohol as a medium for communicating with ancestors continued throughout the early Bronze Age, for Paper[330] maintained that during the late Zhou period, an individual chosen to personify the dead drank alcohol so that the ancestral spirits could enter his or her body.

4.8 COMMON OAT

The common oat, *Avena sativa* L. (Figure 4.21), is a close companion of wheat and barley and a major cereal crop in Old World agriculture. Oats are easily distinguishable from other temperate cereals by the open, spreading panicle bearing large pendulous spikelets. The crop is successful in moist, temperate climates and, in northwest Europe at least, it often thrives better than wheat and is the main grain crop. It has a high nutritive value,

Figure 4.21 *Avena sativa* L. (from *Flora of North America*[387] - by kind permission).

with the grains containing *ca.* 15–16% protein and 8% fat. Oats can serve as a staple in the human diet and as a high-energy supplement for farm animals. Like rye, it may have occurred as a weed amongst other cereals before being cultivated, and it is known to have existed in Europe during the Bronze Age (although probably not established in Britain until the Iron Age). For this reason, oat (like rye) is often considered as a "secondary crop". Oat flour has anti-oxidant properties and is often mixed with other foodstuffs to help delay deterioration.

Of the three cytogenetically distinct stocks of *Avena* L., each with an independent origin, only one species, the hexaploid ($2n = 42$) common oat, *A. sativa*, established itself as a major cereal crop. The other two are crops of little significance, and are represented by the cultivated diploid ($2n = 14$) *A. strigosa* Schreb.

(including the diploid cultivars traditionally named *A. brevis* Roth) and the tetraploid ($2n = 28$) *A. abyssinica* Hochst. The former is grown as a fodder plant in some parts of Western Europe, and originated from wild forms of *A. strigosa*, a species widely spread over the western parts of the Mediterranean basin. The latter is a half-weed, half-crop confined to the highlands of Ethiopia, and is derived from *A. barbata* Pott., a wild and weedy tetraploid oat widely distributed in the Near East and in the Mediterranean basin.[45]

4.8.1 Domestication

The history of oat domestication parallels that of barley and wheat, the primary domesticated cereals of the Middle East, and the primacy of those two grasses in the Neolithic revolution was due to advantages that their progenitor species had over other local candidates for domestication, such as *A. sterilis* L. and *A. longiglumis* Dur. According to Bar Yosef and Kislev,[331] these "advantages" are: local abundance, large seed weight, absence of germination inhibitors and lower ploidy levels. In the archae-ological record, wild oat appears as a weed amongst cultivated cereal crops prior to, and for several centuries after, the Neolithic revolution. Non-domesticated *Avena* species have been identified from many sites in Greece, Iran, Israel, Jordan, Syria and Turkey, all dating from between *ca.* 10,500 and 5000 BC. It is still not possible to give a precise place and date for the domestication of this "weedy crop" but, assuming that it might have spread with wheat and barley, it is likely that oat germplasm held an adaptive advantage over that of wheat and barley in the cloudier, wetter and cooler environments of more northerly Europe. Yanushevich[332] reports of finds of *Avena* spp. in adobe imprints, dated as early as 4700 BC, from Moldavia and Ukraine, although it is not known whether these were cultivated (see below).

The main oat crop comprises numerous contrasting cultivars, all of which are interfertile with one another and share the same hexaploid genomic constitution. All are also characterized by non-shattering panicles. Traditionally, oat cultivars were placed in three "species" (in fact, only morphological types of the same species) according to their response to threshing. Most cultivars are hulled (*i.e.* the grains remain clothed by palea), and in some of these forms (*A. sativa sensu stricto*) threshing results in breakage of the spikelet's rachilla at the base of the upper floret. The rachilla segment then remains attached to the lower floret. In other cultivars (named *A. byzantina* C. Koch), the rachilla segment breaks at the base and remains attached to the upper floret. The basal floret thus exhibits a scar. In several hexaploid cultivars (often referred to as *A. nuda* L.), the grains are free and threshing releases the naked kernels.

As for the archaeological evidence, there is no sign of oat cultivation in Neolithic or Bronze Age sites in the Near East and Mediterranean basin. This is interesting, because it is known that wild *sterilis* oats are widely distributed over these areas and even grow together with wild wheats and wild barley. The few oat remains recovered from Neolithic Near East and European sites seem to represent only shattering wild or weedy *sterilis* or *fatua* forms. Definite indications of domestication (*i.e.* remains of non-shattering *sativa* or *byzantina* plants with their characteristic plump seed) appear first in Europe, but only in second and first millennia BC contexts. Among evidence for early oat cultivation are finds from first millennium BC Czech Republic and Slovakia.

4.8.2 Taxonomy and Genomics

The genus *Avena* comprises 30 different species of wild and domesticated annual grasses that form a distinct polyploidy series, with diploid, tetraploid and hexaploid forms.[333] The basic chromosome number is seven. On the basis of karyotype analysis and observations of meiotic chromosome pairing of inter- and intra-specific hybrids, four *Avena* genomes (A, B, C and D) have been identified and the genome combinations AA, CC, AABB, AACC and AACCDD are found in nature.[334] Structural rearrangements of the chromosomes, and a mechanism that restricts the pairing of chromosomes to homologues, mask the true identity of the genome donor(s) of the polyploids, and no diploid species has been unequivocally identified as being the progenitor of any polyploid *Avena* species.[335]

The cultivated oat is a natural allohexaploid that contains three genomes, A, C and D, and it has been suggested that its evolution involved two distinct steps. The first involved the establishment of the tetraploid AACC by the hybridization of two diploid species

(AA and CC), followed by doubling of the chromosome number. There was then a hybridization of this tetraploid with a third diploid species to form a hexaploid (by the doubling of the chromosomes of the resulting triploid hybrid).[336] The tetraploid species *A. murphyi* Ladiz. and *A. maroccana* Gdgr. (formerly described as *A. magna* Murphy et Terrell) were formerly favoured as the donors of the AC genomes to the hexaploids species. This conclusion has been based on chromosome pairing in hybrids between tetraploid and hexaploid oat and isoenzyme analysis. In 1998, Ladizinsky[337] reported on *Avena insularis*, a new species from Sicily that had greater pairing affinity with the hexaploid and was advanced as the most likely tetraploid ancestor of the hexaploids.

Molecular studies have now provided good evidence that the A and D genomes are very closely related, although it is possible that a D genome species never existed and that the D genome is a derived A genome.[338] No B genome species have been identified and one AABB species (*A. barbata* Brot.) is apparently a near autoploid arising from the AsAs genome (As = *A. strigosa*) genome. It is still unclear, however, which of the diploid species contributed to the polyploidy species. All hybrids between A genome diploids and the hexaploids have been produced,[335] but none of these hybrid combinations have shown the level of chromosome homology that would suggest any of them to be the donor of the A genome.

On the basis of karyotype, the three diploid C genome species have been separated into two genome types, Cv and Cp,[335] and both of them have been proposed as the putative donors of the C genome of the hexaploids. In 2000, Li *et al.*[339] reported the isolation of a satellite DNA sequence from *A. sativa*, and this appeared to identify the diploid and tetraploid progenitors of the crop. The satellite sequence, ASS49, was isolated from a microsatellite-enriched library and Southern hybridization showed that it was a species-specific rather than a genome-specific satellite. ASS49 was able to distinguish species that may be the diploid and tetraploid progenitors of the hexaploid oat. The phylogenetic relationship of *Avena* species was further investigated using 40 microsatellite and four minisatellite primers, and results obtained showed accord with the findings from ASS49. It appeared that the Ac genome diploid species, *A. canariensis* Baum Raj. Et Samp., is the progenitor and A-genome donor of the hexaploid oat rather than the As-genome species *A. strigosa*. Instead, *A. strigosa* seemed to be a member of a separate lineage of diploid and tetraploid species (which includes the tetraploid, *A. abyssinica*).

The primary cultivated species are the hexaploids, *A. sativa* L. and *A. byzantina* C. Koch, although five other species have been cultivated for human food. These are the diploids *A. strigosa* Schreb., *A. brevis* Roth., *A. hispanica* Ars. and *A. nuda* L., and the tetraploid *A. abyssinica* Hochst. The demarcation of the two cultivated hexaploids is based on small morphological differences, and is considered by many to be of more historical than contemporary relevance. *A. byzantina*, the "red oat", was the original germplasm base of most autumn-sown North American cultivars, while *A. sativa* was the germplasm base of spring-sown cultivars. Late twentieth-century breeding populations contain inter-crosses of both species, and this has led to the almost exclusive use of "*A. sativa*" when describing new cultivars.

Oats can be broadly classified as hulled and naked (hull-less), with the latter being nutritionally superior. Some would argue that this classification is an artificial one created by oat breeders, but there is more to it than that. Naked oats contain a higher percentage of β-glucan as compared to hulled oats, and this is by virtue of the fact that the hull contains negligible amounts of this polymer compared to the grain endosperm. Naked oats have a thin, non-lignified husk on the outside of the grain, which falls off during harvesting, resulting in a grain of higher energy, protein and lipid and lower fibre content compared to conventional oats.

Oat cultivars contain 3.0–5.0% of β-glucan,[340] with variations in its content being influenced by cultivar, and environmental and agricultural practices. Food processing profoundly affects the physical nature of oat β-glucan, making it more available at the site of action in the gut. Thus, the measured grain β-glucan content at the point of utilization for food product manufacture will be lower than that which finally reaches the site of action. Factors affecting β-glucan levels during oat cultivation have recently been studied.[341]

An evaluation of accessions of all *Avena* species has recently been presented by Loskutov.[342] The work discussed karyotype structure data, and the relationships of genomes of different species at each ploidy level were discussed. From a study of species diversity and the analysis of geographic distribution of oat forms and species, the A and C genomes were confirmed as being of prime importance in the formation of *Avena* spp. It was then suggested that species in the sub-section Aristulatae had reached their evolutionary climax and did not get involved in the formation of hexaploid plants. Diploid and tetraploid species of sub-section Avenae did participate in the evolution of wild and cultivated hexaploids.

The most likely centre of origin of the genus *Avena* L. was confirmed as lying in the western part of the Mediterranean where *A. byzantina* originated. The secondary centre for the formation of *Avena* L., and the origin of the cultivated oat, is in Asia Minor. Results from analysis of the intraspecific diversity of landraces enabled the centres of differentiation of all cultivated oat species to be ascertained. The centre of origin and diversity of the diploid *A. strigosa* was evidently the Iberian peninsula, for naked forms (*A. strigosa* subsp. *nudibrevis* (Vav.) Kobyl et Rod. – designated by Linnaeus as *A. nuda*) was the UK, for the tetraploid species *A. abyssinica* it is Ethiopia and for hexaploid species of *A. byzantina* it is Algeria and Morocco. The hulled forms of *A. sativa* subsp. *sativa* originated in Iran, Georgia and parts of Russia (Tatarstan), whilst the non-hulled forms (*A. sativa* subsp. *nudisativa* (Husnot.) Rod. et. Sold.) came from Mongolia and China. Zhou *et al.*[343] recently suggested that the two cultivated hexaploid oat species *A. byzantina* and *A. sativa* arose independently from *A. sterilis*.

As Igor Loskutov points out, there is no general agreement among researchers regarding the systematics of *Avena* L. The history of the systematic of this genus commenced about 300 years ago, when Linnaeus described four species: *A. sativa*, *A. fatua*, *A. sterilis* and *A. nuda*. Various taxonomic treatises have been produced since, perhaps the most significant being by the Russian botanist Regel, who, in 1880, developed a classification based on rachilla and lemma tip shape, lemma pubescence and the disarticulation of florets. These traits remain the key criteria in *Avena* L. taxonomy today. The most detailed classification, and still the most cited Avena monograph, was produced by Malzev in 1930.[344] By using lemma tip characters, he divided the genus into two sub-sections, Aristulatae and Denticulatae (= Avenae L.), and, in total, his classification included 7 *Avena* species, 22 sub-species and 184 varieties and forms. Apart from that of Ladizinsky and Zohary,[345] who invoke 7 species, in more modern classifications of oats, one sees an increase in the number of species, *i.e.* 14[346] and 34.[347]

4.8.3 History of Usage

Probably because of their rather late domestication (around AD 100?), oats meant very little to our ancient civilizations; they were despised by the Egyptians, and viewed only as animal fodder by the Greeks and Romans. There is no reference to the plant in the Bible but, by the first century AD, Roman writers began to make reference to it. The vigour and aggressive nature of the oat were noted by Pliny the Elder (AD 23–79), who, in *Natural History* (XXXVII, 3: 183), found it to be a particularly troublesome weed of cereal crops grown in damp environments. He also referred to an autumn-sown, non-shattering "Greek oat" used for forage. He thought that oats were a "diseased form of wheat", and the belief that oats were a degenerate form of another grain flourished well into the nineteenth century. Dioscorides described the medicinal qualities of oats and reported them to be a natural food for horses.

Jack Harlan[348] thought that oat domestication was a separate process for each ploidy level, with the diploids domesticated primarily as fodder crops in the Mediterranean area, and subsequently cultivated more widely throughout northern and Eastern Europe, particularly on poor, upland soils. The tetraploid *A. abyssinica*, found exclusively in the highlands of Ethiopia, was regarded as an intermediate form between a truly wild and a fully domesticated type. Harlan later[349] suggested that it is a tolerated admixture in cereal production because of the belief that it improves the quality of malt. The hexaploid progenitor of the cultivated hexaploid oat has long perplexed experts, and three species have been suggested, *A. sterilis* L., *A. hybrid* Petrem., *A. fatua* L. *A. maroccana* Gdgr. and *A. murphyi* Ladiz., which have a narrow geographic distribution (southern Spain and Morocco) and are believed to be the tetraploid base for cultivated hexaploids.[350]

In Europe, oats have provided human food and animal feed since Roman times and, presently, the crop ranks behind wheat and barley (and in some areas, rye) on that continent. Its importance in the human diet is greatest in Northern Europe, where, in the Middle Ages, a three-year crop rotation was fallow, followed by wheat, followed by oat or barley.

According to Coffman,[361] wild and cultivated oat arrived in the Americas by two routes. The Spaniards introduced *A. byzantina*, and the wild/weedy *A. fatua* and *A. barbata* Pott ex Link to southern latitudes, while the British (and other Europeans) brought in *A. sativa*, and probably *A. fatua*, to the northern colonies. The degree of usage varied in the Americas, often dependent upon the distribution of Scottish colonists, but Sturtevant[362] noted that Native Americans in California gathered wild oats and used them for bread-making. During the twentieth century, oat cultivation in North America has been concentrated in northcentral USA and prairies provinces of Canada, where spring-sown varieties predominate. If oats are grown in the southern and southwestern USA, they will likely be winter (autumn-sown) varieties. Most of the latter is grown for winter forage. Although the oat will thrive in cooler climates than wheat and barley, it is more at risk from freezing conditions than those two crops.

Five heterogeneous cultivars provide the foundation germplasm for spring oats in North America: "Kherson", "Green Russian", "White Russian" (all Russian), "Victory" (Swedish) and "Markton" (Greek). All were introduced during the late-nineteenth/early-twentieth centuries. Two heterogeneous cultivars, "Winter Turf" and "Red Rustproof", form the foundation germplasm for winter oats. The former was introduced from Northern Europe, likely in colonial times, while "Red Rustproof" came from Mexico, and became popular just after the Civil War.

The first written evidence for oat as a human food source is attributable to Pliny, who observed that "the people of Germany are in the habit of sowing oats, and make their porridge of nothing else". In a similar vein, oatmeal was an acknowledged Scottish staple as early as the fifth century AD, and by the Middle Ages the reputation of fondness for oats had become firmly attached to the Scots. The fourteenth century French historian and chronicler Jean Froissart attributed the strength and endurance of Scottish soldiers to "a diet of underdone meat and small, flat cakes made from a paste of oats and water rapidly cooked on a hot stove". He noted that Scottish soldiers carried with them, on their horses, bags of oats and metal plates on which to cook oatcakes. In Froissart's own words:

"These Scottish men are right hardy and sore trauelyng in harneys and in warres. For whan they wyll entre into Ingland within a daye and a nyght they wyll dryue theyr hole host xxiiii myle ... Therfore they cary with them none other purueyau[n]ce but on their horse bitwene the saddyll and the panel they trusse a brode plate of metall and behynde the saddyl they wyll haue a lytle sackefull of ootemele to yᵉ entent that whan they haue eaten of the sodden flesshe than they ley this plate on the fyre and temper a lytle of the ootemele & whan the plate is hote they cast of the thyn paste theron and so make a lytle cake in maner of a crackenell or bysket and that they eate to comfort with all theyr stomaks. Wherfore it is no great mueile [marvel] though they make greater journeys than other people do."

The importance of the crop to Scotland can be judged by the fact that, at one time, little else was cultivated. To combat this situation, on September 30th, 1426, the Scottish parliament enacted a law to promote more variety in crop production: "ilk man telande [tilling] with a pleuch of viij oxin sal saw at the lest ilk yere a firlot of peis & xl benis undir payn of x shillingis."

One of the most interesting early records concerning the relationship of Scots to the oat was penned by one John Major, a Scot who was a leading figure in the University of Paris. In his *History of Greater Britain*, of 1521,[363] the first history of Scotland written in a critical spirit, his defence of oats goes as follows:

"Wheat will not grow in every part of the island; and for this reason the common people use barley and oaten bread. And as many Britons are inclined to be ashamed of things nowise to be ashamed of, I will here insist a little ... Just as bread were Christ and his apostles wont to eat, as may be seen from Matthew xiv and John vi ... I say, for my part, that I would rather eat that British oaten bread than bread made of barley or of wheat. I nowhere remember to have seen on the other side of the water such good oats as in Britain, and the people make their bread in the most ingenious fashion ... Just eat this bread once, and you shall find it far from bad. It is the food of almost all the inhabitants of Wales, of the northern English, and of the Scottish peasantry; and yet the main strength of the Scottish and

English armies is in men who have been tillers of the soil – a proof that oaten bread is not a thing to be laughed at."

This defence of the oat was aimed at counteracting the slurs that he had heard whilst in France, including remarks made by one Frenchman who had returned from Scotland with some oaten bread and described it as a "monstrosity".

Some years later, Dr Samuel Johnson, in his *Dictionary of the English Language*, first published in 1755, says of the oat: "A grain, which in England is generally given to horses, but in Scotland supports the people." Evidence suggests that much of the Scottish diet during the eighteenth century was centred around the oat. The following year, Thomas Salmon wrote: "In the Highlands, I am informed, Oats is almost the only Grain that grows there, of which they make both Bread and Beer."

The Dicksons maintain that the species relevant in Scottish archaeology are common oat (*A. sativa*, cultivated); bristle oat, often called "black oat" (*A. strigosa*, cultivated) and wild oat (*A. fatua*, weedy). As they point out, "sativa" means cultivated, "strigosa" means bristly and "fatua" translates as insipid, or not good. They also aver that almost all discoveries of these oats in Scotland, and all of the cultivated oats, fall within the last 2500 years and most of them within the last 1500.

4.8.4 Beverages

Malted oats have been used sporadically by European brewers over the centuries (often mixed with barley), but are now rarely used, except for some speciality beers. Oats contain a large fraction of extract-poor husk (around 30% compared to 10% in barley), and so malted oats give relatively low extracts, often only about 70–75% of those provided by barley malts. Oat malts are also poor in α- and β-amylase and, since the grain is rich in β-glucans, it is necessary to modify it well to avoid producing highly viscous, slow-draining worts. In addition, oats are rich in oils and this is likely to cause off-flavours in beer – especially during storage. Kilning of oat malt is a slow process because of the extensive husk content.

On the credit side, oat grains take up water rapidly during steeping and malt well, and their high husk content makes oat malt ideally suited for brewing in an infusion mash-tun, where they encourage rapid wort separation. Indeed, in the past, oat husks were sometimes added to barley-malt mashes to aid wort run-off.

The nutritional value of the oat has been recognized for centuries, even in the absence of any "hard" evidence. From medieval times it was known that an oat-based diet was ideal for people required to undergo hard physical labour. We now know that oat starch is highly digestible and that oat is a good source of energy.

The primary constituent is starch, which constitutes from 45 to 62% of the groat by weight. This is lower than in other cereals because of the higher levels of fibre, protein and fat. Oat protein is higher than that of most other cereals (15–20%, groat basis) and contains a better balance of essential amino acids.[364] Having said that, lysine, threonine and methionine are present in less than optimal amounts. Oil content is around 5–9% and oat oil is nutritionally favourable because of a high proportion of unsaturated fatty acids, including the essential linoleic acid. Mineral content is fairly typical for cereal grains, and oat provides a significant proportion of Mn^{++}, Mg^{++} and iron and is also a source of Zn^{++}, Ca^{++} and Cu^{++}. The grain is high in phosphorus, but most of this element is bound as phytic acid, and thus unavailable. Oat is relatively vitamin-rich, with significant amounts of thiamine, folic acid, biotin, pantothenic acid and vitamin E, but contains little or no vitamins A, C and D.

Oatmeal caudle, made of ale and oatmeal with added spices, was a traditional British drink and a favourite of Oliver Cromwell.

For human food, the inedible hull (lemma and palea) must first be removed, leaving the groat (caryopsis) for further processing. A hull-less trait, governed by two or three genes, causes the caryopsis to thresh free of the hull, as does the wheat caryopsis.

The oat has been cultivated in China since early in the first millennium AD, and remains a staple food in northern China and Mongolia.

Odd Nordland,[365] in his classic anthropological exposition on brewing in Norway, explains how oat was generally regarded as a raw material for beer, and how it was used. Although barley was considered the best grain for brewing ale, oats were used (often as part of an admixture), and it was accepted that malting them required "more time" (in fact, people who were slightly slow on the uptake were often called "like oats malt, their reactions are slow"), and brewing with them "does not produce the same quantities of ale as barley". It seems that, in certain brewing districts, even some with a stable crop of barley, malt might be made solely from oat. In some parts of Norway, oats ale was considered inferior in taste, colour and strength. It also seems as though oats ale did not have the same keeping qualities as barley ale.

According to Worlidge,[366] oats "make indifferent good malt" though it was usual to incorporate a small quantity when making a strong beer for keeping (oat flour has antioxidant properties, which helps to delay deterioration when ale is in store).

Malted oats have been widely used as an ingredient for beer production, certainly since medieval times. Nowadays, oat malt is used in the brewing industry mainly as a flavour adjunct for the production of specialized products. The importance of the protein matrix surrounding the starch granules in the cereal endosperm is discussed elsewhere, and protein content and size distribution are of great interest in terms of fermentability, filterability and haze stability of a beer, as well as its propensity to foam. As compared to barley, oats has a unique protein composition and a high protein content (11–15%), of which around one-third passes into the final product.

In general, the classification of cereal proteins has been based on their solubility (Osborne fractionation). Thus albumins are water-soluble, globulins are saltwater-soluble, prolamins are soluble in dilute alcohol and glutelins are soluble in acids or bases. The difference between oats and other cereal grains in the structure of the proteinaceous components is consistent with the differences in the distribution of the protein fractions. Oats lack the protein matrix, characteristic of wheat and some other cereals and, in the latter, the storage proteins are insoluble in salt solutions. In oats, a large proportion of the saltwater-soluble globulins belong to the storage proteins of the endosperm. Oats contain a relatively low quantity of prolamins and a high level of globulins (up to 80% of the total oat protein[367]). The prolamins (avenins) of oats account for about 15% of the total protein and have, similar to other cereal prolamins, mainly a storage function. Glutelins have typically been difficult to completely solubilize and a variety of values have been obtained – according to extraction method. Most of the metabolically active proteins of oats are in the water-soluble albumin fraction, and oat albumins can account for 1–12% of total protein. Oat protein distribution also affects the amino acid composition of the crop, and so the generally higher lysine content of albumins and globulins causes the relatively high lysine content in oats (as compared to other cereals). In contrast, glutamic acid and proline contents are relatively lower.[367] The fundamental protein changes occurring during the malting of oats has been studied.[368] Their study revealed some, hitherto undescribed, low molecular weight proteins from the albumin and glutelin fraction.

Whether oats is gluten-free is not yet clear, but it seems as though it can be tolerated by most people with coeliac disease. In Finland, 69% of patients were reported as being able to use oat products,[369] and facts such as this, plus the fact that beer made from malted oats might be suitable for coeliacs, has stimulated new interest in oats.

Other health-related properties, such as high content of dietary fibre, especially β-glucan, as well as minerals and antioxidants, combine to make oats an increasingly more attractive crop. The most abundant antioxidants are vitamin E (tocols), phytic acid, phenolic compounds and avenanthramides, which are all concentrated in the outer layers of the kernel.[370]

Phytic acid (inositol hexakisphosphate), which is present in substantial amounts in oats, has been implicated as being the major factor responsible for impaired absorption of several essential elements, such as zinc, iron, magnesium and calcium.[371] Phytic acid (phytate) is generally regarded as the primary storage form of both phosphorus and inositol in seeds. Phytate concentration in cereals can be lowered by activation of the enzyme phytase, which hydrolyses phytate to phosphate and myoinositolphosphates, thereby increasing mineral availability. In a study on the changes in phytate content and phytase activity during the germination of some cereal grains,[372] a substantial increase in phytase activity was found. They also found that different grains exhibited differing levels of activity of the enzyme, with the highest being in rye, followed by wheat and barley, with the lowest in oats. It was suggested that high phytate levels combined with low phytase activity resulted in the mineral absorption capability of oats/oat products being less than half that of other cereals.

Larsson and Sandberg[373] investigated the effects of malting on phytate reduction in oats, and found that malting for five days at 11 °C and subsequent incubation for 17 hours at 37–40 °C reduced phytate content of grain by 98%. In a later communication, the same workers[374] examined the effects of heat treatment, storage and soaking conditions on oat phytate reduction.

Russia is the largest producer of oats, followed by Canada, and the USA. Today the grain is generally considered to be "healthy" and, with the discovery of its cholesterol-reducing capabilities, is now much more appreciated as a human food. Scotland 1, England 0!

4.9 RYE

Cultivated rye (*Secale cereale* L.) (Figure 4.22) is a cereal belonging to the tribe Triticae, and is thus closely related to the wheats. The connection is such that, in some schemes, rye has been included within the genus *Triticum*. The close relationship can also be demonstrated by the successful breeding of the hybrid cereal Triticale (*Triticum* x *Secale*; *Triticosecale* spp.), which is still grown today. In contrast to most grain crops, which are self-pollinating, rye flowers are cross-pollinated. *S. cereale* is a grain crop of temperate areas of the Old World, and is particularly appreciated in northern and Eastern Europe because of its winter hardiness, its resistance to drought and its ability to grow on acid, sandy soils. This means that it will survive in conditions where wheat is likely to fail.

4.9.1 Taxonomy

Four species are recognized in the genus *Secale*, and all are diploids with $2n = 14$ chromosomes. The centre of diversity of these plants in their wild form is in Southwest Asia. The four species are:

1. The crop complex *Secale cereale* L., which contains the cultivated varieties, as well as conspecific weedy races and wild forms. All are annual, self-compatible, chromosomally homologous and fully interfertile with one another. In the past, rye taxonomists split this complex into different species, but modern genetics has shown this to be unjustified. Nowadays, the complex can be broken down into four main "races":

   i) Cultivated varieties. These are non-shattering plants with large, plump grains.

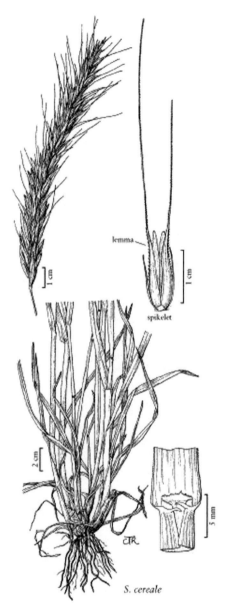

Figure 4.22 *Secale cereale* L. (from *Flora of North America*[387] - by kind permission).

    ii)   Non-shattering weeds. A variable aggregate of obligatory weeds infesting wheat cultivation in Turkey, adjacent areas in Syria, Iraq and Iran, as well as the Balkans, Caucasia and Transcaucasia. The mature ears of these plants do not shatter and, as they mimic wheat in grain size, they are harvested and threshed together with it. Traditional winnowing cannot separate rye from wheat, and so both are to be found in the harvest. Subsequently, the admixture will be sown for another crop. Ancient farmers would tolerate this because in poor growing years (cold and dry), the rye survived when the wheat did not. This sort of situation led to rye being often referred to as "wheat of Allah". When wheat and rye are harvested together, the mixture is said to be a "maslin crop".

iii)   Semi-shattering weeds. Common in northeast Iran, Armenia, Afghanistan and adjacent Central Asian republics. Partially shattering ryes can infest wheat and barley crops. Being semi-brittle, the upper part of the mature ear shatters spontaneously while the lower part remains intact and is harvested with the wheat/barley crop.

iv)   Fully shattering wild types. Here the rachis is fully fragile and the mature ear disarticulates spontaneously into individual spikelets, which serve as seed dispersal devices. The kernels are narrow and fully covered by brittle glumes. The first brittle *S. cereale* forms were discovered in western Turkey, and named *S. ancestrale*. Since this form infests many crops it is now regarded as a weed, rather than a genuinely wild type. Truly wild forms, chromosomally homologous with the crop, are known to exist. First discovered in Armenia, these brittle, annual, cross-pollinated forms have been named *S. vavilovii* Grossh. Subsequently, large populations of them have been found in eastern Turkey.[375] Since they are fully interfertile with cultivated rye, they are now regarded as a sub-species of *S. cereale*.

2.   *Secale montanum* Guss. This comprises a group of perennial, cross-pollinated forms native to elevated ground in the Near East (most notably Turkey), as well as Caucasia, north Iran, south Balkans, south Italy, Sicily and Morocco. Perennial *montanum* forms have a tufted growth habit and adpressed, shattering ears and, in terms of morphology and cytogenetics, they are quite distinct from the rye crop. They differ

from the domestic and wild forms of *S. cereale* by two chromosomal translocations,[376] yet hybrids between annual *S. cereale* and perennial *S. montanum* are easy to make and are only semi-sterile. As Zohary and Hopf[45] state, *S. montanum* grows side by side with *S. cereale* over huge areas of the plateaux of Anatolia. The former is found as a common grass in non-arable steppe habitats, and the latter as a weed in adjacent cultivated fields. Stutz[376] found that spontaneous hybridization between these two cross-pollinated grasses was rather frequent in contact places such as this, particularly at the edges of cultivation. Because the two species are not reproductively fully isolated from each other, they still exchange genes.

3. *Secale iranicum* Kobyl. is a poorly known wild annual, similar to the wild and weedy brittle forms of *S. cereale*. It is, however, divergent from the crop chromosomally. The species is self-pollinated, with small seeds, and it is isolated from *S. cereale* (and other species) by strong hybrid sterility barriers.

4. *Secale sylvestre* Host. is an annual, self-pollinating rye, with long awns, and is otherwise well separated morphologically from other rye species. It differs from *S. cereale* by three chromosomal translocations, and from *S. montanum* by one. The plant is native to the Aralo-Caspian basin.

Both *S. iranicum* and *S. sylvestre* are distant wild relatives of *S. cereale* and almost certainly played little or no part in rye domestication.

On a global scale rye is a minor crop, its production being around 5% that of wheat or rice. In Northern European countries, however, with extreme climatic and poor soil conditions, rye may occupy up to 30% of the acreage. Apart from the advantages mentioned above, rye is also known to have the lowest requirements for chemical treatments, like fertilizers or pesticides, which makes it an ecologically and economically sound crop for certain regions.

Although rye grains contain considerable amounts of proteins and can be baked into a dark-coloured bread, the baking quality of rye flour is inferior to that of wheat. Rye flour, however, has a lower gluten content than wheat flour. Most (more than 50%) of the present-day world production of rye is consumed as bread, but the grains are also used as a high-energy animal feed and for the manufacture of rye whiskey. Other uses for the plant include hay, pasture, cover crop, green fodder and green manure. The standard *schwartzbrot*, or pumpernickel, of Europe was once a major rye product. Despite its relatively low acreage compared to other cereals, rye is of great importance due to its broad tolerance to biotic and abiotic stress, a feature mostly lacking in other cereals. This makes rye an important crop for cool temperate zones.

Since medieval times, rye has been widely cultivated in central and Eastern Europe, and is the main bread cereal in most areas east of the French-German border and north of Hungary. Today, the main rye belt stretches from northern Germany through Poland, Ukraine, Belarus, Lithuania and Latvia into central and northern Russia. It is also grown in Canada and the USA, in Argentina and Brazil, and in Turkey, Kazakhstan and northern China. Production levels are gradually falling, however, with the largest producer, Russia, witnessing a fall from 13.9 million tons in 1992 to 3.4 million tons in 2005.

4.9.2 Genomics

Despite the economic importance of rye, little is known about its genetic make-up at the DNA sequence level, and the crop is underrepresented compared to wheat and barley, and this lack of information is a major limitation for marker development and gene cloning. The monoploid genomic size of rye (1Cx = 7917 Mbp) is the largest among temperate cereals, and almost 40% larger than that of bread wheat. The genome size of some important, closely related crops is given in Table 4.5.

Table 4.5 Genome size of major species of Triticeae.

Species	Chromosome number	1C genome size	1Cx genome size*
Rye	2n = 2x = 14 (RR)	7917 Mbp	7917 Mbp
Durum wheat	2n = 4x = 28 (AABB)	12,030 Mbp	6015 Mbp
Bread wheat	2n = 6x = 42 (AABBDD)	16,937 Mbp	5646 Mbp
Barley	2n = 2x = 14 (HH)	5100 Mbp	5100 Mbp

*Monoploid genome size (see Greilhuber *et al.* (2005)[377] for terminology on genome size). From Bartoš *et al.* 2008.[378]

This is due to the presence of a large amount of highly repetitive sequences, estimated to be 92%.[379] Sequencing the whole rye genome will be a difficult and expensive task, especially when it is realized that genome shotgun sequencing of such a large and repetitive genome currently seems impossible. To partially get around this barrier, molecular biologists are looking at part of the rye genome. The short arm of rye chromosome 1 (1RS) represents only 5.6% of the rye genome and with the molecular size of 441 Mbp is comparable in size to the whole of the rice genome (which has been sequenced, see p. 214). Recently, a method has been developed to dissect large plant genomes into individual chromosomes using flow cytometric sorting Doležel *et al*.[380,381] A protocol for sorting individual rye chromosomes has been established,[382] and Šimková *et al*.[383] created two BAC libraries from flow-sorted 1RS arms. The library represents a valuable tool for map-based cloning, targeted sequencing and marker development.

The short arm of rye chromosome 1 is rich in useful genes and, as it may increase yield, protein content and resistance to biotic and abiotic stress, it has been introgressed into wheat as the 1BL.1RS translocation.[384,385] In fact, some of the most successful wheat cultivars carry the 1BL.1RS translocation and, as well as increasing yield and protein content, it carries a cluster of genes encoding resistance to stem, leaf and yellow rust.

Jan Bartoš and co-workers[378] have provided results from the first survey of rye genome composition through BAC end sequencing of the short arm of chromosome 1R, in full cognizance of the fact that better knowledge of the rye genome will facilitate rye improvement and increase the efficiency of utilizing rye genes in wheat breeding. They provided new sequence data from the rye genome and molecular markers for chromosome 1R.

4.9.3 Some History

Pliny the Elder was not impressed with the crop, which is evident from: "It is a very poor food and only serves to avert starvation." He also reported that spelt was mixed in with it "to mitigate its bitter taste, and even then it is most unpleasant to the stomach" (*N.H.* 18.40).

Rye grains are naked, and when being malted all of the precautions applicable to malting wheat should be observed, although the end-product is quite different. On the plus side, the grains take up water rapidly (because of the lack of a hull and a thin bran layer) and have high potential extract values. Steeping time should be around one-third shorter than is employed for barley, and over-steeping causes the grain to go soft. The grains then are imperfectly germinated. The rye grain is more slender than wheat, with a pointed basal embryo end and a more angular, triangular cross-section. Some newer varieties are plumper and resemble wheat more closely. Little rye is now malted in the UK but elsewhere the high starch-degrading properties of rye malts makes them attractive for distillers employing adjunct-rich mashes.[88] Some extract values may be very high (values over 91% have been recorded), and α- and β-amylase activities comparable to barley malt are known. The grain never really found huge favour with brewers because of poor wort separation and beer haze instability. Although steeping times are short, one weeks' germination is normally necessary to complete modification during malting. Seedlings become detached quite readily during malting and, for this reason, rye and barley were sometimes mixed together for malting because it was believed that barley would protect the delicate rye acrospires! Rye is certainly no friend of the maltster.

According to Vavilov,[386] the progenitor of cultivated rye was a wild species, which entered wheat and barley fields as a weed in Southwest Asia. Through unconscious and conscious selection by man, non-brittle forms with bigger caryopses and suitable maturity time were derived. With the spread of wheat and barley towards the north and higher on the mountains, rye became a crop in its own right due to its adaptation to poorer soils and cooler climate.

Allowing for the fact that there is a wealth of wild and weedy forms of *S. cereale* in Turkey, Armenia and Iran, relatively few remains of *Secale* have been recovered from Neolithic and Bronze Age settlements in the Near East. The earliest remains come from Epi-Palaeolithic sites in northern Syria. Gordon Hillman found charred remains of what he considered to be cultivated rye in Aceramic early Neolithic layers of Tell Abu Hureya in northern Syria[96] and also at Can Hasan III in central Anatolia.[387] These remains are dated to around 6600 BC. At these early sites we don't know whether rye was deliberately sown as a crop in its own right, or whether it was only part of a maslin crop. Whatever, Hillman's

evidence contradicts Hans Helbaek's assertion[388] that rye originated in central rather than western Asia. Sencer and Hawkes[375]
are more specific and postulate that the geographic origin of
cultivated rye was in the Büyük Ağri Daği (Mt. Ararat) and Lake
Van area in eastern Turkey.

Outside of the Near East, early records of rye are fragmentary,
and the situation has been reviewed by Behre.[389] It seems as
though rye reached Europe, probably as a weed, at the dawn of
that continent's Neolithic Revolution. The earliest convincing
finds are from Neolithic Funnel Beaker (TRB) culture sites in
north, central and south Poland (van Zeist *et al.*,[266]). Only limited
numbers of rye grains were found on the Polish sites compared to
plentiful specimens of wheat and/or barley. More abundant rye
retrievals have been obtained from Bronze Age settlements in the
Czech Republic and Slovakia, and records of a similar age exist
from Romania, Moldavia and the Ukraine.[390] Later on, rye
appears in Iron Age settlements in Germany, Denmark, Poland
and the Crimea, and in all of these instances rye was usually
admixed with barley or wheat. Behre[389] compiled a list of the most
ancient finds of pure, or possibly pure, rye cultivated during the
Iron Age in Europe, and this shows concentrations in the eastern
Alps, the countries around the Black Sea and the western and
northern marginal areas of Europe.

There are no records of the crop from the late Neolithic of the
Aegean basin, Greece, Bulgaria and former Yugoslavia, which
would suggest that rye arrived in Europe *via* an easterly migration
from its homeland (*i.e. via* the Caucasus) rather than through the
Aegean basin and south Balkans. It is most likely that rye initially
evolved as a "tolerated weed" and then became adopted as a crop
in its own right. Rye was unable to exhibit its qualities in
Mediterranean climes and the grain was reportedly despised by the
ancient Greeks. There is no mention of rye from ancient Egyptian
sources.

Rye was certainly part of the grain agriculture of the Romans,
and was grown in their cooler northern provinces. This may have
been a response to an increase in population, and the ever-
increasing demand for food. Carbonized grains have been
recovered from several Roman frontier sites along the Rhine and
the Danube, and from the UK. Whether the Romans grew the
crop themselves is a moot point. They would not have been

familiar with the crop in their homelands, because climatic conditions were unfavourable. Behre reckons that in northern Germany, the shift to rye cultivation occurred during the second century AD. He maintains that the expansion of rye cultivation during the Roman period is due to improved harvesting methods beyond simply plucking grains ear by ear. Because all cultivated plants rely on harvesting for seed dispersal, such a thorough method would not have favoured the proliferation of rye. During Roman times, harvesting methods became more sophisticated and rye dispersal became more likely.

Rye became a very important crop in many parts of Europe during medieval times, especially where agriculture was being introduced to "marginal" montane landscapes. In many regions, rye replaced spelt[391] as a major crop. Agronomic considerations apart, there were practical reasons for this. Spelt is a hulled crop, and grains have to be dehusked after threshing. This is unnecessary with rye, or with wheat, but the latter is somewhat easily given to deterioration if it is then stored under unsuitable (damp) conditions. Rye does not suffer as badly, and so the crop was easier to store than wheat and easier to process than spelt. In some areas of Europe (*i.e.* around Lake Constance), rye never replaced spelt. As Behre states: "It is difficult to prove cultivation of rye because rye is a secondary crop plant which expanded originally as a weed in the cornfields and only later attained the status of a cultivated plant ... with archaeological material it is usually impossible to distinguish between the weed and the cultivated form of rye by purely morphological characteristics."

4.9.4   Rye in Beverages and as a Foodstuff

The importance of rye in parts of medieval Europe may be gleaned from the fact that some of the peoples in eastern Germany called themselves *Rugii* (rye-eaters), and this was principally so that the tribe could distinguish themselves from peoples who depended on more "lowly" crops, such as oats. In the first century AD the Rugians lived at the southern shore of the Baltic Sea and, in the late-fifth century AD joined the Ostrogoths. Seemingly related people, the *Rygir* lived in Rogaland, in southwestern Norway. The tribal name is a derivate of the Old Norse word for rye (*rugr*) and translates as "rye-eaters" or "rye farmers". Nordland confirms that rye was used for brewing in Norway, but was not as favoured as barley or oats.

Also, in Anglo-Saxon England, the month of August was called "Rugern" – the time of the rye harvest. The crop must have been important if it gave its name to a month. In England, archaeological evidence of rye cultivation can be found throughout the Anglo-Saxon period, and there is evidence that at least some of it was cultivated with wheat as a maslin crop.[392] With the sensitivity of rye to the fungus ergot (*Claviceps purpurea*), and all that entails, it is likely that this disease would have thrived in Britain's damp climate, and it is possible that this might have been a factor in the demise of the crop in those islands.

Rye reached the New World *via* colonists. In the north, the crop was cultivated extensively in New England and, as Humboldt found when he visited at the beginning of the nineteenth century, further south, in Mexico. In context of the latter, Humboldt observed that rye grew "at heights where cultivation of maize would be attended with no success". He also noted that rye was seldom attacked by a disease that in Mexico "frequently destroys the finest wheat harvests when the spring and the beginning of the summer have been very warm and when storms are frequent".

Julius Thausing[129] records that "rye is only used to a limited extent for brewing purposes, as for instance for the fabrication of some local beers in Belgium and Russia, and it is used as raw fruit [unmalted] as well as in the form of malt. The worts from rye are cloudy and of a mucous quality, the beer produced therefrom does not clarify and has a disagreeable, sharp taste". It was found that considerable amounts of butyric acid were formed in rye flour mixed at summer temperatures, and that the "peculiar sharp taste" of rye beer could be explained by the formation of this compound in rye mashes.

Over a century later, rye has recently been re-evaluated as a brewing ingredient especially amongst the burgeoning micro-brewing fraternity. Rye is now regarded as an "interesting" component contributing a distinctive, refreshing flavour to a beer. The subtlety of rye flavour is due to a variety of factors, including the amount of rye incorporated into the mash (10–20% seemingly being the norm), the form of the grain used (*i.e.* whole grain, malted, flake), the hopping level, type of yeast and other ingredients in the recipe. Experienced brewers of German rye beer (*roggenbier*) find that rye levels above 20% cause problems, not only during processing, but with beer taste.

The high β-glucan content of rye grains causes problems for the brewer because of increased wort viscosity (*i.e.* poor run-off and sparging difficulties). A "set" mash and filtration problems are not uncommon if care is not taken. To improve run-off capability, it is advisable to raise mash temperature a little and use a little more liquor (a "wet" mash). Sparge time should be extended. From a chemical point of view, rye worts contain less maltose and more nitrogen than barley malt worts, and rye malts tend to contain slightly more α-amylase (even though barley worts have slightly more diastatic power). In addition, rye also increases wort pH slightly, so adjustments may be necessary.

During the nineteenth century, *kvass*, a slightly alcoholic drink, was highly regarded in Russia, and became an important part of diet. The drink could be made from barley, oats, rye and wheat, and was most popular in central Russia and Siberia. The Canadians learned how to make whiskey from a rye mash many moons ago. That noted British brewer Lloyd Hind states quite definitely that "malted rye is not used in this country for brewing" … but that "un-malted rye is useful for vinegar brewing and distilling on account of the liquefying power of its diastase". Nowadays, rye is very much thought of as a distillers' grain, principally for the manufacture of whiskey and gin.

According to some Greek commentators, rye beer was a characteristic drink of certain ancient peoples. The late-fifth-century-BC historian Hellanicus of Lesbos reported that "a certain people (as Nelson says,[393] the name has been dislodged; he thinks that it may have been the Phrygians) drink *brūtos* [made] from rye, just as the Thracians drink it made from barley". Nelson also says it is unfortunate that Hellanicus's work on the customs of the barbarians no longer exists. He also reports that another Greek writer, Hecataeus of Miletus, in a work describing Europe from around 500 BC, says that barley *brūtos* was a drink of the Paeonians, a Thracian tribe. Thus *brūtos* was not specifically brewed with rye.

In 1256, the medical writer Aldobrandino of Siena wrote extensively about the effects of alcoholic drinks, and other substances, on the body in a book entitled *Le Régime du Corps*. Aldobrandino was born in Tuscany and, according to some sources, moved to France at an early age to be educated at the University of Paris. Some say he helped to found the medical school there. It seems fairly certain that he lived in Troyes from 1275–1296. From his writings, he was obviously well acquainted with beer, claiming that beer made with oats and wheat was better than that made from barley because "it did not cause as much gas". In his opinion, "beer made with rye, or rye bread, with mint and wild celery as additives, was the best kind". Aldobrandino's work is heavily influenced by previous writers, such as Hippocrates, Galen, Avicenna and Ali Abbas.

Around 300 years later, in 1588, Jacob Theodor von Bergzabern,[394] who was better known as Tabernaemontanus, produced *Neuw Kreuterbuch*, a comprehensive botanical encyclopaedia, which included much information about beer. His principal interest was in the plants used to flavour beer, many of which had known medicinal properties, but he also concerned himself with sources of fermentable material. In his opinion, "any two- or three-part combination of wheat, spelt, rye, or oats was best but that any one of these grains on its own would be acceptable".

Sahti, a full-bodied, turbid, top-fermented beer, is indigenous to Finland, and one of the most characteristic drinks in the world. It is reckoned to be the sole surviving "primitive" beer style in Western Europe, and is still often brewed in an artisanal (rustic) way that has not changed for many centuries. Sahti brewing largely managed to survive the introduction of industrial brewing technology into Scandinavia during the 1840s. The innovations came from Germany, and essentially took the form of an extended wort boil and the use of bottom-fermenting (lager) yeast. Henceforth, many Finns (and other Northern Europeans) became enamoured with the new-fangled, light, clear "Bavarian" beer.

In some rural areas, *sahti* is still the dominant alcoholic drink, and a few commercial samples are available as well. But, because it is a "live" beer, *sahti* does not lend itself to mass-production and reaches peak condition a couple of weeks after being brewed. Typically, its ABV will be in the 7–12% range, but can be stronger for special festive occasions. A less potent, partially fermented (*i.e.* sweet) variety is also brewed, which goes under the name *naisten sahti* (or "*sahti* for women").

Traditional brewing equipment is simple and consists of a wooden mash-tun, a wooden, trough-like device for transferring wort from mash-tun to fermenter and an upright, three- or four-legged wooden fermenter. The latter is fitted with a spigot near its base to enable *sahti* to be dispensed (*i.e.* straight from the fermenter). The wooden trough, which acts as a lautering device, is called the *kuurna* and is lined with a mesh of juniper twigs. It is arranged on a gradient between mash-tun and fermenter and has a bung hole at its bottom end to facilitate wort drainage into the fermenter. In olden times, the *kuurna* was formed from a hollowed-out tree trunk (often aspen, which is fine-grained and easy to work).

There are few documentations of ancient *sahti* brewing methodology, but it is evident that wort was/is not subjected to a boil after being separated from the mash. This is partly responsible for the turbidity and fullness of body of the beer, because proteins are not precipitated out. Turbidity is also caused by the poorly flocculent (baker's) yeast strains used for fermentation. Yeast cells remain in suspension and contribute flavour (fruity esters and phenolics) and haze to *sahti*. The overriding flavour in *sahti* is juniper, which can be imparted in a number of ways, including the addition of crushed berries. Hops are relatively unimportant, certainly in terms of bitterness, because, with the absence of a boil, there will be no means of isomerizing the bitter principles in the hop. If hops are used dry cones are usually placed in with the juniper twig lattice in the *kuurna*, and can only make a contribution towards aroma.

Historically, the main grain used for making *sahti* was rye, often with some contribution from barley, oats and wheat. Both malted and un-malted grain, and even bread, can be used as starter material. Extended step-infusion mashing was traditionally employed – with the wort-making process often taking around six hours. Because there is no wort boiling, lactobacilli invariably multiply during alcoholic fermentation and these impart the typical "lactic sourness" to *sahti*.

Finnish *sahti* has differing characteristics according to exactly where it was brewed. Prominent areas are Sysmä and Lammi, which are both in the region of Tavastia (historically an ancient province in Sweden).

Kvass is a traditional beverage primarily known in Russia, the Baltic States and the Ukraine, where it is predominantly consumed during the summer months. Ingredients and recipes are diverse, as would be expected for a product that had been made domestically for many years. Each family would have used their own ingredients, but there was one common factor; rye was always used in some form, even if it was first baked into bread. Historically, the term "kvass" first appears in the AD 989 document *Die Geschichten vergangener Jahr* (The History of Years Past), which tells us that, as part of the festivities associated with the baptism of Prince Vladimir in Kiev, "food, honey in barrels and bread-kvass" were distributed among the citizens of that town".[395]

Opinions vary as to the alcoholic strength of historical kvass samples, and it seems as though it depended on ingredients, microbial flora and duration and temperature of fermentation. That a wide variety of methods were used to brew kvass was remarked upon by the Swedish physicist and botanist Johann Georg Gmelin in his 1751 work, *Travels across Siberia from 1733 to 1743*.[396] Gmelin, who also wrote *Flora Sibirica*, reported that he was invariably served kvass with meals, and remarked: "In general, kvass refers to a somewhat sour drink made of flour, which one mixes with water and lets ferment, or it can also be made from unleavened bread, to which water is added and which under mildly warm conditions begins to ferment. Also, a very thin beer – made by pouring water over the spent grains left after brewing beer and then letting the water stand on them until it ferments – is served often instead of kvass."

Traditionally, kvass production in its Eastern European strongholds was carried out as part of everyday domestic activity, and this situation appertained well into the nineteenth century. Rye bread would be soaked in water and then fermented spontaneously *via* the activity of yeasts and LAB. After a number of days, the resulting liquid would be coarsely filtered and then consumed. Alcoholic content was usually fairly low (*ca.* 1.5% by volume), and the combination of lactic acid and CO_2 gave rise to a highly refreshing (especially in summer) drink. Flavourings were sometimes added, often in the form of berries, fruits, herbs or honey. Like most artisanally produced drinks, the shelf-life of kvass was limited to a few days.

The situation in Russia changed during the latter half of the nineteenth century when military engagements and large-scale projects, such as the construction of the Trans-Siberian Railway, demanded that victuals be provided over increasingly long periods of time to increasingly large numbers of people. With increasing

Figure 4.23   Kvass trailer in Russia.

industrialization within the Russian Empire, commercial kvass brewers started to emerge. Many kvass brewers tended to specialize in their use of raw materials, and over 150 varieties are recorded. Popular flavourings were apple, pear, mint, lemon, chicory, raspberry and cherry. Flavourings are normally added to the fermentation to effect maximum extraction. Gradually, domestic kvass brewing declined, especially when commercial brewers took to the streets and sold fresh product straight from the barrel.

The first scientific investigations into kvass production were carried out in Kiev during the 1890s, and 70 years later the Russian government – stimulated by the Coca-Cola Company – assigned chemists to look into ways of producing large volumes of kvass commercially. At this time (1961), the *per capita* consumption of kvass in Russia ranged from 60 to 100 litres annually. Large-scale kvass production soon followed and, in the early years, the end-product was put into large containers, usually on trailers (see Figure 4.23), and sold from street corners. Such a sight can still be seen today.

The simplest method developed for industrial kvass production involves the use of wort concentrate, which is a thick, highly concentrated liquid delivered in canisters. The concentrate must be warmed prior to use to reduce its viscosity and render it such that it can be pumped. It is then mixed with water and a sugar solution to create a liquid (wort) with a sugar concentration of 5–7%, at which point it is usually pasteurized to improve stability. The wort is pumped into a fermentation tank at a temperature of 12–30 °C, and fermented for 12–24 hours. During this time, only around 1% of the extract will be fermented out (ethanol, CO_2 and lactic acid). Bakers' yeast and LAB cultures are added to ensure quality. After fermentation, the kvass is cooled to 6 °C, and then clarified through filtration or centrifugation. If necessary, the sugar concentration can be adjusted after fermentation.

Kvass can also be made in a more conventional brewhouse, where the ingredients are milled and mashed. Raw materials often employed include rye, rye malt, barley, barley malt and other cereals such as rice and corn. Potato starch can also be used.

A "special kvass malt", which is based on rye and which gives kvass a unique aroma, is also used for some premium products. The Russian term for this malt translates as "fermented (rye) malt". During the malting of this "special" malt, the temperature is quickly increased by covering the grain towards the end of germination (basically stewing it). This imparts a dark brown colour, a sweet flavour and a bread-like aroma to the malt. To make kvass wort concentrate, such malt is usually mashed when still "green", although for a conventional mash, roasting will occur. Roasting imparts more complex flavours and aromas, and the final product is dark, sour malt with a distinctive "bread" aroma. The malt is very "glassy", and the endosperm highly porous – which can present problems in the brewhouse. The amount of kvass malt in a grist ranges from 20–50%, anything higher than this causes filtration problems because, with their glassiness and low friability, the milled kernels break into very fine particles. The high porosity of kvass malt endosperm results in poor sedimentation of grist particles in the lauter-tun, and the ingress of O_2 into the mash. The latter fact leads to a considerable increase in wort viscosity, caused by the O_2-dependent formation of a network of bonds (gel) among the arabinoxylans (pentosans) in the mash.[88] Because of the sensitivity of rye malt to O_2, it should be used immediately after milling. Failing this, the grist should be stored in an inert atmosphere.

Mashing method will be dependent on raw materials, and both infusion and decoction can be used. Decoction generally enhances the mash aroma, positively affecting the final product, and higher levels of adjunct can be efficiently used with this mashing technique. Appropriate enzymes are available for kvass production. Mash separation is either by lauter-tun or by mash filter, high viscosity being a problem for both methods. Centrifuges are used for producing kvass wort concentrate.

Boiling of kvass wort is for the same reasons as any beer (sterilization, enzyme inactivation, *etc.*), and after boiling hot trub can be removed by whirlpool. Cooling and fermentation are as for using wort concentrate.

REFERENCES

1. H. L. Shantz, *Ecology*, **35**, 1954, 143.
2. W. D. Clayton and S. A. Renvoize, *Genera Graminum*, HMSO, London, 1986.
3. L. Watson and M. J. Dallwitz, Grass genera of the world: descriptions, illustrations, identification, and information retrieval, including synonyms, morphology, anatomy, physiology, phytochemistry, cytology, classification, pathogens, world and local distribution, and references. http://biodiversityunoedu/delta/Version (August 18, 1999).
4. E. A. Kellogg and C. S. Campbell, in *Grass Systematics and Evolution*, ed. T. R. Soderstrom, K. W. Hilu, C. S. Campbell and M. E. Barkworth, Smithsonian Institute Press, Washington, DC, 1987, p. 310.
5. Grass Phylogeny Working Group (GPWP), in *Grasses: Systematics and Evolution*, ed. S. W. L. Jacobs and J. E. Everett, CSIRO Publishing, Collingwood, Victoria, 2000, p. 3.
6. H. P. Linder and I. K. Ferguson, *Grana*, 1985, **24**, 65.
7. B. F. Jacobs, J. D. Kingston and L. L. Jacobs, *Ann. Mo. Bot. Gard.*, 1999, **86**, 590.
8. H. P. Linder, *Kew Bulletin*, 1987, **42**, 297.
9. E. A. Kellogg, *Plant Physiol.*, 2001, **125**, 1198.
10. E. Weiss, W. Wetterstrom, D. Nadel and O. Bar-Yosef, *Proc. Nat. Acad. Sci. USA*, 2004, **101**, 9551.
11. F. J. Grundbacher, *Bot. Rev.*, 1963, **29**, 366.
12. M. A. Blumler, in *The Origins and Spread of Agriculture and Pastoralism in Eurasia*, ed. D. R. Harris, University College Press, London, 1996, p. 25.
13. H. V. Harlan, *The Identification of Varieties of Barley*, Bull. US Dept. Agriculture No. 622, 1918.
14. R. von Bothmer, N. Jacobsen, C. Baden, R. B. Jørgensen and I. Linde-Laursen, *An Ecogeographical Study of the Genus*

Hordeum, Systematic and Ecogeographic Studies on Crop Genepools 7, International Plant Genetics Resources Institute, FAO, Rome, 2nd edn, 1995.

15. R. T. Ramage, in *Barley*, ed. D. C. Rasmusson, Agronomy Series No. 26, American Society of Agronomy and Crop Science Society of America, Madison, WI, 1985, p. 127.

16. K. Kakeda and K. Fukui, *Jpn J. Genet.*, 1994, **69**, 545.

17. M. D. Bennett and J. B. Smith, *Phil. Trans. Roy. Soc. Lond. B.*, 1976, **274**, 227.

18. C. R. Burnham and A. Hagberg, *Heriditas*, 1956, **42**, 467.

19. I. Linde-Laursen, J. S. Heslop-Harrison, K. W. Shepherd and S. Taketa, *Hereditas*, 1997, **126**, 1.

20. M. Koornneef, C. Alonso-Blanco and A. J. M. Peeters, *New Phytologist*, 1997, **137**, 1.

21. E. Nevo, I. Apelbaum-Elkaher, J. Garty and A. Beiles, *Heredity*, 1997, **78**, 373.

22. J. H. Jørgensen, *Euphytica*, 1992, **63**, 141.

23. W. T. B. Thomas, E. Baird, J. D. Fuller, P. Lawrence, G. R. Young, J. Russell, L. Ramsay, R. Waugh and W. Powell, *Mol. Breed.*, 1998, **4**, 381.

24. R. P. Ellis, B. P. Forster, D. Robinson, L. L. Handley, D. C. Gordon, J. R. Russell and W. Powell, *J. Exp. Botany*, 2000, **51**, 9.

25. R. P. Ellis, E. Nevo and A. Beiles, *Plant Breed.*, 1993, **111**, 78.

26. M. J. Erkkila, R. Leah, H. Ahokas and V. Cameron-Mills, *Plant Physiol.*, 1998, **117**, 679.

27. P. M. Hayes, J. Cerono, H. Witsenboer, M. Kuiper, M. Zabeau, K. Sato, A. Kleinhofs, D. Kudrna, A. Kilian, M. Saghai-Maroof, D. Hoffman and the North American Barley Genome Mapping Project, *J. Quantitative Trait Loci*, 1997, **3**, 1.

28. W. T. B. Thomas, *Ann. Appl. Biol.*, 2003, **142**, 1.

29. S. J. Knapp, *Crop Sci.*, 1998, **38**, 1164.

30. J. S. Swanston, W. T. B. Thomas, W. Powell, G. R. Young, P. E. Lawrence, L. Ramsay and R. Waugh, *Mol. Breed.*, 1999, **5**, 103.

31. R. Cook, N. McCaig, J. M. B. McMillan and W. B. Lumsden, *J. Inst. Brew.*, 1990, **96**, 233.

32. S. J. Rae, M. Macaulay, L. Ramsay, F. Leigh, D. Matthews, D. M. O'Sullivan, P. Donini, P. C. Morris, W. Powell, D. F.

Marshall, R. Waugh and W. T. B. Thomas, *Euphytica*, 2007, **158**, 295.

33. M. Ayoub, E. Armstrong, G. Bridger, M. G. Fortin and D. E. Mather, *Crop Sci.*, 2003, **43**, 556.

34. J. R. Harlan and D. Zohary, *Science*, 1966, **153**, 1074.

35. D. Zohary, in *The Domestication and Exploitation of Plants and Animals*, ed. P. J. Ucko and G. W. Dimbleby, Duckworth, London, 1969, p. 47.

36. J. Johnson and E. Åberg, *J. Agron.*, 1943, **35**, 101.

37. R. Takahashi and J. Hayashi, *Berichte des Ohara Institute für Landwirtschaftliche Biologie*, Okayama University, 1964, **12**, 99.

38. R. Takahashi, in *Advances in Genetics*, ed. M. Demerec, Academic Press, New York, 1955, vol. 7, p. 227.

39. G. C. Hillman and M. S. Davies, *Journal of World Prehistory*, 1990, **4**, 157.

40. J. R. Harlan, in *Evolution in Crop Plants*, ed. N. W. Simmonds, Longmans, London, 1976, p. 93.

41. T. Komatsuda and Y. Mano, *Theor. Appl. Genet.*, 2002, **105**, 85.

42. E. Nevo, D. Zohary, A. D. H. Brown and M. Haber, *Evolution*, 1979, **33**, 815.

43. R. A. Nilan and S. E. Ullrich, in *Barley: Chemistry and Technology*, ed. A. W. MacGregor and R. S. Bhatty, St Paul, MN, 1993, p. 1.

44. E. Nevo, in *Barley: Genetics, Biochemistry, Molecular Biology and Biotechnology*, ed. P. R. Shewry, CAB International, Wallingford, Oxon., 1992, p. 19.

45. D. Zohary and M. Hopf, *Domestication of Plants in the Old World: The Origin and Spread of Cultivated Plants in West Asia, Europe and the Nile Valley*, Clarendon Press, Oxford, 3rd edn, 2000.

46. J. McCorriston, *Antiquity*, 1994, **68**, 97.

47. H. Helbaek, *Science*, 1959, **130**, 365.

48. D. Zohary, in *The Origins and Spread of Agriculture and Pastoralism in Eurasia*, ed. D. R. Harris, University College London Press, London, 1996, p. 142.

49. A. Badr, K. Müller, R. Schäfer-Pregl, H. El Rabey, S. Effgen, H. H. Ibrahim, C. Pozzi, W. Rohde and F. Salamini, *Mol. Biol. Evol.*, 2000, **17**, 499.

50. M. A. Saghai-Maroof, Q. Zhang, D. B. Neale and R. W. Allard, *Genetics*, 1992, **131**, 225.
51. L. Petersen, H. Østergård and H. Giese, *Theo. Appl. Genet.*, 1994, **89**, 676.
52. J. M. González and E. Ferrer, *Genome*, 1993, **36**, 1029.
53. I. K. Dawson, K. J. Chalmers, R. Waugh and W. Powell, *Mol. Ecol.*, 1993, **2**, 151.
54. S. Weining and R. J. Henry, *Genet. Resour. Crop Evol.*, 1995, **42**, 273.
55. B. R. Baum, E. Nevo, D. A. Johnson and A. Beiles, *Genet. Resour. Crop Evol.*, 1997, **44**, 147.
56. D. Saisho and M. D. Purugganan, *Genetics*, 2007, **177**, 1765.
57. P. L. Morrell and M. T. Clegg, *Proc. Nat. Acad. Sci. USA*, 2007, **104**, 3289.
58. H. Jones, F. J. Leigh, I. Mackay, M. A. Bower, L. M. J. Smith, M. P. Charles, G. Jones, M. K. Jones, T. A. Brown and W. Powell, *Mol. Biol. Evol.*, 2008, **25**, 2211.
59. A. Turner, J. Beales, S. Faure, R. P. Dunford and D. A. Laurie, *Science*, 2005, **310**, 1031.
60. J. Cockram, H. Jones, F. J. Leigh, D. O'Sullivan, W. Powell, D. A. Laurie and A. J Greenland, *J. Exp. Botany*, 2007, **58**, 1231.
61. P. Azhaguvel and T. Komatsuda, *Ann. Bot.*, 2007, **100**, 1009.
62. E. Åberg, *Ann. R. Agric. Col. Sweden*, 1938, **6**, 159.
63. E. Åberg, *Symbolae Botanicae Uppsalienses*, 1940, **4**, 1.
64. D. Zohary, *Evolution*, 1959, **13**, 279.
65. G. Staudt, *Econ. Bot.*, 1961, **15**, 205.
66. T. Xu, *Barley Genetics*, 1987, **5**, 91.
67. T. Komatsuda, M. Pourkheirandish, C. He, P. Azhaguvel, H. Kanamori, D. Perovic, N. Stein, A. Graner, T. Wicker, A. Tagiri, U. Lundqvist, T. Fujimura, M. Matsuoka, T. Matsumoto and M. Yano, *Proc. Nat. Acad. Sciences USA*, 2007, **104**, 1424.
68. G. Fischbeck, in *Barley Science: Recent Advances from Molecular Biology to Agronomy of Yield and Quality*, ed. G. A. Slafer, J. L. Molina-Cano, R. Savin, J. L. Araus and I. Romagosa, Food Products Press (Haworth Press Inc.), Binghampton, NY, 2002, p. 1.
69. S. A. Palmer, J. D. Moore, A. J. Clapham, P. Rose and R. J. Allaby, *PLoS ONE*, 2009, **4**, e6301.

70. I. S. Hornsey, *A History of Beer and Brewing*, Royal Society of Chemistry, Cambridge, 2003.
71. I. S. Hornsey, *Brewing*, Royal Society of Chemistry, Cambridge, 1999.
72. C. W. Bamforth, *Standards of Brewing: A Practical Approach to Consistency and Excellence*, Brewers Publications, Boulder, CO, 2002.
73. C. W. Bamforth, *Scientific Principles of Malting and Brewing*, ASBC, St Paul, MN, 2006a.
74. C. W. Bamforth, *Brewing: New Technologies*, Woodhead Publishing, Abington, Cambridge, 2006b.
75. C. W. Bamforth, I. Russell and G. G. Stewart, *Beer: A Quality Perspective*, Academic Press, San Diego, 2008.
76. D. E. Briggs, J. S. Hough, R. Stevens and T. W. Young, *Malting and Brewing Science, Volume 1: Malt and Sweet Wort*, Chapman & Hall, London, 2nd edn, 1981.
77. D. E. Briggs, C. A. Boulton, P. A. Brookes and R. Stevens, *Brewing: Science and Practice*, Woodhead Publishing, Abington, Cambridge, 2004.
78. J. S. Hough, D. E. Briggs, R. Stevens and T. W. Young, *Malting and Brewing Science, Volume 2: Hopped Wort and Beer*, Chapman & Hall, London, 2nd edn, 1982.
79. M. Nelson, *The Barbarian's Beverage: A History of Beer in Ancient Europe*, Routledge, London, 2005.
80. J. R. A. Pollock (ed.), *Brewing Science*, Academic Press, London, 1979, vol. 1.
81. J. R. A. Pollock (ed.), *Brewing Science*, Academic Press, London, 1981, vol. 2.
82. J. R. A. Pollock (ed.), *Brewing Science*, Academic Press, London, 1987, vol. 3.
83. F. G. Priest and I. Campbell (ed.), *Brewing Microbiology*, Kluwer Academic, New York, 3rd edn, 2003.
84. F. G. Priest and G. G. Stewart (ed.), *Handbook of Brewing*, Taylor & Francis, CRC Press, London & New York, 2nd revised edn, 2006.
85. A. H. Rose (ed.), *Economic Microbiology*, Academic Press, London, 1977, vol. 1.
86. R. W. Unger, *A History of Brewing in Holland 900–1900*, Brill, Leiden, 2001.

87. R. W. Unger, *Beer in the Middle Ages and the Renaissance*, University of Pennsylvania Press, Philadelphia, PA, 2004.

88. D. E. Briggs, *Malts and Malting*, Blackie, London, 1998.

89. A. De Candolle, *Origin of Cultivated Plants*, Paul, Trench, London, 2nd edn, 1886.

90. N. I. Vavilov, *Chronica Botanica*, 1951, **13**, 1.

91. R. J. Braidwood, *Sci. Am.*, 1960, **203**, 130.

92. J. MacKey, in *Proc. Int. Wheat Genetics Symposium*, Lund. *Hereditus*, Suppl. **2**, 1963. p. 237.

93. M. Nesbitt and D. Samuel, in *Hulled Wheats*. ed. S. Padulosi, K. Hammer and J. Heller, International Plant Genetic Resources Institute, Rome, 1996.

94. V. Jaaska, *Genet. Resour. Crop Evol.*, 1997, **44**, 137.

95. M. Nesbitt, in *Wheat Taxonomy: the Legacy of John Percival*, ed. P. D. S. Caligari and P. E. Brandham, Linnean Society, Special Issue 3, London, 2001, p. 37.

96. G. C. Hillman, *Proc. Prehist. Soc.*, 1975, **41**, 70.

97. G. C. Hillman, S. M. Colledge and D. R. Harris, in *Foraging and Farming: The Evolution of Plant Exploitation*, ed. D. R. Harris and G. C. Hillman, Unwin & Hyman, London, 1989, p. 240.

98. W. van Ziest, *J. Near E. Stud.*, 1970, **29**, 167.

99. M. A. Powell, *Bulletin on Sumerian Agriculture*, 1984, **1**, 48.

100. M. E. Kislev, D. Nadel and I. Carmi, *Review of Palaeobotany and Palynology*, 1992, **73**, 161.

101. M. A. Murray, in *Ancient Egyptian Materials and Technology*, ed. P. T. Nicholson and I. Shaw, Cambridge University Press, Cambridge, 2000, p. 505.

102. M. Hopf, in *Excavations at Jericho*, ed. K. M. Kenyon and T. A. Holland, British School of Archaeology in Jerusalem, London, 1983, vol. 5, p. 576.

103. R. H. Meadow, in *The Origins and Spread of Agriculture and Pastoralism in Eurasia*, ed. D. R. Harris, University College London Press, London, 1996, p. 390.

104. G. C. Hillman, in *Papers in Economic Prehistory*, ed. E. S. Higgs, Cambridge University Press, Cambridge, 1972, p. 182.

105. U. Maier, *Vegetation History and Archaeobotany*, 1996, **5**, 39.

106. E. S. McFadden and E. R. Sears, *J. Heredity*, 1946, **37**, 81, 107.

107. K. Kerby and J. Kuspira, *Genome*, 1987, **29**, 722.

108. E. R. Kerber and G. G. Rowland, *Canadian J. Genetics and Cytology*, 1974, **16**, 145.
109. G. Kimber and E. R. Sears, in *Wheat and Wheat Improvement*, ed. E. G. Heyne, American Society of Agronomy, Madison, WI, 2nd edn, 1987, p. 154.
110. W. van Zeist, *Phil. Trans. Roy. Soc. Lond., Series B*, 1976, **275**, 27.
111. J. M. Renfrew, *Bulletin on Sumerian Agriculture*, 1984, **1**, 32.
112. U. Körber-Grohne, in *L'epeautre* (Triticum spelta)*: histoire et ethnologie*, ed. J.-P. Devroey and J.-J. van Mol, Dire, Treignes, 1989, p. 51.
113. J. Dvorák, M.-C. Luo, Z.-L. Yang and H.-B. Zhang, *Theor. Appl. Genet.*, 1998, **97**, 657.
114. J. Dvorák, M.-C. Luo and Z.-L. Yang, in *The Origins of Agriculture and Crop Domestication*, ed. A. B. Damania, J. Valkoun, G. Willcox and C. O. Qualset, ICARDA, Aleppo, 1999, p. 235.
115. V. Jaaska, *Theor. Appl. Genet.*, 1980, **56**, 273.
116. V. Jaaska, *Theo. Appl. Genet.*, 1978, **53**, 209.
117. M. Rösch, S. Jacomet and S. Karg, *Vegetation History and Archaeobotany*, 1992, **1**, 193.
118. Y. Matsuoka, M. J. Aghaei, M. R. Abbasi, A. Totiaei, J. Mozafari and S. Ohta, *Genet. Resour. Crop Evol.*, 2008, **55**, 1573.
119. N. Stein, N. Ponelies, T. Musket, M. McMullen and G. Weber, *Plant J.*, 1998, **13**, 281.
120. J. Doležel, J. Macas and S. Lucretti, in *Current Protocols in Cytometry*, ed. J. P. Robinson, Z. Darzynkiewicz, J. Dobrucki, W. C. Hyun, J. P. Nolan, A. Orfao and P. S. Rabinovitch, John Wiley & Sons, New York, 1999, p. 5.3.1.
121. J. Doležel, S. Lucretti and I. Schubert, *Crit. Rev. Plant Sci.*, 1994, **13**, 275.
122. J. Doležel, M. Kubaláková, J. Bartoš and J. Macas, *Chromosome Res.*, 2004, **12**, 77.
123. A. V. Carrano, J. W. Gray, R. G. Langlois, K. J. Burkhart-Schultz and M. A. Van Dilla, *Proc. Nat. Acad. Sci. USA*, 1979, **76**, 1382.
124. J. Vrána, M. Kubaláková, H. Šimková, J. Cíhalíková, M. A. Lysák, J. Dolezel, *Genetics*, 2000, **156**, 2033.

125. J. Janda, J. Bartoš, J. Šafář, M. Kubaláková, M. Valárik, J. Číhaliková, H. Simková M. Caboche, P. Sourdille, M. Bernard, B. Chalhoub and J. Doležel, *Theor. Appl. Genet.*, 2004, **30**, 1337.

126. M. Kubaláková, P. Kovářová, P. Suchánková, J. Číhalíková, J. Bartoš, S. Lucretti, N. Watanabe, S. F. Kianian and J. Doležel, *Genetics*, 2005, **170**, 823.

127. M. Erayman, D. Sandhu, D. Sidhu, M. Dilbirligi, P. S. Baenziger and K. S. Gill, *Nucleic Acids Res.*, 2004, **32**, 3546.

128. J. L. Bennetzen, *Plant Mol. Biol.*, 2000, **42**, 251.

129. J. E. Thausing, *The Theory and Practice of the Preparation of Malt and the Fabrication of Beer*, Henry Carey Baird & Co., London and Philadelphia, 1882.

130. F. Delvaux, W. Gys, J. Michiels, F. R. Delvaux and J. A. Delcour, *J. Am. Soc. Brewing Chemists*, 2001, **59**, 135.

131. F. Delvaux, F. J. Combes and F. R. Delvaux, *MBAA Tech. Quarterly*, 2004, **41**, 27.

132. S. Hieronymous, *Brewing with Wheat*, Brewers Publications, Boulder, Colorado, 2010.

133. J. S. Hampl and W. S. Hampl, *J. Roy. Soc. Med.*, 1997, **90**, 636.

134. Y. Matsuoka, Y. Vigouroux, M. M. Goodman, J. Sanchez, E. S. Buckler, J. F. Doebley, *Proc. Nat. Acad. Sci. USA*, 2002, **99**, 6080.

135. D. R. Piperno and K. V. Flannery, *Proc. Nat. Acad. Sci. USA*, 2001, **98**, 2101.

136. J. F. Doebley, *Annu. Rev. Genet.*, 2004, **38**, 37.

137. Y. Matsuoka, *Breed. Sci.*, 2005, **55**, 383.

138. H. H. Iltis, *Science*, 1983, **222**, 886.

139. G. W. Beadle, *J. Heredity*, 1939, **30**, 245.

140. G. W. Beadle, *Field Museum Nat. Hist. Bull.*, 1972, **43**, 2.

141. G. W. Beadle, *Sci. Am.*, 1980, **242**, 112.

142. P. C. Mangelsdorf, *Adv. Genet.*, 1947, **1**, 161.

143. P. C. Mangelsdorf, *Corn: Its Origin, Evolution and Improvement*, Belknap Press, Cambridge, MA, 1974.

144. B. D. Smith, *The Emergence of Agriculture*, Scientific American Library, New York, 1998.

145. B. D. Walden, *Maize Breeding and Genetics*, John Wiley & Sons, Chichester, 1979.

146. M. I. Tenaillon, M. C. Sawkins, A. D. Long, R. L. Gaut, J. F. Doebley and B. S. Gaut, *Proc. Nat. Acad. Sci. USA*, 2001, **98**, 9161.

147. S. I. Wright, I. Vroh Bi, S. G. Schroeder, M. Yamasaki, J. F. Doebley, M. D. McMullen and B. S. Gaut, *Science*, 2005, **308**, 1310.

148. M. Yamasaki, M. I. Tenaillon, I. Vroh Bi, S. G. Schroeder, H. Sanchez-Villeda, J. F. Doebley, B. S.Gaut and M. D. McMullen, *Plant Cell*, 2005, **17**, 2859.

149. J. F. Doebley, *Genetics*, 2001, **158**, 487.

150. J. F. Doebley and A. Stec, *Genetics*, 1993, **134**, 559.

151. J. F. Doebley, A. Stec and C. Gustus, *Genetics*, 1995, **141**, 333.

152. J. F. Doebley, A. Stec and L. Hubbard, *Nature*, 1997, **386**, 485.

153. A. Gallavotti, Q. Zhao, J. Kyozuka, R. B. Meeley, M. Ritter, J. F. Doebley, M. E. Pe and R. J. Schmidt, *Nature*, 2004, **432**, 630.

154. H. Wang, T. Nussbaum-Wagler, B. Li, Q. Zhao, Y. Vigouroux, M. Faller, K. Bomblies, L. Lukens and J. Doebley, *Nature*, 2005, **436**, 714.

155. E. S. Buckler IV, J. M. Thornsberry and S. Kreshovich, *Genetic Res. Camb.*, 2001, **77**, 213.

156. M. A. Hanson, B. S. Gaut, A.O. Stec, S. I. Fuerstenberg, M. M. Goodman, E. H. Coe and J. F. Doebley, *Genetics*, 1996, **143**, 1395.

157. S. R. Whitt, L. M. Wilson, M. I. Tenaillon, B. S. Gaut and E. S. Buckler IV, *Proc. Nat. Acad. Sci. USA*, 2002, **99**, 12959.

158. K. A. Palaisa, M. Morgante, M. Williams and A. Rafalski, *Plant Cell*, 2003, **15**, 1795.

159. W. F. Tracy, S. R. Whitt and E. S. Buckler, *Crop Sci.*, 2006, **46**, S49.

160. J. R. Dinges, C. Colleoni, A. M. Myers and M. G. James, *Plant Physiol.*, 2001, **125**, 1406.

161. C. Correns, *Bibl. Botanica*, 1901, **53**, 1.

162. W. M. Hays, *Improving Corn – Cross Fertilization and Selection*, Bulletin 11, University of Minnesota Agric. Exptl. Station, St Paul, MN, 1890.

163. F. Tian, N. M. Stevens and E. S. Buckler IV, *Proc. Nat. Acad. Sci. USA*, 2009, **106**, 9979.

164. M. M. Rhoades, *Am. Nat.*, 1951, **85**, 105.
165. M. M. Goodman, C. W. Stuber, K. Newton and H. H. Weissinger, *Genetics*, 1980, **96**, 697.
166. J. Wendel, C. W. Stuber, M. Edwards and M. M. Goodman, *Theor. Appl. Gen.*, 1986, **72**, 178.
167. M. D. Gale and K. M. Devos, *Science*, 1998, **282**, 656.
168. G. Moore, K. M. Devos, Z. Wang and M. D. Gale, *Curr. Biol.*, 1995, **5**, 737.
169. B. S. Gaut and J. F. Doebley, *Proc. Nat. Acad. Sci. USA*, 1997, **94**, 6809.
170. N. A. Eckardt, *Plant Cell*, 2003, **15**, 1053.
171. Z. Swigonova, J. Lai, J. Ma, W. Ramakrishna, V. Liaca, J. L. Bennetzen and J. Messing, *Genome Res.*, 2004a, **14**, 1916.
172. F. Wei, E. Coe, W. Nelson, A. K. Bharti, F. Engler, E. Butler, H. Kim, J. L. Goicoechea, M. Chen, S. Lee, G. Fuks, H. Sanchez-Villeda, S. Schroeder, Z. Fang, M. McMullen, G. Davis, J. E. Bowers, A. H. Paterson, M. Schaeffer, J. Gardiner, K. Cone, J. Messing, C. Soderlund and R. A. Wing, *PLoS Genetics*, 2007, **3**, e123.
173. P. S. Schnable, D. Ware, R. S. Fulton, J. C. Stein, F. Wei, S. Pasternak, C. Liang, J. Zhang, L. Fulton, T. A. Graves, P. Minx, A. D. Reily, L. Courtney, S. S. Kruchowski, C. Tomlinson, C. Strong, K. Delehaunty, C. Fronick, B. Courtney, S. M. Rock, E. Belter, F. Du, K. Kim, R. M. Abbott, M. Cotton, A. Levy, P. Marchetto, K. Ochoa, S. M. Jackson, B. Gillam, W. Chen, L. Yan, J. Higginbotham, M. Cardenas, J. Waligorski, E. Applebaum, L. Phelps, J. Falcone, K. Kanchi, T. Thane, A. Scimone, N. Thane, J. Henke, T. Wang, J. Ruppert, N. Shah, K. Rotter, J. Hodges, E. Ingenthron, M. Cordes, S. Kohlberg, J. Sgro, B. Delgado, K. Mead, A. Chinwalla, S. Leonard, K. Crouse, K. Collura, D. Kudrna, J. Currie, R. He, A. Angelova, S. Rajasekar, T. Mueller, R. Lomeli, G. Scara, A. Ko, K. Delaney, M. Wissotski, G. Lopez, D. Campos, M. Braidotti, E. Ashley, W. Golser, H. Kim, S. Lee, J. Lin, Z. Dujmic, W. Kim, J. Talag, A. Zuccolo, C. Fan, A. Sebastian, M. Kramer, L. Spiegel, L. Nascimento, T. Zutavern, B. Miller, C. Ambroise, S. Muller, W. Spooner, A. Narechania, L. Ren, S. Wei, S. Kumari, B. Faga, M. J. Levy, L. McMahan, P. Van Buren, M. W. Vaughn, K. Ying,

C.-T. Yeh, S. J. Emrich, Y. Jia, A. Kalyanaraman, A.-P. Hsia, W. B. Barbazuk, R. S. Baucom, T. P. Brutnell, N. C. Carpita, C. Chaparro, J.-M. Chia, J.-M. Deragon, J. C. Estill, Y. Fu, J. A. Jeddeloh, Y. Han, H. Lee, P. Li, D. R. Lisch, S. Liu, Z. Liu, D. H. Nagel, M. C. McCann, P. SanMiguel, A. M. Myers, D. Nettleton, J. Nguyen, B. W. Penning, L. Ponnala, K. L. Schneider, D. C. Schwartz, A. Sharma, C. Soderlund, N. M. Springer, Q. Sun, H. Wang, M. Waterman, R. Westerman, T. K. Wolfgruber, L. Yang, Y. Yu, L. Zhang, S. Zhou, Q. Zhu, J. L. Bennetzen, R. K. Dawe, J. Jiang, N. Jiang, G. G. Presting, S. R. Wessler, S. Aluru, R. A. Martienssen, S. W. Clifton, W. R. McCombie, R. A. Wing and R. K. Wilson, *Science*, 2009, **326**, 1112.

174. T. Hariot, *A Brief and True Report of the New Found Land of Virginia*, London, 1588.

175. J. Winthorp, Jr., *Phil Trans. Roy. Soc. Lond.*, 1677–78, **12**, 1065.

176. T. Singh and G. S. Bains, *J. Agric. Food Chem.*, 1984, **32**, 346.

177. A. V. Aderinola, *Discov. Innovat.*, 1992, **4**, 103.

178. G. N. Aniche, *Process Biochem.*, 1989, **24**, 183.

179. T. T. Chang, in *Evolution of Crop Plants*, ed. J. Smartt and N. W. Simmonds, Longman, Harlow, 2nd edn, 1995, p. 147.

180. C. Higham and T. L.-D. Lu, *Antiquity*, 1998, **72**, 867.

181. I. C. Glover and F. W. Higham, in *The Origins and Spread of Agriculture and Pastoralism in Eurasia*, ed. D. R. Harris, UCL Press, London, 1996, p. 413.

182. T. T. Chang, *Interdiscipl. Sci. Rev.*, 1987, **12**, 63.

183. D. A. Vaughan and H. Morishima, in *Rice: Origin, History, Technology, and Production*, ed. C. W. Smith and R. H. Dilday, John Wiley & Sons, New York, 2003, p. 27.

184. D. A. Vaughan, B.-R. Lu and N. Tomooka, *Plant Sci.*, 2008a, **174**, 394.

185. D. A. Vaughan, B.-R. Lu and N. Tomooka, *Rice*, 2008b, **1**, 16.

186. J. P. Londo, Y.-C. Chiang, K.-H. Hung, T.-Y. Chiang and B. A. Schaal, *Proc. Nat. Acad. Sci. USA*, 2006, **103**, 9578.

187. A. J. Garris, T. H. Tai, J. Coburn, S. Kresovich and S. McCrouch, *Genetics*, 2005, **169**, 1631.

188. S. Chen, Y. Yang, W. Shi, Q. Ji, F. He, Z. Zhang, Z. Cheng, X. Liu and M. Xu, *Plant Cell*, 2008, **20**, 1850.

189. H. I. Oka, *Origin of Cultivated Rice*, Elsevier, Amsterdam, 1988.
190. T. T. Chang, *Genome*, 1989, **31**, 825.
191. X. Chen, *Bull. Indo-Pacific Prehistory Assn.*, 1999, **18**, 81.
192. S. P. Joshi, V. S. Gupta, R. K. Aggarwal, P. K. Ranjekar and D. S. Brar, *Theor. Appl. Genet.*, 2000, **100**, 1311.
193. H. I. Oka and H. Morishima, *Euphytica*, 1982, **31**, 41.
194. S. Kato, H. Kosaka and S. Hara, *Bull. Sci. Fac. Agric. Kyushu Univ.*, 1928, **3**, 132.
195. G. Second, *Jpn J. Genet.*, 1982, **57**, 25.
196. C. Vitte, T. Ishii, F. Lamy, D. Brar and O. Panaud, *Mol. Gen. Genomics*, 2004, **272**, 504.
197. B. A. Schaal and K. M. Olsen, *Proc. Nat. Acad. Sci. USA*, 2000, **97**, 7024.
198. L. Gao and H. Innan, *Genetics*, 2008, **179**, 965.
199. A. Vicentini, J. C. Barber, S. A. Aliscioni, L. M. Giussani and E. A. Kellogg, *Global Change Biol.*, 2008, **14**, 2963.
200. T. Takeota, *Bot. Gazette*, 1963, **124**, 264.
201. L. Watson and M. J. Dallwitz, *The Grass Genera of the World*, CAB International, Wallingford, UK, 1992.
202. E. A. Kellogg, *Rice*, 2009, **2**, 1.
203. Grass Phylogeny Working Group (GPWP), *Ann. Mo. Bot. Gard.*, 2001, **88**, 373.
204. Y.-L. Guo and S. Ge, *Am. J. Bot.*, 2005, **92**, 1548.
205. G. Second, in *Geographic Origins, Genetic Diversity and the Molecular Clock Hypothesis in the Oryzeae*, ed. P. Jacquard, G. Heim and J. Antonovics, Springer-Verlag, Berlin, 1985, p. 41.
206. J. Yu, S. Hu, J. Wang, G. K. Wong, S. Li, B. Liu, Y. Deng, L. Dai, Y. Zhou, X. Zhang, M. Cao, J. Liu, J. Sun, J. Tang, Y. Chen, X. Huang, W. Lin, C. Ye, W. Tong, L. Cong, J. Geng, Y. Han, L. Li, W. Li, G. Hu, X. Huang, W. Li, J. Li, Z. Liu, L. Li, J. Liu, Q. Qi, J. Liu, L. Li, T. Li, X. Wang, H. Lu, T. Wu, M. Zhu, P. Ni, H. Han, W. Dong, X. Ren, X. Feng, P. Cui, X. Li, H. Wang, X. Xu, W. Zhai, Z. Xu, J. Zhang, S. He, J. Zhang, J. Xu, K. Zhang, X. Zheng, J. Dong, W. Zeng, L. Tao, J. Ye, J. Tan, X. Ren, X. Chen, J. He, D. Liu, W. Tian, C. Tian, H. Xia, Q. Bao, G. Li, H. Gao, T. Cao, J. Wang, W. Zhao, P. Li, W. Chen, X. Wang, Y. Zhang, J. Hu, J. Wang, S. Liu, J. Yang,

G. Zhang, Y. Xiong, Z. Li, L. Mao, C. Zhou, Z. Zhu, R. Chen, B. Hao, W. Zheng, S. Chen, W. Guo, G. Li, S. Liu, M. Tao, J. Wang, L. Zhu, L. Yuan and H. Yang, *Science*, 2002, **296**, 79.

207. S. A. Goff, D. Ricke, T.-H. Lan, G. Presting, R. Wang, M. Dunn, J. Glazebrook, A. Sessions, P. Oeller, H. Varma, D. Hadley, D. Hutchison, C. Martin, F. Katagiri, B. M. Lange, T. Moughamer, Y. Xia, P. Budworth, J. Zhong, T. Miguel, U. Paszkowski, S. Zhang, M. Colbert, W. L. Sun, L. Chen, B. Cooper, S. Park, T. C. Wood, L. Mao, P. Quail, R. Wing, R. Dean, Y. Yu, A. Zharkikh, R. Shen, S. Sahasrabudhe, A. Thomas, R. Cannings, A. Gutin, D. Pruss, J. Reid, S. Tavtigian, J. Mitchell, G. Eldredge, T. Scholl, R. M. Miller, S. Bhatnagar, N. Adey, T. Rubano, N. Tusneem, R. Robinson, J. Feldhaus, T. Macalma, A. Oliphant and S. Briggs, *Science*, 2002, **296**, 92.

208. Y. Harushima, M. Yano, A. Shomura, M. Sato, T. Shimano, Y. Kuboki, T. Yamamoto, S. Y. Lin, B. A. Antonio, A. Parco, H. Kajiya, N. Huang, K. Yamamoto, Y. Nagamura, N. Kurata, G. S. Khush and T. Sasaki, *Genetics*, 1998, **148**, 479.

209. K. Shimamoto and J. Kyozuka, *Ann. Rev. Plant Biol.*, 2002, **53**, 399.

210. Q. Feng, Y. Zhang, P. Hao, S. Wang, G. Fu, Y. Huang, Y. Li, J. Zhu, Y. Liu, X. Hu, P. Jia, Y. Zhang, Q. Zhao, K. Ying, S. Yu, Y. Tang, Q. Weng, L. Zhang, Y. Lu, J. Mu, Y. Lu, L. S. Zhang, Z. Yu, D. Fan, X. Liu, T. Lu, C. Li, Y. Wu, T. Sun, H. Lei, T. Li, H. Hu, J. Guan, M. Wu, R. Zhang, B. Zhou, Z. Chen, L. Chen, Z. Jin, R. Wang, H. Yin, Z. Cai, S. Ren, G. Lv, W. Gu, G. Zhu, Y. Tu, J. Jia, Y. Zhang, J. Chen, H. Kang, X. Chen, C. Shao, Y. Sun, Q. Hu, X. Zhang, W. Zhang, L. Wang, C. Ding, H. Sheng, J. Gu, S. Chen, L. Ni, F. Zhu, W. Chen, L. Lan, Y. Lai, Z. Cheng, M. Gu, J. Jiang, J. Li, G. Hong, Y. Xue and B. Han, *Nature*, 2002, **420**, 316.

211. T. Sasak, T. Matsumoto, K. Yamamoto, K. Sakata, T. Baba, Y. Katayose, J. Wu, Y. Niimura, Z. Cheng, Y. Nagamura, B. A. Antonio, H. Kanamori, S. Hosokawa, M. Masukawa, K. Arikawa, Y. Chiden, M. Hayashi, M. Okamoto, T. Ando, H. Aoki, K. Arita, M. Hamada, C. Harada, S. Hijishita,

M. Honda, Y. Ichikawa, A. Idonuma, M. Iijima, M. Ikeda, M. Ikeno, S. Ito, T. Ito, Y. Ito, Y. Ito, A. Iwabuchi, K. Kamiya, W. Karasawa, S. Katagiri, A. Kikuta, N. Kobayashi, I. Kono, K. Machita, T. Maehara, H. Mizuno, T. Mizubayashi, Y. Mukai, H. Nagasaki, M. Nakashima, Y. Nakama, Y. Nakamichi, M. Nakamura, N. Namiki, M. Negishi, I. Ohta, N. Ono, S. Saji, K. Sakai, M. Shibata, T. Shimokawa, A. Shomura, J. Song, Y. Takazaki, K. Terasawa, K. Tsuji, K. Waki, H. Yamagata, H. Yamane, S. Yoshiki, R. Yoshihara, K. Yukawa, H. Zhong, H. Iwama, T. Endo, H. Ito, J. Hahn, H. Kim, M. Eun, M. Yano, J. Jiang and T. Gojobori, *Nature*, 2002, **420**, 312.

212. Rice Chromosome 10 Sequencing Consortium, *Science*, 2003, **300**, 1566.

213. B. Han, Y. Xue, J. Li, X.-W. Deng and Q. Zhang, *Phil. Trans. Roy. Soc. Lond. B*, 2007, **362**, 1009.

214. International Rice Genome Sequencing Project, *Nature*, 2005, **436**, 793.

215. T. Sasaki and B. Burr, *Curr. Opin. Plant Biol.*, 2000, **3**, 138.

216. J. Yu, J. Wang, W. Lin, S. Li, H. Li, J. Zhou, P. Ni, W. Dong, S. Hu, C. Zeng, J. Zhang, Y. Zhang, R. Li, Z. Xu, S. Li, X. Li, H. Zheng, L. Cong, L. Lin, J. Yin, J. Geng, G. Li, J. Shi, J. Liu, H. Lv, J. Li, J. Wang, Y. Deng, L. Ran, X. Shi, X. Wang, Q. Wu, C. Li, X. Ren, J. Wang, X. Wang, D. Li, D. Liu, X. Zhang, Z. Ji, W. Zhao, Y. Sun, Z. Zhang, J. Bao, Y. Han, L. Dong, J. Ji, P. Chen, S. Wu, J. Liu, Y. Xiao, D. Bu, J. Tan, L. Yang, C. Ye, J. Zhang, J. Xu, Y. Zhou, Y. Yu, B. Zhang, S. Zhuang, H. Wei, B. Liu, M. Lei, H. Yu, Y. Li, H. Xu, S. Wei, X. He, L. Fang, Z. Zhang, Y. Zhang, X. Huang, Z. Su, W. Tong, J. Li, Z. Tong, S. Li, J. Ye, L. Wang, L. Fang, T. Lei, C. Chen, H. Chen, Z. Xu, H. Li, H. Huang, F. Zhang, H. Xu, N. Li, C. Zhao, S. Li, L. Dong, Y. Huang, L. Li, Y. Xi, Q. Qi, W. Li, B. Zhang, W. Hu, Y. Zhang, X. Tian, Y. Jiao, X. Liang, J. Jin, L. Gao, W. Zheng, B. Hao, S. Liu, W. Wang, L. Yuan, M. Cao, J. McDermott, R. Samudrala, J. Wang, G. Ka-Shu Wong and H. Yang, *PLoS Biology*, 2005, **3**, e38.

217. A. H. Paterson, J. E. Bowers and B. A. Chapman, *Proc. Nat. Acad. Sci. USA*, 2004, **101**, 9903.

218. K. M. Devos, *Curr. Opin. Plant Biol.*, 2005, **8**, 155.

219. N. M. Nayar, *Adv. Genet.*, 1973, **17**, 153.
220. R. K. Aggarwal, D. S. Brar and G. S. Khush, *Mol. Gen. Genet.*, 1997, **254**, 1.
221. S. Ge, T. Sang, B.-R. Lu and D.-Y. Hong, *Proc. Nat. Acad. Sci. USA*, 1999, **96**, 14400.
222. X. H. Zou, F. M. Zhang, J. G. Zhang, L.-L. Zang, L. Tang, J. Wang, T. Sang and S. Ge, *Genome Biol.*, 2008, **9**, R49.
223. J. L. Bennetzen, *Curr. Opin. Plant Biol.*, 2007, **10**, 176.
224. H. Tang, J. E. Bowers, X. Wang, R. Ming, M. Alam and A. H. Paterson, *Science*, 2008, **320**, 486.
225. J. S. S. Ammiraju, F. Lu, A. Sanyal, Y. Yu, X. Song, N. Jiang, A. C. Pontaroli, T. Rambo, J. Currie, K. Collura, J. Talag, C. Fan, J. L. Goicoechea, A. Zuccolo, J. L. Bennetzen, M. Chen, S. Jackson and R. A. Wing, *Plant Cell*, 2008, **20**, 3191.
226. A. P. Tikhonov, P. J. SanMiguel, Y. Nakajima, N. M. Gorenstein, J. L. Bennetzen and Z. Avramova, *Proc. Nat. Acad. Sci. USA*, 1999, **96**, 7409.
227. R. Tarchini, P. Biddle, R. Wineland, S. Tingey and A. Rafalski, *Plant Cell*, 2000, **12**, 381.
228. S. Morewood, *An Essay on the Inventions and Customs of Both Ancients and Moderns in the Use of Inebriating Liquors*, Printed for Longman, Hurst, Rees, Orme, Brown and Green, Paternoster Row, London, 1824.
229. C. W. Bamforth, *Food, Fermentation and Micro-organisms*, Blackwell, Oxford, 2005.
230. V. A. Casal, *Trans. Asiatic Soc. Japan*, 2nd series, 1940, **19**, 1.
231. K. H. Steinkraus (ed.), *Handbook of Indigenous Fermented Foods*, Marcel Dekker Inc., New York, 2nd edn, 1996.
232. G. Campbell-Platt, *Fermented Foods of the World*, Butterworths, Sevenoaks, 1989.
233. Y. Teramoto, S. Yoshida and S. Ueda, *World J. Microbiol. Biotechnol.*, 2002, **18**, 813.
234. Y. L. Nene, *Asian Agri-History*, 2005, **9**, 85.
235. J. R. Harlan and J. M. J. de Wet, *Taxon*, 1971, **20**, 509.
236. W. C. MacLean, G. Lopez de Romaña, R. P. Placko and G. G. Graham, *J. Nutrition*, 1981, **111**, 1928.
237. S. Haggblade and W. H. Holzapfel, in *Industrialization of Indigenous Fermented Foods*, ed. K. H. Steinkraus, Marcel Dekker, New York, 1989.

238. H. M. Schwartz, *J. Sci. Food Agric.*, 1956, **7**, 101.
239. B. S. Platt and R. A. Webb, *Proc. Nutr. Soc.*, 1946, **4**, 132.
240. T. G. Watson, *J. Appl. Bacteriol.*, 1975, **38**, 133.
241. M. C. J. Aucamp, J. T. Grieff, L. Novellie, B. Papendick, H. N. Schwartz and A. G. Steer, *J. Sci. Food Agric.*, 1961, **12**, 449.
242. C. C. Ng'andwe, A. N. Hall and J. R. N. Taylor, *J. Inst. Brew.*, 2008, **114**, 343.
243. D. E. Briggs, A. Wadeson, R. Statham and J. F. Taylor, *J. Inst. Brew.*, 1986, **92**, 468.
244. S. M. Fors and H. Nordlöv, *J. Inst. Brew.*, 1987, **93**, 496.
245. C. I. Owuama, *J. Inst. Brew.*, 1999, **105**, 23.
246. N. J. De Mesa-Stonestreet, S. Alavi and S. R. Bean, *J. Food Sci.*, 2010, **75**, R90.
247. R. Zhao, S. R. Bean, B. P. Ioerger, D. Wang and D. L. Boyle, *J. Agric. Food Chem.*, 2008, **56**, 946.
248. N. E. Yousif and A. H. El Tinay, *Plant Foods Hum. Nutr.*, 2001, **56**,175.
249. S. A. Odunfa, in *Microbiology of Fermented Food*, ed. B. J. B. Wood, Elsevier, Amsterdam, 1985, vol. 2, p. 155.
250. B. Demuyakor and Y. Ohta, *Food Microbiol.*, 1991, **8**, 183.
251. B. Demuyakor and Y. Ohta, *J. Sci. Food Agr.*, 1993, **62**, 401.
252. S. Konlani, J. P. Delgenes, R. Moletta, A. Traoré and A. Doh, *Food Biotechnol.*, 1996, **10**, 29.
253. H. Sawadogo-Lingani, V. Lei, B. Diawara, D. S. Nielsen, P. L. Møller, A. S. Traoré and M. Jakobsen, *J. Appl. Microbiol.*, 2007, **103**, 765.
254. S. Sefa-Dedeh, A. I. Sanni, G. Tetteh and E. Sakyi-Dawson, *World J. Microbiol. Biotechnol.*, 1999, **15**, 593.
255. J. A. Ekundayo, *Int. J. Food Sci. Tech.*, 1969, **4**, 217.
256. E. O. Onaghise and Y. S. Izuagbe, *Acta Biotechnologia*, 1989, **9**, 137.
257. A. H. Paterson, J. E. Bowers, R. Bruggmann, I. Dubchak, J. Grimwood, H. Gundlach, G. Haberer, U. Hellsten, T. Mitros, A. Poliakov, J. Schmutz, M. Spannagl, H. Tang, X. Wang, T. Wicker, A. K. Bharti, J. Chapman, F. A. Feltus, U. Gowik, I. V. Grigoriev, E. Lyons, C. A. Maher, M. Martis, A. Narechania, R. P. Otillar, B. W. Penning, A. A. Salamov, Y. Wang, L. Zhang, N. C. Carpita, M. Freeling, A. R. Gingle, C. T. Hash, B. Keller, P. Klein, S. Kresovich, M.

C. McCann, R. Ming, D. G. Peterson, M.-ur-Rahman, D. Ware, P. Westhoff, K. F. X. Mayer, J. Messing and D. S. Rokhsar, *Nature*, 2009, **457**, 551.

258. J. M. J. de Wet, in *Small Millets in Global Agriculture*, ed. A. Seetharama, K. W. Riley and G. Harinarayana, Oxford & IBH Publishing, New Delhi, 1989, p. 19.

259. R. K. Arora, *Econ. Bot.*, 1977, **31**, 358.

260. D. Q. Fuller, *A Millet Atlas – Some Identification Guidance*, Institute of Archaeology, University College London, 2006. http://www.homepages.ucl.ac.uk/tcrndfu/Abot/Millet%20 Handout06.pdf

261. J. Zhang, H. Lu, N. Wu, X. Yang and X. Diao, *PLoS ONE*, 2011, **6**, e19726.

262. K. F. Kiplas and K. C. Ornelas, *The Cambridge World History of Food*, Cambridge University Press, Cambridge, 2000, vol. 1, p. 112.

263. M. Kitagawa, *Botanical Magazine of Tokyo*, 1937, **51**, 150.

264. H. Scholz, *Plant Syst. Evol.*, 1983, **143**, 233.

265. T.-K. Cheng, *Antiquity*, 1973, **47**, 197.

266. K. Wasylikowa, M. Cariumaru, E. Hajnalová, B. P Hartyáni, G. A. Pashkevich and Z. V. Yanushevich, in *Progress in Old World Paleoethnobotany*, ed. W. van Zeist, K. Wasylokowa and K.-E. Behre, Balkema, Rotterdam, 1991, p. 207.

267. D. R. Piperno, *Phytolith Analysis. An Archaeological and Geological Perspective*, Academic Press, San Diego, 1988.

268. H. Lu, J. Zhang, K. Liu, N. Wu, Y. Li, K. Zhou, M. Ye, T. Zhang, H. Zhang, X. Yang, L. Shen, D. Xu and Q. Li, Proc. Nat. Acad. Sci. USA, 2009b, **106**, 7367.

269. H. Lu, J. Zhang, N. Wu, K.-b. Liu, D. Xu and Q. Li, *PLoS ONE*, 2009a, **4**, e4448.

270. X. L. You, *Agric. Hist. China*, 1993, **12**, 1.

271. Y. Li and S. Z. Wu, *Euphytica*, 1996, **87**, 33.

272. L. Barton, S. D. Newsome, F.-H. Chen, H. Wang, T. P. Guilderson and R. L. Bettinger, *Proc. Nat. Acad. Sci. USA*, 2009, **106**, 5523.

273. K. Chang, *Archaeology of Ancient China*, Yale University Press, New Haven, CT, 4th edn, 1986.

274. D. J. Zhang, F. H. Chen, R. L. Bettinger, L. Barton, D. X. Ji, C. Morgan, H. Wang, X. Cheng, G. H. Dong and T. P. Guilderson, *Chin. Sci. Bull.*, 2010, **55**, 1636.

275. G. W. Crawford, in *The Origins of Agriculture: An International Perspective*, ed. C. W. Cowan and W. J. Watson, Smithsonian Institution Press, Washington, 1992, p. 7.

276. G. W. Crawford, in *Archaeology of Asia*, ed. M. T. Stark, Blackwell, Oxford, 2006, p. 77.

277. A. P. Underhill, *J. World Prehistory*, 1997, **11**, 103.

278. G. Shelach, *J. World Prehistory*, 2000, **14**, 363.

279. Z. Zhao, *Antiquity*, 1998, **72**, 885.

280. X. Liu, H. V. Hunt and M. K. Jones, *Antiquity*, 2009, **83**, 82.

281. J. R. Harlan, *Crops and Man*, American Society of Agronomy and Crop Science Society of America, Madison, WI, 2nd edn, 1992.

282. A. N. Doust and E. A. Kellogg, *Am. J. Bot.*, 2002, **89**, 1203.

283. A. N. Doust, K. M. Devos, M. D. Gadberry, M. D. Gale and E. A. Kellogg, *Proc. Nat. Acad. Sci. USA*, 2004, **101**, 9045.

284. N. I. Vavilov, *Inst. Appl. Bot. Plant Breeding*, 1926, **16**, 1.

285. S. Sakamoto, *Jpn Agr. Res. Q.*, 1987, **21**, 84.

286. A. N. Doust, K. M. Devos, M. D. Gadberry, M. D. Gale and E. A. Kellogg, *Genetics*, 2005, **169**, 1659.

287. B. S. Gaut and J. F. Doebley, *Proc. Nat. Acad. Sci. USA*, 1997, **94**, 6809.

288. K. M. Devos, Z. M. Wang, J. Beales, T. Sasaki and M. D. Gale, *Theor. Appl. Genet.*, 1998, **96**, 63.

289. Z. M. Wang, K. M. Devos, C. J. Liu, R. Q. Wang and M. D. Gale, *Theor. Appl. Genet.*, 1998, **96**, 31.

290. K. E. Prasada Rao, J. M. S. de Wet, D. E. Brink and M. H. Megesha, *Econ. Bot.*, 1987, **41**, 108.

291. T. L.-D. Lu, *Antiquity*, 1998, **72**, 902.

292. P.-T. Ho, in *Origins of Agriculture*, ed. C. A. Reed, Mouton, The Hague, 1977, p. 413.

293. Z. An, in *Foraging and Farming: The Evolution of Plant Exploitation*, ed. D. R. Harris and G. C. Hillman, Unwin & Hyman, London, 1989, p. 643.

294. Y. Li, S. Z. Wu and Y. Cao, *Euphytica*, 1995, **83**, 79.

295. K. Fukunaga, Z. Wang, K. Kato and M. Kawase, *Genet. Resour. Crop Evol.*, 2002, **49**, 95.

296. K. Fukunaga and K. Kato, *Euphytica*, 2003, **129**, 7.

297. K. Fukunaga, K. Ichitani and M. Kawase, *Theor. Appl. Genet.*, 2006, **113**, 261.

298. K. Hammer and K. Khoshbakht, *Genet. Resour. Crop Evol.*, 2007, **54**, 907.

299. H. V. Hunt, M. V. Linden, X. Liu, G. Motuzaite-Matuzeviciute, S. Colledge and M. K. Jones, *Vegetation History and Archaeobotany*, 2008, **17**(1), S5.

300. M. K. Jones, in *Traces of Ancestry: Studies in Honour of Colin Renfrew*, ed. M. K. Jones, Oxbow Books, Oxford, 2004, p. 127.

301. G. W. Crawford and G.-A. Lee, *Antiquity*, 2003, **77**, 87.

302. J. R. Harlan, *Science*, 1971, **174**, 463.

303. L. Marchais, S. Tostain and I. Amoukou, in *Le mil en Afrique diversité génétique et agrophysiologique: Potentialités et contraintes pour l'amélioration génétique et agriculture*, ed. S. Hamon, ORSTOM, Paris, 1993, p. 119.

304. V. Poncet, F. Lamy, J. Enjalbert, H. Joly, A. Sarr and T. Robert, *Heredity*, 1998, **81**, 648.

305. J. N. Brunken, J. M. J. de Wet and J. R. Harlan, *Econ. Bot.*, 1977, **31**, 163.

306. J. K. Pritchard, M. Stephens and P. Donnelly, *Genetics*, 2000, **155**, 945.

307. I. Oumar, C. Mariac, J.-L. Pham and Y. Vigouroux, *Theor. Appl. Genet.*, 2008, **117**, 489.

308. V. Poncet, F. Lamy, K. M. Devos, M. D. Gale, A. Sarr and T. Robert, *Theor. Appl. Genet.*, 2000, **100**, 147.

309. V. H. Techio, L. C. Davide and A. V. Pereira, *Genet. Mol. Biol.*, 2006, **29**, 353.

310. A. R. Opuku, S. O. Ohenhen and N. Ejiofor, *J. Agric. Food Chem.*, 1981, **29**, 1247.

311. G. Ejeta, M. M. Hassen and E. T. Mertz, *Proc. Nat. Acad. Sci. USA*, 1987, **84**, 6016.

312. M. Nesbitt, in *The Cultural History of Plants*, ed. G. Prance, Routledge, New York, 2005, p. 45.

313. E. J. Nkanga and N. Uraih, *Acta Biotechnologica*, 1981, **2**, 153.

314. Y. O. U. Iwuagwu and Y. S. Izuagbe, *J. Appl. Bact.*, 1985, **59**, 487.

315. A. Opoku and N. Uraih, *Acta Biotechnologica*, 1983, **3**, 53.

316. L. A. M. Pelembe, J. Dewar and J. R. N. Taylor, *J. Inst. Brew.*, 2002, **108**, 13.

317. L. A. M. Pelembe, J. Dewar and J. R. N. Taylor, *J. Inst. Brew.*, 2004, **110**, 320.
318. A. Ganava, Millet beer brewing and its socio-economic role in the lives of Mafa women in the far north province of Cameroon, Masters Thesis, Faculty of Social Science, University of Tromso, 2008.
319. G. Muller-Kosack, *The Way of Beer: Ritual Re-enactment of History Among the Mafa; Terrace Farmers of the Mandara Mountains (North Cameroon)*, Mandaras Publishing, London, 2001.
320. R. Jaouen, l'Euchariste du Mil: Langage d'un People, Expression de la Foi Karthala, Paris, 1995.
321. H. C. Nzelibe and C. C. Nwasike, *J. Inst. Brew.*, 1995, **101**, 345.
322. R. C. Agu, *Process Biochem.*, 1995, **30**, 311.
323. S. O. Ogbeide, *J. Eng. Appl. Sci.*, 2007, **2**, 1161.
324. J. R. N. Taylor, T. J. Schober and S. R. Bean, *J. Cereal Sci.*, 2006, **44**, 252.
325. M. Loewe and E. L. Shaughnessy (ed.), *Cambridge History of Ancient China: From the Origins of Civilisation to 221 BC*, Cambridge University Press, Cambridge, 1999.
326. A. P. Underhill, *J. East Asian Archaeol.*, 2000, **2**, 93.
327. C. K. Maisels, *Early Civilisations of the World: The Formative Histories of Egypt, The Levant, Mesopotamia, India and China*, Routledge, New York, 2001.
328. D. N. Keightley, in *Ancient Mortuary Traditions of China*, ed. G. Kuwayama, Far Eastern Art Council and Los Angeles Museum of Art, Los Angeles, 1991, p. 12.
329. K.-C. Chang, *Art, Myth, and Ritual: The Path to Political Authority in Ancient China*, Harvard University Press, Cambridge, MA, 1983.
330. J. D. Paper, *The Spirits are Drunk: Comparative Approaches to Chinese Religion*, State University of New York Press, Albany, 1995.
331. O. Bar-Yosef, and M. E. Kislev, in *Foraging and Farming: The Evolution of Plant Exploitation*, ed. D. R. Harris and G. C. Hillman, Unwin Hyman, London, 1989, p. 632.
332. Z. V. Yanushevich, in *Foraging and Farming: The Evolution of Plant Exploitation*, ed. D. R. Harris and G. C. Hillman, Unwin & Hyman, London, 1989, p. 606.

333. J. M. Leggett, in *Oat Science and Technology*, ed. H. G. Marshall and M. E. Sorrells, American Society of Agronomy, Madison, WI, 1992, p. 29.

334. T. Rajhathy and H. Thomas, *Misc. Publ. Genet. Soc. Canada*, 1974, **2**, 1.

335. J. M. Leggett and H. Thomas, in *The Oat Crop: Production and Utilization*, ed. W. Welch, Chapman & Hall, London, 1995, p. 121.

336. H. Thomas, in *Oat Science and Technology*, ed. H. G. Marshall and M. E. Sorrells, American Society of Agronomy, Madison, WI, 1992, p. 473.

337. G. Ladizinsky, *Genet. Resour. Crop Evol.*, 1998, **45**, 263.

338. J. M. Leggett, in *Proceedings of the V International Oat Conference*, University of Saskatchewan Saskatoon, 1996, p. 128.

339. C.-D. Li, B. G. Rossnagel and G. J. Scoles, *Crop Sci.*, 2000, **40**, 1755.

340. A. A. Chernyshova, P. J. White, M. P. Scott and J.-L. Jannink, *Crop Sci.*, 2007, **47**, 2330.

341. U. Tiwari and E. Cummins, *J. Cereal Sci.*, 2009, **50**, 175.

342. I. G. Loskutov, *Genet. Resour. Crop Evol.*, 2007, **55**, 211.

343. X. Zhou, E. N. Jellen and J. P. Murphy, *Crop Sci.*, 1999, **39**, 1208.

344. A. I. Malsev, *Wild and Cultivated Oats: Section* Euavena Griseb., Works of Applied Botany and Plant Breeding, Supplement No. 38, 1930.

345. G. Ladizinsky and D. Zohary, *Euphytica*, 1971, **20**, 380.

346. G. Ladizinsky, in *Proc. 3rd International Oat Conference*, Lund, Sweden, 1989, p. 76.

347. B. R. Baum, *Oats: Wild and Cultivated. A Monograph of the genus* Avena *(Gramineae)*, Monograph No. 14, Biosystematics Research Institute, Canada Dept. Agriculture, Ottawa, 1977.

348. J. R. Harlan, in *Origins of Agriculture*, ed. C. A. Reed, Mouton, The Hague, 1977, p. 357.

349. J. R. Harlan, in *Foraging and Farming*, ed. D. R. Harris and G. C. Hillman, Unwin Hyman, London, 1989, p. 79.

350. G. Ladizinsky, *Evolution*, 1969, **23**, 676.

351. F. A. Coffman, *Oat History, Identification and Classification*, Technical Bulletin No. 1516, Washington, DC, 1977.

352. E. L. Sturtevant, *Sturtevant's Notes on Edible Plants*, ed. U. P. Hedrick, Report of the New York Agricultural Experiment Station for 1919, part II, J. B. Lyon Co., Albany, NY, 1919.

353. J. Major, *Historia majoris Britanniae, tam Angliae quàm Scotiae ... veterum monumentis concinnate*, Paris, 1521.

354. G. S. Robbins, Y. Pomeranz and L. W. Briggle, *J. Agr. Food Chem.*, 1971, **19**, 536.

355. O. Nordland, *Brewing and Beer Traditions in Norway*, Universitetsforlaget, Oslo, 1969.

356. J. Worlidge, *Systema Agriculturae*, T. Dring, London, 3rd edn, 1681.

357. R. Lásztity, *The Chemistry of Cereal Proteins*, CRC Press, Boca Raton, FL, 2nd edn, 1996.

358. C. Klose, B. D. Schehl and E. K. Arendt, *J. Cereal Sci.*, 2009, **49**, 83.

359. M. Peraaho, K. Kaukinen, K. Mustalahti, N. Vuolteenaho, M. Mäki, P. Laippala and P. Collin, *Scand. J. Gastroenterol.*, 2004, **39**, 27.

360. D. M. Peterson, *J. Cereal Sci.*, 2001, **33**, 115.

361. E. R. Morris, in *Phytic Acid: Chemistry and Applications*, ed. E. Graf, Pilatus Press, Minneapolis, 1986, p. 57.

362. M. Bartnik and I. Szafranska, *J. Cereal Sci.*, 1987, **5**, 23.

363. M. Larsson and A.-S. Sandberg, *J. Food Sci.*, 1992, **57**, 994.

364. M. Larsson and A.-S. Sandberg, *J. Cereal Sci.*, 1995, **21**, 87.

365. H. A. Sencer and J. G. Hawkes, *Biol. J. Linn. Soc. London*, 1980, **13**, 299.

366. H. C. Stutz, *Am. J. Bot.*, 1972, **9**, 59.

367. J. Greilhuber, J. Doležel, M. A. Lysák and M. D. Bennett, *Ann. Bot.*, 2005, **95**, 255.

368. J. Bartoš, E. Paux, R. Kofler, M. Havránková, D. Kopecký, P. Suchánková, J. Šafár, H. Šimková, C. D. Town, T. Lelley, C. Feuillet and J. Doležel, *BMC Plant Biol.*, 2008, **8**, 95.

369. R. B. Flavell, M. D. Bennett, J. B. Smith and D. B. Smith, *Biochem. Genet.*, 1974, **12**, 257.

370. J. Doležel, M. Kubalakova, J. Bartoš and J. Macas, *Chromosome Res.*, 2004, **12**, 77.

371. J. Doležel, M. Kubalakova, E. Paux, J. Bartoš and C. Feuillet, *Chromosome Res.*, 2007, **15**, 51.

372. M. Kubaláková, M. Valárik, J. Barto, J. Vrána, J. Cíhalíková, M. Molnár-Láng and J. Dolezel, *Genome*, 2003, **46**, 893.

373. H. Šimková, J. Safar, P. Suchankova, P. Kovárová, J. Bartos, M. Kubaláková, J. Janda, J. Cíhalíková, R. Mago, T. Lelley and J. Dolezel, *BMC Genomics*, 2008, **9**, 237.

374. R. Schlegel and V. Korzun, *Plant Breed*, 1997, **116**, 537.

375. S. V. Rabinovich, *Euphytica*, 1998, **100**, 323.

376. N. I. Vavilov, *Bulletin of Applied Botany, Genetics and Plant Breeding*, 1917, **10**, 561.

377. G. C. Hillman, *Anatolian Studies*, 1978, **28**, 157.

378. H. H. Helbaek, in *Plant Life of South-West Asia*, ed. P. H. Davis, P. C. Harper and I. G. Hedge, Botanical Society of Edinburgh, Edinburgh, 1971, p. 265.

379. K.-E. Behre, *Vegetation History and Archaeobotany*, 1992, **1**, 141.

380. W. van Ziest, K. Wasylikowa and K.-E. Behre, *Progress in Old World Palaeoethnobotany*, Balkema, Rotterdam, 1991.

381. M. Rösch, S. Jacomet, and S. Karg, *Vegetation History and Archaeobotany*, 1992, **1** 193.

382. A. Hagen, *A Second Handbook of Anglo-Saxon Food and Drink: Production and Distribution*, Anglo-Saxon Books, Hockwold, Norfolk, 1995.

383. M. Nelson, *The Barbarian's Beverage: A History of Beer in Ancient Europe*, Routledge, London, 2005, p. 91.

384. Jacobus Theodorus Tabernaemontanus, *Neuwe Kreuterbuch*, 2 volumes, Frankfurt, 1588–1591.

385. P. Bahns, R. Michel, T. Becker and M. Zarnkow, *Brauwelt International*, 2010, **28**, 96.

386. J. G. Gmelin, *Reise durch Sibirien von dem Jahr 1733 bis 1743*, Abram Bandenhoect, Göttingen, 1751.

387. M. E. Barkworth, K. M. Capels, S. Long and L. K. Anderton (ed.), *Flora of North America: North of Mexico, Vol. 24, Poaceae Part 1*, Oxford University Press, New York, 2007.

388. M. E. Barkworth, K. M. Capels, S. Long and M. B. Piep (ed.), *Flora of North America: North of Mexico, Vol. 25, Poaceae Part 2*, Oxford University Press, New York, 2003.

389. T. Komatsuda, P. Maxim, N. Senthil and Y. Mano, *Theor. Appl. Genet.*, 2004, **109**, 986.

CHAPTER 5

Other Sources of Sugar

5.1 HONEY

"Honey falls from the air principally about the rising of the stars and when a rainbow rests upon the earth"

Aristotle, *Historia animalium*

Rig-Veda 1: 90, 6–8 has:

"Let every wind that blows drop honey
Let the rivers and streams recreate honey
Let all our medicines turn honey
Let the dawn and evening be full of honey
Let the dark particles be converted to honey
Our nourisher, this sky above, be full of honey
Let our trees be honey
Let the Sun be honey
Let our cows secrete honey"

The honey bee (*Apis mellifera* L.) preceded *Homo sapiens* on our planet by around 10–20 million years and, from an anthropomorphic viewpoint, is one of the most important forms of animal life, having served man since prehistoric times. It is highly likely that honey formed the basis for mankind's first encounter with ethanol, since honey and water left together in a pot (or any

Alcohol and its Role in the Evolution of Human Society
Ian S. Hornsey
© Ian S. Hornsey 2012
Published by the Royal Society of Chemistry, www.rsc.org

recess) will ferment, and this drink, what we would probably call mead, would have often been flavoured with wild herbs and fruits.

Honey was undoubtedly the earliest sweetener available to man, and was certainly the most concentrated source of sugar in prehistoric temperate Europe, but exactly how much was available is a moot point. Judging by a number of examples of ancient art, it would certainly appear that early man often experienced difficulty recovering his prize; the example illustrated (Figure 5.1) is from a 15,000-year old rock painting from Cueva de la Arana, near Valencia, Spain, and shows prehistoric man gathering honey from a highly inaccessible vertical site. The enormous effort made to secure honey meant that it was destined to occupy an exalted position in the legend of mankind, and in his art, literature, mythology, religion and medicine. Honey is one of mankind's oldest medicines and it is mentioned in that capacity in ancient Sumerian texts.

Forbes[505] maintained that it was a much-traded commodity, although collection and movement must have been rather difficult.[1] As Crane[2] explained, there were a number of ways in which "natural" honey production could be increased by human intervention but, generally speaking, the amount of honey at the disposal of early peoples was undoubtedly small, with few ways of increasing yield. This meant that honey became a prestige commodity.

Figure 5.1   Raiding the wild bees' nest.

There is some conjecture as to where the bee might have been tamed and, to me, there is still no better lead-in than that given all those years ago by Grahame Clark:[1] "On the general question of the origin and diffusion of apiculture, the antiquity of the art can be proved for Egypt, and is rendered likely for Mesopotamia by the fact that the wild honey bees are known to have flourished in neighbouring regions during antiquity."

5.1.1 Early Records

As a number of scholars have opined,[3] it is sometimes difficult to ascertain which sweet substance is being designated as "honey" in some ancient writings. In many cases, it has been presumed that bee honey was the substance being ingested. As Crane[4] has pointed out, the Greek μλι, like *mel* in Latin, can refer equally to honey produced by bees or to any number of other sweet substances, including dates, figs and pods, sap and/or gum from carob or other trees. As she says about interpretations of ancient texts: "Unless the context makes clear a connection with hives, bees, or honeycomb, caution is warranted."

While wild bees are attested as having been abundant in Palestine there is no reference there to beekeeping *per se* until late-Hellenistic times, and scholars have sought to clarify whether the Hebrew Bible's numerous references to "a land flowing with milk and honey" refer to the product of honey bees or from some sweet substance emanating from trees (*i.e.* dates, figs or sap). Jewish scriptures, while containing no specific references to apiculture, do mention honey – presumably wild honey. "Wild honey", as opposed to that arising from apiculture, has added significance because of some of the references to John the Baptist in the New Testament gospels of Mark and Matthew. Mark 1:6, for instance, states that while in the wilderness: "John was clothed with camel's hair, and with a girdle of a skin about his loins; and he did eat locusts and wild honey."

Edward Neufeld,[5] in his extensive account of apiculture in Early and Middle Iron Age Palestine, presented a thorough analysis of the extensive, but sometimes conflicting, evidence available. As he says, there is no allusion to the practice in the Ugaritic texts or in the Amarna Tablets, and follows with: "The earliest specific reference to beekeeping comes from Philo, and from the Mishnah.

Philo, a contemporary of Jesus, mentions beekeeping in his account of a semi-monastic, ascetic Jewish sect (somewhat resembling the Essenes), and called "Therapeutae", believed to have been located in the vicinity of Alexandria."

Bodenheimer,[6] however, took the bull by the horns when coming to the following conclusion: "We do not find in the Bible one definite reference to beekeeping, but all passages indicate the hunting of wild honey ... Beekeeping seems to have penetrated into Palestine and Persia at a late period and only under Hellenistic influence ... no hint of any knowledge of the horizontal bee-pipes, which have been in use in Egypt for at least 5000 years, can be traced in the Bible, or even in the Talmud ... The honey which was delivered in the midst of the second millennium BC to Egypt from Syria and Palestine was certainly honey collected from wild bees." He further maintains that: "In Israel of Biblical times wild honey-hunting only was known, whereas, at the same time, real and extensive beekeeping was carried on in Egypt and Anatolia."

Tova Forti, in her excellent article written nearly 50 years later,[7] agrees that the Bible does not offer any definite evidence of beekeeping as an organized agricultural activity. There is no textual evidence of people building hives, and everything points to the practice of collecting honey from the hives of the wild bee that naturally abides in rock clefts. *Deuteronomy* xxxii 13, for example, says: "He made him ride on the high places of the earth, that he might eat the increase of the fields; and he made him to suck honey out of the rock, and oil out of the flinty rock." As Forti says, in Biblical literature the word *děbāš* denotes not just honey, but various types of fruit syrup (as does the Akkadian *dišpu*), and the absence of testimony to hives has meant that the most frequent interpretation of *děbāš* has been sweet fruit syrup made from the pulp of grapes, dates, figs and carob. How does this square with the aforementioned "land flowing with milk and honey"? Perhaps the most conclusive statement is found in *Hebrew Encyclopaedia Biblica*, where at 2:909, Deutscher says: "This almost certainly refers to fruit syrup, which is the most commonly found honey in Palestine, apparently bartered in commerce with neighbouring countries (*c.f. Gen.* xliii 11; *Ezek.* xxvii 17; *M. Bik.* 1: 10)."

In the Bible, honey was likened to "wisdom" by Solomon, *viz.*:

"My son, eat thou honey, because it is good: and the honeycomb, which is sweet to thy taste:

So shall the knowledge of wisdom be unto thy soul: when thou has found it, then there shall be a reward, and thy expectation shall not be cut off." (*Proverbs* 24: 13–14)

Thus, like honey, which is natural sweetness, natural wisdom is accumulated slowly, carefully and (sometimes) painfully, and should be distinguished from spiritual wisdom which is bestowed, not acquired.

In Forti's paper,[7] an overview of the literary adaptation of honey in biblical narrative and poetry led to an amazing assemblage of honey metaphors in the wisdom books of Proverbs and Job. Her paper identified four rhetorical categories of honey imagery: 1. "Honey" as metaphor of internalization wisdom and attaining good reputation; 2. "Honey" as a symbol of restraint and moderation against over-indulgence; 3. "Honey" as a metaphor for temptation; 4. "Honey" in the context of the two strongly contrasted idiomatic expressions: "honey under the tongue" and "venom under the tongue". Forti's work was aimed at assessing the various qualities of honey that might have been responsible for penning the various metaphors.

The Mishnah is rather more enlightening about honey, and contains references to honeys from both bees and dates. There is also more specificity about honey type, with that from bees generally being considered superior to date honey.[8] A little earlier, the Romano-Jewish historian, Josephus ben Matthias (Titus Flavius Josephus) (37–c. 100 AD) described the abundance of honey in the Bible Lands: "Of the date-palms watered by bit [a nearby spring] there are numerous varieties differing in flavour and in medicinal properties; the richer species of this fruit when pressed under foot emit copious honey, not much inferior to that of bees. And so the region is abundant in honey." Josephus noted that dates and bees are plentiful around Jericho.

5.1.2 "Melisso-madness"

Wild honey was undoubtedly widely consumed by the ancients, but its popularity was moderated by the fact that partaking of it could be dangerous, if bees had made it with pollen from poisonous plants. Writings by such as Aristotle, Xenophon, Strabo and Pliny attest to the dangers of ingesting "maddening honey". In *De mirabilibus auscultationibus* ("On marvellous things heard"; Hett[9]), attributed to Aristotle, we learn: "At Trapezus in Pontus, honey from boxwood has a strong scent; and they say that healthy people go mad, but that epileptics are cured by it immediately."

Strabo reports the intentional incapacitation of foreign troops by drugging them with "wild honey" (see Pompeii's escapade below), and there are a number of instances of ancient wars being lost or prolonged by the injudicious ingestion of suspect honey.

One of the earliest examples was penned by Xenophon in *Anabasis* and he describes how, in 401 BC, after Cyrus the Younger's failed attempt to conquer the Persian Empire, he was assigned to take command of 10,000 dispirited Greek soldiers. On their way back to Greece, he led them through the mountains of Kurdistan, through Georgia to Armenia ("the march of the ten thousand"), and pitched tent in the ancient kingdom of Colchis, on the Black Sea coast. Xenophon (*Anabasis*, 4.8, 18–21) tells us that some of the soldiers went off pillaging and found a source of honey; Xenophon explains: "The swarms of bees in the area were numerous, and the soldiers who ate of the honey all went off their heads, and suffered from vomiting and diarrhoea, and not one of them could stand up, but those who had eaten a little were like people exceedingly drunk, while those who had eaten a great deal seemed like crazy, or even in some cases dying men ... On the next day, however, no one had died, and at approximately the same hour as they had eaten the honey they began to come to their senses; and on the third or fourth day they got up, as if from a drugging."

Perhaps one of the best known examples of honey toxicity in warfare occurred some three centuries after Xenophon's unfortunate experience, and befell a Roman campaign against Mithridates VI, the king of Pontus. Led by Pompeii the Great in 67 BC, the troops camped near Trebizond on the southern shores of the Black Sea (not far from where Xenophon's troops had camped). Knowing that the local honey was toxic, the Heptakometes ("people of the seven villages"), allies of Mithridates, placed the deadly honeycombs in Pompeii's path. True to form, soldiers from three of the squadrons ate the honey and suffered the expected consequences, after which, in their state of stupor, they were slaughtered by the Heptakometes. This episode was recorded by the geographer Strabo and, in *Geographica*, XI, ii, 17, he refers to "the bitter honey of Colchis". Dioscorides, in *De Materia Medica*, wrote about the "peculiar nature" of the honey produced in the Black Sea region, which confirms the fact that King Mithridates's land south and east of the Black Sea was notorious for its poisonous plants and minerals.

Some people never learn, for, in AD 946, Russian enemies of Olga of Kiev fell for the same trick, in the same region, when they gladly accepted a large quantity of fermented honey from her followers. The inevitable happened and they were all (around 5000) slaughtered as they lay in a stupor. Then, in 1489 in the same area, around 10,000 Tatar soldiers were slaughtered by Russians after stopping to sup huge casks of mead in an abandoned camp. These, and other episodes, are related by Mayor.[10]

As one would expect, the "poisoned honey" notion did not escape the attention of Pliny the Elder, for, in *Natural History*, XXI. 45, we read: "In the country of Sanni, in the same part of Pontus, there is another kind of honey, which, from the madness it produces, has received the name of *meli mænomenon* ('maddening honey'). This evil effect is generally attributed to the flowers of the rhododendron, with which the woods there abound; and that people, though it pays a tribute to the Romans in wax, derives no profit whatever from its honey, in consequence of these dangerous properties." Some authorities refer to this as "Pontic honey". The Greek version of "mad honey" was *meli chloron*.

The cause of the toxicity was determined by Plugge in 1891,[11] and called andromedotoxin, a name that was subsequently changed to acetylandromedol, or even rhodotoxin. Also known as "rhododendron poisoning", it is one of the small group of diterpenoid grayanotoxins (Figure 5.2). Grayanotoxin I (acetylandromedol) is the principal toxic isomer.

Ott[12] has made a case for the fact that, in olden times, toxic honeys may well have served as pointers for people searching for medicinal and other useful plants. As he puts it: "I suggest that immemorial pursuit of wild honey, the only concentrated sweet which occurs naturally, could have led inexorably to the discovery of psychoactive and other toxic honeys, while subsequent observation of bees' foraging habits could easily have led preliterate shaman/pharmacognosists to single out toxic plant

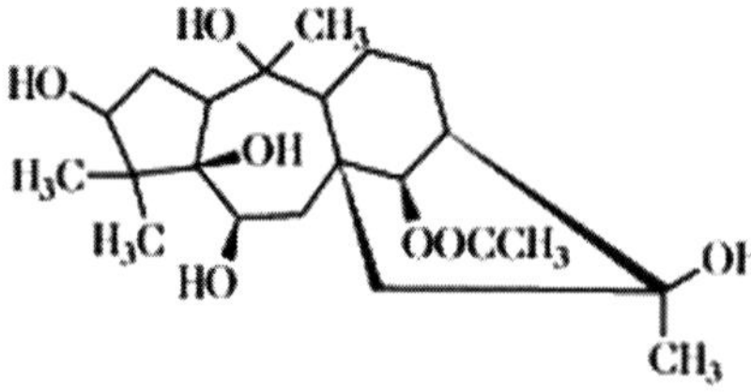

Figure 5.2 Acetylandromedol.

species even against a background of extreme biodiversity, as in Amazonia."

5.1.3 Ancient Near East

The earliest written references to honey from bees come from Nippur *ca.* 2100–2000 BC, while, if we take note of Hoffner,[13] "honey was the sweetening agent in widest use in the ancient Near East. It was known and prized in Mesopotamia, although bees are not native to that land and had to be imported by the kings". He also affirms that the Akkadian language possessed a variety of words for worker-bees, honey and beeswax. Elsewhere, Hoffner advises that "the land of the Hittites was a bee-keeping country and had been since the earliest times of recorded history", adding that "the bee (NIM LÀL) also figures in the oldest known myths from the Hattian culture (the vanishing god myths such as Telepinu)". Most Hittite apicultural terms are known only in ideographic form (*e.g.* the hive, É NIM LÀL, and the bee-keeper, LÚ NIM LÀL), but the word for honey is known (*milit*). Honey was also a constituent of beer in the ancient Near East, either as a source of yeast inoculum, or as a flavouring. The Hittites are known to have drunk "beer-honey", which was either mead or regular grain-based beer flavoured with honey.

In terms of Hittite mythology, in Beckman[14] we read: "Hannahanna sent a bee: You go search for [my son] Telepinu. When you find [him], sting his hands and feet and make him stand up. Then take wax and wipe him off. Then purify him and make him holy again. Then conduct him back here to me."

The earliest laws relating to bees come from fragments of the archives of Hittite kings, *ca.* 1300 BC, which were written in Babylonian cuneiform.[15] One states: "If anyone steals bees out of a hive, formerly he had to pay 1 mina of silver, now he need only pay 5 shekels, then he is set free." Another says: "If anyone steals two or three beehouses he formerly had to have his own hives destroyed, now he needs to pay 6 shekels of silver. If anyone steals a hive and there are no bees in it, he needs to pay 3 shekels of silver." These laws indicate that there were even more ancient laws in existence, and suggest that the offences mentioned were not regarded as being as heinous as they once were. Hilda Ransome[16] reckons that the lesser penalty might show that the number of hives was greater than in past times and that, proportionally, they were not such valuable items.

By and large, however, the evidence for apiculture in Mesopotamia is pretty scarce – especially when one considers the plethora of extant cuneiform tablets detailing nearly every aspect of life, including agriculture. Some scholars have suggested that this situation might have arisen because the Mesopotamian words for "honey" (*dišpu*, Akkadian; *làl*, Sumerian) refer to date syrup as well (similar to the situation regarding the Hebrew word for honey). The bee does not feature conspicuously in Mesopotamian texts, and not at all in the region's art. Most of the Akkadian words for "bee" appear solely in lexical texts, and so there is no specialized vocabulary associated with beekeeping. The first recorded mention of beekeeping in cuneiform appears on the stela of Šamaš-reš-usur, a regional governor on the Syrian Euphrates in the middle of the eighth century BC.[17] He claimed to have brought bees down from the mountains (presumably the Taurus Mountains, where beekeeping was a tradition), and that he was the first to do so: "I, Šamaš-reš-usur, the governor of the land of Suhu and Mari, I brought bees – that collect honey and which from the time of my fathers and forefathers no-one had seen nor brought to the land of Suhu – down from the mountains of the Habna-people and settled them in the gardens of the town of Algabbaribani. They collect honey and wax and so can the gardeners."

Without the benefit of modern technology, Mesopotamia would have been an alien environment for bees. The Iraqi Plain is far too hot and the flowering season too short to encourage apiculture; only in the mountainous north of the region were native honey bees found. It seems, therefore, that in Mesopotamia most honey was imported, and that it was an expensive commodity. This can

be judged from a record of the Ur III period (twenty-first–
twentieth century BC) when one shekel of silver bought only two
pounds of *làl* ("honey"), but the same amount of silver procured
300 litres of dates!

Our knowledge of the history of apiculture in the ancient Near
East has been revolutionized by the recent discovery of an apiary
at the huge earthen mound called Tel Rehov.[18] Seemingly
constructed on an industrial scale, the artefact was found in an
Iron Age IIA context (tenth– early-ninth centuries BCE) in what is
known as "Area C" of the excavation. Tel Rehov is located in the
heart of the Beth-Shean Valley in northern Israel, some 5 km south
of Beth-Shean and 6.5 km west of the River Jordan (Figure 5.3).
The mound is composed of two equal parts – a lower city on the

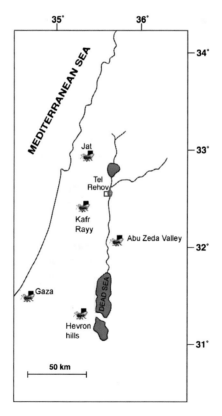

Figure 5.3   Schematic map of Israel and Jordan. Ethnographic hives are
marked with a bee. After Mazar *et al.*[18] - by kind permission.

north and an upper city to the south. The whole area covers some ten hectares. Tel Rehov is not mentioned in the Bible, but it figures in other ancient sources. Several texts dating to the Egyptian New Kingdom indicate that Rehov was an important city-state in the region during the Late Bronze Age.

The apiary, which dates to the time of the Biblical accounts of King David and King Solomon, was surrounded by high mud-brick walls, and the initial excavations yielded 25 cylindrical containers, each 80 cm long and with an external diameter of 40 cm. It is estimated that there were at least 75, and maybe as many as 200, cylinders in the apiary, which may have housed around one million bees. Annual production capacity for the apiary was calculated at 500 kg honey and 70 kg beeswax.

In addition to the honey production facility, there was clear evidence of accompanying cultic activity, for Prof. Amihai Mazar's team found a four-horned clay altar that features carved figures of two naked female goddesses flanking an incised tree (Figure 5.4). Other paraphernalia was recovered, including a large, elaborately painted chalice. This correlation between industrial and ritual

Figure 5.4 Horned clay altar found in association with the apiary at Tel Rehov. After Mazar and Panitz-Cohen[20] - by kind permission.

activity is not unique and it is known from other situations, such as olive oil manufacture.[19] Mazar is of the opinion that honey was burnt during rituals, even thought there is a Biblical prohibition against burning honey on an altar: "… ye shall burn no leaven, nor any honey, in any offering of the Lord made by fire" (*Leviticus* 2: 11). The aversion to the use of honey in Jewish sacrificial offerings was probably due to the fact that it was widely used as such in "foreign" religions (such as in Egypt and Babylon), as explained by Maimonides in his *Guide of the Perplexed*. It may also have been proscribed because of its fermenting properties!

Although the Bible does not specifically mention apiculture, the find at Tel Rehov shows that it was practised at that time in Israel, and apparently functioned very much like that of the ancient Egyptians. As Mazar has said about this ancient Biblical location,[20] "It is the land of honey".

In a later paper, Guy Bloch *et al.*[21] reported on the discovery of remains of honey bee workers, drones, pupae and larvae inside the Tel Rehov hives. Morphometric analyses of these remains indicated that the bees differed from the local sub-species, *A. mellifera* subsp. *syriaca*, and from all other sub-species except *A. mellifera* subsp. *anatoliaca*, which can presently be found in parts of Turkey. Bloch's team concluded that either the Western honey bee sub-species distribution has undergone rapid change during the last 3000 years, or that the ancient inhabitants of Tel Rehov imported bees that were presumably superior (*e.g.* in honey yield) to anything found locally. Ancient attestations indicate that long-distance transport of bees was technically feasible.[22] Honey is still used to flavour (sweeten) traditional Middle Eastern sherbets.

Thanks to palynology, and that science's ability to signature "fossilized honey", we now know more about ancient apiculture and the role of honey a little further to the east, in southern Georgia.[23] In the middle of the third millennium BC, a period known as the Culture of the Early Kurgans developed in the eastern Causcasus, and both settlements and cemeteries have been uncovered from a wide area. Analysis of the organic content of ceramic pots from the Kodiani burial mound indicated that they contained honey, which is dated to twenty-seventh–twenty-fifth centuries BC, has told us much about the way of life of these people. The principal new element of culture was a distinctive burial ritual, for the deceased were buried in kurgans, graves characterized by stone or soil mounds. The palynological spectra obtained from the Kodiani mound were dominated by insect-pollinated plants, especially members of the Rosaceae. *Tilia* pollen was also well represented. It is known that agriculture, especially wheat cultivation, was established in Georgia during this Early-Bronze Age period, and that the landscape and climate of this period were probably quite different from those of today.

A little further north, we find that in the diet of early Russians, honey was a prized sweetener and was often fermented into mead (hives of forest bees were raided). We also know from nineteenth-century travellers in that country that mead drinking had gradually declined and had become a rarity by mid-century.

In Eastern Asia, where there are many competing sources of sweetness, honey has largely been prized for its medicinal properties, and it is in that context that we find it mentioned in Chinese literature around 2000 BC.

5.1.4 Ancient Egypt

In ancient Egypt the bee was thought to have been of divine origin and, according to one myth, bees were the tears of the Sun-god Ra, the following passage coming from the Salt Magical Papyrus: "When Ra weeps again the water which flows from his eyes upon the ground turns into working bees. They work in flowers and trees of every kind and wax and honey come into being."

To the ancient Egyptians, the Sun-god Ra framed the Earth and sea, and was responsible for causing the Nile to flood. He was usually represented as a hawk-headed human figure wearing a Sun-disc headdress, but in the underworld he appeared with a ram's head. According to cult, his right eye was the Sun, his left the moon, and when he opened his eyes there was light.

As well as being the ancient Egyptians' major sweetening agent, honey was perhaps the most effective of all their medicines (chiefly on account of its *ca.* 80% sugar content, which is desiccatory to most microbes). Other sweeteners used by the ancient Egyptians at various times were dates and raisins. There are documentations of vast quantities of honey being produced by ancient Egyptian beekeepers, perhaps the most amazing being the 15 tons that Rameses III (1184–1153 BC) reputedly offered to the god of the Nile.

The hieroglyph symbol of the bee (*bit*) played an extremely important role in ancient Egypt and, being the "throne name", was part of one of the five names (the "five-fold titulary") held by the pharaohs from early Dynastic times. The throne name "He of the sedge and bee" (*nesw-bit*) was first used in the reign of Den (*ca.* 2950 BC), and the practice continued until the Alexander the Great-inspired demise of the Late Kingdom in 332 BC. The practice of beekeeping in Egypt continued through the Ptolemaic and Roman periods and beyond.

Bee-keepers are graphically represented in ancient Egypt as early as the Fifth Dynasty, as witnessed on a relief of Nyuserra (2445–2421 BC) in his splendid Sun temple at Abu Gurab. This record indicates that apiculture was well organized by the middle of the Old Kingdom, although there is some evidence of the practice during that country's Neolithic period. Indeed, Crane[24] believes that Egyptian apiculture was initiated in the bee-rich Nile delta during the Predynastic period. French Egyptologist Gaston Maspero discovered honey cakes and pots of honey at the Eleventh-Dynasty site of Gebelein, and bee-keeping is also depicted in the Eighteenth-Dynasty tomb of Rekhmira[25] and the Twenty-sixth Dynasty tomb of Pabasa (Figure 5.5), both at the

Figure 5.5   Tomb of Pabasa – bees.

Theban necropolis. The latter depicts bees kept in pottery apiaries, although mud and other materials were also known to have been used. The walls of Pabasa's tomb (TT279) are literally covered with rows of bees. In a scene from the tomb of Rekhmira, we see bees being smoked out by one man, and their honey being extracted by another; a third man is storing the bounty in large jars. Honey was certainly traded at this time and it is likely that many communities kept their own hives.

Most apiaries were seemingly made of cylindrical pots arranged horizontally in stacks, a system that is known from many parts of Africa, although some tomb paintings indicate rolls of papyrus being used. Tristram[26] maintained that the method of keeping domesticated bees in hives made of large tubes of sun-dried mud had probably not changed from the earliest times "because of the ecologic and geologic conditions". Typical pipe hives made of mud or clay are about one metre in length and are stacked together, thus imitating logs where bees naturally swarm. The ends are sealed except for small apertures that allow bees to pass through.

Honey from wild bees was gathered by professional collectors called *bityw*, and the harvest took place at the end of autumn. The apiary was smoked in order to remove the honeycomb, the contents of which were then decanted into large, spherical jars that were then hermetically sealed. Different grades of honey are attested for, including a light, "pure" variety, and a darker blended variety emanating from the desert.

In the Old Kingdom, honey was scarce and costly and was under the Pharaoh's control. It was still an expensive item by the Middle Kingdom (2040–1640 BC), but was now regularly being used as a tomb offering and extensively employed in temple rituals.[27] Honeycombs, honey cakes and sealed jars of honey were placed next to the sarcophagi as food for souls of the dead. The importance of honey during the Middle Kingdom can be judged by the instigation of an unusual Office of State, "The Overseer of the Beekeepers".[28]

Apart from being used as a flavouring, honey was a constituent of many ancient Egyptian ointments, particularly one known as *kyphi* (Dioscorides, 1.24). Manniche[29] reports that a honey solution was incorporated into an eye remedy, which was dropped onto the eyelids (Ram III A 24–5), and that in its natural form it was used to cleanse hands and pots when preparing certain ointments. The Ebers papyrus (*ca.* 1555 BC), the largest preserved medical document (with around 700 formulae, mostly employing food and herbs), gives 147 prescriptions for which honey may be used in external applications, and the Edwin Smith papyrus (*ca.* 1600 BC – but most of the work is based on information written around 2640 BC[30]) tells us how to use honey to deal with a wound: "Now after you have stitched it, thou shouldst bind with fresh meat the first day. If you find the stitching of the wound is loose, draw it together and treat afterwards with grease, honey every day until he recovers."

Egyptian medicine had a good reputation in the Classical World and such luminaries as Hippocrates and Galen freely admitted that part of their information came from Egyptian works that they had studied at the temple of Imhotep at Memphis. Imhotep (the architect of the first step pyramid, at Saqqara – the necropolis of Memphis) was considered to be a god of wisdom and medicine, and is regarded by some to be the founder of Egyptian medicine and the author of the Smith papyrus. He was certainly the first architect and physician known by name to written history.

5.1.5 Greeks and Romans

The Greeks identified Imhotep with their own god of medicine, Asklepios, and Imhotep's cult centre at Saqqara became a centre of pilgrimage for those wanting to be healed. The Greeks called the site the "Asklepion" and its fame lasted for a couple of millennia after Imhotep's demise. In ancient Greece, the bee was a sacred symbol of Artemis (Figure 5.6), and was a design on Ephesian coinage for several centuries. The first description of the

Figure 5.6   Ephesian bee coin.

production of honey was given by Aristoteles, and Hippocrates, who mentions the foodstuff sparingly, reported that it: "… cleans sores and ulcers, softens hard ulcers of the lips, heals carbuncles and running sores." He also refers to "oxymel", a mixture of honey and vinegar, as a topical application for aches and pains, and "hydromel" (honey and water) for curing fever-induced thirst. Dioscorides, in his *Materia Medica*, often praised the healing properties of mead.

Aristotle wrote at length about honeybees in his ten-book *History of Animals*, and the information was used as a source of information for beekeepers for many centuries. There are contributions on the insect in books IV, V, VII and IX. Not all of his writings are technically correct, but are the result of much observation and are a great insight into how our ancestors viewed the ancient world around them. For example, on the bee (Book VII, Chapter ix), he says: "The bee will live for six years, some have lived for seven, and if a swarm lasts nine to ten years, it is considered to have done well." On honey he offers: "… the gold-coloured honey is also good. The white honey is not formed of pure thyme, but is good for the eyes, and for wounds. Weak honey always floats on the surface, and ought to be separated. The pure honey is beneath." Aristotle recognized that the quality of honey was dependent upon the amount of rainfall and the species of flowers present in different regions and seasons.

Aristotle lived during the time of Alexander the Great, and helped to educate him. He maintained a close relationship with the great leader and this allowed him to study what was afoot in the Greek empire. After his death, in 323 BC, Alexander the Great was embalmed in a coffin filled with honey, as it was not unknown for Greek citizens who had died far from home to have their bodies preserved in this medium.

For the ancient Greeks, apiculture was a craft associated with the Minoans, and they regarded mead as a Cretan intoxicant, which pre-dated wine. Homer never regarded bees as anything else but wild creatures, and offers several similes in his great works. In Book II of the *Iliad*, for example, he likens the Achaeans disgorging from their ships before battle as being "like buzzing swarms of bees that come out in relays from a hollow rock".

Honey is mentioned by several Roman writers, most notably Virgil, Varro and Plinius. To my mind Virgil's treatment of bees and their honey is quite remarkable, because Book IV of his *Georgics* is devoted to apiculture. The seven line "Introduction" sets the scene with: "Next I'll speak about the celestial gift of honey from the air ..." Virgil follows with sections on: "Location and Maintenance of the Apiary", "The Fighting Swarms", "The Surrounding Garden", "The Nature and Qualities of Bees", "Gathering the Honey", "Disease in Bees" and "Autogenesis of Bees".

Honeybees and honey are also dealt with by Marcus Terentius Varro (116–27 BC) in his major surviving work, *De Re Rustica*, and are given reverential treatment by Pliny the Elder in Book XI of his *Natural History*. Book XI deals with insects generally, and Pliny considers that: "Among all insect species, the pride of place is reserved for bees, and they are particularly admired – rightly so, because they alone have been created for man's benefit. Bees collect honey: the sweetest, finest, most health-promoting liquid ..." The above, and other writers, often give details of the plant sources of honey, but it is significant that most Greco-Roman accounts deal with apiculture, and have little to say about "wild" honeys.

In the first century AD, Apicius, a wealthy gourmet, produced a series of books on Roman cuisine in which over half of the recipes made use of honey. Apicius's main source of information is *De Re Coquinaria*, which contains around 500 recipes, and details of much of his work can be found in Flower and Rosenbaum's classic, *The Roman Cookery Book*.[31] Dalby and Grainger[32] have written a more readily available text on the same subject.

Stacked horizontal hives are known to have been employed in Classical times in Greece and, according to Andrew Sherratt,[33] are still to be found in the Cyclades. A little later, Columella describes wooden or wicker horizontal hives in Roman Italy, which were apparently typical of unforested areas. Further north in Europe, there was a tradition of forest bee-keeping (*i.e.* managing wild bee populations in trees), and this lasted down to historic times. This practice was certainly recorded in Medieval Britain, where mead was the drink of the elite during the Early Medieval period. In some parts of northern Germany, hives were constructed of vertical logs so that they mimicked trees. Two beech tree trunk hives have been recovered from peat bogs near Oldenburg in northern Germany, and dated to the first century AD, and another similarly dated trunk was dredged from the River Oder in Poland.

The wicker skep (a woven basket with an aperture), the typical manifestation of a Medieval Western European beehive, seems to emerge during the early centuries AD (an early example being from Feddersen Wierde), although there is a hint that it may have been used as far back as the Neolithic. As Sherratt says: "The keeping of domestic bees, therefore, although of some antiquity in the Mediterranean, was largely a development of the historical era in temperate Europe."

5.1.6 Anglo-Saxon Times

Ann Hagen[34] confirms that honey (*hunig*) was the most important sweetening agent in Anglo-Saxon Britain, and attests to the universality of mead in Britain during that time. She finds no evidence of beekeeping at the start of the Anglo-Saxon period, even though the Romans had introduced the art many years before. Anyone finding honey and wax was entitled to keep them. The first attempts at apiculture involved making incisions into trees that contained bees' nests and making a portion of the trunk into a door through which honey could be collected. The next evolutionary step was to make bark or log hives (*ruscae*), which were recorded from Suffolk in the Domesday Book, before wicker hives (skeps), the typical medieval form of West European beehive, were developed.

Pollington,[35] in his excellent *Leechcraft*, reckons that honey was the only source of sweetness in the Anglo-Saxon diet other than ripe berries, and was often used to counteract the sharp, bitter taste of many herbal concoctions (*wyrtdrencas*). Honey gets numerous mentions in the major books of Old English remedies, notably the *Old English Herbarium*, the *Lacnunga* manuscript, *Fragmenta H Omont*, and Bald's *Leechbook III* (where, in one recipe, "English honey" (*huniges...englisces*) is specified). Pollington also confirms that meadowsweet was used as a flavouring for mead, and that "Welsh Ale" was sweetened with honey, spiced, and was a highly regarded drink in medieval Wales.

5.1.7 Honey Taxonomy

Contemporary honey taxonomy takes into account how it is produced (*i.e.* extracted, pressed or drained), whether it is "organic" and whether it is "unifloral". There are also designations for honeys that are in a liquid/crystalline state ("normal"), those that are sold in the whole comb ("comb") and those that contain one, or more, pieces of comb ("cut comb", or "chunk").

Because there are numerous plants that can produce enough nectar, or honeydew, from which beekeepers can produce honey,[36] no two honeys are completely alike. This can be problematical in an age where importance is attached to "uniformity". There are dozens of plants from which unifloral honey can be made, and these can be attested by their pollen (melissopalynological) and physico-chemical characters. Most unifloral honeys are made and marketed in Europe, where over 100 species are used for their production.[37] The sources of the main unifloral honeys in the world are shown in Table 5.1.

Table 5.1 The sources of the main unifloral honeys in the world.

Plant name	Place of Harvest
Acacia (*Robinia pseudoacacia*)	Temperate Europe, Asia, America, Oceania
Eucalyptus (*Eucalyptus* spp.)	S. Europe, Oceania, Africa, S. America
Fir (*Abies alba*)	Central and Southern Europe
Heather (*Calluna vulgaris*)	Europe
Lavender (*Lavendula* spp.)	Temperate Europe, Asia, N. America
Lime (*Tilia* spp.)	Temperate Europe and Asia, temperate and sub-tropical North America
Orange blossom (*Citrus* spp.)	Europe, temperate and sub-tropical North America, S. America
Pine (*Pinus* spp.)	Temperate Europe, Asia, Oceania
Rape (*Brassica napus*)	Europe, N. America
Rosemary (*Rosmarinus officinalis*)	Temperate Europe, Asia, Africa
Sunflower (*Helianthus annuus*)	Temperate Europe, S. and N. America, Asia, Africa, Oceania
Sweet Chestnut (*Castanea sativa*)	Europe
Thyme (*Thymus* spp.)	Mediterranean and temperate Europe, N. America, Oceania
White Clover (*Trifolium repens*)	Europe, N. America

5.1.8 Honey Composition

The precise composition of honey varies according to the plant species on which the bee forages, but the main constituents are the same in all honeys. Most reports deal with honey from *A. mellifera*, but Minh *et al.*[38] reported the chemical composition of honey produced from *A. dorsata*. Carbohydrates constitute more than 95% of honey solids, and there is a highly complex mixture of sugars. As Table 5.2 shows, the monosaccharides fructose (levulose) and glucose (dextrose) are the two main sugars, and account for around 85% of honey solids. The following disaccharides have been identified:[39] sucrose, maltose, isomaltose, maltulose, nigerose, turanose, kojibiose, laminaribiose, α-, β-trehalose, gentiobiose and the trisaccharides: melizitose, 3-α-isomaltosylglucose, maltotriose, 1-kestose, panose, isomaltotriose, erlose, theanderose, centose and isopanose have also been identified.[40] Two more (complex) sugars, isomaltosyltetraose and isomaltosylpentaose, have also been identified. Many of these sugars are not present in nectar, but are formed during the ripening and storage effects of bee enzymes and honey acids. "Maltose" in Table 5.2 is in reality a mixture of several disaccharide sugars, while "higher sugars" represents those compounds sometimes referred to as "honey dextrins". Cotte *et al.*[41] used the chromatographic analysis of sugars to characterize monofloral honeys.

Table 5.2 Composition of a "typical" honey (after various sources).

Water content, %	17.20
D-Fructose, %	38.38
^{d}-Glucose, %	30.31
Fructose : Glucose ratio	1.23
Sucrose, %	1.31
"Maltose", %	7.31
Other sugars, %	8.81
Reducing sugars, %	76.75
"Higher sugars", %	1.50
Lactone as gluconolactone, %	0.14
Nitrogen, %	0.041
Minerals (ash), %	0.169
Proteins, mg/100 g	168.60
Free acid as gluconic acid, %	0.43
Total acidity, meq/kg	29.12
pH	3.91

Although acids represent less than 0.5% of honey solids, they make important contributions to flavour and aid microbial stability. They can also be used as indicators of deterioration and to measure purity and authenticity.[42] The predominant acid in honey is gluconic acid,[43] and its presence is largely attributable to the activity of glucose oxidase, which bees add at ripening. Some is due to bacterial activity, as has been shown.[44] Other acids that have been identified are: formic, acetic, butyric, lactic, oxalic, succinic, tartaric, fumaric, maleic, pyruvic, pyroglutamic, α-ketoglutaric, glycolic, citric, malic and 2- or 3-phosphoglyceric. The situation regarding the dicarboxylic acids fumaric and maleic acids is interesting because, despite their identical molecular weights, their pH relationships in honey differ. Suárez-Luque *et al.*[45] found that total fumaric acid had a correlation with honey pH, whereas there was no correlation between honey pH and total maleic acid. They also found that the *cis-* form of maleic acid (*cis*-2-butenedioic acid) was predominant in all honey analysed, whereas for fumaric acid *trans*-2-butenedioic acid was the major form in most samples. The predominant acid form was dependent on honey pH value. The total fumaric acid content is known to be related to the citric acid content of a honey, while the maleic acid content is not correlated.[46]

Of the low nitrogen content of honey, around 50% will consist of amino acids, with only proline, phenylalanine and aspartic acid generally being found in quantities exceeding 200 ppm. Others commonly present are: glutamic acid, alanine, tyrosine, leucine and isoleucine. In addition, there are several enzymes (proteins) present, the most important of which are introduced by the honey bee during the conversion of nectar to honey. These are invertase, amylase (most) and glucose oxidase (hence the presence of gluconic acid). Other enzymes are yeast- or plant-derived, such as catalase, acid phosphatase and a small amount of amylase. Glucose oxidase is of special interest because it causes the release of hydrogen peroxide, which stabilizes nectar and possesses antimicrobial activity.

Giri[47] was the first to show the presence of acid phosphatase in honey, and he also observed that fermented honeys showed higher acid phosphatase activities than unfermented honeys. It has also been shown that the enzyme might be correlated with the botanical origin of the honey. Pollen is the main source of the enzyme, with some contribution from nectar. There is also evidence that acid phosphatase activity decreases with honey (unfermented) storage time. Alonso-Torre *et al.*[48] examined acid phosphatase levels in Spanish honeys from two different climates (continental and oceanic), and found that activities of the enzyme were higher in the latter. All samples showed a decrease in activity with time, and induced granulation did not affect the trend of this activity. Honey pH had a considerable effect on the activity of acid phosphatase, such that the higher the pH, the lower the increase in activity. Other proteins have been poorly studied, but it is known that they lower the surface tension in honey, which increases its propensity to foam (especially noticeable in buckwheat honey, which has high protein content).

The phenolic content of honeys of differing floral origin were investigated by Andrade *et al.*,[49] who found that there were clear differences in the phenolic profiles of the electrophoretograms. Heather honeys had higher contents of phenolic derivatives and lower levels of flavonoids. In contrast, citrus and rosemary honeys contained low levels of phenolic derivatives. In some instances, an individual phenolic proved to be unique to a unifloral honey type and could be considered as a potential marker for that honey. In this context, rosmarinic acid appears to be characteristic of thyme honey, while naringenin defines lavender honey. Hesperitin would seem to be characteristic of citrus honey.

Honey contains varying amounts of mineral matter, ranging from 0.02–1.03%.[50] The major mineral element is potassium, which normally comprises around one-third of the total. With the proviso that some heavy metals can be of anthropogenic origin, several groups of workers, including Feller-Demalsy *et al.*,[51] have reported that the trace element content of honey depends on its botanical origin. It has also been possible to distinguish between honeys of different botanical origin. For example, Nozal Nalda *et al.*[52] showed that it was possible to differentiate between ling (*Calluna vulgaris*), heather (*Erica* spp.), rosemary (*Rosmarinus officinalis*), thyme (*Thymus vulgaris*), lavender (*Lavandula* spp.) and oak (*Quercus* spp.) honeydew honeys by measuring Mg, Ca, Al, Fe, Mn, Zn, B, Cu, Co, Cr, Ni, Cd and P and subsequent discriminant analysis. The mineral content of honey can also be correlated with its geographical origins (*e.g.* [53,54]). As Bogdanov's group confirmed,[55] the electrical conductivity of a honey sample can be correlated with mineral content, and this parameter is frequently used for the characterization of the botanical origin of honey.

A certain amount of water is naturally present in the honeycomb after ripening, the actual amount depending upon the original moisture of the nectar and the weather. After honey is extracted, moisture content may change according to conditions of storage. Moisture content is all important in determining granulation, body and keeping quality, and can vary between 13 and 25%. Honeys may be graded by their water content; in the USA, for example, honey must not have a moisture content of more than 18.6% in order to qualify for Grades A and B. Grade C samples may contain up to 20% water, while anything over that value constitutes Grade D (sub-standard). It is possible for a honey sample to have too little water, in which case it becomes difficult to handle and can only really be used to blend in with samples that are too moist.

The microbial load of a honey is also important and, as a rule of thumb, a sample with a substantial microbe count and a water content of 17% is likely to auto-ferment and spoil, and should be pasteurized. Honey with a water content of 19% or over would be expected to ferment. Fermented honey usually tastes sour owing to the subsequent acetification of ethanol. Honey that has been fermented can sometimes be reclaimed by heating it to 150 °F for a short time. A recommended pasteurization regime for honey is 145 °F for 30 minutes.

5.1.9 Antimicrobial Activity of Honey

The reasons for the antibacterial activity of honey have been well documented, but there has always been some controversy about them. The first *in vitro* demonstration of the phenomenon was published in 1937,[56] and the active principle was named "inhibine", without any suggestion as to its composition, although the substance was defined (assayed by its effect on the growth of *Staphylococcus aureus* on nutrient agar plates containing graded amounts of honey) by Dold and Witzenhausen.[57] One of the earliest suggestions as to the cause of the antibacterial activity of honey was hydrogen peroxide,[58] when it was found that both the antibacterial activity of honey and hydrogen peroxide were destroyed by light and by catalase. The following year, White and Subers reported that hydrogen peroxidise, which is produced by honey glucose oxidase, could be the inhibitory substance. They also demonstrated that the inhibine number (as defined by Dold[56] and described by Schade *et al.*[59]) of honey is linearly related to the logarithm of the hydrogen peroxide accumulation.

Non-dissociated organic acids also play a role in the antibacterial activity of honey,[60,61] and these compounds are known to be very soluble in cell membranes and induce alterations in cell permeability.[62] Bogdanov,[63] during an investigation into the antibacterial content of honeys, extracted a flavonoid compound, while Molan and Russell[64] showed that the botanical source of honey can make a contribution to its antibacterial capacity. Flavonoids exist in many plants and are the largest group of compounds in the resinous propolis, which possesses considerable antimicrobial activity. The most significant flavonoids in honey are the flavanones pinocembrin and pinobanksin, and the flavone chrysin, according to Ferreres *et al.*,[65] who developed an extractive technique for analysing honey flavonoids. Using HPLC and GLC, caffeic acid and ferulic acid were identified from honey,[66] and these phenolic acids are also known to have antimicrobial properties.

Inhibine came to be regarded as a recognized property of honey, and it was soon discovered that it was sensitive to heat and light, retained by bacterial filters and capable of passing through a dialysis membrane. In addition, inhibine was found to be active against numerous Gram-negative and Gram-positive bacteria. In the USA, Jonathan White Jr. and co-workers made investigations into the nature of the active principle, and were the first to demonstrate the presence of a glucose oxidase in honey[67] – it having been previously detected in the pharyngeal gland of the honey bee. By producing gluconolactone and hydrogen peroxide from glucose, it was a typical glucose oxidase and it was suspected that the antimicrobial activity of honey might be explained along these lines. This notion was confirmed when it was found that all of the properties of inhibine could be explained by the production of hydrogen peroxide by honey glucose oxidase. It was then shown that if the Dold-Witzenhausen assay is carried out in the presence of either catalase or peroxidise plus an acceptor, then antibacterial activity disappears. If honey glucose is removed by dialysis, then activity is also destroyed (this can be reversed by glucose addition). It was also found that glucose oxidase is only "active" in diluted honey. "Undiluted" honey inhibits bacterial growth *via* its high sugar concentration but, being hygroscopic, it becomes "diluted" and then glucose oxidase becomes active and liberates hydrogen peroxide.

Some of the properties of honey glucose oxidase were described by White *et al.*,[68] and the enzyme was later partially purified and found to be different from samples isolated from other sources, particularly in respect of its very high substrate concentration requirement, and its high specificity for D-glucose.[69] It was later reported that the honey enzyme oxidized the β-form of D-glucose around six times faster than the α-form,[70] and further kinetic characteristics were unravelled.[71]

White and Subers[72] developed a rapid assay for peroxide accumulation in honey and, when applied to about 90 honeys, they were able to show that the Inhibine Number was linearly related to the logarithm of the hydrogen peroxide accumulation (see Table 5.3).

Cotton honey was found to be consistently very high in peroxide accumulation. The peroxide accumulation (inhibine) value of a honey is affected not only by glucose oxidase content but also by a number of minor peroxide-destroying components in honey from sources such as yeasts, nectar, pollen and various enzymes. It may also be affected by handling, storage and processing.

Fundamental work on the diverse nature of the antibacterial activity of honey has been presented by Peter Molan,[73,74] and it was concluded that the broad spectrum antibacterial activity, although multi-factorial, could only be attributed with any certainty to hydrogen peroxide and osmolarity (*ca.* 80% w/v sugar content). Some honeys contain specific compounds in detectable

Table 5.3 Relationship between Inhibine Number and hydrogen peroxide accumulation (after White and Subers[72]).

Inhibine No.	Hydrogen peroxide accumulation (mg H_2O_2 accumulated gm honey^{-1} in 1 hr under assay conditions)
0	<3.4
1	3.4–8.7
2	8.8–20.5
3	20.6–54.5
4	54.6–174.0
5	>174.0

antimicrobial levels, perhaps the best example being methylglyoxal (MGO) in Manuka honey from *Leptospermum scoparium* (see below). One conclusion to be drawn from this work was that the potency of antibacterial activity can vary markedly, and that it is impossible to predict with any certainty that a particular honey sample will have this or that level of activity. Attention should also be paid to how honey is processed if it is intended for sale as an antibacterial product. Pasteurization is often employed to eradicate yeasts that are a potential hazard in honey with a high water content, and/or to dissolve sugar crystals that might initiate granulation in a liquid honey. Such treatment will normally inactivate the antibacterial activity of a honey. Loss of antibacterial activity on exposure to light is another consideration.

Some clinicians have grave doubts about the efficacy of honey in medicine, especially in the realm of wound healing, and Molan has written an erudite and thorough summation supporting the use of honey as a wound dressing. Other areas where honey has been advocated as being beneficial include: cancer (reviewed by Bardy *et al.*[75] and Orsolic[76]), eye diseases; Aristotle in *Historia animalium* says: "Honey is good as a salve for eye sores"), gastroenterology (the Roman physician Celsus used honey as a cure for diarrhoea and, more recently, it shows potency against *Helicobacter pylori*; Al Somal *et al.*[77]) and cardiovascular health.[78]

In the never-ending search for new therapies, it has recently been reported that the antimicrobial activity of a Chilean honey made from the ulmo tree (*Eucryphia cordifolia*) was greater against certain pathogenic bacteria than that exhibited by manuka honey.[79]

Propolis ("bee glue") has a long history of medical use – from Hippocrates onwards – but that is another matter. For an account of the diversity of propolis and biological activity see Bankova,[80] and for an account of the functional properties of honey, propolis and royal jelly, see Viuda-Martos *et al.*[81]

Because of the presence of high levels of MGO in manuka honey (the concentration is up to 100-fold higher than in conventional honeys), it has been claimed to have therapeutic advantages over other honeys. MGO is a potent protein-glycating agent and an important precursor of advanced glycation end-products. The latter, known as AGEs, play a role in the pathogenesis of impaired diabetic wound healing and can modify the structure and function of target molecules. Matjan[82] regards MGO as a potential risk factor when manuka honey is used in diabetic patients.

The enhanced antibacterial activity of manuka honey is not purely down to peroxide levels, and the product has been reported as being effective against a wide range of microbes, including those of clinical significance. Cooper *et al.* reported on the sensitivity of Gram-positive cocci, such as methicillin-resistant *Staphylococcus aureus*,[83] and burn-infecting *Pseudomonas aeruginosa*[84] to honey. With the rise in prevalence of antibiotic-resistant bacteria, medical-grade honeys have evolved with known antibacterial ingredients.[85] The characterization of the antibacterial activity of medical-grade honey has been reported by Kwakman *et al.*,[86] and the same team[87] used the same product enriched with antimicrobial peptides to give enhanced inhibition of the growth of some antibiotic-resistant pathogens, including the troublesome *Burkholderia cepacia*. Richard Jones[88] has written a concise history of honey and healing through the ages, and aspects of the medical uses of honey have been summarized by Jeffrey and Echazarreta.[89]

5.1.10   Micro-organisms in Honey

As we have intimated, honey is essentially a supersaturated sugar solution, which provides an environment sufficiently hostile to prevent the growth of most bacteria and yeasts. This is primarily due to honey having a low water activity (a_W), a unit that is proportional to its free water content. As in many foodstuffs, a proportion of the absolute water content is bound and unavailable for microbial growth, and only that comprising a_W is available for potential spoilage organisms. Water activity is defined as the relation of the water vapour pressure of the food (p) to the water vapour pressure of pure water (p_0) at the same temperature. From this, the a_W of pure water is 1, and each addition of water-fixing substances causes $p < p_0$ and so a_W becomes <1. The a_W values of honey vary between 0.55 and 0.75, and honeys with a value <0.60 are microbiologically stable. As a rule of thumb, bacteria require an a_W of 0.90, yeasts 0.80 and moulds 0.70. Osmophilic yeasts have a requirement for high sugar levels, and an a_W of *ca.* 0.60.

Some water in honey is molecularly fixed to sugars, mainly *via* hydrogen bonding, and this will have no influence on a_W. Glucose and fructose, the main honey sugars, behave rather differently during the crystallization of honey, because the latter is more soluble and stays in solution for a longer period of time. Glucose, however, crystallizes by forming glucose monohydrate[90] and the water "fixed" to glucose, while in solution, is released, thus increasing a_W. Honeys with high water content are liable to separate out into a crystallized phase at the bottom and a liquid phase above, which is susceptible to fermentation. Hydrogen bonding between water and fructose is *via* low-energy bonds, and water surrounding a fructose molecule will normally be "available" and contribute towards a_W (hence encourage microbial growth). Thus, it has been found that water activities of crystallized honeys are higher than those of liquid honeys having the same water content, and that liquid honeydew honeys show higher water activities than liquid floral honeys with the same water content.[91]

As Snowdon and Cliver[92] attest, the very nature of the product and control measures within the industry determine that low numbers of only a few types of micro-organism are going to be found in a sound honey. The natural antimicrobial activity of honey discourages the growth and persistence of many micro-organisms. Some organisms (*e.g.* coliforms) may be introduced through insanitary practices, and can be used as indicators of the same; others are introduced by other means. There are primary and secondary sources of micro-organism, and in the former category we should include air, dust, pollen, the honey bee digestive tract and nectar, which are all difficult to control. Secondary sources include factory buildings and air, equipment, operatives and cross-contamination, and these should be controlled by good manufacturing practice.

The microbes of concern in honey are primarily yeasts and spore-forming bacteria, although most samples will contain low numbers of viable yeasts (<100 cfu g^{-1}). Because of this, standard industry practices are aimed at restricting the proliferation of yeasts. Bacteria do not readily multiply in the honey environment, and high bacterial counts suggest a recent secondary contamination. Spore-forming bacteria are the most likely prokaryotes to cause spoilage, but some vegetative cells can survive in honey for extended periods.

Because of its unique composition, microbial load specifications for honey will be at variance with many other food specifications.

Most purchasers tend to specify a total plate count, coliform count, yeast count, mould count and specified potential pathogens. Examples of some recent honey specifications are shown in Table 5.4.

In an investigation into the microbiology of ten Nigerian honeys from different geographical locations, Omafuvbe and Akanbi[93] could not detect yeasts and moulds but obtained total aerobic mesophilic bacterial counts of between 1.0×10^3 and 5.0×10^3 cfu g^{-1}. Six Bacillus species were identified: *B. cereus*, *B. megaterium*, *B. polymyxa*, *B. licheniformis*, *B. firmus* and *B. pumilis*. The samples exhibited antibacterial activity against a range of organisms including *E. coli*, *Salmonella* spp., *Shigella* spp., *Proteus vulgaris* and *Clostridium sporogenes*, while they had no effect on *Staphylococcus aureus* and *Pseudomonas aeruginosa*.

5.1.11 Honey Bee

One cannot talk about honey and mead without making some reference to the honey bee, and the first record of any insect that could be identified as a modern bee is a fossilized, female stingless bee (*Trigona prisca*) found in coniferous resin and originally thought to date back some 80 million years.[94] The specimen actually came from late-Cretaceous (Late Maastrichtian; *ca.* 65–70 Mya, although it was originally thought to be Campanian; *ca.* 80 Mya) amber, and was almost twice as old as previously found specimens, which were from late-Eocene Baltic amber. Found during the 1920s–1930s by Alfred C. Hawker in Kinkora, New Jersey, the specimen was given to Columbia University, where it remained for several decades before being critically examined.

This New Jersey specimen was a worker bee, which means that social organization had arisen by this time. The first record of an insect more closely allied to our honey bee (*i.e.* a species of *Apis*) comes from the Eocene period, around 50 Mya. Bee fossils are relatively rare in amber but, as Engel[95] remarks: "The Baltic-amber fauna, recently dated approximately 45 million years in age (middle Eocene), is the earliest diverse fossil record of bees ... and actually harbours a remarkable array of taxa representing most higher clades of the Apoidea" (see page 354). The previous year, Engel[96] published a new interpretation of the original work and transferred the Kinkora fossil into a new genus, *Cretotrigona*.

Table 5.4 Some recent honey specifications.

Company	Total plate count	Coliforms	E. coli	Assay (cfu g⁻¹) Yeasts	Moulds	Yeasts and moulds	Staphylococcus	Salmonella	Spore- and rope-forming bacilli
1	1000	10	0	100	100		0	0	50
2	5000	10	<10	100	100		100	0	
3ᵃ	10,000	<3	<0.3	100	100		<3	0	
4	10,000	10	0				0	0	
5	10,000	10	0			10	0	0	
6	10,000	100	0			200	0	0	
7	1000	10	10	100	100		100	0	
8	10,000	10	0	100	100		0	0	50
9	10,000	10	0	100	100		0	0	50
10ᵇ	5000	10	0	10	10				

ᵃAlso negative for Clostridium botulinum and Cl. perfringens.
ᵇAlso, 10 cfu g⁻¹ of mesophilic and thermophilic aerobes and anaerobes.

In the insect world, bees fall into the order Hymenoptera (which includes 100,000 species of sawflies, wasps, ants and bees), where they are placed in the sub-order Aculeata (bees, wasps and ants). Within this sub-order, we find the superfamily Apoidea, which, itself, contains the family Apidae. The Apidae, with over 5600 described species, is the largest family of bees and includes our honey bees, which are placed in the sub-family Apinae, and then, finally, in the Old World honey bee tribe, Apini, which contains a single genus, *Apis*.

Hymenoptera exhibit haplodiploid sex determination, where males arise from unfertilized haploid eggs and females arise from fertilized diploid eggs. Haplodiploidy has distinct sex-determination mechanisms compared with other organisms because Hymenoptera lack sex chromosomes.[97] Hymenoptera is one of the 11 orders of insects that undergo a metamorphic moult (called "homometabolous"). The honey bees' last common ancestor with humans was 600 Mya, while the group diverged from flies (Diptera) and butterflies and moths (Lepidoptera) some 300 Mya.[98] The genus *Apis* evolved in tropical Eurasia, and migrated northwards and westwards, in context of the latter reaching Europe by the end of the Pleistocene (10,000 years ago). The place of origin of *A. mellifera* has been variously given as Asia,[99] the Middle East[100] or Africa[101] – which now seems most likely – but, wherever it was, mankind eventually carried the honey bee worldwide.

Overall, there are around 20,000 species of bee showing great diversity, and they are known for their dominant role as pollinators, thus being the cornerstone of many ecological systems. Most bees are solitary, and advanced social behaviour (called "eusocial") is only found in one group, the "corbiculate Apinae", thus named because of the modification of their hind tibia into a corbicula ("pollen basket"). The corbiculate bees consist of four extant tribes, orchid bees (Euglossini), bumble bees (Bombini), stingless bees (Meliponini) and honey bees (Apini), and, of these, only the honey bees and the stingless bees exhibit advanced eusocial behaviour. Bumble bees are eusocial, but the only difference between worker and queen is size.

Numerous versions of the species breakdown of the genus have appeared, one containing 24 species. The biodiversity of *A. mellifera* was first assessed using morphometrics, and it was

Ruttner *et al.*[102] who proposed a classification that was long recognized as being definitive. This was based on multi-variate analyses, and they hypothesized that northeastern Africa and the Near East would be the centre of origin of the species. They also proposed that *A. mellifera* invaded Africa and Europe in three distinct branches: a South and Central African branch, a North African and West European branch and a North Mediterranean branch. A fourth branch, which included Near and Middle Eastern sub-species, was later added.[99] Then Franck *et al.*,[103] using microsatellite and mitochondrial DNA data to assess the genetic diversity of the honey bee in Africa, found an important differentiation of Ethiopian and Egyptian honey bees from those of other areas in Africa. Accordingly, two new lineages, probably originating in the Near East, were described.

There are currently ten species assigned to *Apis*, with one (*A. laboriosa*) being questioned on the basis of scant morphological differentiation from *A. dorsata* (the giant honey bee), despite differences in ecology and behaviour. Nine of these species are confined to Asia, the exception being the "Western honey bee", *A. mellifera* L.[104]

According to Ruttner's classification,[99] there are 24 sub-species of *A. mellifera* (Engel[105] has more than 25), each occupying locations in the Near East, tropical Africa and the west and central Mediterranean regions. These sub-species are delimited primarily on morphological characters, and they typically exhibit reduced gene flow with other groups due to water, mountain or desert barriers. As such, they have been called "geographic races" to reflect their adaptation to specific areas. It is reckoned that *A. mellifera* separated from its nearest relative (*A. cerana* Fabr.) around 2.5 Mya. *A. cerana* (syn. *A. indica*) is very closely related to *A. mellifera* in both morphology and ethology, and some have doubted whether they are separate species.[106]

Arias and Sheppard[104] PCR amplified, cloned and sequenced two different genomic regions of the ten *Apis* species collected within their natural range distribution. DNA sequences were then analysed using parsimony, distance and maximum likelihood methods to investigate phylogenetic relationships within the genus. These analyses grouped the honey bees into three major clusters: giant bees (*A. dorsata*, *A. binghami* and *A. laboriosa*), dwarf bees (*A. andreniformis* and *A. florea*) and cavity-nesting bees (*A.*

mellifera, A. cerana, A. koschevnikovi, A. nuluensis and *A. nigrosincta*).

Apis mellifera has a native distribution that extends from the Caspian Sea in the east to Ireland in the west, and from the southern tip of Africa to southern Russia and Scandinavia in the north.[24]

The Americas have no endemic Apidae, and it is known that *A. mellifera* subsp. *mellifera* arrived *via* European settlers during the early seventeenth century. There are other families of honey bees that are indigenous in the Americas, the Meliponae and the Trigonae, and when Mexico and Central America were conquered by the Spaniards, they found the natives engaged in apiculture. Most authorities would agree that the Inca and Aztec civilizations generally settled at altitudes too high for apiculture, but the Maya people mention the honey bee in some of their ancient Codices. In particular, the Maya of Yucatan were renowned for their beekeeping abilities, and their aptitude for collecting honey from the wild.

The honey bee is reported as having been imported into Virginia in 1622, but is not reported (as a swarm) from Texas until 200 years later! Subsequent introductions of *A. mellifera* subsp. *ligustica* (the "Italian" bee) commenced in 1859, and this was followed by the introduction of at least seven other sub-species from Europe, the Near East and northern Africa – the descendants of which are collectively referred to as "European".[107,108] Mitochondrial DNA evidence for the nineteenth-century introduction of African honey bees into the USA has been obtained.[109]

From what is known about introductions into South America, the first were the Western European *A. mellifera* subsp. *mellifera* and *A. mellifera* subsp. *iberiensis*, and these were followed by Eastern European sub-species.[107] In 1956, *A. mellifera* subsp. *scutellata* from the African savannas was introduced into Brazil, where it became established and spread through South and Central America.[110,111] This sub-species reached its southern limit, in Argentina, in the 1970s, and had penetrated into the USA by 1990. This "Africanized" sub-species has largely replaced the longer-established European sub-species, *mellifera*. Schneider *et al.* have examined the factors that enabled sub-species *scutellata* to be so successful in the Americas, concentrating on its ability to replace European honey bees. In particular, they considered: the genetic composition of the expanding population and the symmetry of gene flow between African and European bees, the mechanisms that favour the preservation of the African genome and the possible range and impact of the African bee in the USA. Without wishing to be dramatic, the team say of *A. mellifera* subsp. *scutellata* that: "Since its introduction [1956], the African bee has spread throughout Latin America and the southwestern United States. The colonization of much of the western hemisphere in less than fifty years by a single race of insect is one of the most rapid and spectacular biological invasions known."

Using modern molecular techniques, Whitfield *et al.*[101] characterized over 300 *A. mellifera* individuals from both native and introduced ranges, and were able to conclude that the species originated in Africa and expanded into Eurasia on at least two occasions. This resulted in populations in Eastern and Western Europe that are geographically close but genetically distinct. They were also able to discern a third expansion – into the New World. The latter has seen the near-replacement of the previously introduced "European" honey bee by descendents of the more recently introduced ("African") sub-species *scutellata*. In a subsequent piece of work on North American honey bees, Whitfield was able to show that Africanized bees exploited the genetic diversity present from previous introductions in an adaptive way.[112]

A. mellifera is a social species that was one of the first to build a highly advanced cavity nest with multiple parallel combs, and this contributed greatly to the evolutionary success of the insect. For example, during the Pliocene in Europe, low temperatures proved to be an important selective force on bees with nests open to the air.[99] Having closely packed multiple combs and concomitant aggressive behaviour, these honey bees were able to migrate from their tropical African "homelands" and radiate into temperate Europe and Asia. Cold periods could be survived by forming clusters within nests.

Michael Engel[95] doubts whether the advanced eusociality[113] of certain bees is the sole factor responsible for their ecological success. His work on Baltic amber fossils indicates that there were far more eusocial bee lineages during the middle Eocene period than we see today, where there are only two eusocial tribes comprising less than 2% of total bee diversity. He maintains that this is a trend analogous to that of hominid evolution, and that eusociability arose only once in the bees. As Engel says: "Advanced eusocial behaviour is not in its nascent stages in the bees but presently is a rudiment of what it once was and seems to have been unimportant in the greater diversification of bees." In his view, the permeating factor in the success of the bees as a whole has been their mutualistic association with flowering plants, a link that developed during the early mid-Cretaceous.[114] For further reading see O'Toole and Raw,[115] Grimaldi[116] and Michener.[117]

Being a key model for social behaviour and essential to global ecology through pollination, the sequencing of the genome of *A. mellifera* was an important event (Honeybee Genome Sequencing Consortium[118]). A number of significant points emerged, among which we may mention: 1) it is closer, in some respects, to vertebrate genomes (than to fruit fly (*Drosophila*) and malaria mosquito (*Anopheles*) genomes) for genes involved in circadian rhythms, RNA interference and DNA methylation; 2) it is distinguishable from other sequenced insect genomes by high A + T content, greater spatial heterogeneity of A + T content, high CpG content and an absence of most major families of transposons; 3) the honey bee evolved more slowly than *Drosophila* and *Anopheles* and 4) *A. mellifera* has fewer genes than *Drosophila* and *Anopheles* for some traits (*e.g.* innate immunity), and more genes for others (*e.g.* odorant receptors).

Behura[119] showed that at least 0.08% of the *A. mellifera* nuclear genome contains sequences that originated from mitochondria. These nuclear copies of mitochondrial sequences (numts) are scattered throughout the honey bee chromosomes and phylogenetic analysis shows that they arose *via* multiple insertions of mitochondrial DNA (mtDNA). Transfer of mitochondrial sequences to the nuclear genome is known from many other organisms, including several insect species,[120] where mtDNA is extensively used as molecular markers in ecological and evolutionary studies.[121] Once an mtDNA fragment is transposed to a nuclear genome, it remains there as a "molecular fossil", and numts collectively form a "natural library" of valuable information on the population and speciation history of the mtDNA of an organism. As Behura says: "Comparative analysis of numts of European, African, and Africanized honeybees suggests that numt evolution in *A. mellifera* is probably not demarked by speciation time frame but may be a continuous and dynamic process."

5.1.12 Mead

> Mead can be regarded as the ancestor of all fermented drinks ... antedating the cultivation of the soil.
>
> Maguelonne Toussaint-Samat

The first paragraph in Chapter 1 of Robert Gayre's classic 1948 work[122] evocatively intimates how highly mead was regarded by our forefathers: "Although we may little realise it today, when the world was young, when gods walked the earth and communed with men, and when men had more ready access to paradise than they have had since, mead was the liquor drunk by gods and men alike."

Mead was regarded as having magical, even sacred, properties, and this resulted in its raw material (honey) and the creature that creates the raw material (the bee) being held in high esteem as well. All three entities occupy high places in ancient sacred mythologies. Honey was considered to be a "giver of life" and the bee was a "messenger of the gods". According to the four Vedas (emanating from 1500–200 BC), honey was a dew that came down from Heaven and was gathered from flowers by bees. A similar sort of theme is recounted in Hesiod (eighth century BC), Aristotle (fourth century BC), Virgil (70–19 BC) and the Bible.

The Sanskrit word for honey is *mádhu*, which is etymologically related to *methu* (Greek) and *medu*, the Anglo-Saxon word for mead. The gods Vishnu, Krishna and Indra were called *Madhava*, "the nectar-born ones", and their symbol was the bee.[16] Vishnu, for instance, is represented by a blue bee resting on a lotus flower (Figure 5.7). The lotus flower is an ancient symbol of life, of resurrection and of Nature whose power slumbers until the warmth and light of the Sun calls it to life. As an old poem says:

> "When the Sun rises, the lotus flower opens And frees the bees from their prison"

Figure 5.7   Blue bee resting on lotus flower.

In the Rig-Veda, Vishnu is said to have three "steps" – sunrise, zenith and sunset – and the highest step is connected with life after death. Here, the pious are happy and there is a spring of mead:

> "In the wide-striding Vishnu's highest footstep, There is a spring of mead"

> [*Rig-Veda* 1: 154, 4–6]

5.1.12.1   Archaeology of Mead

Mead, of all the alcoholic beverages used by our ancestors, leaves us direct archaeobotanical evidence with which to work: pollen. Because pollen is normally only preserved in an un-carbonized condition, suitable environments must be provided for pollen preservation, and these are: waterlogged conditions, xeric conditions and/or the presence of copper salts. There are known marker compounds for the presence of honey in archaeological samples, such as the beeswax n-alkanes $C_{23}H_{48}$, $C_{25}H_{52}$, $C_{27}H_{56}$ and $C_{29}H_{60}$.[123] Under certain circumstances, ancient wine can be verified by the presence of pollen (from grape skin), as long as its production technology made no attempt to remove it. Similarly, wine must enriched with honey can be detected by palynology, as evidenced by Rösch,[124] who found some of the Coptic wine amphorae from the early medieval period of Šaruna, Middle Egypt, contained residues implying just that practice.

At present, the first archaeological evidence of any beverage that might qualify as being mead has been presented by McGovern *et al.*,[125] who showed that organic material absorbed into pottery jars from the early Neolithic village of Jiahu in China's Henan Province originated from a fermented beverage containing rice, honey and fruit. The latter was likely to have been from hawthorn, although grape was considered a possibility, and the findings indicate that honey was being used in the making of alcoholic beverages in China by the seventh millennium BC. A seemingly similar mixed beverage was detectable in vessels recovered from Liangchengzhen, a late prehistoric site from the (Longshan) period of northern China.[126] This is seen as a significant finding because it suggests that, even at this time, the technique of producing alcoholic beverages by initially saccharifying cereal starches, by mould enzyme activity, had not yet reached this part of China. The Longshan Yellow River Valley culture covers the period *ca.* 3000–1900 BC (Neolithic and Chalcolithic). The earliest documentary evidence of a fermented honey beverage in China emanates from the Warring States period (*ca.* 480–221 BC; Morohashi, 1987 – quoted in [126]). A little later, during the Tang period (AD 618–906), honey and rice were known to have been fermented together.[127] For readable accounts of these periods of Chinese history see Loewe and Shaughnessy[128] and Ebrey.[129]

It is possible that in Bronze Age Greece mead may have been a major competitor of wine, and may have even preceded it as an everyday drink[130] but, further north in Europe, the honey supply is unlikely to have sustained substantial mead production. This is borne out by the fact that in Medieval British bartering mead was four times as valuable as ale! When compared to their southern neighbours, the northern Europeans had few opportunities for using simple sugars for their fermentations, and honey was probably the most accessible.

From European archaeology, there are several attestations of mead, or at least honey containing alcoholic beverages, from Bell Beaker ceramics. Some highly regarded authorities, the late Andrew Sherratt for example,[131] have suggested that the characteristic vessels of the Bell Beaker culture were primarily for imbibing such alcoholic beverages. The Bell Beaker culture refers to a pottery style widespread in Europe and North Africa during the third millennium BC, a period that approximates to the end of the Neolithic. The culture covers a huge geographical area, which has prompted a number of different interpretations. Bell Beaker territory extends from the British Isles to the North African coast, as its north-south extension, and from Hungary to Portugal as its east-west extension.

Figure 5.8   Bronze Age mead beaker from Ashgrove Farm, Methilhill, Fife (from Dickinson[133] - by kind permission).

Perhaps the most enlightening evidence in support of Sherratt's notion, certainly in terms of honey composition, comes from the cist-burial on Ashgrove Farm, Methilhill, Fifeshire. The burial, fully documented in Henshall's paper,[132] contained a spilled beaker (Figure 5.8) associated with a fine metal dagger. Pollen samples from vegetation adjacent to the prostrate beaker gave a pollen spectrum that matched that of the material impregnated on the inner wall of the beaker – probably reflecting its original contents. The pollen profile[133] was dominated by small-leaved lime (*Tilia cordata* Mill.) at 54%, and this was followed by meadowsweet (*Filipendula* (presumably *F. ulmaria* (L.) Maxim.)) at 15%. Heather (*Calluna* sp.) and ribwort plantain (*Plantago lanceolata*) were both at around 7% of the sample. The huge presence of the lime (and, to a lesser extent, meadowsweet) presented a problem, because the former is not native to this part of Scotland and there is no evidence of any planting nearby. Why so much pollen from a largely insect-pollinated tree?

Meadowsweet, while common throughout much of the UK, also had a high presence, and gave food for thought. A number of possibilities for the observed pollen profile were considered, but Dickson considered that the liquid spilled from the beaker had been mead (which removed the need to explain why there was so much local small-leaved lime). To support this, it is known that bees are very keen on *Tilia* nectar. Similarly, meadowsweet pollen is avidly gathered by honey bees, and both of these plant species are known constituents of British honey[134] and, even in those days, it would have been possible to transport honey (or mead) over large distances (indeed, the Ashgrove honey could have come from northern England, where *Tilia cordata* is recorded!).

Meadowsweet crops up numerous times as an ancient honey component, as a couple of its ancient names, *medesweete* (= "mead-sweet") and *medewurte* ("meadwort"), would suggest. One of the most celebrated Bronze Age records involving meadowsweet comes from a burial mound at Egtved, Denmark.[135] The tumulus contains an oak-coffin burial and associated with it was a birch-bark bucket, which contained the remnants of a mixed beverage (in the form of a dark brown crust). The grave was that of a young woman (18–20 years old), and the pail, sewn together with bast, was placed near her feet.

Analysis of the dregs in the bucket suggested that the original drink had three components: 1. honey, as indicated by the pollen of *Tilia* sp., white clover (*Trifolium repens* L.) and meadowsweet; 2. fruits and leaves used respectively as a source of sugar, and for flavouring. The fruits were identified as being from either *Vaccinium oxycoccus* L. (cranberry), or *V. vitis-idaea* (cowberry), and the leaves were those of sweet gale (*Myrica gale* L.); and 3. cereal grains (identified as wheat and, therefore, most likely to be emmer). Thus, superficially, this beverage would appear be a mixture of ale and mead with, perhaps, a hint of a fruit wine. Similar dregs were recovered in Denmark, from the grave of a Roman Iron Age woman, who was buried with a rich set of grave goods at Juellinge on the island of Lolland. The dregs were found in a bronze bowl, which had an accompanying ladle and strainer. Also present nearby were a glass vessel, and two drinking horns.

Two other Danish Bronze Age beakers contained botanical remains that yielded a pollen profile similar to those encountered at Egtved and Juellinge; these were at Nandrup, on the island of Mors in Jutland, and Bregninge on Zealand.[133]

This sort of mixed beverage is not a million miles away from what had been reported by Tacitus, whose work was contemporary with the Danish early Iron Age. He rather ruefully describes a Germanic beer or cider-like drink as a bastard type of wine, and something that was highly distasteful to a Roman palate.

That honey was transported over wide areas in order to ferment was confirmed by Manfred Rösch,[136] who performed pollen analyses on the organic contents of bronze vessels from royal burial sites dating to the Iron Age in southern Germany. Rösch examined artefacts from fortified hill-top sites at Heuneburg, Hochdorf and Glauberg, all dating to the late-Hallstatt/early-La Tene period, and all within 100 km of each other. The Glauberg barrow dates to the fifth century BC, while the burials at Heuneburg and Hochdorf are from the previous century. All pollen assemblages observed were dominated by non-arboreal pollen from non-wind pollinated species, a characteristic feature of honey. Furthermore, beeswax could also be detected by chemical analysis. Rösch describes the type of drink found in the samples as "fresh mead unfiltered", or "drink sweetened by honey". One beverage was listed as "unknown". He also maintained that the high diversity of the pollen assemblages differed from recent honeys and "points to a high biodiversity in the Iron Age landscapes, and to the use of honey mixtures that originate from a large area that included the surrounding uplands". In a later paper, Rösch[124] maintained that these mixed honeys were characteristic of Central Europe at this time.

The first evidence for pure mead in temperate Europe comes from the Iron Age royal burial site at Hochdorf, near Stuttgart, where a large bronze cauldron set in one corner of the grave chamber contained an organic deposit consisting largely of pollen grains and beeswax. The double-walled tomb at Hochdorf is dated to around 525 BC and contained the body of a middle-aged male laid out on a bronze couch. Also in the tomb were prestige goods, such as a four-wheeled wagon (laid out with table settings for nine people), rugs on the floor and fine textiles lining the walls. The 500-litre capacity cauldron, which was seemingly three-quarters full of liquid when placed in the tomb, was of Greek design. Pollen analysis of the dried remains strongly suggested that the liquid was mead (mostly), and that the honey from which it was made contained the pollen from around 60 different plants, including wild thyme, lime and willow.[137] It was impossible to say whether the Hochdorf cauldron liquid was solely mead, or whether it was a mixed, honey-containing, beverage, the sort of thing that has been called "grog" by Patrick McGovern.[138] The pollen content of the residue was indicative of a substantial honey content (15–60%), suggesting a very high-quality beverage. One thing that is fairly certain though is that the artefacts and the beverage in the royal prince's tomb were strongly indicative of a communal feast that had been held in his honour.

Among the Germanic tribes mead was the drink of the gods – often being consumed straight from vats! To the Celts, there were rivers of mead in paradise. By noting the quantities of mead that were apparently consumed on Earth by these folk, it is pretty evident that apiculture must have been widespread at the time.

In the British Isles, where the honey bee is indigenous, mead has long been a favourite drink (see above), and especially so during Anglo-Saxon times.[34] Mead features prominently in the old English poem Beowulf (which translated as "bee-wolf"), in which the "mead-hall" and the "mead-bench" are frequently mentioned. The following is from the fragment which tells of Beowulf being bidden by King Hrothgar at Hart Hall:

Site nú tó symle [Sit now to banquet]
and onsœl meodo [and unseal with mead]
sige hréther secgum [brave breast with warriors]

5.1.12.2 Making Mead

Andrew Boorde,[139] quite simply, says of mead: "Meade is made of ho*n*ny and water boyled both togyther; yf it be fyned and pure, it preseruth helth; but it is not good for them the whiche haue the Ilyacke or the colycke" … and of metheglyn: "Metheglyn is made of ho*n*ny and water, and herbes, boyled and sodden togyther; yf it be fyned and stale, it is better in the regyment of helth than meade."

Some of the very first written practical details for making mead were penned by Pliny the Elder in *Naturalis historia*, one translation being: "A wine is also made of only water and honey. For this it is recommended that rain-water should be stored for five years. Some who are more expert use rain-water as soon as it has fallen, boiling it down to a third of the quantity and adding one part of old honey to three parts of water, and then keeping the mixture in the Sun for forty days after the rising of the Dog-star. Others pour it off after nine days and then cork it up. This beverage is called in Greek 'hydromeli'; with age it attains the flavour of wine. It is nowhere more highly prized than in Phrygia."

One of the most amazing books that I have perchanced upon regarding mead is Sir Kenelm Digby's *The Closet of Sir Kenelm Digby Knight Opened*, first published in London in 1669.[140] Digby, amongst other accolades, a Fellow of the Royal Society, provides a series of recipes for mead (meath, meathe), metheglin and hydromel and in his introduction "To The Reader" explains that "There is herein a sufficiency of Solids as well as Liquids for the sating of Curiosities of each or the nicest Palate; and according to that old Saw in the Regiment of Health, *Incipe cum Liquido*, &c. The Liquids permitted to the Solids. These being so Excellent in their kinde, so beneficial and so well ordered, I think it unhandsome, if not injurious, by the trouble of any further Discourse, to detain thee any longer from falling to; Fall to therefore, and much good may it do thee."

In all, there are 50 mead/meathe recipes, including "white" and "strong" versions, 48 recipes for metheglin, including "white" and "small", and two ways to make hydromel. If there had been a seventeenth-century *Who's Who*, then Digby's entry would have included "Scholar, traveller, archaeologist and sometime pirate"; he is also credited with making a relatively cheap, sturdy wine bottle.

Under the heading of "To Make Excellent Meathe", an example of a Digby mead recipe is as follows: "To every quart of Honey, take four quarts of water. Put your water in a clean Kettle over the fire, and with a stick take the just measure, how high the water cometh, making a notch, where the superficies toucheth the stick. As soon as the water is warm, put in your Honey, and let it boil, skimming it always, till it be very clean; then put to every Gallon of water, one pound of the best Blew-raisins of the Sun, first clean picked from the stalks, and clean washed. Let them remain in the boiling Liquor, till they be thoroughly swollen and soft; Then take them out, and put them into a Hair-bag, and strain all the juice and pulp and substance from them in an Apothecaries Press; which put back into your liquor, and let it boil, till it be consumed just to the notch you took at first, for the measure of your water alone. Then let your Liquor run through a Hair-strainer into an empty Woodden-fat [vat], which must stand endwise, with the head of the upper-end out, and there let it remain till the next day, that the liquor be quite cold. Then Tun it up into a good Barrel, not filled quite full, but within three or four fingers breadth; (where Sack has been, is the best) and let the bung remain open for six weeks with a double bolter-cloth lying upon it, to keep out any foulness from falling in. Then stop it up close, and drink not of it till after nine months."

The above seems to be a relatively straightforward Digby recipe, certainly in terms of herb additions. Many of his mead recipes are flavoured as well as those for metheglin (which should have added herbs). Here is a metheglin recipe; it is announced as "Sir Thomas Gower's metheglin for health":

"First boil the water and scum it; then to 12 Gallons put 6 handfuls of Sweet-bryar leaves, of Sweet-marjoram, Rosemary, Thyme, of each one handful: Flowers of Marigold, Borrage, Bugloss, Sage, each two handfuls. Boil all together very gently, till a third waste. To eight Gallons of this put two Gallons of pure honey, and boil them till the Liquor bear an Egge, the breadth of threepence or a Groat, together with such spices as you like (bruised, but not beaten) an ounce of all is sufficient.

You must observe carefully. 1. Before you set the Liquor to boil, to cause a lusty Servant (his Arms well washed) to mix the honey and water together, labouring it with his hands at least an hour without intermission. 2. That when it begins to boil fast, you take away part of the fire, so as it may boil slowly, and the scum and dross go all to one side, the other remaining clear. When you take it off, let none of the liquor go away with the dross. 3. When you take it from the fire, let it settle well, before it be tunned into the vessel, wherein you mean to keep it: and when it comes near the bottom, let it be taken carefully from the sediment, with a thin Dish, so that nothing be put into the vessel, but what is clear. 4. Stop it very close (when it is set in place, where it must remain) cover it with a cloth, upon which some handfuls of Bay-salt and Salpeter is laid, and over that lay clay, and a Turf. 5. Put into it, when you stop it, some New-laid-eggs in number proportionable to the bigness of the vessel, Shells unbroken. Six Eggs to about sixteen Gallons. The whole Egg-shell and all will be entirely consumed."

Fermentation times seem to vary with recipe but, as a rule of thumb, "the longer you keep it, the better it will be".

It is evident from works like Digby's and others, such as William Harrison,[141] that the best honey was that which was allowed to drip out of the combs and, after this had been collected, the remainder was pressed out. The combs were then washed in water to remove any residual honey, and the "washings" were used as a basis for a run-of-the-mill mead. This drink, which was apparently common in Essex, was called "swish-swash" by Harrison. For details of a variety of meads and metheglins (and hydromels), see Renfrow.[142]

Traditional mead fermentations are protracted affairs, often taking several months and, for this reason, commercial manufacturers have the opportunity to employ immobilized cell technology. Aspects of the use of such technology on the food industry have been reviewed by Norton and Vuillemard.[143] Initial methodology for hastened (continuous) mead production employed calcium alginate immobilized yeast cells.[144,145] Calcium pectate gel beads have proved to be more efficient,[146] and the beads are less sensitive ions and chemical agents that destroy calcium alginate beads. The stability constant of calcium pectate is almost one order of magnitude higher than of calcium alginate. Mead production time can be reduced even more by employing the two-column packed-bed system[145] with an entrapped ethanol-tolerant strain of *S. cerevisiae.*[147]

Gayre informs us that there are five basic categories of mead, "traditional mead" being the drink that is made from honey and water, in the proportion of three pounds of honey to one imperial gallon of water. What he refers to as "sack mead" is brewed in a similar way to the above, but with 20–25% more honey. The drink should not have a honey "nose". "Metheglin" is basically mead with added herbs (*gruit* in olden times). "Sack metheglin", as expected, is metheglin with a higher proportion of honey. Gayre's final category comprises those beverages prepared by mixing honey with fruit juice, and fermenting the result. There are several types, of which "pyment" ("clarre") is perhaps the most widely known. It is a mixture of grape juice and honey, and was popular in medieval times. "Cyser" is a ferment of honey and apple juice, and the term "cyder" (or "cider") is seemingly derived from it. Cyser is credited with sherry-like characters. "Melomel", also known as "mulsum", describes any fermented drink containing honey and the juice of any fruit apart from grape and apple. Some authorities would say that mulsum is not a mead, because it is unfermented honey blended into a wine. Some of these honey-fruit juice mixtures have accredited names, "morat", for instance, being a ferment of honey and mulberry juice, and "rhodomel" one of honey and rose hips/petals.

For modern producers, there are set mead recipes, but our ancestors often used rule of thumb. Basically, any unprocessed honey sample will contain a yeast population capable of initiating fermentation once water is added – as long as there is sufficient mixing (aeration) when one is added to the other. Such products will, by definition, be highly variable. A low strength, artisanal mead was made by rinsing out honeycombs with water after honey had been harvested, and this tends to confirm the notion that making mead was a by-product of apiculture.

Commercial producers will usually pasteurize their honey, or add sodium metabisulfite to the must. A yeast culture would then be added. Fermentation is usually a protracted process and, when sugars have been exhausted, moribund yeast cells tend to settle out. After clarification, the product is bottled and left to mature. The alcoholic strength of meads fall in the range 8–18% ABV and, rather like wines, can be categorized as dry, sweet, still and sparkling.

Although ancient peoples would have used any honey that came to hand, lighter honeys (say from alfalfa and clover) are generally considered to be superior for making mead. A survey of wines made from the fermentation of honey has been provided by Morse and Steinkraus,[148] and Ken Schramm has written a useful book[149] on the practicalities of mead-making.

5.1.12.3 *Honey Wine*

Mead is essentially honey wine, but it is only one appellation of a series of alcoholic beverages made from honey in various parts of the world. Early scientific work on honey wines was carried out by Sarin,[150] who analysed around 40 products from Eastern Europe, and found that ABV varied from 6.4 to 16.6%. Related products that spring to mind are aguamiel, medovukha and ogol. The latter, a traditional Ethiopian honey wine, has been the subject of scientific study.[151] The beverage, characteristic of Majangir villages in southwest Ethiopia, is made from wild honey, rainwater and pulverized bark of the mange tree (*Blighia unijungata* Bak). The latter, one of the floral sources of the honey, imparts antimicrobial compounds to the must, and may also contribute nutrient material. The initial step is the production of a yeast starter culture (called *bite*), which involves incubating some ground bark, wild honey and fresh well water for around a fortnight. The mixture is then introduced into a wild honey/water mixture in a large earthenware pot and left next to a dying fire to ferment. The honey sample should contain wax and even the bodies of dead bees in order to contribute micro-nutrients for fermentation. By the third night, fermentation should be complete and the wine is then strained and is ready to drink.

By floating to the top of the must during fermentation, the ground bark, wax and any debris would help to create anaerobic conditions. The ground bark *etc.* is recovered after being strained off and, being impregnated with yeast cells, is used as an inoculum for a subsequent batch of ogol (*i.e.* acts as *bite*) – in which case only a couple of days are needed before it is seeded. The "natural" yeast was isolated and identified as *S. cerevisiae* (named strain ET99), and the final alcoholic content was found to be 16.5% by volume. Other chemical analyses are shown in Table 5.5.

Interestingly, when wine and sake strains of *S. cerevisiae* were inoculated into ogol must, final ethanol content increased to 17.5 % by volume. Except for pH readings, the final levels of all other determinands also varied.

Balché, another honey wine/mead, this time from the central Americas, was a favourite ritual drink of the ancient Maya living

Table 5.5   Some characteristics of ogol (after Teramoto *et al.*[151]).

Determinand	Value
Ethanol (%, v/v)	16.5
2-Methylpropanol (mg/l)	89.0
3-Methylbutanol (mg/l)	154.0
Propan-1-ol (mg/l)	43.0
Ethyl acetate (mg/l)	30.0
Iso-amyl acetate (mg/l)	0.4
CO_2 output (g)	11.5
Acidity (ml)	6.0
Initial pH	4.5
Final pH	3.8
Reducing sugar (mg/ml)	9.7

in the area of Yucatan, northern Guatemala and Belize. Traditionally, the drink was made by mixing stingless-bee honey and water in a large trough (or even a canoe), and then adding lumps of bark and root from the psychoactive leguminous tree (Fabaceae), *Lonchocarpus violaceous* (Jacquin) DC.[152] Thorough mixing of these ingredients causes fermentation to start immediately, and the resulting drink was imbued with magical properties.

In the pre-Hispanic period, the sign of *balché* can be found in hieroglyph manuscripts and in a number of archaeological objects containing astronomical and historical data. An early glyph is shown in Figure 5.9, and explanations of this, and other, signs are to be found in Thompson.[153]

The drink (or something akin to it) was first documented in a sixteenth-century text by Diego de Landa,[154] who makes a number of references to the liquid obtained from fermentation of a solution of honey in water with added root of a plant called *balché*. He noted that people used to "drink until they get drunk", and that "this wine is healthy for them because it acts as a purgative and it induces them to vomit worms; for these reasons they grow up strong and healthy and the old men live much and in good health".

The same tome gives details of the preparation of *balché*, and describes it as "a wine, rather strong with a bad smell". He continues: "I will say about the wine that the Indians like it very much; for this reason they plant [the tree] in their orchards and in the spaces near to their houses. The tree is ugly and has no other fruit than its roots in order to make wine with honey and water". Thompson[155] asserts that *balché* was associated with sacred ceremonies, and also reported on the contemporary production

Figure 5.9 Maya glyph, *balché*.

of the drink; an activity that he witnessed in 1930 in what was then British Honduras. The groups that he observed used tree bark, rather than root, which they washed and dried "for a certain time; the longer the better". Many Indians kept the bark for many months, and when they came to use it, they crushed it with stone hammers and placed it in a vessel that contains honey and water "in the same proportions". The resulting solution was fermented for "four to six days", at which point it was ready for use.

Some of the Maya peoples keep a store of empty tree trunks in a sacred hut in order to obtain the wild honey needed for *balché*, and honey obtained from this source could never be used for trivial (*i.e.* non-ritual) purposes. Thompson also gives a more detailed description of the making of *balché* in the "little town" of Cham-Kom: "Four pieces of bark, about 33 cm long, are cut into pieces, crushed and put in a jar with two cups of water and one cup of honey. This mixture is left at room temperature for three days. At this point the beverage is tested in order to ascertain its quality. If necessary, more honey is added, and the mixture is kept fermenting until it acquires a yellow colour. The technique in the latter case, is characterized by the progressive addition of honey in order to increase the alcoholic content. This technique is also called "ripening" because it is characterized by the appearance of the desired yellow colour."

It is said that *L. violaceous* became a cultivated plant because it was indispensable for Maya ritual.[154] The name *balché* also means "hidden tree", presumably thus named because it was not native, and had been used by a "foreign culture". The plant eventually became distributed to all Maya-inhabited areas.

Comparisons have been made between *balché* and other ritual beverages,[152] and with *pulque* in particular, where the use of *ocpatli* (*oc* = wine, *patli* = medicine) is used in its preparation. Unfortunately, during the early period of the Conquest, the Spanish authorities assigned hallucinogenic properties to *ocpatli* and prohibited its use (and even the making of *pulque*), and so there is a paucity of technical information. One thing is certain: the use of an aromatic, bittering plant with antimicrobial (preservative) capabilities is representative of a common theme in the preparation of ancient alcoholic beverages, the hop being only one of a long line.

The psychoactivity of *L. violaceous* was attributed four hydroxyl (or methoxy) prenyl-stilbenes, called longistyline A, B, C and D.[156] Longistyline A is isometric with longistyline C, and longistyline B is isometric with longistyline D. *Lonchocarpus longistylus* Pittier is a synonym of *L. violaceous*, hence the name "longistyline".

The Christmas vine, or "*ololiuhqui*" (*Turbina corymbosa* (L.) Raf.), a species of morning glory, was also used for making *balché* by the Maya (Ott, 1998). The plant, "*xtabentún*" in Mayan, is apparently not now in contemporary use by those people,[157] but is almost universally used by indigenous peoples in Oaxaca, and occurs elsewhere in Mexico.[158] It is reported that the Maya used to place all of their hives adjacent to natural stands of *xtabentún* so that all of their honey came from that flower. Thus, it seems as though *T. corymbosa* honeys were produced intentionally, and were very highly regarded because of their ergoline alkaloid (ergine) content which has effects on the uterus as well as on the mind. It was Hofmann[159] who found that the main constituents of *T. corymbosa* are D-lysergic acid amide (ergine) and D-iso-lysergic acid amine (isoergine), both of which are closely related to the ergot alkaloid derivative D-lysergic acid diethylamide (LSD).

Ololiuhqui is the least known in the outside world of the major Mexican divinatory agents, yet is probably the best known and most widely used among the Indians of that country. Its properties are as mind-blowing as those of peyotl. According to Wesson, *ololiuhqui* is the name of the seeds, not of the plant itself (the word translates as "round thing"). Other materials known/thought to have been included in Maya *balché* recipes include mushrooms of the genus *Psilocybe* and the toad *Bufo marinus*, whose venom contains a psychedelic component! Some workers have rejected *B. marinus* as being the toad responsible as the source of the ritual intoxicant because of the extreme toxicity of its venom, and have suggested that a desert toad, *B. alvarius*, is a more likely candidate.[160] *B. alvarius* secretes large amounts of the potent known hallucinogen, 5-methoxy-N,N-dimethyltryptamine (5-MeO-DMT).

One of the most popular honey-based fermented beverages in Ethiopia is *t'ej*.[161] It is a yellow, sweet, turbid (due to yeast) and effervescent drink, which varies with area of production. Honey can be collected from wild sources, or from hives, and proportions of honey to water can vary from 1 : 2 to 1 : 5 v/v; ethanolic content may vary between 7 and 14% ABV (although Desta's survey[162] of traditional *t'ej* samples ranged only from 13.18 to 13.73 ABV). Most production is artisanal and, since honey has always been a relatively expensive commodity in Ethiopia, was originally consumed only by the higher echelons of society. Honey wines (and honey itself) play an important role in feasts and ceremonies (particularly weddings). With roots that stretch right back to the fourth century AD, *t'ej* is regarded unofficially as the national drink of Ethiopia.

To counteract its inherent sweetness, bitter plant material is often infused into the must. The most commonly used plant is *Rhamnus prinoides* ("African dogwood"; a buckthorn, but often referred to as "hops"), known locally as *gesho*. In many regions of Africa all parts of the plant may be used for nutritional, medicinal or religious purposes. In the making of *t'ej*, the stems are boiled in water and the extract is added to aqueous honey. *Gesho* is also used in the preparation of *tella*, an Ethiopian beer. In many instances, the inner surface of the fermentation pot is smoked in order to impart an appropriate flavour. Spices may also be added. A flow sheet for the production of *t'ej* is given in Figure 5.10.

Experimental work with *t'ej* is not exactly bounteous, but Bahiru *et al.*[163] examined the microflora of 200 different samples from ten production units at different production times. The pH of the *t'ej* samples varied from 3.2 to 4.3 (mostly <4.0). As would be expected, yeasts were the dominant microbes in all samples giving mean counts of 6 log cfu ml^{-1} in all samples, with *S. cerevisiae* always contributing at least 25% of the yeast flora. Samples from all production units yielded at least two yeast species. All yeasts identified were ascosporogenous and fermentative, and other notable species present were *Kluyveromyces marxianus* (16%), *K. thermotolerans* (10%) and *Debaryomyces polymorphus* (14%). Other species encountered (all <10%) were *Zygosaccharomyces rouxii, Pichia subpelliculosa, Kluyveromyces lactis* and *Pichia burtonii*. It is assumed that these other yeasts make some contribution to *t'ej* fermentation, as they are known to do in some wine fermentations.[164]

Counts of mesophilic bacteria were too low to be of significance (mean counts generally being less than 3 log cfu ml^{-1}), while coliform numbers were below detectable levels. Aerobic bacterial spores were also calculated as being less than 3 log cfu ml^{-1}. When

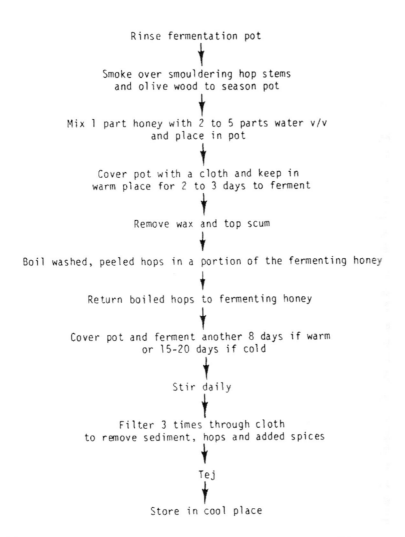

Rinse fermentation pot

↓

Smoke over smouldering hop stems
and olive wood to season pot

↓

Mix 1 part honey with 2 to 5 parts water v/v
and place in pot

↓

Cover pot with a cloth and keep in
warm place for 2 to 3 days to ferment

↓

Remove wax and top scum

↓

Boil washed, peeled hops in a portion of the fermenting honey

↓

Return boiled hops to fermenting honey

↓

Cover pot and ferment another 8 days if warm
or 15-20 days if cold

↓

Stir daily

↓

Filter 3 times through cloth
to remove sediment, hops and added spices

↓

Tej

↓

Store in cool place

Figure 5.10 Flow sheet for *t'ej* production (after Steinkraus[417] - by kind
permission).

the lactic acid bacteria (LAB) were selectively isolated, it was
found that at most sites heterofermentative forms outnumbered
homofermentative ones. LAB had counts of 6 log cfu ml^{-1} with a
significant variation within samples of a production unit. The
lactic flora consisted of species of *Lactobacillus*, *Leuconostoc*,
Streptococcus and *Pediococcus*, with lactobacilli being the most
prevalent.

We started this section with a quote from Toussaint-Samat,[165] and it is pertinent to end it with one from the same source: "Practically no mead is brewed today; try looking for it in the off-licence or on the supermarket shelves. Despite some efforts by farmers to popularize it, it remains a small folk industry, perhaps drunk occasionally at an ecological gathering, or as a conscious celebration of the past, or out of amused curiosity. One enjoys it and then forgets it, which is a pity, when it used to signify so much that it also now forgotten. Perhaps the gods really are dead."

5.2 THE GRAPEVINE

5.2.1 The Origins, Taxonomy and Biogeography of the Grapevine

The grapevine, *Vitis vinifera* subsp. *vinifera* L. (Figure 5.11), is a perennial, woody, heliophilous liane, which climbs by means of coiled tendrils. Like many plants that have been established in cultivation for a prolonged period of time, the grapevine is difficult to accommodate easily into any orthodox botanical taxonomy. What is generally accepted is that grape-bearing vines for making

0   20   40 mm

Figure 5.11   Wild grapevine *Vitis vinifera* subsp. *sylvestris* (after Zapriagaeva[166]).

wine belong to the genus *Vitis* L., a member of the Angiosperm family Vitaceae. Several other names have been proposed for the grapevine family, including Ampelideae and Ampelidaceae, but the accepted name under the International Code for Botanical Nomenclature is Vitaceae. The Vitaceae comprises some 1000 species in 12–14 genera (depending on taxonomy), and most of these are tropical, or sub-tropical, climbing plants. There are about 60 inter-fertile wild *Vitis* species distributed in Asia, North America and Europe under sub-tropical, Mediterranean and continental-temperate climatic conditions. *V. vinifera* subsp. *vinifera* is the sole European representative of genus, with two-thirds of species being native to North America, and one-third being distributed over Central and Eastern Asia. Other temperate members of the Vitaceae are to be found in the genus *Parthenocissus*, where Virginia creeper (*P. quinquefolia*) and Boston ivy (*P. tricuspidata*) will be familiar to many readers.

Systematics within the genus *Vitis* has created considerable controversy over the last century or so, and there are many more names in the literature than there are accepted species! Owing to the long history of grapevine, the possibility of vegetative propagation and the movement of plant material between geographical locations around the world, the exact number of cultivars is difficult to judge, and this gives rise to numerous homonyms and synonyms. Cultivar estimates vary from 6000 to 11,000,[167] with many authorities electing for the lower figure, and this rather grey area has been partially attributed to the lack of information on the relative importance of wild vines in Europe.

Currently, most botanists regard the wild ancestral grape (was *Vitis sylvestris* C.C.Gmelin) as the primitive form of the cultivated grape because of the close morphological resemblance and free gene flow between them,[168] and have accordingly reduced its taxonomic status to sub-species level within the *V. vinifera* crop complex.[169] Thus we have the modern domesticated vine as *Vitis vinifera* L. subsp. *vinifera*, with the wild Eurasian plant being *V. vinifera* subsp. *sylvestris* (C.C.Gmelin) Berger. Some works split *V. vinifera* L. into two sections (varieties), rather than sub-species. The status of the sub-species (variety) *sylvestris*, however, is somewhat controversial, not least because some authorities maintain that there is no difference between that and *vinifera*. A widespread synonym for the cultivated plant has been *V. sativa* Beger (or *V. vinifera* subsp. *sativa*).

Wild grapevines occur in disjunct populations from the Atlantic coast of Europe eastwards to Tadzhikistan and the western Himalayas,[170] and occasionally they come into contact with cultivated vines. This gives rises to complex, introgressive hybrid swarms, which form in distinct transition zones between the two types. The disjunct pattern of wild grape distribution has long occupied biogeographers (see below).

Negrul,[171] for example, who travelled widely and examined thousands of samples, concluded that the *Vitis sylvestris* of Gmelin is taxonomically equivalent to the cultivated forms of *V. vinifera*. In many regions, the distinction between the wild and cultivated forms of *V. vinifera* is often obscured by a continuum of intermediate weedy forms, some of which are the result of hybridization. As Jansen *et al.*[172] have observed, although the Vitaceae have great economic importance for man, the phylogenetic placement of the family is still unresolved. Modern techniques, such as the use of complete chloroplast genome sequences, may well alter this situation (see below).

One of the first classifications of the genus *Vitis* was by the French ampelographer Planchon,[173] who regarded American and Asiatic species as sufficiently different that they warranted placement in separate series. The main problem with Planchon's scheme, and with some subsequent schemes, is that only those species with practical uses in propagation are dealt with. In particular, species found to be useful as germplasm in the "war" against phylloxera, are given priority status, whilst other species are, more or less, ignored. The result is a somewhat blinkered and unbalanced view of the genus. Another factor contributing to the difficulty sometimes experienced in delimiting wild *Vitis* species is the fact that some of the geographical and ecological barriers that would normally ensure that different species were reproductively isolated from each other are incomplete and permit hybridization. Such a phenomenon, which occurs in some parts of North America and Eastern Asia, causes a great deal of confusion, with one species merging into another. This would not occur, of course, where two species were reproductively isolated from each other, either by incompatibility or hybrid sterility.

Fossil evidence suggests that archaic vines can be traced back to Tertiary sediments some 60 million years old, and that by the beginning of the Pleistocene (*ca.* 2 million years BP) two sub-genera had evolved, *Euvitis* and *Muscadiniae*, both of which were distributed over the mid-latitude regions of North America and Eurasia. Glaciation decimated muscadine grapes, and eliminated them from all but an area extending around the Gulf of Mexico into southeastern USA and, as a result, there are now only three extant species assigned to this sub-genus. In the above-mentioned region of the Americas, one muscadine species, *V. rotundifolia*, has been commonly employed to make sweet wines, known locally as "scuppernong". All other remaining species of *Vitis* are assigned to the section *Euvitis* (some taxonomies would have "*Vitis* sub-genus *Vitis*"). As we shall see, the two sub-genera are sufficiently distinct that some authorities have elevated muscadine grapes to genus level. Biogeographically, wild *Euvitis* species in North America were concentrated east of the Rocky Mountains, and of the 30-plus species identified, *V. lambrusca* is by far the most important oenologically, its genes being traceable to most of the major grape cultivars in the USA. The current distribution of species in the *Euvitis* sub-genus includes North America, Central America, northern South America (the Andean chain in Colombia and Venezuela), Europe and Asia, whereas the *Muscadiniae*, as indicated above, are restricted to northeastern Mexico and the southeastern USA. Across Eurasia, wherever wine-drinking is a part of tradition, one species of *Euvitis*, *Vitis vinifera* L., dominates, suggesting that its superiority for making wine must have caused it to replace other competing species. This is a reasonable assumption because, from Central Asia eastwards to Japan and Southeast Asia, where drinking wine made from grapes is not ingrained into the culture, and therefore the attributes of *V. vinifera* are not put at a premium, a number of different species can still be found. So, what are the oenologically important characters of *V. vinifera*? Firstly, the fruit achieves a fine balance between fermentable sugar content and acidity and, secondly, with human assistance, the plant has evolved into a vigorous, self-reproducing hermaphrodite, capable of producing a virtually limitless number of stable cultivars with subtle variations in taste, colour and aroma. Apart from being the result of some of its biological characteristics, the dominance of *V. vinifera* in Europe and

Western Asia may well be partly attributable to the physiography of parts of that area, which could have played an important role during past glacial and interglacial periods. It is fairly evident that the genus *Vitis* appears to have established its present geographical ranges by the end of the last major Quaternary glacial period (*ca.* 8000 BC), and it is thought that repeated glacial advances and retreats during the Quaternary, and the alignment of the major mountain ranges, dramatically affected the evolution and distribution of *Vitis* species, particularly *V. vinifera*. In the Americas, for example, the mountain ranges are to be found running predominantly north-south, whereas in Europe and Western Asia, they are mostly aligned east-west. Such arrangements would have allowed the displacement of North American and Eastern Chinese taxa southward in advance of the expanding ice-sheets, but prevented the same happening to species in Europe and Western Asia. In the latter areas upland masses, such as the Himalayas, Caucasus, Alps and Pyrenees, would have provided effective barriers to any southward movement of *Vitis* in the face of advancing ice. Thus, we have the situation where there are numerous species of *Vitis* in North and Central America, several species in China but only one, *V. vinifera*, in Eurasia. It is a sobering thought that the one surviving Eurasian taxon has, during the last 500 years, been broadcast all over the temperate regions of the planet by *Homo sapiens*, and is now responsible for almost all of the world's wine! The present distribution of the wild Eurasian grapevine is almost certainly a fraction of what it would have been some 50 million years ago, when there was a warmer climate. This ancient, more widespread distribution would have appertained until the advent of the most recent Quaternary Ice Age, which can be dated to some 2.5 million years ago. More latterly in the geological time-scale, man has carried *Vitis vinifera* outside of its natural range, and the plant can now be found growing in the tropics, as well as its favoured mid-latitude haunts. Other *Vitis* species were undoubtedly present in Eurasia immediately prior to the most recent ice age, but only *V. vinifera* actually survived the glacial advances, thanks to it finding shelter in certain areas that provided a suitable micro-climate (known as refugia). Evidence suggests that pockets of *V. vinifera*, apart from being able to merely survive the glacial advances, were vigorous enough to be able to expand the range of the plant (*i.e.* re-colonize) during interglacial times. In Europe, refugia occurred around the Mediterranean basin, and south of the Black and Caspian Seas, areas that played an important role in the overall evolution of the grapevine. In some areas the vine started to expand its range immediately after the ice retreated. This is certainly true in some regions of southern France, where, after displacement, *V. vinifera* was again apparently flourishing by 8000 BC, from which time onward the climate in Europe started to warm gradually. Partial domestication of the vine (as indicated by seed remains having properties intermediate between wild and domesticated forms) seems to have occurred around 5–6000 BC in Europe. This is several thousand years before the agricultural package is thought to have spread into Western Europe.

5.2.2 The Early Days

Homo sapiens would have probably encountered the wild vine, *Vitis vinifera* subsp. *sylvestris* (Figure 5.11), at a very early date, when the first groups of humans migrated from East Africa some 2 million years ago. The most likely venues for making the acquaintance were the upland areas of eastern Turkey, northern Syria or western Iran, or, maybe, the hilly lands of Palestine and Israel, or the Transjordanian highlands. Significantly, most eastern Mediterranean myths pinpoint the origin of viniculture to somewhere in northeastern Asia Minor. In theory, however, the event could have occurred anywhere within the distribution range of the wild grapevine (Figure 5.12). As we have already intimated, Palaeolithic man was probably the first to become familiar with wine, purely by the accidental "spoilage" of stored, or over-ripe, grapes. Wine may, of course, have been the result of unsuccessful attempts to store grape juice, which is a particularly unstable beverage. Because our Stone Age forebears did not dwell in permanent, year-round settlements, they had no opportunity to investigate what was actually happening when grapes/grape juice spoiled (fermented), and so they were unable to learn about and perfect the process. It was not until man adopted a sedentary way of life, and had developed the need for a continuous supply of wine, that he put his mind to a rudimentary form of viticulture. One thing that is indisputable is that, like mead, production of a basic grape wine from its raw material requires relatively little human interference; it is practically unavoidable.

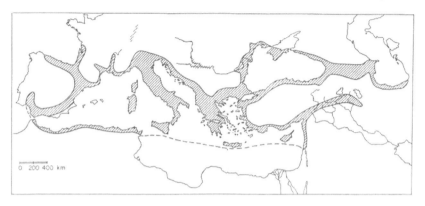

Figure 5.12   The distribution range (shaded area) of *Vitis vinifera* subsp.
sylvestris. Note that toward the east the wild vine extends beyond
the boundaries of this map, and reappears in a few places in
Turkmenistan and Tadzhikistan (from Zohary and Hopf[170] - by
kind permission).

From what we know at the present time, the Neolithic period in
the Near East, somewhere between 8500 and 4000 BC would seem
to provide all prerequisites for the intentional manufacture of
wine. Not only was mankind flirting with agriculture and
beginning to lead a settled existence, but he was starting to
manufacture items such as pottery vessels, essential items if liquids
were to be stored for any period of time.

The world's earliest known pottery, from Japan, dates to the
eleventh millennium BC, but it seems to have been "invented"
independently in many places throughout Anatolia, Mesopotamia
and the Levant almost simultaneously around 7000–6000 BC. The
earliest pots were simple cups, vases and dishes, probably
originally sun-dried. Later, more sophisticated vessels were fired
in kilns. Although probably used by nomadic peoples, pottery was
a real benefit for those with a sedentary lifestyle. With the
domestication of plants and animals, clay pots provided a sound
way to store liquids (*e.g.* milk, fruit juice) and, most importantly,
the wherewithal for cooking (heating and boiling) and fermenting.
The inventor of the first clay pot is unknown, but its creation was
most likely attributable to women who carried most of the burden
for domestic chores. The ability to heat food, and to ferment,
opened up a whole new world of potential food sources that were
safer, tastier and easier on the digestive tract.

Although there is some controversy, the cultivation and domestication of the vine seems to have occurred between the seventh and the fourth millennia BC. If we are looking for a single site for the origin of the domestic grapevine (the Noah "hypothesis"), then, according to the archaeological, archaeobotanical and historical information that we have at present, some northern mountainous region of the Near East would appear to be the most likely. With the ingenuity of *Homo sapiens*, and the vast natural range of the wild vine, it is tempting to suggest that there might have been multiple domestications of the plant in different places, and at slightly different times. This is not impossible, but all the available evidence does seem to point to our upland site in the northern part of the Near East.

As we shall see, the earliest confirmed traces of wine have been found in sediments in a pottery jar from the Neolithic village of Hajji Firuz in northeastern Iran. Relatively little is known about wine making at Neolithic sites further north and at higher altitudes in the Taurus and Caucasus mountains, where the wild sub-species of *V. vinifera* still thrives today. Domesticated grape seeds have been recovered from Chokh in the Dagestan mountains of the eastern Caucasus, dating from the beginning of the sixth millennium BC, and from Shomutepe and Shulaveri along the Kura River in Transcaucasia, dating from the fifth through early fourth millennia BC. Should confirmatory residue analysis from pottery fragments from these areas ever be possible, then we may be able to locate the ultimate origins of viticulture. Unfortunately, few Neolithic sites in this vast region have, thus far, been excavated. It has long been thought that the domestication of *V. vinifera* occurred in Transcaucasia, or in neighbouring Anatolia, around 4000 BC. The region between and below the Black Sea and the Caspian Sea, where the grapevine is indigenous, represents the closest the vine comes to the Near Eastern origins of plant and animal agriculture. As Jackson[174] says: "Therefore it seems reasonable to assume that grape domestication may have first occurred in the Transcaucasian Near East." Certainly, as far as Vavilov's "centres of genetic diversity" are concerned, it is in this Transcaucasian region that wild grapevines are both the most abundant and the most variable (see p. 384). Before this, of course, wild vines would have been used as a source of grape juice and, as is the case for most domesticated plants, they would have required constant attention from man in order to give him satisfactory yields.

The notion that viticulture and viniculture originated somewhere in the Near East, and were disseminated from there, is supported by the fact that there is a remarkable similarity in most Indo-European languages in the words for "vine" and "wine"; indeed, Renfrew[175] goes further and maintains that the spread of agriculture into Europe resulted from the dispersal of peoples speaking Proto-Indo-European languages. Conversely, there is little resemblance between the words for "grape" in the same group of languages, and this is taken to signify that the appreciation of, and use of, the grape occurred a long time before the advent of viticulture and viniculture, and the above-mentioned dispersal of Indo-European languages into Europe. In contrast, Semitic languages, which evolved in regions where the vine was not indigenous, have often adopted words for "vine" and "grape" that are very similar to those for "wine", and are thought to be related to *woi-no*, an ancestral term for wine. According to some authorities, this lexical similarity implies that knowledge relating to grapes, viticulture and viniculture was obtained concurrently. If the above Noah hypothesis is correct, then it is possible that from Transcaucasia the vine (and wine) became transplanted into the Indian sub-continent, and then back to the Mediterranean basin. Support for such a notion comes from the fact that, in Sanskrit, the oldest of the Indo-European family of languages, "vena" translates as "favourite".

5.2.3 Dissemination of Viticulture

In simplistic terms, it is still commonly believed that domesticated vines were carried westwards by humans, along with the shift of agriculture generally, into the Mediterranean Basin, and from thence into different parts of Europe by Greeks and Roman colonists. The Greeks are credited with the transport of cultivated vines into Italy, and the Romans took them to Spain, France and Germany. It should be stressed, however, that movements such as these are not as certain as parts of the historical record might suggest, for changes in *V. vinifera* seed and pollen morphology indicate that domestication was occurring in some parts of Europe before the agricultural revolution supposedly reached that continent! Work by Stevenson,[176] for example, has indicated that an extensive system of viticulture was already in existence in southern Spain several centuries before colonization by the Phoenicians. Until many wild grapevine populations were annihilated by "foreign" pests and diseases in the mid-nineteenth century, the sub-species *sylvestris* remained abundant throughout its natural range, from Spain to Turkmenistan. Large populations of wild vines still exist in certain regions within their indigenous range, such as southwestern Russia, where they were afforded protection from the above-mentioned scourges, and have been used for wine-making over several millennia. It is thought that these locally occurring and adapted vines might be the progenitors of most of today's European cultivars, and it is notable that the Pinot group of cultivars, for instance, possesses many traits resembling *V. vinifera* subsp. *sylvestris*.

From whatever its exact place of origin, the domesticated vine seems to have diffused in two directions. One route was to Assyria, and thence to the Mesopotamian city states of Kish and Ur, and, in later times, Babylon, where wine, because of its rarity, was mostly regarded as a drink for the upper classes, and the priesthood. The hot, dry climate and soil salinity in Mesopotamia were not conducive to growing vines, and so most of the wine consumed by the civilizations there was imported from further north, from Assyria, for example, where there was higher rainfall. From Mesopotamia, under the influence of man, wine and the vine reached the Jordan Valley, somewhere around 4000 BC. The vine never grew in this part of the world, and from there it reached ancient Egypt, where, again, the grapevine is certainly not indigenous, but where there were certain areas favourable for the growth of *Vitis*. Vines became established in the Nile Delta by at least 3000 BC, at the beginning of the Early Dynastic period and, by New Kingdom times, a very sophisticated form of viticulture was being practised. Viticulture followed the Nile upstream into Nubia, where it flourished until the advent of Islam. By 1000 BC wine drinking was widespread over much of the Near East, and was practised by, amongst many others, the Hebrews, the Canaanites and the closely related Phoenicians, even if it was largely restricted to the higher strata in society.

The second route carried wine across Anatolia to the Aegean, where, by 2200 BC, it was being enjoyed by the Minoan civilization on Crete, and in Mycenaea, where viticulture and viniculture became highly specialized, and wine was an important commercial commodity. Of great significance is the fact that wine (and vines) became an integral part of Greek culture and, wherever they established a colony around the Mediterranean, wine drinking, viticulture and viniculture were sure to follow. Indeed, Sicily and southern Italy became very important vine-growing areas to the Greeks. Another important site was Massilia (now Marseilles), from where the Greek methods of viticulture and viniculture spread inland, following the valley of the River Rhône. The suggested major diffusion routes of viticulture in Southwest Asia and Europe are shown in Figure 5.13.

The possibility of two independent domestication events was voiced by Arroyo-García *et al.*,[178,179] who employed chloroplast microsatellite markers (DNA polymorphisms). The team were addressing the existence of morphological differentiation between grape cultivars from the eastern and western ends of the modern distribution of the Eurasian grape, which suggests different genetic contributions from local wild (*sylvestris*) populations or multi-local

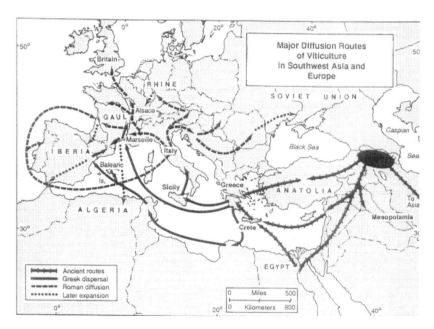

Figure 5.13   Likely diffusion routes of viticulture in Southwest Asia and Europe (after de Blij[177]).

selection and domestication of *sylvestris* genotypes. The analysis of 1200 samples of wild and cultivated genotypes suggested the existence of at least two important origins for the cultivated germplasm, one in the Near East and another in the western Mediterranean region. The latter appears to have given rise to many of the extant Western European cultivars (over 70% of cultivars on the Iberian Peninsula show chlorotypes that are only compatible with their having been derived from western *sylvestris* populations).

5.2.4 Chemical Evidence for Ancient Wine

In 1996, McGovern *et al.*[180] reported on the re-investigation of some residues found on the inner surfaces of pottery sherds, the analysis of which showed, beyond doubt, that the original jars had contained wine. The sherds had been recovered in 1968 from Hajji Firuz Tepe, a Neolithic site southwest of Lake Urmia, in the northern Zagros Mountains. The site is on the eastern fringes of the "Fertile Crescent". Wild *Vitis* still grows in this region, and pollen cores taken from the deposits of Lake Urmia showed that it grew there during Neolithic times. The site was excavated as part of the University of Pennsylvania Museum's Hasanlu project. The jars had been found embedded in the earthen floor of the kitchen of a mud-brick building dated to *ca.* 5400–5000 BC. During the original excavations, a yellow residue was noted on the inside lower half of a jar fragment. At the time, the deposit was assumed to be from some sort of dairy product, even though chemical analysis yielded no positive results. After a gap of 25 years, some sherds (one with a reddish residue) were re-examined using more sophisticated methods. Results showed that the deposits contained tartaric acid, calcium tartrate and the oleoresin of the terebinth tree (*Pistacia atlantica* Desf.). Tartaric acid occurs naturally in large amounts only in grapes, and was converted into its insoluble calcium salt in the calcareous environment of the site. *P. atlantica* grows abundantly throughout the Near East, and was widely used as a medicine and a wine preservative in antiquity. Judging by their long, narrow necks, the fact that the residues were confined to their bottom halves and the presence of clay stoppers of approximately the same diameter as the necks, it was evident that these jars once contained some sort of liquid, and had been sealed. All the evidence supported the conclusion that the Hajji Firuz jars originally contained resinated wine, the like of which has been found in other, more recent contexts in the ancient Near East and ancient Egypt. If all the six jars found contained wine, then it was estimated that there would have been around 14 gallons in all, quite a significant production for household use. As McGovern commented:[181] "If the same pattern of usage were established across the whole of the site's Neolithic stratum, only part of which was excavated, one might conclude that the grapevine had already come into cultivation." He also tentatively asked the question (somewhat tongue-in-cheek) whether the difference in colour (yellow and red) of some of the deposits might have represented early attempts at making white and red wine.

The first published chemical evidence for ancient Near Eastern wine[182] emanated from excavations carried out on the Period V site at the Late Chalcolithic village of Godin Tepe, in the Kangavar Valley in west-central Iran, during the years 1967–1973. The site is located high in the Zagros Mountains, and Period V dates from 3500–2900 BC, which is contemporary with the Late Uruk period (as the Late Chalcolithic in lowland Greater Mesopotamia is known) in southern Mesopotamia, and sits alongside the "High Road", or the "Great Khorasan Road", which later on became part of the famous "Silk Road", leading from Lower Mesopotamia to Iran. Godin Tepe controlled the most important east-west route through the Zagros Mountains between Baghdad and Hamadan, and may have been a Sumerian or Elamite trading post. Certainly, Godin Tepe was well positioned to participate in trade and, especially, to protect the trade route, being situated, as it was, some 2 km above the alluvial lowlands. It was adjacent to the eastern edge of Lower Mesopotamia, where some of the world's earliest literate cultures had formed themselves into city-states, such as Lagash, Ur, Uruk and Kish. In addition, the Elamite capital of Susa (modern Shush, in Iran) was some 300 km to the south, and the proto-Elamites at this time were already well on their way toward developing an urban way of life. The city-states in the Tigris-Euphrates valley were based on the irrigation culture of cereals, dates, figs and other plants.

From evidence of imported items, these adjacent lowland cultures appear to have been in contact with several other parts of the Near East, such as Anatolia, Egypt and Transcaucasia.

There was much evidence of an import trade in precious commodities, such as gold, copper and lapis lazuli, and even in everyday requirements, such as wood and stone, which were not readily available in the lowlands. Many resources, essential to urban lowland life, were available in the Zagros Mountains, and so Godin Tepe was ideally placed to take account of, and assimilate, developments occurring elsewhere in the world.

In 1988, some years after the original excavations at Godin Tepe had been completed, a reddish residue found on some pottery sherds from unusually shaped jars was re-examined, using infrared (IR) spectroscopy. The spectra obtained clearly showed the predominant presence of tartaric acid, and its salts, principal components of grapes, and it was presumed that the jars had once contained wine, which had evaporated and left a residue. The wine jars proved to be of a type not found elsewhere or, at least, imported from a region not explored archaeologically (or not reported). The jars, which had been stored on their sides, and had been stoppered, were unique in several respects, all of which were consistent with them being used to store liquid. Firstly, a rope design had been applied as two inverted "U" shapes along opposite exterior sides of each vessel (termed *rope-appliqué*). The specific rope pattern on the jars implied that actual rope had once been laid underneath, in an effort to stabilize them once they had been laid on their sides. The red, organic deposits on the interior of the one whole vessel available for examination were confined to the base and to the side wall directly opposite that which had the rope decoration on its outer surface – exactly the position that one would expect if precipitation of sediment had occurred inside the jar after it had been placed on its side. Secondly, the jars had tall necks, and relatively narrow mouths, certainly when compared to other known jars of this period. Thirdly, the interior of one of the sherds appeared to be lined with a slip (a compact, fired fine layer of clay), which would have helped to make the vessel resistant to leakage. All in all, these jars, with their elongated, narrow necks, were ideally suited for storing, transporting and pouring wine. The question now arose as to whether the wine was produced locally, and probably destined for export, or whether it was imported from elsewhere. As Badler and co-workers reported,[182] it is quite possible that the wild Eurasian grapevine grew around Godin Tepe in antiquity, since grapevines are common today throughout parts of the Zagros Mountains, especially some 500 km north of the village. The ecology and climate of the Godin Tepe region, with a perennial supply of water provided by its river, the Gamas Ab, is well suited to growing wild or domesticated *Vitis*, and there have been occasional reports of the vine growing there. Moister, milder conditions in ancient times might have permitted viticulture during the Late Uruk period. Excavations from Period V yielded no unequivocal evidence of grapes, but grape seeds dating to the much later Period II (first millennium BC) have been identified. In the words of Badler *et al.*:

"Given the beneficial upland environment for grape cultivation, and the evident contacts that existed with lowland horticultural areas, it is quite conceivable that viticulture and wine trade already existed at Godin Tepe and other key sites in the hill country in the Late Uruk period."

Also found with the jars was an unusually large funnel, and circular lid, which might have been used as a strainer of some sort (*e.g.* for extracting juice from grapes), and might be indicative of viniculture. If, as seems the most likely scenario, the wine was indeed imported, rather than being indigenously produced, then the most likely source was Transcaucasia, where Godin Tepe had contacts by the fourth millennium BC. Another possibility is that the wine came from slightly further west, in southeastern Anatolia. The wild grapevine grows prolifically along the northern and southern boundaries of Anatolia, and it is known that consumption of fresh grapes, or raisins, extends at least as far back as the Neolithic period, as exemplified by the ninth millennium BC site at Çayönü. Cultivation and processing of *Vitis* may date to as early as the late-fourth millennium BC in southern Turkey in, for example, Kurban Höyük.

The hot, dry climate in most of Mesopotamia made it difficult to cultivate the grapevine. Vineyards were mostly to be found in the northern, hilly, region. There is little evidence of vineyards prior to the second millennium BC, but one example comes *via* documents from the time of King Gudea of Lagash (2125–2110 BC), that make reference to irrigated terraces, protected by trees, for planting vines. Records from the royal palace at Mari, dating to the eighteenth century BC have yielded clay tablets containing information on viticulture and viniculture.

5.2.5 The Holy Land

According to *Genesis*, 9: 20, one of the first things that Noah did after emerging from the Ark was to plant a vineyard: "And Noah began to be an husbandman, and he planted a vineyard." As we know, he then went on enthusiastically to sample the products of the vine, with inevitable results. The Bible also mentions the various steps necessary to care for a vineyard (*Isiah*, 5: 1–6) and, as well as other things, prohibits the crossing of strains in a vineyard, lest "the fruit of thy vineyard be defiled".

The kings of Judah and Israel are said to have owned vast vineyards and huge stores of wine. King David's vineyards were so numerous that he had to appoint special officials to supervise them; one "over the vineyards" and the other "over the produce in the vineyards for the wine cellars" (*1 Chronicles*, 27: 27). The land that is now Israel lies in one of the oldest wine producing areas in the world, there being evidence of viticulture in Jericho, Lachish and Arad, during the period 3500–3000 BC. In antiquity, some of the wines shipped to Egypt were apparently so bad that they had to be flavoured with honey, pepper or juniper berries, in order to make them more palatable.

The Moslem conquest of 636 brought a halt to commercial viticulture for around 1200 years, and during the period of Ottoman rule, from 1516–1917, Moslem law prevented Jews and Christians from manufacturing all but "house wine", which enabled non-Moslems to use wine during their religious ceremonies. It was not until 1870 that Jews began to make wine again on any scale, when Baron Edmond de Rothschild imported some French grape varieties. The wines produced in those days were mostly red, sweet and unsophisticated, such that, in 1875, when the then British Prime Minister, Benjamin Disraeli, was given a bottle of kosher red from Palestine, he commented that it tasted "not so much like wine, but more like what I expect to receive from my doctor as a remedy for a bad winter cough".

The modern wine industry in Israel owes its existence to the importation of French cultivars by Baron de Rothschild. Prohibition of the consumption of wine under Islamic law had the effect of fostering the cultivation of table varieties in the Middle East and North Africa. This, in turn, led to the introduction of table grapes into Spain, from where they spread to France and Italy, and eventually to the New World. The growing of table grapes also spread to the Balkans during the Ottoman occupation, but the Turks, unlike the Arabs, allowed their Christian and Jewish subjects to cultivate and use wine grapes.

5.2.6   Ancient Egypt

As James[183] so forcefully put it:

> "A great deal of nonsense has been written about the origins of the vine and of wine in Egypt, much of it based on the fanciful ideas purveyed by classical and post-classical writers, whose authority in other matters of early Egyptian culture is known to be wholly unreliable."

For a start, the vine is not a component of the native flora of Egypt and, also, until effective irrigation techniques were developed, the vine remained rare in ancient Egypt, especially in Upper Egypt, where the climate was hot and dry. From the evidence that we have at present, it seems most likely that the domesticated grape was introduced into Egypt from the Levant at least as early as the Predynastic period[170,184]. Archaeobotanical evidence for the grapevine in ancient Egypt manifests itself in many forms, and includes both charred and desiccated whole grapes (or fragments), grape seeds, stems, leaves and woody material. At present, the earliest finds of grape seeds from Egypt are from the Predynastic settlements of Tell Ibrahim Awad and Buto (now Tell el-Fara'in; "mound of the Pharaoh's) in the Nile Delta and also from Tomb U-j at Umm el-Qa?ab at Abydos, on the middle Nile. The numerous grape seeds from the latter site, which was the burial chamber of King Scorpion I,[185] are, to date, the earliest to be ascribed to wine-making from Egypt (*ca.* 3150 BC), and they were found inside storage jars, which had patently been used to store and transport wine. It should be stressed, at this point, that presence of the grape does not necessarily indicate viniculture. The Abydos find consisted of three rooms stacked with around 700 wine jars. McGovern's team[186] calculated that if each jar had been full, then there would have been some 1200 gallons of wine at the site. Of the 207 jars that were recovered, 47 of them (*ca.* 23%)

contained grape seeds, and several contained completely preserved grapes. The seeds were most closely related to the domesticated *V. vinifera* subsp. *vinifera*. Eleven of the jars (*ca.* 5.5%) contained the remains of sliced figs, which had been perforated and strung together. It is surmised that the figs were suspended in the liquid in order to impart sweetness, and the addition of whole grapes may have served the same purpose. This appears to be the sole example of figs being used as an additive in this way. In addition to the jars, many clay sealings with rim and string impressions were recovered. The sealings had probably been pressed onto leather or cloth covers and tied over the jar mouths with string, an arrangement that would have been particularly convenient for storing a young alcoholic beverage from which fermentation gases needed to escape. In addition, most of the Abydos tomb vessels were bottle-shaped, with narrow mouths, which would have been easy to stopper and, therefore, well-suited for long-distance travel.

The volatile constituents of the jars had long since disappeared, leaving yellowish deposits. Chemical analysis of these deposits confirmed the presence of tartaric acid, which strongly suggested that the liquid had been wine. Analysis also showed the presence of aromatic hydrocarbons derived from the resin of the terebinth tree. The presence of terebinth resin is consistent with the ancient practice of resinating wines in order to prevent oxidation (to vinegar) during storage.

McGovern's team speculated as to where the wine might have originated from, in view of the fact that the wild grapevine never grew in ancient Egypt, and that the domesticated sub-species was not transplanted to the Nile Delta before the end of the fourth millennium BC. The clay sources of the jars were characterized by neutron activation analysis (NAA), and comparison of results with a database showed that the pottery originated in more than one place, but that none of the jars were of Egyptian origin. Of the jars that could be clearly identified by their clay composition, it was apparent that they had originated in southern Levant, an area where there is also archaeobotanical evidence for grapevine cultivation. It was, therefore, presumed that the wine from Abydos came from this region. Interestingly, the clay seals were made of Nile alluvial clay, suggesting that the jars were sealed on entering Egypt (somewhere in the Nile Delta), or at the tomb itself. Archaeological investigation has established that the use of the overland trade route between the southern Levant (the lowlands of Israel, the Palestinian uplands, the Jordan Valley and Transjordan) and the eastern Nile Delta – "the Ways of Horus" as it was known to the ancient Egyptians – intensified during Early Bronze Age 1. There were two-way exchanges in both goods and technologies and, during the latter stages of this period, the Egyptians took control of the route, and established a number of trading stations in the southern Levantine coastal plain. The most important of these centres was at 'En Besor (in modern Palestine).

A two-stage process in Early Bronze Age 1 interactions between ancient Egypt and the southern Levant may account for the presence of the Abydos wine jars and the start of an indigenous Egyptian winemaking industry in the wide alluvial plains of the Nile Delta soon after. In the first phase, increasing Egyptian demand stimulated the trade in horticultural products, particularly grapes. Once a sufficient market for wine had been established in Egypt, the transplantation to, and cultivation of, grapevines in the Nile Delta could be accomplished, and this would have been quickly followed by vinification itself, probably under the auspices of foreign specialists. The earliest indications so far, for ancient Egyptian wine itself, are from the very start of the historic period, the beginning of the First Dynasty (*ca.* 3000 BC). The first appearance of the hieroglyph for what is generally considered to be a wine press (Figure 5.14) occurs at around this time, and the ancient Egyptian term for wine (*irp*) is known from the Second Dynasty onwards. Representations of vineyards date as far back as the Old Kingdom, the third millennium BC.

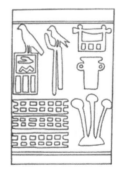

Figure 5.14 First-dynasty seal impressions, with wine press in top right hand corner (after Kaplony[186]).

A thriving royal winemaking industry had probably become established by the First and Second Dynasties (*ca.* 3050–2700 BC), but certainly by the Third Dynasty, which was the beginning of the Old Kingdom (*ca.* 2700 BC). That much can be confirmed by results of organic analyses carried out on residues from thousands of jars found in the tombs of the pharaohs at Abydos and Saqqara. Over a period of time, the Egyptian elite were able to call upon a variety of locally produced wines that could satisfy both daily needs and religious purposes. New Kingdom (1550–1069 BC) tomb reliefs and paintings depict elaborate storehouses and wine cellars in the palaces and temples at Amarna and Thebes, with row upon row of amphorae awaiting consumption during royal celebrations. The bulk of the information regarding Pharaonic wine production comes from the artistic and linguistic records of the New Kingdom, particularly from Thebes, where there are 42 tombs in the necropolis illustrating the wine-making process. Most tomb-paintings show grapes being harvested and pressed, perhaps the most notable being those in the tomb of Nakht at Thebes. Figure 5.15 shows Middle Kingdom winemaking scenes in the tomb of Amenemhet at Beni Hasan (dated about 2050–2000 BC).

From the evidence available in the pictorial record generally, the juice from crushed grapes was collected in vats and fermented. When partly fermented, the juice/young wine was decanted into amphorae and left to mature. It might then be filtered, and have spices and/or honey added before being transported to its destination in amphorae. The shoulders of amphorae (and, sometimes, their mud stoppers) are frequently inscribed with hieroglyphics listing the pharaoh's regnal year, the variety of

Figure 5.15 Winemaking scenes in the tomb of Amenemhet (after Newberry[187]).

grape, the vineyard, its owner and the person responsible for production. Such information is the forerunner of the modern wine bottle label, and enables us to pinpoint the location of certain vineyards.

It is predominantly the multiple funerary uses of wine (*e.g.* the presence of wine jars in tombs) that have helped to preserve the record of ancient Egyptian viniculture that comes to us today. Other finds of grape remains include those from the First Dynasty graves at Abydos and Nagada, from the Third Dynasty Step Pyramid of Djoser at Saqqara, from Twelfth Dynasty Kahun, from Thirteenth Dynasty Memphis and from Second Intermediate Period Tell el-Dab'a. Several Eighteenth Dynasty sites have also yielded grape remains: these include Amarna, Deir el-Medina, Memphis and the tomb of Tutankhamun. There are numerous post-Pharaonic finds of grape. In Egypt, viticulture and wine consumption became much more widespread during the Ptolemaic period, mainly due to the presence of a large number of Greek immigrants, but also due to improved irrigation techniques at that time.

As Powell[188] has discussed, in both ancient Egypt and Mesopotamia, wine was reserved for the elite, and for special occasions. Grapes and raisins were always more expensive than barley or dates. During the Ramesside period (Nineteenth–Twentieth Dynasties; New Kingdom) at Deir el-Medina, for example, wine was five to ten times more expensive than beer.[189] Palmer[190] has queried whether the population at large in ancient Egypt, and other parts of the Near East, would have actually preferred wine to beer if they had obtained greater access to it. No entry on winemaking in ancient Egypt would be complete without referring the reader to Mary Ann Murray's[191] superb chapter on the subject.

5.2.7   China

One has to be careful when writing about the history of wine in China for, since prehistoric times, the Chinese have made an alcoholic beverage called *chiu*, which still plays a central part in their cuisine. *Chiu* has always been translated as "wine", even though it is made with cooked grains. It is all really a case of "what's in a name?", as Huang[127] splendidly explains. The same author also offers the following as part reason for any failure to discover when and how Chinese processes arose: "This is a mystery in the history of food science and technology in China because the traditional process for making Chinese wine (*chiu*) is quite different from that developed for making wine or beer in the ancient Mediterranean world." In the Chinese lexicon, any wine that is not made from grain is considered a "medicated" wine, and this would include grape wine.

If the tartaric acid-bearing fruit in McGovern's "Jiahu grog" was indeed grape rather than hawthorn (see p. 361) then this represents the earliest known use of this fruit in an alcoholic beverage. Such a find is quite understandable because around a half of all extant wild *Vitis* species are to be found in China (according to the revised Flora of China,[192] there are 36 species of *Vitis* that have a natural distribution in China). Huang,[127] however, tells us that vine cultivation did not commence in China until seed of the wine grape (*phu thao*) was brought back from Central Asia around 126 BC by the military envoy Chang Chhien (Zhang Qian).

The aggressive sixth emperor of the Han Dynasty, Wu-ti, in an effort to expand Chinese territory and influence, sent out General Zhang Qian on a mission to explore China's western frontier. The knowledge obtained allowed Wu-ti to conquer part of Western Asia (now XinJian Province) around 120 BC. One of the lands visited by Zhang Qian was the ancient country of *Da wan* (Ferghana; today's Ferghana Basin of Uzbekistan) where the indigenous people made grape wine and then stored it before drinking it. Seeds were brought back and given to the emperor, and were eventually planted near the emperor's palace at the capital of ancient China, *Chang'an* (now Xi'an),[193,194] the plants ostensibly being grown for fruit and medicament. Thus, the Chinese derive great pleasure when they claim that the arrival of *V. vinifera* in the imperial city of Xi'an predates its arrival in France by a couple of centuries! Given these kind of dates, it seems as though viticulture in China would have commenced around 100 BC.

The Pharmacopoeia of the Heavenly Husbandman, a Han compilation, noted that wine can be made from grapes and, around AD 220, an emperor pronounced of grape wine that "it is sweeter than the wine made [from cereals] using ferments and sprouted grain (*chhü nieh*) ... one recovers from it more easily when one has taken too much".

There is little in the literature to indicate how grape wine was made, and it has been assumed that it was brewed in a similar way to grain wine. One of the very few examples of anything resembling a recipe is to be found in *Compendium of Food and Drink* (*Yin Chuan Fu Shih Chien*) of 1591, which says: "Take one *tou* of pressed juice from grapes and blend in four ounces of *chhu ferment* in a jar. Cover the mouth and [the juice] will turn into wine as indicated by the presence of an exotic fragrance." Even by this time, there is no hint of the use of yeast (*chiao*), and methodology seems to be similar to that used in rice wine manufacture.

During the Han Dynasty (202 BC–AD 220) and for around five-hundred years afterwards the grape was a delicacy and its wine a rare and exotic drink in China. During the Thang Dynasty, with the conquest of Kao-chhang (Turpan), grape wine became more popular, as the following from Thai Phing Yü Lan (AD 983) avers:

"Grape wine was always a great thing in the Western countries. Formerly they sometimes presented it (as tribute) but it was not until the capture of Kao-chhang that the seeds of the "mare teat" grape were obtained, and planted in the imperial gardens. The method of winemaking was also obtained, and the emperor himself took a hand in preparing it. When finished it came in eight colours, with strong perfumes like those of springtime itself; some types tasted like a kind of whey." Further details of Chinese viticulture and winemaking can be obtained from Huang's tome.[127]

In common with other aspects of daily life, the increasing prosperity of China has resulted in a greater demand for "Western-style" grape wines. Currently, most of China's wineries are located in milder climes southeast or southwest of Beijing, although moves are afoot to grow the grapevine in more northerly parts of the country, particularly the Yellow River basin and Inner Mongolia. Up there, of course, plants must be able to withstand harsh winters and a short growing season.

One remarkable site of long-term viticulture in China is at the ancient city of Turpan in the autonomous region of Xinjiang Uyghur, Northwest China. Situated on the Silk Road, Turpan played a key role in the cultural exchange between East and West, and it provided a surprise oasis for travellers because of its system of underground irrigation (*karez*), which is supplied by snowmelt water from surrounding high ground. The city is in a depression, some 500 feet below sea level and is one of China's hottest places (July average is 90 °F). Winters are cold! Vines are thought to have been grown here since at least 300 BC, the famous variety being "Wuhebai", a white, seedless grape with a normal sugar content of 20–24%w/w. This variety was transported to California in 1870, where it became known as "Thompson's seedless" – a classic example of homonymy.

In the last couple of years, archaeological evidence has suggested that viticulture is more ancient in China than previously thought.[195] The discovery of a *V. vinifera* stem in the Yanghai Tombs, Turpan district, has been calculated (by AMS, then calibration) to be 2245 ± 35 ^{14}C years BP (*ca.* 2300 calendar years old), in accordance with the above-mentioned date of 300 BC. The Yanghai Tombs are located in the Turpan Basin and lie at the base of the Flaming Mountains, at the base of the Heavenly Mountains. The Tombs belong to the Subeixi Culture, which prospered between 3000 BP and 2000 BP. A review of the taxonomic research on Chinese wild grapes has recently been presented.[196]

5.2.8 Differences between Wild and Domesticated *Vitis*

In some ways, the wild grapevine proved to be an ideal candidate for bringing into cultivation, because it required relatively little modification before it ultimately became domesticated. Its mineral and water requirements are not excessive, allowing it to grow on impoverished sites (*e.g.* hill-sides) bearing soil unsuitable for other crops. Its tendrils gave it the ability to climb on other plants, and so it would have required relatively little attention if grown with other crops, or in a scrub, or woodland environment. In addition, the wild vine possessed considerable regenerative capabilities, and so was indifferent to the intense pruning regimes, which converted a liane into a shorter, shrubbier, more manageable plant. The shorter stature of the cultivated vine reduced the necessity for structural support, and probably reduced water-stress, thus enabling it to survive in warmer climes. The regenerative powers and general hardiness of the grapevine allowed the plant to survive in cooler conditions than it would have liked, and so permitted the spread of viticulture into Central Europe. The variation of seed morphology in wild and domesticated vines is such that they can be effectively distinguished on that criterion. The most obvious difference is that wild vine seeds have a prominent beak, and are more elongated than those of *V. vinifera* subsp. *vinifera*, which lack a beak, and are, by definition, more rounded. The degree of "roundness" of a seed can be expressed by its "seed index", which is simply seed width / seed length. In wild grape seeds the index is, on average, 0.64, whilst in the more rounded, cultivated seed the index is approximately 0.55. Why this should have happened is unclear, because there is no obvious biological advantage for the domesticated vine in having more rounded seeds. Pollen of wild and domesticated vines is also distinguishable, and in *V. vinifera* there are differences between fertile pollen (produced by male and bisexual flowers) and sterile pollen (produced by female flowers).

As already indicated, a major difference between wild and domesticated grapevines is the sexuality of their flowers, and the size of their fruit. Wild vines are dioecious, that is there are separate male and female plants, bearing the appropriate sex organs. As a result of domestication, *V. vinifera* has become hermaphrodite (monoecious), with functional anthers (male) and a pistil (female) on the same flower. Such a condition permits self-fertilization by wind (or even by gravity), which has obvious biological benefits for the plant itself, and for the viticulturalist aiming to maintain plants with a standard fruit size. In addition, nearly every hermaphrodite flower produces fruit of a consistently larger size than would be found in wild *Vitis*. If the ancient, dioecious habit had persisted into the cultivated form, then viticulturalists would have been forced to grow male and female vines, with only the latter bearing fruit. So, in effect, genetic events that gave rise to a biological advantage for the vine (*i.e.* hermaphroditism) have also been a boon for the grower. It is almost certain that biologically useful mutations in wild *Vitis* would have occurred very slowly in nature, until man began to encourage selection of desired (domestic) traits in the plant, whence change was accelerated somewhat. Early attempts at cultivation would have involved selecting fruit-bearing (female) vines from their woodland habitat, thus separating them from their sources of pollen (male vines). Favouritism would almost certainly have been shown towards vines with large fruits, and/or those with enhanced flavour and aroma characters. By selectively propagating such plants away from male vines, unisexuality would have been progressively discouraged. Nowadays, nearly all cultivars bear bisexual flowers and set fruit after self-pollination. One of the unheralded consequences of the shift from being a cross-pollinating plant to a self-pollinating one is that the nectarines in the monoecious flowers are poorly developed and non-functional, so that they are no longer attractive to pollinating insects. As is the case with most other fruit trees, viticulture is based on the "fixation" and maintenance of vegetative clones, and vines are propagated vegetatively by rooting winter-dormant twigs, or by grafting. Although wild vines are essentially dioecious, male plants sometimes possess non-functional pistils, and some female plants may contain non-functional anthers. Such a phenomenon hints that the "archetypal *Vitis*" may have been bisexual (monoecious), but that they reverted to being dioecious at some stage during evolution. Indeed, primitive forms of *Vitis* from the Tertiary have been shown to have been hermaphrodite, just like contemporaneous, cultivated *V. vinifera* subsp. *vinifera*. This means that at either end of its long existence on planet Earth the grapevine has been capable of self-fertilization. Somewhere in between, for reasons not yet fully apparent, but probably related to the harsh environment during the last Ice Age, the wild grape became dioecious throughout its range. The segregated male and female plants each had an anther and a pistil, but a mutation suppressed the development of one, or the other, on different plants, so that, functionally, flowers on an individual plant can be considered to be either male or female. In what was to become the male plant, a dominant mutation of one of the nuclear chromosomes (designated Su^F) suppressed the development of the female reproductive apparatus. In the nascent female, a recessive mutation (designated Su^m) prevented the development of the male apparatus. The upshot of all this is that, in the wild, male vines seldom yield any grapes, and female plants produce fruit that is highly variable. As we find it today, the Eurasian wild grape is hardly likely to be a source of good wine. It is small, many-seeded, has a relatively low sugar content, is of high acidity and has a tough skin. Colour-wise, they are usually black or dark red, very rarely white.

It would appear that the modifications that occurred to transform the wild vine into a domesticated form were effected at an early stage in history. Certainly, by the time that we have the first depictions of vines from ancient Egypt, it would appear that the wild vine had been transformed into its domestic counterpart. There are no known illustrations of non-grape-bearing, male vines from ancient Egypt, even though a fairly high percentage of plants would have been fruitless in the days when the sexes were separated (dioecy). Let us envisage how the transformation from wild (dioecious) to domesticated (monoecious) grapevine might have occurred. Unfruitful male vines would have been grubbed up, thus depriving the fruitful female vines of their source of pollen. This would eventually result in the female vines ceasing to bear fruit. Meanwhile, there would have been a few hermaphrodite vines among the wild population, bearing both pollen and female organs (ultimately fruit). Their fruit size would have been small, certainly smaller than that from female vines, but better than nothing, so they would have been persevered with. Because pollen was obtainable from these hermaphrodite vines, male, non-fruiting plants could be dispensed with and, as a result of selection by man, fruit yields on hermaphrodite plants were improved (in size and, presumably, taste). Dioecious vines grown from seed are notoriously heterozygous, and unlikely to bear much resemblance to their parent plants. Young vines from self-pollinated hermaphrodite (monoecious) plants are very weak, because a number of harmful recessive genes are expressed and, without the intervention of man, natural selection would lead to the elimination of hermaphroditism within a population. As it has turned out, man has been responsible for the hermaphrodite condition being synonymous with cultivated *Vitis*. It is interesting to note that, by back-crossing, viticulturalists can obtain the primitive dioecious condition in *Vitis* plants within a few generations.

5.2.9   The Grape Berry

The mature grape is the result of complex physiological and biochemical phenomena, which, themselves, are intimately linked to environmental conditions. Botanically, the grape, being a multi-seeded fleshy fruit, is classified as a berry and, when compared to other types of fruit, it does not lend itself easily to detailed study. This is partly due to the fact that berry growth and development are the result of a long and complex reproductive cycle, and partly due to the fact that berries are aggregated into a cluster. The study of grape maturation is greatly complicated because of the variability of individual berries within a bunch. That said, it is possible to use biochemical markers to assess fruit maturation. Probably the most important outward manifestation of berry maturation is *véraison*, which marks the beginning of berry ripening. In red cultivars, this point is marked by a change in grape colour from green to red.

The grape, being a true berry, is a simple fruit with a pulpy pericarp, and it is this fleshy region that is of major significance to the winemaker. In the simplest of terms, the grape berry contains three major tissue types: flesh, skin and seed. All three tissues vary considerably in composition, and therefore contribute differently to overall wine composition. The skin typically represents around 15% of the berry weight, and is the principal source of aromatic compounds and flavour precursors. It also contains flavonoid phenolic compounds (including flavonols, anthocyanins and the large polymeric flavonoids called tannins). Phenolic compounds in skins represent some 30% of the total berry phenolics. The seeds, which represent *ca.* 4% of berry weight, contain both non-flavonoid and flavonoid phenolics, including relatively large amounts of tannin. Seed phenolics approximate to 60% of those compounds found in the berry. Seeds also contain significant levels of nitrogenous compounds, minerals and oils (primarily oleic and linoleic acids). The pulp accounts for around 78% of the berry weight, and its primary constituents are hexose sugars (notably glucose and fructose), organic acids (mainly tartaric and malic), mineral cations (especially K^+), nitrogenous compounds (soluble proteins, ammonia and amino acids), pectic substances (cell wall structural material composed of galacturonic acid polymers) and non-flavonoid phenolics (primarily benzoic and cinnamic acid derivatives). Phenolic compounds in the pulp represent around 10% of the total phenolic content of berries. Division of the grape berry into its constituent regions has engendered a number of different systems, but most authorities would agree that the pericarp is divisible into three main regions: the exocarp (skin), the mesocarp (flesh) and the endocarp. The latter is, by definition, the tissue that surrounds the seeds, and is scarcely distinguishable from the rest of the pulp. Again, most authorities would agree that the "skin" of the grape should include the epidermis, and its waxy covering, and the hypodermis, in which the cell walls are thickened with cellulose (called collenchyma cells). The cells in the hypodermis are separable into two regions, with the outer cells being rectangular in shape, and the inner ones being polygonal. The cells in this dermal area contain a high percentage of the substances responsible for pigmentation, flavour and aroma. The mesoderm is largely composed of large, un-thickened, polygonal cells, and this tissue forms the bulk of the berry flesh. On the outside, the cuticle is secreted by the epidermis, and is responsible for the "bloom" of grapes. It is composed of overlapping platelets of wax, which look fascinating under the electron microscope. Wax thickness is relatively constant throughout berry development, and is usually in the order of 100 μg wax per cm^2 berry surface. Platelets are hydrophobic, and help to prevent water loss from the fruit. Chemically, epicuticular wax consists of two-thirds hard wax (mainly oleanolic acid) and one-third soft wax – a complex mixture of organics, including fatty alcohols, such as lauryl, stearyl and cetyl.[197]

The grape is richly supplied with vascular tissue, which after it enters the fruit *via* the pedicel, branches out to supply the developing seeds (*via* an ovular network), the flesh (*via* a central network) and the skin, by a peripheral system of veins, which have been likened to a mesh of "chicken wire".

From an oenologists point of view, the grape berry is an independent biochemical factory because, in addition to the primary metabolites essential for plant survival, it has the ability to synthesize all the other secondary components, such as flavour and aroma compounds, that go toward defining a particular wine. There is tremendous potential for variability in ripening berries within a cluster and, therefore, within a vineyard, and it is often difficult to determine when a vineyard, with a large discrepancy in berry maturity, is at its optimum stage for harvest. One of the main aims of modern viticulture, and certainly one of its greatest challenges, is to be able to produce a uniformly ripe crop.

The transition from ovary to ripe grape can be conveniently divided into three phases and, irrespective of the way in which the fruit is set, the growth of the berry follows a double-sigmoid pattern, which essentially represents two successive sigmoidal growth periods separated by a lull in growth. If we just consider the biological reasons for the production of the grape berry, *i.e.* that the vine is trying to reproduce itself, then all of the changes that occur in grape maturation make sense. During what is called

Stage I, the vine seed is immature and requires protection from what are essentially its dispersal agents (birds and other animals). The plant protects its young seeds by making them as unappealing as possible to likely predators, *i.e.* by synthesizing organic acids, tannins and pyrazines, which taste pretty disgusting. Such compounds are protective agents, as far as the vine is concerned. Stage II is essentially a "lag phase", with little berry expansion. During this stage, however, there is considerable growth of the embryo, and the seeds mature. Having produced a viable seed, Stage III, the second period of berry growth, is all about making the seed covering (the juicy berry) as appealing as possible to birds and mammals, so that the seeds will be dispersed. It's as simple as that.

As can be seen from Figure 5.16, the initial branch from the vine cane is called the peduncle, which itself is furcated into rachi (sing. rachis). Each rachis bears numerous pedicels on which the berries themselves are borne. The berry is supplied *via* the pedicel with a vascular system composed of xylem and phloem elements. Xylem is responsible for transporting water, minerals, growth regulators and nutrients from the root system, whilst the phloem transports photosynthate from the leaves. Berry formation, of course, is preceded by flowering and then sexual reproduction and, botanically, the arrangement of the flower head in *V. vinifera* is called a panicle. Each berry is lavishly supplied with vascular tissue, and the way in which berries are spatially arranged into a cluster is determined by the length of the pedicels. If they are long and thin, then the cluster will be loose, whilst if pedicels are short, bunches will be more compact.

Under temperate conditions, flowering in the mature vine is usually a three-step process, the first of which is the formation of uncommitted primordia, sometimes referred to as *anlagen* (sing. *anlage*). *Anlagen* are formed by the apices of latent buds on shoots of the current season, and will develop either as inflorescence primordia, or as tendril primordia, at which point they enter dormancy. Flowers are formed from inflorescence primordia in the spring of the following season – at the onset of "bud burst". Using scanning electron microscopy, the origin of inflorescence primordia in latent buds, and the subsequent development of the inflorescence, was studied extensively and reported on by Srinivasan and Mullins,[198] who discerned 11 developmental stages.

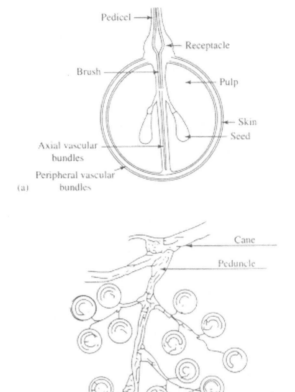

Figure 5.16   The grape (a) sectioned berry at maturity; (b) composition of grape cluster (reproduced by kind permission of John Wiley & Sons Ltd).

As we have noted, in nature, most *Vitis* species are dioecious, but nearly all important commercial cultivars are hermaphrodite, their flowers containing functional stamens and functional pistils. According to Negi and Olmo,[199] the sexuality of such a flower is determined by three alleles: Su^+, Su^F and Su^M and, depending on matters of dominance, flowers become functioning males or functioning females. The primitive hermaphrodite state is

Su^+Su^+, and a dominant mutation, Su^F, suppresses ovary development to give functional male flowers. Conversely, in the presence of the recessive Su^M, sterile pollen is produced and the flower is effectively female. The dominance relationships of these three alleles are: $Su^F > Su^+ > Su^F$.

The exact means of pollination in *V. vinifera* is imprecisely known, and our knowledge is beset by anomaly. Most authorities agree that the vine is primarily anemophilous (wind-pollinated), even though the flower structure does not fundamentally imply this method. For example, at the base of the ovary is a whorl of nectaries, called the "disc", and this is thought to be responsible for the odour of the vine flower. Production of nectar, however, has not been unequivocally proven and, in any case, nectaries are not compatible with anemophily. Bud-pollination, involving dehiscence of the anthers before anthesis, is fairly common, and it is also thought that some insects may play a role in pollination. Similar question marks exist in the debate about whether *V. vinifera* is cross-pollinated or self-pollinated, but most evidence points to the latter mechanism as being the norm. The dioecious progenitors of the European vine would, of course, have been cross-pollinated. Where cross-pollination is necessary with *V. vinifera* plants, this is effected either by hand or by machine.

Whatever the case, pollination is normally followed by fertilization, which signals the onset of berry formation. Grapevine embryology has not been the subject of extensive study, and information about the embryo, endosperm and seed formation is fragmentary, to say the least. When the seed is formed, its outer layers are formed from the integuments that supported the egg cell (ovule). The hardening of the grape seed is due to lignification of part of the outer integument. The inner integument remains unlignified, and closely aligned to the seed's food reserve (endosperm). Grape seeds are very resistant to decay and are, therefore, of considerable archaeological and palaeobotanical interest. Poor levels of fertilization can lead to the formation of rudimentary (aborted) seeds, a condition called stenospermocarpic seedlessness, whilst a lack of fertilization leads to totally seedless berries, a condition known as parthenocarpic seedlessness. In the latter situation, there is a degeneration of the embryo sac, and pollination alone is a trigger for fruit development. With stenospermocarpy, both pollination and fertilization are triggers for fruit development, but the embryo and endosperm fail to develop. As a consequence, the outer integument around the ovule contains little, or no, lignin and the aborted ovules appear as small, white specks in the flesh of the fruit.

In general, not all of the flowers in a cluster are successfully pollinated, and non-pollinated stigmas lead to non-fertilized ovaries, and these eventually give rise to small green berries. It has been shown that the fertilization/berry-setting ratio decreases as the number of flowers in the cluster increases. After berry-set, a variable proportion of ostensibly fertilized young berries cease to grow, and fall from the vine. This is called "shatter" (Fr. *coulure*), and it is caused by the formation of an abscission layer at the base of the pedicel. Under temperate growth conditions, shatter is linked to the carbohydrate status of the vine, but in more northerly vineyards, climate plays a part in the phenomenon, and shatter is worst when conditions are cold and wet. There is also a varietal variation in propensity to shatter, with Chardonnay, for example, being much more likely to shatter than Riesling. Some *coulure* is beneficial, as a vine might have difficulty in ripening a full, heavy crop, resulting in loss of overall grape quality. Heavy shatter will result in a very small grape crop.

Sugars, of ultimate interest to the winemaker, have been shown to play an important role in grapevine flowering. The process, commencing with inflorescence initiation and ending with fruit maturity, takes two seasons, which is typical of woody perennials. Much of the available data concern the role of sugars during the formation of the reproductive organs, from initiation of the inflorescence during the summer of the first year, until flower opening the following spring. The sugars concerned originate from photosynthate[202] or from reserve material in wood[201,202] – depending on the stage of the reproductive cycle. In some species, sugars are known to be regulators of source-sink interactions under standard or stress conditions.[203,204] Carbon allocation in plants has been the subject of intense study for many years, and many ways of assessing the phenomenon have been invoked including L- systems,[205,206] which have been applied to woody plants.[207] A model for sink feedback regulation in *Vitis* has been provided by Quereix *et al.*[208]

The control of plant development and gene expression by sugar signalling has recently been reviewed.[209] Not surprisingly, sucrose and its cleavage products glucose and fructose are the central molecules for carbohydrate translocation, metabolism and sensing in higher plants. Plants possess three types of invertases, which are located in the apoplast, the cytoplasm and the vacuole. Extracellular and vacuolar invertase isoenzymes are key metabolic enzymes because their substrates and reaction products are both nutrients and signal molecules. The role of apoplastic solute accumulation in grape ripening was reported by Wada *et al.*,[210] who found that high concentrations of solutes accumulate in the mesocarp apoplast prior to véraison and that apoplast solutes play an important role in cellular metabolism prior to the onset of ripening. For a discussion of the function and regulation of plant invertases see Roitsch and Gonzalez,[211] and for information about sugars and grapevine flowering do look at Lebon *et al.*[212]

5.2.10 Microbiology of the Grape Surface

It was Louis Pasteur who first showed that microbes found in wine could also be identified on the grape surface and, since that time, the presence of certain wine yeasts[213] and bacteria[214] on grape surfaces has been well documented. The enormous diversity of organisms present makes exhaustive identification difficult, especially those microbes present in low numbers. In addition, the grape berry surface is a dynamic habitat, with the microflora varying according to ripeness,[215] environmental factors[216] and even agrochemicals and pesticides.[217] Traditional microbiological methods (*e.g.* isolation and enumeration with selective media) on their own are unreliable, but when combined with modern molecular methods the composition of complex microbial communities can be unravelled.[218–220]

An extensive investigation into the microbial ecosystem on the grape berry surface was carried out by Renouf *et al.*,[221] who conducted experiments on the varieties "Merlot", "Cabernet Sauvignon" and "Cabernet Franc" grown in several areas of Bordeaux. They confirmed that species and their populations varied according to berry development stage. Microbial numbers were at their peak at the time of harvest, when berry surface area was at its maximum (as long as no chemical applications had been made for some weeks!). The most widespread yeast at berry set was *Aureobasidium pullulans*, but this was never present at harvest.

Numbers of this yeast started to decline at véraison, and it started to be ousted by fermentative yeasts. Conversely, the bacterium *Gluconobacter oxydans* was detected mainly at harvest. *Oenococcus oeni*, on the other hand, could be detected on immature and mature berries. Véraison, a key stage during grape ripening (see above), was seen to be an important stage in the development of the microbial community, as well. Species such as *S. cerevisiae*, *Pichia anomala* and *Candida stellata* became progressively more abundant on the grape surface, but they were never as prolific as species of the Basidiomycetous yeasts, *Rhodotorula*, *Cryptococcus* and *Sporobolomyces*. The French team saw this work as being a foundation for the establishment of a grape surface "biofilm" flora.

Although not prevalent on grapes, *Brettanomyces bruxellensis* plays an important role in winemaking, especially in the development of wine aroma. It interacts with other yeast species during fermentation and is often the dominant yeast at the end of alcoholic fermentation. Using molecular methods, the interactions between *B. bruxellensis* and other yeast species during the initial stages of winemaking have been unravelled.[222] It was found to be the most adapted of all non-*Saccharomces* yeasts to the grape must environment, and appeared to be more ethanol-tolerant than *S. cerevisiae* itself.

5.2.11 The Grape Genome

In 2007, the French-Italian Public Consortium for Grapevine Genome Characterization published what they described as a "high-quality draft" of the genome sequence of *V. vinifera*.[223] The researchers chose a variant derived from a Pinot Noir cultivar that was bred to have high homozygosity (it was a special inbred variety, which was about 93% homogeneous), such that both copies of each chromosome pair carried the same version of the majority of the plants genes. The genome was unravelled by means of "shotgun" sequencing, which is a method that literally fragments DNA and then enables the fragments to be read (over six million fragments were produced!). The information obtained can be pieced together in jigsaw-fashion. Low homozygosity (shown by some *V. vinifera* cultivars) impedes shotgun sequencing.

The genome has 19 chromosomes amounting to 487 Mb of DNA, and the team reported 30,434 protein-encoding genes and 600 tRNA genes (rRNA genes were not included in the paper), genes comprising 46% of the genome (37% introns, 7% exons). It was evident that the *V. vinifera* genome had not undergone recent genome duplication, something that facilitated the discovery of ancestral traits. Analysis revealed that three ancestral genomes had contributed to the haploid content of the plant, an arrangement that has been demonstrated from other dicotyledonous plants (but not from the monocotyledon, rice).

As with most other crop plants, molecular genetics is helping us to understand much more about the grapevine, and the genetic resources that are available to the community studying these plants have increased enormously over the past decade. The genetic diversity that exists today in the vine is largely attributable to the vineyard practices of the last few centuries. The grapevine lends itself to genetic research because it is diploid and has a small genome size (475–500 Mb) relative to other plants.

Recent molecular studies have demonstrated that considerable genetic variation exists within grapevine varieties, on both a global[179,224] and a local scale (*e.g.* France,[225] Turkey[226], Portugal[227]). Such studies have also intimated that part of the genetic variation in cultivated grape is geographically restricted, and this has led some workers to suggest that the wild grape could have been taken into cultivation in several places at different times, thus capturing different subsets of the total variation of the wild ancestor.[179]

In an examination of chloroplast DNA diversity in a geographically wide set of 113 grapevine cultivars, including 40 from the Republic of Georgia, Schaal *et al.*[228] observed four plastid haplotypes – each with a three-letter acronym. These were designated by their character states at each of the three polymorphic positions: (AAA) – 23 samples, (ATT) – 29 samples, (GTA) – 26 samples and (ATA) – 35 samples. Each plastid haplotype was accorded an informal group name (*e.g.* (AAA) = "Rkatsiteli" group). The (AAA) haplotype was found only in the cultivars from Georgia, and 25 of them (63%) showed it – all originating from eastern Georgia. The other three haplotypes were also exhibited by some Georgian specimens, with nine cultivars of the "Chkhaveri-Pinot noir" group (GTA), most of which are cultivated in western Georgia near the Black Sea. Six of the Georgian cultivars showed the "Saperavi-Cabernet Sauvignon" group (ATT) haplotype (Saperavi is believed to have originated in southwest Georgia), and two exhibited the "Chardonnay-Mtsvane Meskhuri" group (ATA), which are mainly represented by French cultivars. That the Georgian cultivars exhibited both unique chloroplast DNA variation (AAA) and all other observed plastid haplotypes was consistent with previous work that has suggested both unique and high levels of genetic variation in wild grapes in the greater Caucasus region.

Chloroplast markers have also been used to establish a phylogeny of the Vitaceae,[229,230] which has demonstrated that the family shows complex multiple inter-continental relationships within the northern hemisphere and between northern and southern hemispheres. In another study,[172] complete chloroplast sequences provided strong support for the position of the Vitaceae as the earliest diverging lineage of the rosids (a large clade of flowering plants, with *ca.* 20,000 species).

The first attempt to study the genetic diversity of Georgian wild grapes has been made by Pipia *et al.*,[231] who found that, by sample analysis at four polymorphic microsatellite loci, a slightly higher level of polymorphism was exhibited in the wild population than had been observed in cultural varieties. This would seem to indicate that the wild sub-species *sylvestris* has easily crossed with the cultivated sub-species *vinifera* producing a gamut of inter-mediate forms (as originally suggested by Negrul). Study of the genetic diversity of wild grape populations is important from an ethnobotanical standpoint, since it aids our understanding of the evolution of the cultivated vine. As Zohary and Spiegel-Roy[184] reported, wild forms of *V. vinifera* are abundant in the south Caucasus, with some carrying characters associated with cultivated forms, such as white berries, hermaphrodite flowers and sizeable seeds.

An early step towards germplasm management of the rich ampelography of Transcaucasia and Anatolia was made by Vouillamoz *et al.*,[232], who genotyped over 100 traditional cultivars (using 12 nuclear microsatellite markers) from Armenia, Georgia and Turkey. Their work detected 17 identical genotypes and 6 examples of homonymy, mainly within each national germplasm. Neighbour-joining analysis of genetic distance showed that each germplasm might have multiple origins – although they are now separated. To support the case for Georgia being the ancient origin of some distant cultivars, it was shown that four varieties from western Europe, included as outgroups in the study, proved to be more related to Georgian cultivars than other germplasms. The "foreign" cultivars were "Chasselas", "Pinot noir", "Syrah" and "Nebbiolo", and they were primarily included as standards in order to have consistent allele sizes.

For almost two decades, microsatellites have been used as a reliable means of characterizing grape cultivars, as can be ascertained from Sefc *et al.*,[233] Sanchez-Escribano *et al.*,[234] Vouillamoz *et al.*[235] and This *et al.*[236] – the latter developing a standard set of microsatellite reference alleles. In addition, Sefc *et al.*[237] have penned a definitive chapter on grapevine microsatellites and cultivar identification.

Conscious of the fact that the identity of ancient grapevine cultivars and mechanisms of domestication are still largely unknown, seed variation morphology has been used to address this particular problem. Although the analysis of genetic diversity of living material has been successful,[238] the first analyses carried out on ancient DNA using a few microsatellite markers[239] did not produce a means of identifying ancient grapevines. Analysis of variation in seed morphology has now been used to provide detailed criteria on which to distinguish between wild grapevines and modern cultivars and to understand changes in traits related to vine domestication.[240] Geometrical analysis (elliptic Fourier transform method) were applied to grapevine seed outlines from modern wild individuals, cultivars and well-preserved archaeological material from southern France. Results seemed to confirm the complexity of human contact, exchanges and migrations that spread vine cultivation in Europe and in Mediterranean areas. There was also cause for arguing for the local domestication of the vine in the Languedoc region of southern France during Antiquity.

In an attempt to explain the aforementioned disjunct distribution of *Euvitis* (*Vitis* subg. *Vitis*) in the northern hemisphere, Péros *et al.*[241] assessed 30 species and hybrids of chloroplast DNA polymorphism and found the sub-genus to be a monophyletic group. The four main haplotypes in *V. vinifera* corresponded to two different origins, and nuclear microsatellites indicated that genetic variation was especially large in North America. Some evidence suggested that sub-genus *Vitis* had an Asian origin and then dispersed to Europe and North America. It was also noted that, in contrast to chloroplast DNA, nuclear DNA showed a larger than expected genetic variation.

Using 20 microsatellite markers, Riahi *et al.*[242] evaluated the genetic diversity, structure and differentiation of *V. vinifera* cultivars traditionally grown in parts of the Maghreb region of North Africa, which has a long history of varied viticulture owing to the settlement of many populations and civilizations, and lies within the range of distribution of the wild vine. They observed a high level of differentiation between cultivated and wild accessions indicating that cultivated forms may not derive directly from local wild populations but could mostly correspond to imported material introduced over the years.

2010 saw publication of the results of the first integrated investigation of both ancient DNA and proteins in archaeological cultivated grape seeds,[243] a study that demonstrated how ancient bio-molecules allow us to understand the past diffusion of the grape. Samples, which were subjected to nuclear microsatellite locus analysis, came from an early medieval (seventh–eighth century AD) Byzantine rural settlement in the Lecce area of southern Italy and a late-fourteenth–fifteenth century site at York (UK). Results indicated that the grapes sampled from the two disparate areas had interesting origins: the York material was not grown locally, whilst the remains from the small settlement of Supersano, near Lecce, hinted at contact with the eastern Mediterranean. It is envisaged that similar work will allow the historical profiling of modern grape cultivars.

There is another reason why molecular methods can be of great benefit. Anthropogenic pressure on their natural habitats and pathogens introduced from North America during the second half of the nineteenth century has seen wild grapevine populations decline. According to some, a number of wild populations are threatened with extinction,[244] and modern technologies will be essential in the battle to preserve *Vitis* biodiversity. Having said that, it is nice to know that some long-standing methods can still be of use, and a recent work has reported the use of ampelometry to test for genetic diversity in Tunisian grape ecotypes.[245]

In 2007, research institutes and grapevine collections from 17 European, Caucasian and North African countries decided to work together for the conservation, characterization and management of *Vitis* genetic resources. The first job was to make an inventory of all the partners' accessions using standard morphological, agronomical and molecular descriptors, and to link the information into a unique database. Today, the database contains passport data from 25,000 living accessions, of which 2500 are linked to photographs and 1600 are characterized for 48 OIV descriptors and microsatellite markers. Information about the current activities of the project can be obtained from Bacilieri *et al.*[246]

There are many books available on winemaking, and to conserve space in this text the reader is directed to my 2007 effort.[247]

5.3 THE APPLE

God Almighty first planted a garden. And indeed it is the purest of human pleasures

Francis Bacon, *Essays*, 1597

5.3.1 Introduction

Many people will be aware that Eve was said to have been cast out of the Earthly Paradise forever because she ate an apple. In fact, the Bible does not specify which fruit that Eve consumed, and the implication of the apple emanates from the fifth century AD. Toussaint-Samat argues[165] that the first translators of the Old Testament into Latin thought that if there was only one fruit in Paradise it must have been the apple, because of its abundance. The Latin for "fruit" is *pomum*, and for "apple" at that time was *malum*, and so the apple then took on the identity of *THE* fruit, or the "fruit of fruits", *pomum.*, from which we get the modern French *pomme*. Maybe, with the connotation of sin in *Genesis*, it was not coincidental that *malum* was a homonym of the Latin for "evil", "crime", "damage", *etc.*

Apples are among the most popular and important fruit trees in the world, and would rank as the fourth most economically important fruit crop worldwide after citrus, grapes and banana.

Apples are produced in all temperate and sub-tropical countries of the world, and are the main fruit crop of temperate regions of the world. A small crop can be gained from high altitude sites in some tropical countries. The popularity of the apple is in some part due to the fact that the fruit has multiple uses, and the fact that some cultivars can be stored for long periods. The apple has a juvenile period of six to ten years (occasionally more), is self-incompatible and highly heterozygous – characteristics that have seriously hampered apple breeding.

This fruit of legends is classified as a pome fruit and, because of its origin, has a compartmented core, which consists of five leathery chambers (loculi) each containing two seeds. The emergence of the pome, which is essentially the swelling of the flower receptacle (and thus a "false fruit", because it does not arise from the ovary), seems to have been some 50 Mya and is characteristic of a small number of apple-related fruits, such as pear, rowan and medlar. Cultivated apples, of which there are around 7000 varieties, are descendents of wild "crab" apples, and evidence suggests that the crop was being cultivated and stored at least 5000 years ago, which is some while after cereals such as wheat and barley. Apples were grown by the ancient Egyptians, the Greeks and the Romans and, as with many other foodstuffs, were spread around Europe by the latter. Some cultivated apples almost certainly reached the British Isles *via* the Romans, and before their introduction only crabs would have been consumed. Whether apples were grown in the UK at that time is a moot point; most would argue that they arrived with William the Conqueror! The crop reached the Americas during the seventeenth century with the early colonists (today, there are around 2500 varieties grown in North America!). Native North American apples are very astringent and indigenous peoples do not seem to have exploited them, although Hendrick[248] relates that one species might have been used for cidermaking.

In Greek mythology, Mother Earth (Gaia) gave a tree with golden apples to Zeus and his bride Hera on their wedding day. The tree was in the Garden of the Hesperides[249] (Figure 5.17), daughters of the Evening Star, and was guarded by Ladon, a serpent who never slept. These golden apples were implicated in many ancient Greek tales, including the abduction of Helen of Troy. On *terra firma*, the Greek historian Xenophon (*ca.* 400 BC)

Figure 5.17 The Garden of the Hesperides – supposedly situated near Mount Atlas, or, according to some, near Cyrenaica. They were described by the sixth-century BC geographer, Scylax. The gardens contained the golden apples that Juno gave to Jupiter on the day of their nuptials.

was so taken by the beautiful walled gardens that he had seen while on duty in Persia that he constructed one himself on his Greek estate. He named it after the Persian *pairidaeza* ("walled garden") and this later evolved into the Latin, paradises, and from there our "paradise".

Being self-incompatible plants, production of fruit is dependent upon a cross-pollinating mechanism, which is an insect in this case. Commercial growers would usually introduce the relevant species, or at least take steps to make sure that it is present in order to effect pollination. Honey bees, or orchard mason bees, are the most commonly used in commercial orchards. There are a number of pollination groups in apples (usually labelled A–H), and these are climate-dependent. Fruit set results when one type is pollinated by a different genotype.

5.3.2 Taxonomy

Apples are to be found in the genus *Malus* which is a member of the dicotyledonous order Rosales Bercht. & J.Presl. Within this order, the domesticated apple, *Malus pumilla* Mill. (syn. *Malus* × *domestica* Borkh.), belongs to the family Rosaceae Juss., which contains over 100 genera and some 3000 species, most of which are perennial trees, shrubs and herbs.[250] Division of the Rosaceae is into four sub-families, with *Malus* being in the sub-family Maloideae (but see Potter *et al.*[251] below). Sub-families may be broken down into tribes, and the apple falls in the tribe Maleae. The Plant List[252] has 35 accepted species names in the genus *Malus*, although up to 79 have been described at various times. Part of the problem of *Malus* taxonomy is the intimate association that humans have had with apples over the years, and this has tended to blur the distinction between "wild" and "cultivated" forms. As has been pointed out, caution is needed when the term "wild apple" is applied.[253] In practice, the term has been applied to all *Malus* species apart from *M. pumila.* In Europe, this would refer to *M. sylvestris*, whereas in the USA the name might apply to several species, including *M. angustifolia* and *M. coronaria.*

There are several species of wild fruit trees in the genus *Malus* that produce the coloured fruit known as the crab apple, and these have been used as a food by humans since prehistoric times. This is despite them being harsh, acidic fruits, which are a challenge for many consumers when raw. Most domesticated apples have a high sugar/carbohydrate content, are low in tannins and are slightly acidic. Fruit is graded by its gastronomic characters into categories such as "dessert" ("eating"), "cooking" and "cider". Cooking apple fruits are harder and usually larger than their eating counterparts, while cider apples are wanted for their juice.

For such an important crop species, it is puzzling why there should have been so much conjecture surrounding its scientific name. Having said that, classification down to species is somewhat difficult by traditional means because of the apple's great genetic diversity, its hybridization potential and its propensity for apomyxis and polyploidy.[254] Historically, Linnaeus's classification (1767) combined apple, pear and quince into the genus *Pyrus* (in his family Pomaceae), and he proposed *Pyrus malus* L. for the cultivated form. This appellation has been used by some workers well into the early-twentieth century. Mainly because of the distinctive character of its fruit, the botanist and fruit-grower Philip Miller (1691–1771) separated apple from the genus *Pyrus* and described it as having the characteristics of the genus *Malus*, writing: "Apple should be distinguished as a separate genus and this difference is included in the Nature of these plants, as they are compatible on budding and whip grafting. Therefore, I separate apple from pear as it was always done by the botanists before Linnaeus." After Miller's contribution, the plant became *Malus pumila* Mill. Another enduring contribution to apple taxonomy was forwarded in 1803 by the German botanist M. B. Borkhausen (1760–1803), who invoked the name *Malus* x *domestica* Borkh. Even the name *Malus malus* was suggested at one stage! In a typically erudite contribution, Mabberley *et al.*[255] concluded that the correct Latin binomial for the common cultivated apple is *Malus pumilla* Mill. This nomen is adopted by Juniper and Mabberley,[256] who regard "*M. domestica*" as illegitimate.

In contrast, Luby[257] firmly plumps for *Malus* x *domestica* Borkh. (Korban and Skirvin[258]) or *M. domestica* Borkh. (Phipps *et al.*[259]) in his discourse. When first describing "*Malus* x *domestica*" in 1803, he believed that it had originated as a hybrid from *M. sylvestris* Mill., *M. dasyphyllus* Borkh. (a synonym for *M. pumilla*) and *M. praecox* Borkh. (a synonym for *M. sylvestris* var. *praecox*). This taxon would appear to be illegitimate, in view of what we now know, and what has been hypothesized, about the wild apple *M. sieversii*. Phylogenetic reconstruction of the tribe Pyreae and the genus *Malus*, relative to major Rosaceae taxa, enabled the Italian-led international research consortium[260] to identify the progenitor of cultivated apple as *M. sieversii*.

In an attempt to develop a new phylogenetically based infrafamilial classification of the Rosaceae, Potter *et al.*[251] investigated the phylogenetic relationships among 90-odd genera in that family. They used nucleotide sequence data from six nuclear and four chloroplast regions, which were examined separately and in various combinations, with parsimony and likelihood-based Bayesian approaches. The new classification that they proposed consisted of three sub-families: Rosoideae; Dryadoideae and Spiraeoideae. Following work by Campbell *et al.*,[261] the Spiraeoideae were broken down into supertribes (Pyrodae), tribes (Pyreae) and sub-tribes (Pyrinae), with the parentheses denoting those applicable to the apple. The sub-tribe Pyrinae corresponds to the long-recognized sub-family, Maloideae. Descriptions of the various groupings are provided in the paper.

All species in the sub-family Maloideae have a base chromosome number of 17, which is to be contrasted with all other members of the family Rosaceae, which have $x = 7$, 8 or 9. Because of this, for many years it was hypothesized that the origin of the Maloideae had involved hybridization between ancestors of two other sub-families. When this hypothesis was tested using cloned and sequenced granule-bound starch synthase genes,[261] it was shown that it was untenable, and that the results were consistent with a polyploidy Maloideae origin involving only members of a lineage that contained the ancestors of *Gillenia*, a genus of the south-eastern USA (with $x = 9$). As the authors concluded: "Under this hypothesis, the sub-family originated in North America, and the high Maloideae chromosome number arose *via* aneuploidy from $x = 18$."

Many cultivars have a diploid ($2n = 34$) chromosome constitution, but there are several triploid ($3n = 51$) clones and a few tetraploids ($4n = 68$). The normal means of cultivating apples is *via* vegetative methods, with grafting being the norm. Most cultivars cannot be reproduced from cuttings or suckers and when they are grown from seed, they segregate widely making most progeny economically worthless. Zohary and Spiegel-Roy[184] saw this shift to vegetative propagation under domestication as a "pre-adaptation for domestication early in the development of agriculture".

5.3.3 Origins of the Apple

Exactly how the domesticated apple evolved has tested the minds of great naturalists for hundreds of years, and men like de Candolle and Vavilov spent some time addressing the problem. It is only in the last decade or so that the situation has been unequivocally resolved, and in *The Story of the Apple*[256] the following possible reason for this is put forward: "... paradoxically – probably because of the relative inaccessibility of its homeland, both for geographical and political reasons – the origin of *Malus pumila*, unlike that of many other fruits and vegetables, long remained unknown."

The Central Asian wild apple, *M. sieversii* (Ledeb.) M.Roem., is a wild apple tree native to the mountains of Central Asia, notably in southern Kazakhstan, eastern Uzbekistan, Kyrgyzstan, Tajikistan, northern Afghanistan and Chinese Xinjiang. In Kazakhstan this apple in known as alma, and the region where it may have originated as Alma-Ata ("father of apples"). It was Vavilov who first suggested that this wild apple and its close relatives were the progenitors of the domesticated apple.[263] Fieldwork carried out in this region over the past couple of decades has confirmed the similarity between wild and cultivated apples.[264] Janick *et al.*,[265] talking of Central Asia, and heeding Vavilov's idea that the centre of diversity is the centre of origin, offered up: "This area is the centre of greatest diversity and the centre of origin of the domesticated apple."

This vulnerable species grows to *ca.* 5–12 m high and its fruit is the largest of any wild apple species, and can reach up to 7 cm diameter (very similar to cultivated fruits). The plant is widespread at elevations between *ca.* 1200 and 1800 m, where the tree is often the overstorey species. Man has inhabited and practised nomadic agriculture in this region for millennia.

M. sieversii is closely related to a group of apple trees bearing fruits of varying size. One of these, *Malus baccata*, the Siberian crab apple, has small (cherry-sized), red fruits which hang in clusters and are dispersed by birds. As Harris *et al.*[253] said: "*Malus baccata* might have had a wider distribution than that of the present day, and we think that populations became "trapped" as the Tien Shan began to rise out of the Tethys Ocean. Over seven million years, perhaps up to ten million years, mammals, such as bears, acted as selection vehicles by selecting the largest and juiciest fruits; a small, bird-distributed, cherry-like delicacy giving way to a large mammal-distributed form."

When humans began to occupy this area, somewhere between 5000 and 8000 years ago, they would have found that the early evolution of the apple was almost complete, and it had started to be dispersed. Beside human influence, migration of the apple was aided by the horse, which was a fairly recent domesticate.[266] Incidentally, the domestication of the horse has profoundly affected the course of man's civilization. Archaeological and molecular data intimate that late Neolithic/early Bronze Age travellers, on the trade routes from China to Eastern Europe, carried *M. sieversii* seed (either intentionally, or in their horses intestines) westward. *Via* Persia, apples probably spread to the ancient Greeks, and then passed to the Romans.

The leaves of *M. sieversii* redden in autumn, but this trait has been lost during domestication.[267] The biochemistry and the physiology of autumn colours are well known, but their adaptive value still baffles scientists. Several hypotheses have been forwarded, including, interestingly, one that suggests that autumn colours are the result of coevolution.[267,268]

Volk *et al.*[269] found that the genetic diversity of a wild *Malus* population collected near Arslanbob in Kyrgyzstan, when compared with seedlings of *M. sieversii* collected in Kazakhstan, contained genotypes assigned to a common genetic linkage with *M. sieversii*. The Arslanbob site is around 200 km from the nearest Kazakhstan collection sites, and is separated from them by the rugged Karatau and Kyrgyz mountain ranges, and it was felt that these allele variations could prove to be valuable complements to *Malus* germplasm collections.

The wild European (crab) apple, *M. sylvestris* (L.) Mill., is distributed widely in Eastern, Central and Western Europe, but is nowhere a common plant. In some areas it is regarded as endangered, and is certainly not particularly common in the UK. The northern limit of all wild apple species would seem to be 60–62 °N, while sweet cultivars seem to inhabit a zone between 25 °N and 55 °N. As Juniper and Mabberley say,[256] there is a great deal of confusion between genuine *M. sylvestris* and feral seedlings of imported *M. pumila*. The sole obvious floral difference is observable during flowering, where the calyx lobes of the latter are hairy on both surfaces, whereas in *M. sylvestris* only the inner one is hairy (the outer being smooth). *M. sylvestris*, a well-studied species, is now thought to have played little part in the evolution of the domesticated apple, although in parts of pre-agricultural Europe it was an important small-tree element in open woodland and semi-natural grassland. At one point, it was believed that the domesticated apple had arisen as a result of chance hybridizations among various wild species, but DNA evidence has refuted that notion.

China, being the centre of origin of the genus *Malus* Mill., can claim to be home to around 80% of all species. A list of all of these, together with their distribution, characteristics and uses, has been provided by Zhou[270] and, a little later, the wild apples of Central Asia were similarly treated.[271]

5.3.4 Apple History through Grafting

In many ways, one can trace much of the history of the apple through a short discourse on grafting, for the two are inextricably linked. It seems as though the ancient Chinese may have known about this method of fruit tree propagation, for there is mention of the art by a Chinese diplomat called Feng Li around 5000 BC. This technique was quite possibly used by the ancient Mesopotamian culture at Mari, which was founded in the early third millennium BC, and pillaged and destroyed *ca.* 1760 BC. Situated on the Euphrates, Mari was the centre of a kingdom controlling trade and river traffic between Babylonia and Syria, and a feature of which was a huge palace. Clay tablets recovered from the site document receipt of shipments of apples from the mountainous north, and there is evidence that the intricacies of grafting grapevines were known at Mari.[272] Further north, in second millennium BC Nuzi, in northeastern Iraq, a tablet records the sale of an apple orchard, which could signify grafting, although it is more likely that early domesticated apples were grown from seed (with all the attendant problems). Better evidence for grafting of apples comes from records from around 2500 BC Persia. Mudge *et al.*[273] are of the opinion that the evidence that grafting was used in ancient Mesopotamia is "murky at best". Reasons why these early cultures are likely to have used grafting are discussed by Juniper and Mabberley.[256] In ancient Egypt, apples were part of a "second wave" of domestication, because of the need to propagate them by grafting, and this art does not seem to have been perfected until Greco-Roman times.

During Sir Leonard Woolley's excavations at Ur (1923–1934), food debris was recovered from the "Royal Cemetery" and sent back to the British Museum for subsequent examination. The material was found on saucers or in other pottery vessels, usually accompanied by other grave goods. When the material was finally analysed, on saucers at one location were the charred remains of perforated, sliced crab apples. It appears that they were cut transversely in half when fresh and threaded on a string and dried before being placed in the tomb.[274] At the time, this was the first find of crab apples in the Near East, although they were commonly halved and dried in prehistoric times,[275,276] albeit not threaded on a string.

The Hebrew Bible, written over a 1000-year period (*ca.* 1400–400 BCE), alludes to the practice of grafting numerous times, without being too specific. In *Leviticus* 19: 19 we read: "Thou shall not sow thy field with two types of seed", but the Mishna (third century CE) is more specific: "It is unlawful to graft tree on tree, vegetable on vegetable, tree on vegetable, or vegetable on tree." As has been stated, Greek and Roman sources indicate that grafting was well known and widely practised in the Mediterranean region by the fifth century BCE and during the Talmudic-Hellenistic times, when the Mishna was composed.[273]

Grafting, although not apple-related, is mentioned in the New Testament, where in *Romans* 11: 24 it is used as an interesting analogy for how Gentiles can become one with Israel: "For if you were cut off from what is by nature a wild olive tree, and grafted, contrary to nature, into a cultivated olive tree, how much more will these natural branches be grafted back into their own olive tree?"

The earliest *bona fide* written account of grafting is from the Hippocratic tract, *On the Nature of the Child*, written by a follower of Hippocrates around 424 BC. A passage in the work makes it obvious that grafting was a common technique at this time and that it has been established for some considerable time:

> "Some trees, however, grow from grafts implanted into other trees: they live independently on these, and the fruit which they bear is different from that of the tree on which they are grafted. This is how: first of all the graft produces buds, for initially it still contains nutrient from its parent tree, and only subsequently from the tree in which it was engrafted. Then, when it buds, it puts forth slender roots in the tree, and feeds initially on the moisture actually in the tree on which it is engrafted. Then in course of time it extends its roots directly into the earth, through the tree on which it was engrafted: thereafter it uses the moisture which it draws up from the ground."[277]

The author(s) of the above recognized at this early stage that a grafted plant produces fruit true to the scion cultivar rather than the stock, a view reiterated by Theophrastus, who, around 300 BC, described grafting techniques for a number of fruit crop plants and noted that seed propagation of these plants always produced inferior specimens to those obtained by grafting. He referred to grafting as "propagation in another tree", and discussed grafting techniques. Theophrastus took notice of dwarf fruit trees, which were thought to have been introduced to Greece by Alexander the Great after his rummaging in Asia Minor.

Since Persia was at the crossroads of importations of fruit from the East, it is unsurprising that horticulture was an important entity in that ancient land. Based on pottery evidence, the origins of Persian gardens probably date to the fourth millennium BCE, and remnants of the gardens of Cyrus the Great are dated to 500 BCE. It is quite likely that grafting originated in Persia.

The importance of the vine to the Romans ensured that grafting received plenty of attention from their writers. Cato, for example, in *De agri cultura*, describes methods that are still used today and, a little later, Varro wrote extensively on stock/scion relationships. On compatibility, he proffers: "… you cannot, for instance, graft a pear on an oak, even though you can on an apple." Later still, Virgil wrote *Georgics*, essentially an almanac for gentleman farmers, which contained information about grafting. As one would expect, Pliny the Elder had plenty to say about these matters, even though some of his material is dubious, especially relating to compatibility. The oldest pictorial evidence of grafting comes from a mosaic of the Roman period, which was discovered in St. Roman-en-gal, Vienne, France (Figure 5.18).

From all this, it is evident that grafting was well known and established around the Mediterranean area by at least the fifth century AD. After the fall of the Roman Empire, it is likely that the art survived in Christian monasteries and similar establishments. With the rise of Islam, gardening and horticulture reached new heights. As has been said, the Arabs took special interest in their gardens and gardening, probably because their ancestors originated from desert countries.[273] Islamic gardening was inspired by the traditional Persian fruit tree and ornamental gardening, and many new crop plants were introduced during the Islamic period.[278]

The invention of the printing press was to lead to increased literacy and the demand for more books on horticulture. One of the first English treatises was John Fitzherbert's *Boke of*

Figure 5.18 Detached scion grafting – a third-century AD mosaic from St. Roman-en-gal, Vienne, France.

Husbandry of 1531, which contains accurate details on grafting. In 1558, the Italian Gambattista della Porta published a four-volume work (in Latin) called *Natural Magick*, and in 1584 this was expanded into 21 volumes and translated into English. The work was widely read in Europe for over a century, and was very influential. Another, albeit written in the vernacular, tome widely used on the continent was Johann Domitzer's *Ein Neues Pflantzbüchlin* (1531). One of the first books to deal specifically with fruit tree gardening was Leonard Mascall's 1572 work (Figure 5.19), *A Booke of the Arte and Manner How to Plante and Graffe All Sorts of Trees* ..., although this was evidently a translation of a French work. Mascall's treatise was the forerunner of many subsequent, related efforts during the sixteenth and seventeenth centuries, and one should mention: Markham, 1635; Meager, 1688; Langford, 1699. Robert Sharrock's *History of the*

Figure 5.19 How to graft trees – Leonard Mascall, 1572.

Propagation and Improvement of Vegetables of 1672 makes it clear that most of our modern grafting methods had been developed by the seventeenth century.

Some authors paid more specific attention to the apple than others. John Beale, for example, accorded the title of "The Father of Herefordshire orchards", extolled these areas as the "Orchard of England". He goes into great detail about the apples grown there, and comments on their cyder-making qualities: "Some apples make a cool "small wine" with water added at grinding." One of Dr Beale's favourite apples was "Redstreak", of which he speaks of its "mordant sweetness". Beale contributed to John Evelyn's classic *Pomona*[279] (Figure 5.20), which is a discourse on fruit trees in relation to cyder, and an appendix to his masterly *Sylva* of 1664. In his contribution, Beale tells us how to plant apple trees on grazing land; each should be planted on a mound surrounded by prickly bushes and circular ditch to protect the trees from cattle. Herefordshire was known for the longevity of its inhabitants at least as early as the middle of the seventeenth century, and Beale attributed this to apple trees purifying the air![280]

At around this time, there was a group of learned men, mostly Fellows of the Royal Society (one being John Beale), who corresponded with Evelyn on cyder matters. Two facets of cyder exercised the minds of these men: firstly the fuel savings in its manufacture as compared to beer-brewing and, secondly, the fact that grain supplies (needed for bread-making) are not depleted during cyder-making. Thus, there were economic reasons for drinking cyder, certainly instead of imported wine!

Evelyn was one of the group that founded the Royal Society, and became noted for his knowledge of trees, and for his concern about the growing air pollution problem in our cities, notably London. *Sylva* was written with the aim of making the populace aware of the importance of wood to the needs of the Royal Navy for shipbuilding.

One of the major works on apple and cyder of the seventeenth century was John Worlidge's *Vinetum Britannicum* (1676, plus subsequent editions, Figure 5.21).[281] Worlidge was a central figure in the seventeenth century English renaissance of cyder and one of his claims to fame was the invention, in the 1670s, of the "Ingenio" cyder mill. Designed to replace the horse-powered cyder mill, with

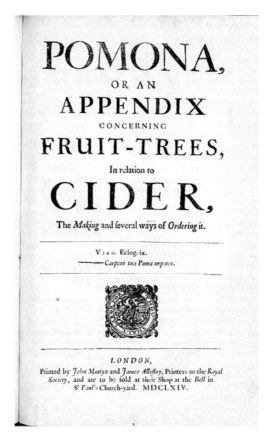

Figure 5.20   Title page of John Evelyn's *Pomona* of 1664.

its expensive associated stonework, it consisted of two rollers in a wooden frame surmounted by a hopper (Figure 5.22), and achieved some success. Robert Hooke, a contemporary of Evelyn and Curator of the Royal Society, devised a cyder mill that would prevent apple solids from descending with the juice (Figure 5.23). Thus, this acted as a press as well as a mill. Worlidge invented several presses, one of which was illustrated in his *Two Treatises* of 1694 (Figure 5.24).

Worlidge often eulogized about the apple (and cyder!), and went into great detail about the intricacies of cydermaking. He was particularly interested in prolonging the shelf-life of cyder and, spurred by the fact that his bottle drink kept much longer that his casked version, he paid considerable attention to the production

Figure 5.21   Title page of *Vinetum Britannicum* (1676).

processes, which he thought might improve cyder's longevity. Apple-pressing was obviously important, and so was what happened to the fruit prior to pressing. Worlidge said that if murc (pomace) was allowed to stand for two days, fermentation would

Figure 5.22   Worlidge's "Ingenio" cider mill.

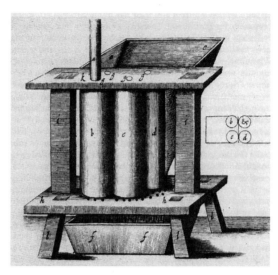

Figure 5.23   Hooke's cider press.

ensue and "juice is transferred from the large particles". He also
maintained that, by remaining on the ground apple skins, the juice
acquired an enhanced colour, as would happen in winemaking.
With the availability of microscopes, Worlidge carried out
investigations into apple micro-structure and concluded that the
"pores" of the fruit were much smaller just under the skin. These

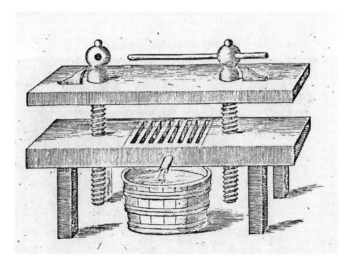

Figure 5.24   Worlidge's improved cider press of 1694.

finer pores were easily disrupted during milling and pressing and yielded the best (ripest) juice. Hardly any pressure was required to harvest this juice, which was prized for producing vintage cyder. Worlidge was always prepared to incorporate the new advances in scientific theory that resulted from the impetus provided by the recently formed Royal Society (for example, Robert Hooke's work on the structure of cork).

Grafting has engendered some controversy over the years and, while some naturalists have regarded it as "instant domestication", others have labelled it "a kind of adulteration". For accounts of grafting methodology see Garmer,[282] du Breuil[283] and Rosenfeld.[284] A fascinating account of apple tree architecture and tree size has recently been published.[285]

5.3.5 Some Apple Biochemistry

Apples are a reasonable source of vitamin A and a good source of vitamin C, the latter particularly in the peel. They can also contribute carotenoids and fibre to a diet, and they are also a fair source of minerals, especially potassium. The long-held view that "apples are good for you" is borne out by the fact that the fruit is rich in antioxidants. Two compounds associated with the uniqueness of the apple are malic acid and sorbitol. Malic acid is the predominant organic acid in pome fruits generally, although many others are present in small amounts. In apple, malic acid is present as 3–19 milliequivalents per 100 g fresh weight. Quinic acid (1,3,4,5-tetrahydroxycyclohexanecarboxylic acid) is often the second most abundant acid in apples, and has even been reported to be the major acid in the cultivar "Worcester Pearmain" stored for several days in nitrogen.[286] It was proposed that one of the functions of this compound was to combine with caffeic acid to form chlorogenic acid (see below). Quinic acid can be formed synthetically by the hydrolysis of chlorogenic acid. Phillips, Pollard and Whiting[287] showed that the amount of quinic acid present in juices of English cider apples is second only to that of malic acid. Phenolic acids are not found in free forms in fruits because their carbonyl groups are very active and easily form esters or amides when combined with aliphatic alcohols and phenols (or amino compounds). Table 5.6 illustrates the main nutritional value of 100 g of unpeeled apple tissue.

Retention of flavour is of prime importance in the harvesting and processing of apples, and with over 300 volatile components[288] contributing to flavour profile, it is essential that there are appropriate methods for assessing the effects of handling and storage on the fruit. These volatile compounds include alcohols, aldehydes, carboxylic esters, ketones and ethers,[289] and around 20 of them are "character impact" compounds, which have a range of flavour thresholds. Some are present in very low concentrations and contribute to the "typical" apple aroma and flavour (*e.g.* ethyl-2-methyl butanoate), while others contribute to aroma intensity (*e.g.* trans-2-hexenal) or are related to aroma quality (ethanol). The major group of volatiles is the esters (78–92%), with alcohols coming in second place (6–16%). As Paillard[290] says, the most abundant compounds are even-numbered carbon chains including combinations of acetic, butanoic and hexanoic acids with ethyl, butyl and hexyl alcohols. Volatile compounds in apples, produced by lipid and amino acid catabolism, are primarily synthesized in the skin.[291] Important apple volatile compounds and their sensory descriptions, as well as their aroma threshold values have been presented by Dixon and Hewett.[292]

Apples are a rich source of phytochemicals, including quercetin, catechin, phloridzin and chlorogenic acid, all of which are strong

Table 5.6   The main nutritional value of 100 g of unpeeled apple tissue.

Water	85.56 g
Carbohydrate	13.81 g (sugar, 10.39 g)
Fibre	2.4 g
Fat	0.17 g
Protein	0.26 g
Vitamin C	4.3 mg
Vitamin A (equiv.)	3 µg
Vitamin B_6	0.041 mg
Thiamine	0.017 mg
Riboflavin	0.026 mg
Niacin	0.091 mg
Pantothenic acid	0.061 mg
Folate	3 µg
Calcium	6.0 mg
Iron	0.12 mg
Magnesium	5.0 mg
Phosphorus	11.0 mg
Potassium	10.7 mg
Zinc	0.04 mg

antioxidants. The phytochemical composition of apples varies greatly between varieties, and there are small changes during fruit ripening and maturation. Apple peel contains higher levels of most antioxidants than does the flesh (from two to six times more phenolic compounds, according to variety). Chlorogenic acid is more prevalent in apple flesh. Vitamin C, although a powerful antioxidant, contributes less than 0.4% of total antioxidant activity in apples.

Lee *et al.*[293] studied the major phenolics in six cultivars and found average concentrations of: quercetin glycosides, 13.2 mg/100 g fruit; procyanidin B, 9.35 mg/100 g fruit; chlorogenic acid, 9.02 mg/100 g fruit; epicatechin, 8.65 mg/100 g fruit; and phloretin glycosides, 5.59 mg/100 g fruit. Vitamin C content was 12.8 mg/100 g fruit.

Vitamin C is present in two forms in apple, L-ascorbic acid and its oxidized form L-dehydroascorbic acid.[294] The total quantity of the two forms is constant per unit weight during fruit growth, although the ascorbic acid to dehydroascorbic acid ratio increases to at least 95 : 5 at fruit maturity.[295] Apple phytochemicals and their health benefits have been superbly summarized by Boyer and Liu.[296]

The presence of fructose, glucose and sucrose has long been recorded from apple juice samples, with fructose being the most common and glucose the least common. Early work indicated that pentose sugars were absent until the presence of xylose was confirmed by Aso and Matsuda.[297] Because of the problem of juice adulteration, many laboratories have devised precise methods for sugar analysis in pure apple juice. Adulteration, though a problem, can be detected fairly easily because a true apple juice should contain a minimum fructose/glucose ratio of 1 : 6, and a maximum sucrose content of 3.5%. Sugars, together with sorbitol, can be used as indicators of fruit maturity because the relative ratio between the different major sugars changes during ripening. For example, fructose is synthesized from sorbitol *via* the activity of sorbitol dehydrogenase and, later on, glucose levels may start to increase due to starch hydrolysis as the fruit reaches maturity. A variety of analytical techniques have been employed to separate and identify apple juice sugars, one of the most widely used being HPLC.[298]

In many plants, sucrose is a major photosynthetic product and a major phloem-translocated component. Some plants, however, can synthesize carbohydrates other than sucrose in source leaves and translocate them to sink organs, such as fruit. A number of plants, including Rosaceae, can synthesize sugar alcohols (polyols) and use them as phloem-translocated compounds.[299,300] In many members of the Rosaceae sorbitol (D-glucitol) is the polyol synthesized (Rosaceae produce sorbitol, sucrose and starch as primary products of photosynthesis). Polyols are reduction products of sugars, and among the angiosperms mannitol is the most widely distributed.[301]

5.3.5.1 Sorbitol

Sorbitol (molecular formula $C_6H_{14}O_6$) is found in woody members of the Rosaceae, such as apple, pear and cherry, and is (unsurprisingly!) also a constituent of members of the genus *Sorbus*, in which the rowan and whitebeam are to be found. Sorbitol seems to play a variety of roles in protecting metabolism during periods of stress, as do other polyols. Several studies have suggested that polyols may play a role in low-temperature tolerance, drought and salt-stress.[302]

It has been known for some time that translocation from the apple leaf is in the form of sorbitol and sucrose, with the former outweighing the latter;[303] indeed, later it was found that around 65–70% of translocated carbon in apple can be attributed to sorbitol.[304] The fate of these two, and other sugars resulting from photosynthesis, including their interconversions, was studied by Priestley[305] as part of the extensive work carried out at the much-missed East Malling Research Station in Kent. It was shown that tissue samples excised from orchard-grown apple trees during a growing season exhibited different capabilities of transferring the ^{14}C-label from sucrose, fructose and sorbitol to other soluble carbohydrates. All tissues incorporated fructose ^{14}C into sucrose, but only leaves incorporated noteworthy amounts of label from sucrose into sorbitol. The work confirmed that sorbitol and sucrose are the major translocated photosynthetic products of apple leaves, now generally reckoned to be: sorbitol 60–70% of total sugar, with sucrose 16–24%.[306] It was suggested, however, that whereas sorbitol seemed to be an end-product of photosynthesis, sucrose could be regarded as a substrate rather more implicated in carbohydrate utilization. In a previous paper, Priestley[307] had found that approximately 80% of the label from $^{14}CO_2$ fed to apple leaves was rapidly incorporated into sorbitol. Sorbitol is also a major component of apple shoot carbohydrates.[308]

Sorbitol is synthesized *via* the reversible reduction of glucose-6-phosphate to sorbitol-6-phosphate by aldose-6-phosphate reductase (A6PR), also called sorbitol-6-phosphate dehydrogenase (S6PDH) (Figure 5.25). Together with sucrose, sorbitol is translocated in the phloem to sink tissues *via* a complex process, which is regulated by rate of photosynthesis, phloem loading, long-distance translocation and unloading, post phloem transport and sink tissue metabolism.[309,310]

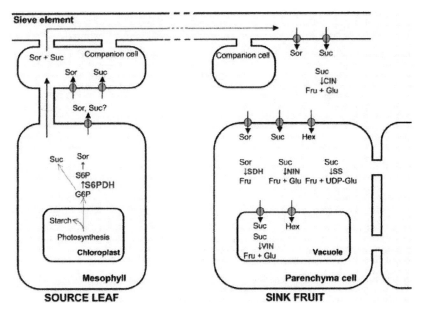

Figure 5.25   Carbon metabolism in apple. Sorbitol (Sor) and sucrose (Suc) are synthesized in the leaf from G6P. Sorbitol is synthesized by S6PDH and sorbitol dehydrogenase. S6PDH is the regulatory step. After translocation to the fruit sink, sorbitol is converted to fructose (Fru) by SDH and sucrose is converted to fructose and glucose (Glu) by invertases (NIN, VIN). Starch synthesis and breakdown (SS) also occurs in fruit. Hex = hexose. (After Teo *et al.*[340] - by kind permission).

The production of sorbitol in Rosacean leaf tissue and its conversion to other sugars in terminal sink tissue, such as fruit, has been the subject of much investigation.[299] According to Bieleski,[311] sorbitol is extensively converted to other carbohydrates (fructose, glucose, sucrose, cellulose and starch) in sink tissues, and it has been shown that most sorbitol breakdown to fructose in sink tissue is dependent upon the enzyme sorbitol dehydrogenase (SDH), with NAD^+ as cofactor:

$$\text{Sorbitol} + NAD^+ \rightarrow \text{Fructose} + NADH + H^+$$

Expression of SDH in the developing apple fruit is at its maximum some 2–3 weeks after bloom, a period sensitive to carbon availability.[312] Rates of SDH activity, and thus sorbitol utilization by fruit, may be reduced as sorbitol availability declines, but why this should be is imprecisely known at present.

Sorbitol can also be metabolized by sorbitol oxidase (SOX), which converts it to glucose in the absence of a cofactor but, to date, the levels of SOX encountered have been too low to be able to explain *in vivo* sorbitol interconversions. The activity of SOX is reported as being 10-fold lower than SDH during apple fruit development.[313]

Very young apple leaves cannot deal with sorbitol at all, and only when they are expanding can they synthesize and metabolize it.[314] Even when older, leaves, which are net sorbitol importers for only a short time, lose the ability to metabolize sorbitol when they are fully expanded (*i.e.* mature).[315]

Sorbitol metabolism and sink-source interconversions in developing apple leaves were the subject of a study by Loescher *et al.*,[316] who investigated changes in the activities of the sorbitol enzymes, rates of photosynthesis and levels of carbohydrates in leaf tissues as leaves underwent sink-source interconversions. In spring, they found a decrease in sorbitol dehydrogenase activity and an increase in A6PR activity as leaves underwent the sink-to-source transition. Sorbitol dehydrogenase activity reached a minimum as A6PR peaked, and these changes were related to increases in leaf carbohydrate levels (especially sorbitol) and to increases in rates of net photosynthesis. In autumn, when leaves were senescing, sorbitol dehydrogenase activity increased, and A6PR activity, leaf carbohydrates and rates of photosynthesis all decreased. Together with other results, this suggested that the metabolism of sorbitol in apple is tightly controlled and related to source and sink activity.

In spite of the importance of sorbitol in Rosaceae, evidence of sorbitol transporters was for some time limited to sink organs, notably fruit.[317,318] Chlorophyll DNAs encoding sorbitol transporters PcSOT1 and PcSOT2 were identified from the related sour cherry (*Prunus cerastus*)[319] from both fruit and leaf (source) tissues, and then Watari *et al.*[320] cloned and characterized sorbitol transporters from apple leaf source leaves. They isolated three cDNAs (*MdSOT3*, *MdSOT4* and *MdSOT5*), which are homologous to plant polyol transporters. *MdSOT3* was expressed only in source leaves, whereas *MdSOT4* and *MdSOT5* were expressed in source leaves and in some sink organs, expression being greatest in flowers. Fruits exhibited little or no *MdSOT* expression. When yeasts were transformed with these *MdSOT*s, they took up significant amounts of sorbitol and, in view of other results as well, the group were able to conclude that these *MdSOT*s are involved in sorbitol loading in Rosaceae.

Within five years, a sucrose transporter, MdSUT1, and a sorbitol transporter, MdSOT6, both localized in the cell membrane, had been identified in apple fruit.[321] The sucrose and sorbitol transport functions of these two transporters were assayed by yeast transformation; MdSUT1 used strain SUSY7[322], MdSOT6 employed strain RS453.[323] It was also shown that the two transporters interacted physically with an endoplasmic reticulum-anchored cytochrome *b*5 (MdCYB5) *in vitro* and *in vivo*. Yeast (*S. cerevisiae*) cytochrome *b*5 (ScCYB5) had no function in the regulation of the two sugar transporters, which suggested that there is a novel mechanism by which the apple plant can adjust sugar uptake according to surrounding availability.

5.3.5.2   Malic Acid

L(–)-Malic acid (the biologically active form) is found in all living organisms as a key intermediate in the citric acid cycle, but is especially plentiful in green apples, something which explains its German name, *Äpfelsaure* ("apple acid"). Being the major acid in apple, the level of this compound is crucial for maintaining a balance between sweetness and acidity in the fruit, since overly "sweet" or "sour" fruit is of little use to the grower.

In all, Hulme and Rhodes[324] detail eleven organic acids in apple pulp with five additions in whole fruit. The most commercially successful dessert apple cultivars are in the medium acid/medium sugar, medium acid/high sugar and low acid/medium sugar categories. The nature and inheritance of sweetness and acidity in the cultivated apple was reported on by Brown and Harvey.[325] The inheritance of sweetness showed a quantitative pattern with a normal distribution of progenies about the mean that is very close to the mean value of the two parents. The inheritance of acidity, however, was shown to be more complex with two patterns being involved. Acidity (malic acid content) of fruit is strongly additively inherited, as shown by Visser *et al.*,[326] but a role is also played by a single dominant gene, designated *Ma*, "superimposed on a quantitative pattern". Thus, low acidity behaves as an acidic character and, because most cultivars appear to be heterozygous, a quarter of the seedlings in progenies may be of the low-acid kind. The very low acid progeny are often called "sweets", and have 0.1–0.3% malic acid in their fruits.

To expand this slightly, we know that in progenies where one or both parents are homozygous (*Ma Ma*), all offspring will have normal acidity. In other progenies (*Ma ma* × *Ma ma* or *Ma ma* × *ma ma*) a quarter or half of the offspring will be "sweet", and regarded as undesirable. In progenies where one or both parents are homozygous dominant, the distribution of acidity is normal with the progeny mean approximating the parental mean.

Malate dehydrogenase (MDH) is a highly conserved, ubiquitous enzyme which catalyses the reversible reaction from oxaloacetate to malate – associated with oxidation/reduction of dinucleotide co-enzymes. In apple, malic acid is usually accumulated in the cytosol during apple development, and is then stored in vacuoles. Such synthesis involves cytosolic phosphoenolpyruvate carboxylase (PEPC) and cytosolic NAD-dependent cyMDH. The apple *MdcyMDH* gene encoding the cyMDH enzyme has been cloned and functionally analysed.[327] The protein was sub-cellularly localized in the cytoplasm and plasmalemma, and it was found primarily to catalyse the reaction from oxaloacetate to malate *in vitro*. The level of expression of *MdcyMDH* was positively correlated with MDH activity throughout fruit development, but not with malate content, especially in ripening fruit. It was also shown that over-expression of *MdcyMDH* contributed to malate accumulation in apple callus tissue.

The same group[328] investigated the role of *MdcyMDH* in growth and stress tolerance. They characterized the role of the gene in growth and tolerance to salt and low temperature, and found that *MdcyMDH* expression is positively correlated with the growth vigour of the apple plant, and that it was sensitive to abiotic stresses. It was also found that *MdcyMDH* over-expression enhanced tolerance to cold and salt stresses.

In another contribution, the team[329] looked at the relationship between the activity of some key enzymes related to malic acid metabolism and the accumulation of malic acid and soluble sugars. Their results demonstrated that PEPC and cyMDH, which we have said are implicated in cytosolic malic acid synthesis, and vacuolar H^+-pyrophosphatase (VHP), implicated in malic acid storage in the vacuole, showed no significant difference between high- and low-acid fruits. Vacuolar H^+-ATPase, however, involved in malic acid storage in the vacuole, exhibited much higher enzyme activity in high-acid fruit. Malic acid-degradation-related cytosolic NADP-dependent malic enzyme (cyME) and ATP-dependent phosphoenolpyruvate carboxylase kinase (PEPCK) showed high enzyme activity in low-acid fruit. The effects of these key enzymes on the accumulation of soluble sugars were discussed in terms of the transformation of malic acid into sugars.

5.3.6 Photosynthate and Apple Fruit Development

Apple fruit quality can be judged by size, colour, flavour and sugar and acid content, and with over 90% of the final dry weight of the fruit being composed of carbohydrates, fruit quality and yield must be heavily dependent on the import of these compounds. Once in fruit, sorbitol, the major sugar export from leaves, is converted to fructose. Unfortunately, fruits are weaker sinks than shoots and roots, and will suffer if there is stiff competition for translocated carbohydrate.

After pollination and fertilization, apple fruit develops *via* cell division for 4–6 weeks, and this is followed by an extended period (50–180 days, depending on cultivar) of cell expansion until fruit is harvested.[330] Fruit in the cell division phase grows exponentially, and when cells are in their expansion phase, growth is linear.[331]

The final size of an apple is a function of cell number produced during the first 4–6 weeks[332] and, as one would expect, fruit relative growth rate (*i.e.* sink activity) is highest during this time. Aspects of carbon supply and demand in apple fruits have been studied by Lakso *et al.*[333]

As Berüter[334] reported, in developing fruit, most of the accumulated, soluble carbohydrate is manifested as fructose (45–60%), with smaller contributions from glucose (20%), sucrose (10%) and sorbitol (3–8%). From the onset of cell expansion until ripening commences, much of this carbohydrate is stored as starch and, when ripening is underway, starch is converted to fructose and glucose.

Because fruits are weaker sinks than growing shoots, steps are taken by commercial growers to maximize their fruit yield, and some fruit is invariably removed ("thinning") to ensure adequate development of what remains. This often involves chemical application and this is usually undertaken during the first month of fruit growth. Some fruit is shed naturally during this early period ("June drop"). If fruits were more competitive for photosynthate, more young material could be left on the tree and more with commercial properties could develop. Experiments on pear have shown that fruit growth is limited by its sink strength rather than by the capacity of the transport pathway,[335] strongly suggesting that fruit yield and quality are dependent on the way (essentially the rate) in which carbohydrates are actually metabolized.

As has been intimated, sorbitol levels in apple phloem are high, but are much lower in fruit, and this, together with the fact that fructose is the most abundant soluble carbohydrate in fruit, would seem to suggest that the sorbitol-to-fructose conversion may be the rate-limiting step in sorbitol utilization by the apple fruit. This demonstrates the importance of SDH, and it is interesting that four highly homologous *SDH* genes (*SDH1*, *SDH2*, *SDH3* and *SDH4*) have been found to be expressed in apple fruit.[336]

The presence of these multiple genes suggested tissue-specific expression, and it was found that the expression of three of these genes was confined to sink tissues (immature leaves, stems, roots maturing fruit), while the fourth was expressed in both immature and mature leaves. Not all expression patterns could be resolved, but the presence of two isomers, SDH1 and SDH2, was detected in

90-day old fruit, and the other two isomers (SDH3; SDH4) detected in 120-day old fruit. None of the isomers could be identified from fruit 30 days after bloom.[336] A considerable volume of work has been conducted at the University of Kentucky into the activity of SDH during the early stages of apple fruit growth,[312,337] and other valuable contributions to this field have been reported since.[338–341] Phloem loading, the starting point for the export of carbohydrates from leaves, and all that this entails, has been reviewed recently,[342] and those interested in transport mechanisms for assimilated products should consult Lalonde *et al.*[343] For comparison, Shiratake[344] has provided a review of the genetics of sucrose transporters in plants.

The high concentration of sorbitol in phloem is directly correlated with a combination of the activity of S6PDH and the inhibition of sucrose synthesis catalysed by sucrose-phosphate synthase (SPS). S6PDH activity is inhibited by inorganic phosphorus, which may be the basis of a control mechanism in plants. Sorbitol concentrations in apple leaves therefore vary considerably, gradually increasing after dawn, reaching a peak in the late afternoon, and declining to their lowest level at the end of the photosynthetic dark period.

Because of its function in translocation and storage, sucrose plays a pivotal role in plant growth and development. In addition, there is gathering evidence that sucrose (or some derived metabolite) may play a non-nutritive role as a regulator of cellular metabolism, possibly by acting at the level of gene expression.[345] Sucrose synthesis is catalysed by two completely different enzyme systems in higher plants: SPS and sucrose synthase (SuSy). The reaction for the former is:

$$\text{UDP-glucose} + \text{fructose-6-phosphate}$$
$$\leftrightarrow \text{sucrose-6}'\text{-phosphate} + \text{UDP} + \text{H}^+$$

and for the latter:

$$\text{UDP-glucose} + \text{fructose} \leftrightarrow \text{sucrose} + \text{UDP} + \text{H}^+$$

With cell production being an essential facilitator of fruit growth and development, Malladi and Johnson[346] used expression profiling of cell cycle genes, and identified key potential regulators of cell production during different phases of fruit production (carpel development, fruit set and fruit growth).

5.3.7 The Apple Genome

Four years after releasing the complete genome sequence of the grapevine, a consortium led by scientists based in Trento, Italy, concluded their task of sequencing the apple genome. The work, which used the diploid cultivar "Golden Delicious", one of the most widely dispersed varieties in the world, was a collaboration between 18 research institutions worldwide, and lasted two years, during which time some 13 billion sequenced nucleotides were produced. Sequencing and assembly of the genome followed the whole-genome shotgun approach, and the work produced a high-quality draft genome sequence.[260] The apple genome was assembled using a combination of traditional and next-generation sequencing technologies that produced approximately 600 Mb of the estimated 743.2-Mb genome. The team demonstrated that most of the cloned apple genes (>90%) are found in this sequence, and that the majority of the missing sequence is highly repetitive (as is the case for most sequenced genomes) and likely to contain relatively few functional gene sequences.

Evolutionary analyses of sequence duplications suggested that the apple genome underwent a relatively recent (*ca.* 50–65 Mya) genome duplication, which contributed to the 1-chromosome karyotype of the sub-tribe (Pyreae) to which apples and pears belong. There was also evidence to support the notion that the modern apple genome resulted from the autopolyploidization of a 9-chromosome progenitor to 18 chromosomes, followed by the loss of a single chromosome, resulting in the current 17-chromosome constitution. The timeframe for the genome duplication, an important evolutionary change, in apple suggests that it might have been coincident with the major environmental event that resulted in the demise of the dinosaurs, and that it was part of survival strategy.

Phylogenetic reconstruction of Pyreae and the genus *Malus*, relative to major Rosaceae taxa, enabled the identity of the ancestor of today's cultivated trees to be confirmed as *M. sieversii* (Ledeb.) M.Roem., which had been previously suggested,[347] but challenged because of molecular similarities between the domestic apple and *M. sylvestris*, which were reported by Coart *et al.*[348] It seems as though the cultivated apple was domesticated around 3000–4000 years ago from *M. sieversii*. On the progenitor of the cultivated apple, the consortium offer: "Our study fully supports the proposal that *M. x domestica* and *M. sieversii* are the same species, for which the more appropriate nomenclature of *M. pumilla* Mill. could be adopted."

The work also threw some light onto the possible origin of the pome, a fruit which is found only in the tribe Pyreae, and a structure that has enabled the apple to be a "storable" commodity. The pome is unique because the bulk of the fruit tissue is formed from the expansion of tissue at the base of the floral organs (the receptacle). It appears that a sub-clade of MADS-box genes, normally involved in flower and fruit development, is expanded in apple to include 15 members. MADS-box genes determine the eventual fate of floral tissues in all plant species thus far examined,[349] and it has been shown that an apple MADS-box gene that is a member of the *AP1* clade, common to all flowering plants,[350] is differentially expressed during pome development.[351] The *MCM1*, *AGAMOUS*, *DEFICIENS* and *SRF* (MADS)-box genes encode for a family of highly conserved transcription regulators involved in a diverse range of important biological functions, among which are embryogenesis and flower and root development in plants. The evolution of plant MADS-box transcription factors has been dealt with by Shan *et al.*,[350] and Jack[352] has summarized the molecular and genetic mechanisms of the control of flowering.

Using microarrays to study the developmental processes occurring in fruit formation from pollination to full ripeness, it has been possible to understand the regulation of the events required to produce a complex apple fruit, in which both the core (ovary) and cortex (hypanthium) tissues expand, and it was evident that genes in different functional classes are expressed at different times during fruit development.[351]

With photosynthesis-derived carbohydrates being transported mainly as sorbitol in Rosaceae,[311,353] it is to be expected that apple has more copies of key genes related to sorbitol metabolism. Thus, we find enhanced numbers of *sorbitol transporter PcSOT2* (specific to Rosaceae fruit[319,320]), *aldose-6-P reductase* (*A6PR*), which is rate-limiting for sorbitol biosynthesis (see below), and *sorbitol dehydrogenase* (*SDH*), which converts sorbitol into fructose in fruit.[316] Overall, 71 sorbitol metabolism genes in apple were identified.

The domesticated apple has the highest number of genes (putative number; 57,386), of any plant genome studied to date, and a complete set of 992 genes responsible for gene resistance were identified, which will be highly useful for breeding purposes. With the apple genome now available, its sequence can be harvested for DNA markers to assist in breeding and for explaining the genetic basis of the domestication of the crop. Genome size variation in *Malus* species has been studied by Höfer and Meister,[354] whose results indicated that species found far from the centre of origin in Asia had a greater DNA content.

The beauty and fragrance of an apple orchard when the trees are in flower have inspired many great minds, and the beauty of plants generally has attracted the attention of mathematicians for centuries. With the publication of *The Algorithmic Beauty of Plants* (now on-line) we had the first account of algorithms behind the creation of "virtual plants", mostly based on the concept of Aristid Lindenmayer's L-systems.[205] Mathematics is used to explore the beauty of plants. The apple inflorescence is an example of a closed raceme, whereby the main apex eventually terminates its development and produces a terminal flower – everything is fully explained!

5.3.8 Cider or Cyder?

Traditionally, the alcoholic drink cyder was, by definition, made from the first pressings of the apple, and so the juice would contain no added water. As a result, it was strongly alcoholic, and would seldom need preservative herbs added to it to stabilize it. Cider, on the other hand, was an ale-strength brew, which was often made from the final pressings of the apple, or even from the re-pressed pomace, and this was often "watered down" to make it go further. The shelf-life of cider was consequently much shorter than its more robust sister drink (especially when diluted with barely-potable water!) and, like beer, was often flavoured with aromatic plants. Roger French, in his classic tome,[355] said quite categorically:

"Cyder is no longer made. This book shows why not ... It is a living wine of some subtlety, matured in cask and bottle in the manner of champagne. At its seventeenth-century height, cyder was a wine, not a long drink. At its best it was often preferred to good French white wine."

The geographical northern limit of the apple is roughly the same as the grapevine and so it is not unreasonable that wine and cider should essentially be drinks of the median and southerly latitudes of Europe. It is likely that the wandering Celts took cider-making technology northwards with them. In a British context, French puts it thus: "Within historical times cyder emerged as preeminently a Celtic drink, associated with them from their profoundest antiquity, and no doubt brought by them from their southern origins. The Celtic 'Britons' came to Britain long before the Romans are said to have introduced the apple tree, and the poetry of their descendents, the Welsh, contains ecstatic descriptions of the apple tree, and no one interested in folklore need be reminded of druids, sacred groves and mistletoe, which depends on the apple tree for its existence."

Pliny the Elder mentions cider, and gives descriptions of cider- (and perry-) making around the Mediterranean basin, but gives relatively few details. In some instances it is not certain whether he was describing an alcoholic drink, or an unfermented mixture of apple juice and water. Toussaint-Samat mentions that in the Roman Empire a kind of cider was made using an excellent apple, the *matiana*, which was sweetened with honey to compensate for the paucity of sugar in the fruit. It would seem that cider was a more popular drink among the ancient Greeks (where it was known as *sikera*) than the Romans. It is tempting to invoke the ancient Hebrew word *sicera* as being the root of our modern "cider" but, although it is a word often paired with wine in the Old Testament (cider being "apple wine"), it has long been used as a term to denote any intoxicating drink (from *shekar*, Old Testament Hebrew for "strong drink"). This seemingly emanates from Oxford scholar John Wyclif, who translated the "strong drink" of the Bible as "cider", and this may be a sign of the strength of the drink in England at that time.

Nearer our own time, Charles the Great (Charlemagne), who reigned from AD 768 to 814, employed *siceratores* on his estates, and these "makers of sicera" were subject to many regulations. It is clear from the regulations that the duties of the *siceratores* encompassed the production of cider, perry and beer. Grape wine and beer, however, seem to have been the most popular drinks in Carolingian Gaul.[356] It is likely that, in olden times, in the absence of specialized equipment, pear would have been a better fruit for alcoholic fermentation (perry), because it liberates its juice far more freely, and contains more fermentable material.

As Juniper and Mabberley say,[256] "cyder is a word that occurs in almost all the Indo-European languages, but it may have existed longest in those languages in the warmer regions". Cyder, widely used until the nineteenth century, and cider originate from the Middle English *sidre*, almost certainly emanating from the identical Old French word. There are a number of now scarcely used variants, such as *sidir*, *sydar* and *seider*.

From a UK point of view, there are no references to cider-making during the Roman occupation of Britain, although it is likely that the drink was imported during the occupation from northern France. This would have undoubtedly been an expensive commodity. Some authorities[357] believe that cider-making had been introduced to England from Normandy well before the 1066 Conquest, and there is some evidence from Anglo-Saxon times. One story involves Harold Godwinson, a patron of Edward the Confessor, created Earl of Anglia in 1045 (and Earl of Hereford in 1058), and later to become King Harold, who at a banquet in Hereford was known to have served cider, as well as other fermented drinks. Otherwise, as Hagen says, historians have presumed that cider was made in Anglo-Saxon England, but have not produced an Old English word for it. It is possible that the strong drink, *beor*, may have been used for cider – it certainly did not equate to "beer".[358] Having said this, most authorities would agree that the art of making cider was introduced to Britain from Normandy in the middle of the twelfth century.

At first cider-making was confined to the Kent and Sussex area, but it soon spread to East Anglia and Yorkshire and obviously fairly quickly westwards, for there is a record of cider being sold from the Earl of Gloucester's manors in Gloucestershire in 1184.[359] Most of the earliest official references to the drink, in fact, come to us from thirteenth-/fourteenth-century documents from Norfolk and Yorkshire and, from these locations, cider-making made a broad sweep westwards to the more renowned counties of Devon, Somerset, Dorset, Worcestershire and Herefordshire. One early Norfolk record, dated 1205, is from "The Lordships of Redham and Stoksley", who were "held by the Petty Sergeantry by the tenure of paying yearly on Michaelmas Day [29th September] ... 200 Pearmain apples and 4 Hogsheads of wine made from Pearmains". In Yorkshire, crab apples were being gathered to make cider in the manor of Wakefield in 1296.

William of Malmesbury (*c.* 1096–*c.* 1143), one of England's greatest early historians, on one of his travels, reported that Thorney (near Peterborough) in the see of Ely was "so fully cultivated with apples and vines that it is like an earthly paradise".

From the Middle Ages we have numerous attestations as to the popularity of cider in England, one such being by William of Shoreham, a fourteenth-century poet who was at one time vicar of Chart in Kent. Worried about the sanctity of religious rites, he reports a Church edict stating that: "young children were not allowed to be baptised in cider"! Cider is mentioned several times in William Langland's *Piers Plowman*.

Cider gets several mentions in English texts during the sixteenth century, although one gets the impression that the country was in the throes of recovering from the ale/beer transition. William Harrison,[360] who had lots to say about the benefits of English ale, penned the following around 1577: "In some places of England there is a kind of drinke made of apples, which they call cider or pommage, but that of peares is named pirrie, and both are ground and pressed in presses made for the nonce. Certes, these two are verie common Sussex, Kent, Worcester, and other steads where these sorts of fruits do abound, howbeit they are not their onelie drinks at all times, but referred unto the delicate sorts of drinke."

Andrew Boorde (1490–1549), physician, priest and traveller, was ever eager to expound upon the medical effects of what could be taken in by mouth, and penned the following in relation to cider, and was obviously somewhat confused about its distinction from perry:[139] "Cyder is made of the iuce of peeres, or of the iuce of aples; & other whyle cyder is made of both; but the best cyder is made of cleane peeres, the which be dulcet; but the beest is not praysed in physicke, for cyder is colde of operacyon, and is full of ventosyte, wherfore it doth ingendre euyll humours, and doth swage to moche the naturall heate of man, & doth let dygestyon, and doth hurte the stomacke; but they the which be used to it, yf it be drunken in haruyst, it doth lytell harme". Boorde's *Breviary of Health* (1547)[139] was the first medical book in the English language to be written by a medical man, and was a companion volume to *Compendyous Regyment* or *Dyetary of Health*, in which the above quote appeared.

By the seventeenth and eighteenth centuries, cider-making had become an art in England and, as we saw above, the seventeenth century saw the publication of several specialist books about cider-making, with Worlidge's *Vinetum Britannicum*[281] being perhaps the most influential to "the man in the street". Evelyn's *Pomona*[279] was probably a must for the "coffee table" of many a grand house. By some, comparisons were drawn between cider and fine Rhenish wine. As the nineteenth century progressed, the popularity of cider gradually diminished and it became regarded as a cheap source of alcohol. The decline was probably associated with cider quality – cider-making was still largely a cottage industry – and the improving quality of beer (especially in bottle), which was due to brewing becoming more industrialized (and science-based). By the turn of the twentieth century, consumers began to demand a clear, sweet cider instead of the "rough" and dry farm-made product. Unfortunately, sweet ciders are far more likely to develop "cider sickness".

The cider made in the eastern and western counties differed in as much as the latter only used selected "cider" varieties, such as "Kingston Black" and "Yarlington Mill", whereas East Anglian producers would regularly use culinary varieties, such as "Bramley" and desserts like "Worcester". Production in eastern England had apparently ceased by the end of the eighteenth century, for Nathaniel Kent in his survey of Norfolk agriculture[361] wrote: "Orchards very few, and much neglected, consequently no cider." If this was the case, then the situation changed dramatically within 50 years, for in White's 1845 Directory we read: "Orchards are numerous in Norfolk especially on the south side of the county, where farmers make cider for their own consumption and a little for sale ... there are a great variety of apples peculiar to, or cultivated in Norfolk, some few of which are rarely to be seen elsewhere." This snippet indicates that cider-making was still a rural enterprise and largely for domestic consumption. Compare this with the contemporary brewing industry in Britain, which by then had started to become industrialized.

In his admirable 1972 review paper, Beech[362] noted: "Over the last twenty years there has been a progressive improvement in the fermentation methods employed by the cider industry. The mixed microflora, typical of traditional cider making and derived in the main from the processing equipment, was first controlled by the rational addition of sulphur dioxide to the juice. This, together with the maintenance of a rigorous standard of hygiene in the pressing equipment, led to the need for pure yeast cultures. The process has been further improved by the increasing adoption of fermenting vessels made with impervious surfaces and the use of sterile concentrated juice which can be stored until required for dilution and fermentation. These changes have been based largely upon chemical, biochemical and microbiological research into all aspects of the process."

Cider was an important drink to the early settlers in North America, with both "hard" and "soft" cider being made from an early date. Hard cider was alcoholic, soft was unfermented. Much was written about the rough, variable nature of the alcoholic drink in America – and its effects! Apart from the obvious, some of the enthusiasm of these settlers for cider may well have had something to do with its antiscorbutic effects, as cider had been known to help keep scurvy at bay for some time. Lore has it that after the Mayflower berthed in America, one of the first acts of those in charge was to plant the apple seeds that they had brought over with them, and records show that, in 1630, the recently formed Massachusetts Bay Company planted numerous orchards in the vicinity of Boston.

One of the outstanding characters of early American pomology was John Chapman, alias "Johnny Appleseed" (b. 1774), who was responsible for spreading the apple over large areas of the land. Starting in New York and Pennsylvania, he journeyed through Ohio and Indiana, and parts of Kentucky and Illinois, and his itinerant lifestyle allowed him to visit countless cider-mills every autumn, where he would filch around for apple seeds in spent pomace. He would wash the seeds and plant them the following spring in carefully selected, hand-cleared land. He has sometimes been portrayed as a "scatterer of seeds", and been charged with the claim that his trees produced apples only fit for cider-making. The latter is probably true, but he was actually very methodical in his planting and tending of trees. His orchards were always created in wildernesses in promising locations, and it is amazing how many towns subsequently arose adjacent to his plantings. It is reckoned that he planted some 100,000 square miles in all. Some of his trees still bear fruit!

There is a possibility that Basque fishermen and mariners spread cider-making knowledge around other parts of Europe, including Normandy. Both the Basque country and Normandy were great apple-growing areas. Toussaint-Samat avers that, early in the seventh century AD, the Merovingian king Thierry II served *sidre* at a banquet at which St Columba was a guest, and that it was apparently Queen Radegonde's favourite drink.

5.3.8.1 Cider-making and Cider Apples

As with brewers, cider-makers (ciderists) have a terminology all of their own, and this was especially so prior to the industrialization of cider-making, which began in the late-nineteenth century in the UK. Before the commercialization of cider the drink was only produced after the annual apple harvest but, with modern technology, it can be manufactured over the entire year. Decent cider cannot be made from any old apple, and there are certain, selected varieties that should be used (Morgan and Richards[363] describe 72 English cider cultivars, most of which are from the West Country). Cider apples should be bitter (tannins) with plenty of sugar, and the best varieties have a russet skin, the "juiciest" having a smooth skin. Cider apples are divided into four categories: sweets (*e.g.* "Sweet Alford"), bittersweets (*e.g.* "Dabinett"), bittersharps (*e.g.* "Kingston Black") and sharps (*e.g.* "Bramley's Seedling"), with the last two being far too astringent for most human palates. Apples can also be classified according to when they were picked, with those picked latest in the year (December) usually producing cider of the highest alcoholic content. Fully ripe cider apples are generally stored for a few weeks after harvest to allow all starch to be converted into sugars.

True cider apples have several advantages for the cider-maker: a) potentially high sugar levels (up to 15%); b) a range of acidity (0.1–1.0%); c) a fibrous cortex to facilitate pressing; d) main- tenance of texture during storage while starch converts to sugar;

e) relatively high tannin (polyphenol) level for body and mouthfeel in end-product. Some ciders are made from a single apple variety, usually a bittersharp or bittersweet cultivar, and these may be sold as a defined year "vintage cider". Only certain cultivars are capable of being thus used for it is difficult to achieve a balance of sugar, acid and tannin in a single variety. "Kingston Black" and "Dabinett" are notable exceptions.

Much cider is made from a blend of bittersweet and other traditional cider apple cultivars, from a blend of culinary or dessert apple cultivars or from blends of both cider and culinary ("cull") fruit. Examples of some characteristics of juices from different cultivars are given in Table 2 in Jarvis *et al.*[365] From the table, it will be appreciated how a wide range of product flavours can be achieved. To the consumer, the primary characterization of a cider taste is a "spicy, aromatic, appley note", and this is what differentiates cider from all other alcoholic beverages.[364]

Some key cider flavour compounds are listed in Table 5.7.

Marks *et al.*[366] identified and quantified 17 phenolic compounds from around 20 English ciders, and found that total phenolic content varied from 44 to 1559 mg L^{-1}. Four groups of compounds were identified: hydroxycinnamates; flavan-3-ols; flavonols and dihydrochalcones, with the first-named being predominant in most

Table 5.7 Key cider flavour compounds.

Alcohols	Ethanol, heptanol, hexan-1-ol, 2 and 3 methyl-butan-1-ol, 2-phenylethanol
Acids	Malic, acetic, lactic, butyric, hexanoic, nonanoic, octanoic, succinic
Aldehydes	Acetaldehyde, benzaldehyde, butylaldehyde, hexanal, nonanal
Carbonyls	Pyruvate, decalactone, decan-2-one
Esters	Amyl acetate, butyl acetate, diethyl succinate, methyl pentanoate, ethyl acetate, ethyl benzoate, ethyl butyrate, ethyl hexanoate, ethyl guaiacol, ethyl lactate, ethyl-2- and ethyl-3-methylbutyrate, ethyl octanoate, ethyl octanoate, ethyl decanoate, ethyl dodecanoate, diethyl succinate
Sulfur compounds	Methanethiol, ethanethiol, methyl thioacetate, dimethyl disulfide, ethyl methyl sulfide, diethyldisulfide
Others	Diacetyl, 2-methyl propanol, ethyl benzoate (after Jarvis *et al.* 1995[365])

Some chemical contributors to this "cidery note" have yet to be identified.

samples. A "substantial variation in the profile and quantity of the phenolics" was found, and analysis of the single-variety ciders emphasized the importance of using a cultivar rich in phenolics for producing a phenolic-rich cider, with possible health benefits.

Much mass-produced cider, which has to be manufactured throughout the year, is made (wholly or partially) from apple juice concentrate. The composition of a juice concentrate is affected by the concentration process, with some volatiles being lost and a few being thermally introduced. An example of the latter is hydroxyl-methyl furfural (HMF), which is believed to hinder fermentation. Another consequence of juice concentration is the development of "cooked" or "biscuity" notes.[367]

On reaching the cidery, sound apples are separated from any gross extraneous materials (*e.g.* orchard debris and stones), washed and then conveyed to a mill for breaking into fragments. The result of milling is a pulpy mixture of the solid and liquid parts of the apples, which, by definition, is known as "pomace" ("pummice"), or "murc" in some areas. Olden means of pulping apples involved stone or wooden rollers. There is no intention to release juice at this juncture; that is effected at the next stage, pressing. Again, in olden times, this would have often meant the construction of a "cheese" in a screw-driven "rack and cloth" press, but modern presses, which are economically suited to large operations, are often horizontal cage presses. The Bucher-Guyer "HP" series presses seem to be almost universal. A decent press would be expected to recover 600 L juice from one tonne of fruit, but this can be increased to 800 L if pectinase enzymes are added to the pomace. Spent pomace is either de-watered and sent for pectin manufacture or ensiled for cattle food.

In France, a traditional juice-making process called *maceration et cuvage* still survives, whereby pulp is packed into barrels to stand at 5 °C for a couple of days.[368] This allows large amounts of pectin to be solubilized and leached out into the juice, and permits enhanced colours to develop *via* the action of polyphenol oxidase on tannins. Once the cuvage has finished, juice is pressed out conventionally.

Once run off from the press, juice (must) is screened to remove gross matter and placed in a receiving vessel, where it may be blended, and mixed with other fermentables (*e.g.* glucose syrup), if necessary, prior to fermentation. With the generally low soluble nitrogen content of apple juice, especially FAN, it is usual to raise the value to around 100 mg N L^{-1} by addition of an ammonium salt. Vitamins and other yeast growth factors may also be required in order to facilitate completion of fermentation.

Apple juice differs from brewers' wort in having a lower pH (generally 3.0–4.4), a much lower soluble nitrogen (TSN) content and a virtual absence of any sugars except mono- and disaccharides. Because of the variable composition of apple juice (according to cultivar, orchard husbandry, climate and fruit maturity at pressing) the cider-maker has less control over his main raw material than the does the brewer. The lower TSN in apple juice means that cider fermentations are much more protracted than beer fermentations.

The soluble nitrogen content of cider apple must largely consists of asparagine, aspartic acid and glutamic acid, with lesser amounts of α-alanine, γ-amino butyric acid, isoleucine, methylhydroxyproline, serine and valine. Arginine, β-alanine, glycine, lysine, methionine, phenylalanine, pipecolinic acid, proline and threonine can also be present in minute amounts. Protein, purines, nucleosides and nucleotides can also usually be detected.[369,370]

In traditional (artisanal) cider-making, no external source of yeast is added to the must and alcoholic fermentation is initiated by wild yeasts from apples and/or equipment. As collected apple must is laden with microbes and inherently unstable, and if must is kept at 10 °C or above, spontaneous fermentation ensues within a few hours. The apple surface yeast microflora may well contain 5×10^4 cells g^{-1} fruit.[371]

If clarified juice concentrate is being fermented, then it will be necessary to incorporate some insoluble solid material into the fermentation in order to provide surfaces for yeast cell attachment and foci for ethanol and CO_2 liberation. Bentonite serves this purpose, as it does in white-wine making. Pectolytic enzymes may have to be added prior to fermentation, especially if must has been stored for any period of time, but apart from that the only other necessary additive need be controlled doses of SO_2. The original aim of this addition (it replaced the old burning sulfur candle!) was to restrict the growth of acetifying bacteria, unwanted lactic acid bacteria and "wild" yeasts, but it also prevents the excessive oxidation of must components. Generally, the use of SO_2 in cider-making parallels that in winemaking.[247] Wedzicha[372] has summarized the chemistry relating to the use of SO_2 in foods. Most countries prescribe a maximum permitted level of SO_2 in the final product – in the UK this is 200 mg kg^{-1}.

With improved hygiene in the cidery, it is now usually necessary to inoculate must with a pure yeast culture, certainly for large-scale manufacture. Most cider fermentations are carried out in the absence of any method of temperature control, certainly those on a small scale. This means that temperature can vary enormously (*ca.* 4 °C in winter to *ca.* 30 °C in summer), and the nature of the metabolites produced will vary accordingly. Cider fermentations at between 20 and 25 °C normally require 3–4 weeks to attenuate, and most UK producers ferment at slightly higher temperatures (22–25 °C) than do French ciderists (15–18 °C), the latter taking longer, of course. Most cider fermentations are taken to "dryness" (*i.e.* PG 1000–995°), rather than leaving residual sugars, and this is possible because most of the juice sugar consists of fructose, sucrose and glucose.

The production of cider involves two fermentations, the "obvious" alcoholic fermentation, and the rather less obvious malo-lactic fermentation (MLF). The former is yeast-driven, while the latter is the result of lactic acid bacteria (LAB) activity. In French cider-making, where alcoholic fermentation is very slow, the malo-lactic fermentation may occur concurrently but, when alcoholic fermentation is more rapid, it is most likely to occur after attenuation (as in the winery). The malo-lactic fermentation results in the decarboxylation of L(−)-malic acid to L(+)-lactic acid, which serves to reduce the acidity of the end-product. Fortunately, the L-form of lactic acid is the usual end-product of the malo-lactic conversion, but D-lactic acid and acetic acid can appear – with undesirable effects on cider flavour (called *piqûre lactique* in France).

Cider fermentation yeasts have been dealt with by Beech,[373] and a consensus would list the essentials of a good cider-making yeast as follows: must have a good rate of fermentation, even at lower temperatures; must settle out rapidly at fermentation end; must be ethanol- and SO_2-tolerant; should produce polygalacturonase (to convert demethylated pectin to galacturonic acid); must have a low oxygen requirement; must produce sulfite-binding compounds (*e.g.* pyruvate and acetaldehyde) in low amounts, and should not produce hydrogen sulfide or acetic acid. Most yeast starter cultures used in cideries have been isolated from traditional cider fermentations, or taken from wine yeast culture collections. Frequently encountered yeasts in cider apple juice are given as genera of *Saccharomyces*, *Pichia*, *Torulopsis*, *Hansenula* and *Kloerckera apiculata*.

Some yeasts produce considerable amounts of diacetyl in cider, especially when SO_2 has been overused (>30 mg L^{-1} "free" SO_2) in the must. A low level of diacetyl is desirable in cider, but once it rises to *ca.* 0.6 mg L^{-1}, it starts to become unpleasant (compare this with the 0.15–0.2 mg L^{-1} desirability threshold for beer).

When the karyotype of a cider yeast (called "CID1") was compared to the karyotypes of some known yeast species, it was found that its genome contained *S. cerevisiae*-like and *S. bayanus*-like chromosomes,[374] thus making it likely that it was a hybrid. Such a phenomenon has also been shown to be relatively widespread in breweries and wineries,[375] and a rapid identification and differentiation method for these hybrids by multiplex PCR was developed by Torriani *et al.*[376]

When fermentation ceases, ciders are left to stand for a period, usually around a fortnight, to facilitate yeast sedimentation (if in a hurry, centrifuge), before being racked into holding/storage tanks, where blending can now occur – as desired. Cider made from cider apples has a greater storage potential than that made from cull fruit. Many commercial makers produce a "base" cider from bittersweet fruit (stored at 5 °C and purged periodically with SO_2), and blend it with other material (cider, apple juice, sugar) to give different products. Factors affecting the development of cider flavour have been discussed by Jarvis *et al.*[365]

Where it occurs, the malo-lactic fermentation results from the activity of a number of bacteria, including *Lactobacillus mali*, *L. plantarum*, "*L. pastorianus* var. *quinicus*",[377] *Leuconostoc mesenteroides* and *Oenococcus oeni*, the latter (syn. *Leuconostoc oenus*) of prime importance in malo-lactic fermentations of wine.[378] In fact, Salih *et al.*,[379] during an investigation of some microbiological aspects of French cider-making, found that out of 165 LAB strains isolated, 79 were identifiable as *O. oeni*. They also found that LAB cells were present throughout fermentation, but that a critical mass of cells (10^6 CFU ml^{-1}) had to be reached before the malo-lactic conversion would ensue. Yeast-wise, they found that the juice flora at T_0 comprised species of *Saccharomyces*, *Kloerckera/*

Hanseniaspora and *Candida*, with the apiculate yeasts being in greater number. By day 7, *Candida* sp. had disappeared, and by day 21 *Saccharomyces* (as *S. uvarum*) was totally dominant, and fermenting well. *Kloerckera/Hanseniaspora* numbers increased slightly after alcoholic fermentation had been completed (day 105).

Original studies on LAB in apple juice and cider suggested that heterofermenters outnumbered homofermenting types in apple juice and ciders, but an analysis of musts running from the presses of a large cider-maker showed that almost 50% of LAB, *L. plantarum* for example, were homofermentative.[380] Further analysis of these organisms indicated that certain strains were able to metabolize quinic acid, and that the end-products of this metabolism were different from those of heterofermenters, where the major end-product is dihydroshikimic acid.[381] Because of this, and other biochemical properties, a new species, *L. mali*, was proposed.

The same workers,[382] on the basis that most heterofermentative lactic rods found in the cider industry formed a fairly coherent group, and that they were sufficiently different from other heterofermentative lactobacilli, proposed a new name for these organisms, *L. collinoides*. It was noted that these characteristic bacteria occurred most frequently in cideries where SO_2 was used sparingly, or not at all. The kinetics, stereospecificity and expression of the malolactic enzyme from *L. collinoides* were later reported by Arthurs and Lloyd.[383]

Of importance for spontaneous cider fermentations is the yeast flora of cider apples, the subject of a study by Bowen and Beech.[384] Their study monitored the yeast distribution on apples from the main cider-making counties of England, and an overall conclusion was that the overall surface flora of apples picked aseptically from the tree consisted mainly of small numbers of weakly fermentative, non-sporing yeasts of the genera *Torulopsis*, *Candida* and *Rhodotorula*. No single yeast was found to be characteristic of a particular apple variety, but a difference could be detected in the yeast floras of orchards in different counties. Invariably present, irrespective of apple cultivar, was *Candida pulcherrima* and this was usually the predominating species. Where it did not predominate, it was usually replaced with some species of *Rhodotorula*, which was also present in all areas (as was *Kloerckera apiculata*). *Debaryomyces kloeckeri* was widely distributed, except in Somerset, and in Herefordshire it was a major microflora component. In one Somerset orchard, *Sporobolomyces roseus* was a principal yeast – but not found elsewhere. *Candida krusei* was common, but not universally spread, while also encountered were *Torulopsis* spp. (*e.g. T. famata*), other *Candida* spp. (such as *C. catenulata*) and, from one Herefordshire orchard, *S. delbrueckii*, which was unusual for a strongly fermenting yeast. The nearest thing to yeast-apple cultivar specificity was *T. famata*, which was found on three (out of four) samples of Kingston Black and one, isolated, sample of Dabinett.

The same workers[385] investigated the yeast flora of cider-making plants and found that commercially harvested and stored fruit showed larger numbers and a wider range of yeast species than did that carefully picked. They also found that the ratios of species changed during storage. Factories with strict hygiene regulations did not contain detectable *Saccharomyces* species, or any other spore-formers, whereas "artisanal" makers' equipment obviously harboured these organisms, and they started to appear (with numerous others) after the milling and pressing operations. Some years later, Laplace *et al.*[386] documented the incidence of indigenous microbes from utensils and the surrounding air in a traditional French cidery.

An early report of the yeast flora of apple juices and ciders was provided by Beech[387] at Long Ashton, and the importance of some of the work carried out there at this time cannot be overestimated. A superb retrospective account of some of the activities of this world-renowned research station has been written by Carr.[388]

In another study of French ciders, Laplace *et al.*[389] found a similar sort of overall pattern, with the exception that *Saccharomyces* spp. were not detected in freshly-pressed juice (day 0). Their experiments were aimed at examining the effect of soil (silt, siliceous clay and green clay) on the growth of a single apple variety, and the fermentation of the resulting apple juices (with a view to being able to provide a guarantee of origin (AOC). From days 0–18 the yeast flora consisted of *Rhodotorula rubra* and *Kloerckera apiculata* (+ *Candida famata* from silt-grown apples); from days 18–50 they recovered *Rhodotorula rubra* and *S. cerevisiae*, and from days 50–85 solely *S. cerevisiae*. LAB during the same experiments also showed a consistent pattern with *O. oeni*, *Lactobacillus brevis* and *L. acidophilus* (days 0–50) and *O.*

oeni and *L. brevis* (days 50–85). All in all, the ciders produced showed differing organoleptic characters, which could be attributed to the microbial floras of the juices, which, in turn, were determined by must composition and, ultimately, soil type. The apparently low populations (or even the absence) of *Saccharomyces* yeasts in these studies is in accordance with the situation reported on undamaged grape berries.[390]

Over the past few decades a number of investigations have been conducted, notably in northern Spain, into the effects of traditional and modern cider-making techniques on the end-product (*sidra*). When traditional and modified cider must-making and fermentation methods were compared for their effect on the flora of cider fermentation,[391] alcoholic fermentation was shown to be carried out mainly by *S. cerevisiae* and *K. apiculata* in both instances. In the case of the "traditional" run, malo-lactic fermentation proceeded concurrently with alcoholic fermentation and the *piqûre lactique* occurred but, in the "modern" method, no malo-lactic fermentation took place.

In the Basque Country of northern Spain, where cider production doubled between 1987 and 2002, cider is usually made in the centuries-old tradition of milling, pressing and re-pressing, natural clarification of juice and spontaneous fermentation to dryness, mainly in chestnut barrels. With increased demand for their cider, manufacturers in this region have been replacing their old presses and wooden barrels with modern equipment, including vertical stainless-steel tanks with means of attemperation. In an exhaustive study,[392] pressing (traditional (T) or pneumatic (P)) and vat material (chestnut wood (W) and stainless-steel (S)) were compared for their influence on the fermentation microflora (yeasts, LAB and acetic acid bacteria) and the production of volatile and acidic compounds.

It was shown that alcoholic and malo-lactic fermentations developed later and were slower in cider made from pneumatically pressed juice and fermented in stainless-steel tanks. Decrease in phenolic compounds was also slower and these differences were transcribed into flavour and aroma differences in cider. Of the volatiles analysed, there were observed differences in methanol, *n*-propanol, *n*-butanol, iso-butanol, iso-amyl alcohol and ethyl acetate. Methanol is not a direct product of fermentation, but arises from the de-methylation of methyl esters in pectin *via* pectinase activity. Pressing method did not affect methanol production, but there was more found in S ciders than W ciders. The major volatile in these experiments, *n*-butanol, was found at similar levels (5.8–6.0 mg L^{-1}) regardless of mode of pressing or vat used. The production of higher alcohols is known to increase in the presence of finely divided pulp particles and by oxygenation,[362] and musts are known to be aerated more as a result of traditional pressing methods. The content of iso-amyl alcohol and total higher alcohols was certainly higher in T must fermentations.

The major organic acids monitored were malic, lactic, acetic, pyruvic and succinic, together with glycerol, a product of alcoholic fermentation whose yield is generally related to that of succinate. Pyruvic acid content was much higher in PS ciders than the others, although it was not present in the initial musts. Not unexpectedly, pyruvic acid content was found to be highly correlated with log yeast population (*i.e.* log turbidity). In terms of malic/lactic acid conversions, it was evident that the malo-lactic conversion was influenced by type of press and vat, along the lines TW > TS > PW > PS.

Must turbidity was significantly higher in pneumatic pressings than traditional ones, and this high turbidity was maintained during the early days of fermentation, now mainly due to high yeast numbers. Post-fermentation racking showed that PS cider was still the cloudiest and required most clarification. The wooden cider press had a significant effect on the LAB population, and this effect was enhanced by using wooden vats. PS cider had the lowest acetic acid content.

In Asturias, Spain, traditional cider is produced by the spontaneous fermentation of apple juice by yeasts originating from fruit and cider-making equipment. Studies on the population dynamics of fermentations in this region have indicated that *S. cerevisiae* usually dominates alcoholic fermentation, with non-*Saccharomyces* genera such as *Kloerckera*, *Candida*, *Pichia*, *Hansenula*, *Hanseniaspora* and *Metschnikowia* being found during the initial stages of fermentation.

It is known that geography, climate, apple variety and, as we have seen, cider-making technology can influence the diversity of yeasts present in a fermentation.[392,393] Prior to the advent of modern molecular methods, identification of yeasts had involved their morphological and physiological traits. With the introduction of PCR-based methods,[394,395] it became possible to provide an accurate definition of the must microflora during fermentation. One such study involved the yeasts associated with the spontaneous fermentation of cider in Asturias.[396] Yeasts associated with apple juice before and throughout fermentation were identified using RFLP analysis of the 5.8S rRNA gene and the two flanking ITSs. The juice was prepared by both traditional and pneumatic pressing methods.

The work showed that a wide variety of yeast species were encountered during the cider-making processes, with the strongly fermentative *S. cerevisiae* and *S. bayanus* both being prominent. Non-Saccharomyces yeasts included apiculate yeasts with low fermentative activity (*Hanseniaspora uvarum*, *H. valbyensis*, *H. osmophila*) and oxidative species (*Metschnikowia pulcherrima*, *Pichia guillermondii*). In general, non-*Saccharomyces* yeasts flourish during the early stages of fermentation, but are later replaced by *Saccharomyces* species. In pneumatically pressed must, *Hanseniaspora* species comprised 66–84% of the isolates, with either *H. valbyensis* or *H. uvarum* dominating – according to year. From the fourth day of fermentation, *Saccharomyces* species take over, and this coincides with the beginning of alcoholic fermentation proper. Interestingly though, *S. bayanus* was dominant at the start of fermentation while *S. cerevisiae* dominated the end stages of fermentation. The presence of *S. bayanus* here may well be related to its cryotolerance.[397,398] since Asturias is a somewhat cold area, and cider fermentations were conducted in a cellar (12–15 °C).

When must originated from a "time-honoured" press, fresh must was populated with larger numbers of *Saccharomyces* yeasts (20–50%, with *S. bayanus* constituting 18–46% of these populations). *Hanseniaspora* species were present throughout the fermentations, with their numbers representing between 40 and 66% of the total yeast count. Different *Hanseniaspora* species dominated through the course of fermentation and the pattern varied year on year. In one fermentation, *H. valbyensis* was present at all stages and increased its numbers throughout until it reached 94% of viable yeasts at fermentation end. Previous work had suggested that this organism was characteristic of the early stages of fermentation only,[399] but in this work it survived right to the end. It is known that apiculate yeasts can tolerate more than 6% ethanol.[400]

In terms of biochemical events, it was evident that the major must sugars were fully utilized after 20–28 days from pressing (regardless of method). Sorbitol, which is not metabolized, was present in musts at levels of 6.5–6.6 g L^{-1}, and glycerol (produced from pyruvic acid) levels varied between 3.6 and 4.8 g L^{-1}. In all fermentations, alcoholic and malo-lactic fermentations were undertaken, with must malic acid level never exceeding 0.5 g L^{-1}. Lactic acid, the main organic acid in most ciders, varied from 4.0–4.8 g L^{-1}. The chemical composition of Asturian natural ciders, including the influence of geographic origin, was presented by Picinelli *et al.*,[401] who took into account pH, titratable and volatile acidities, total polyphenol, acidic polysaccharides, non-volatile acids, polyalcohols, residual sugars and major volatile compounds.

Of 21 strains of *Saccharomyces* isolated from cider-making processes in Brittany and Normandy, and then subjected to electrophoretic karyotyping, PCR-RFLP and Y' occurrence, 18 were identified as *S. bayanus* var. *uvarum*, and three came out as *S. cerevisiae*.[402] It was noted that *S. bayanus* var. *uvarum* has several physiological and biochemical advantages over *S. cerevisiae*, in terms of usefulness in cider-making, including good fermentation at low temperature, production of higher quantities of glycerol and malic and succinic acids, production of lower quantities of acetic acid and the production of pectinolytic enzymes.

Traditional Irish cider fermentations were the subject of a study by Morrissey *et al.*,[403] who, by using selective growth media and PCR-RFLP methods, were able to assess the role of indigenous yeasts. They concluded that such fermentations were divisible into three sequential phases, according to the predominant yeast present: a) "fruit yeast phase", dominated by *Kloerckera/Hanseniaspora*-type yeasts, b) "fermentation phase", dominated by *Saccharomyces* spp. and c) the "maturation phase", which is dominated by *Dekkera/Brettanomyces*. Of interest was the fact that *Brettanomyces* origin could be traced back to the fruit and to the press house, which also harboured high levels of *Saccharomyces* yeasts. *H. uvarum* definitely came from the fruit.

In reality, cider is an apple wine, and accordingly its manufacture has more in common with winemaking than with brewing. Strange, therefore, that cider has usually been marketed to compete with beer in most bars, and is sold "by the pint" rather than "by the glass". Much of the early scientific work on cider-making emanated from the Long Ashton Research Station, near Bristol, UK, which was founded in 1903, but, generally speaking, research into this industry has lagged behind that relative to brewing and winemaking.

5.4 AGAVES

The monocotyledonous *Agave* L. species (long placed in the Agavaceae), "century plants", are important xerophytic plants of semi-arid to arid regions whose geographic centre of origin is Mexico. From this centre, populations spread from the south-western USA, through Central America, the Caribbean and into northern South America. The family is comprised of some of the most advanced monocotyledonous plants and these have been subjected to a variety of taxonomic schemes. The genera associated with Agavaceae have been dispersed among other families in past taxonomies, one of the earliest putting them (as a tribe) in with the lilies (Liliaceae) – then a huge family. Agaves have an inferior ovary and were consigned to the Amaryllidaceae on that basis. Hutchinson[404] invoked the family Agavaceae to include *Agave* and its relatives and genera such as *Yucca*. In addition to their having vegetative similarities, *Agave* and *Yucca* both have chromosome numbers based on $x = 30$, with 25 small chromosomes and 5 large ones. In Hutchinson's scheme there were 19 genera and around 550 species in the Agavaceae. An interesting text on agaves has been contributed by Nobel.[405]

A major problem in monocotyledon concerns the classification and phylogeny in the Liliaceae *sensu lato* and smaller families that have been split off from it, such as the Agavaeae. Two distinct systems of classification were commonly used at the end of the twentieth century,[406,407] with the former scheme being typically followed by North American floras. Molecular evidence now gives us a narrower view of the family, and a number of works have appeared. A chloroplast DNA study of the Agavaceae by Bogler and Simpson[408] resolved some problems and showed that there were two major lineages in these taxa, and the same workers[409] produced a phylogeny of the family, based on ITS rDNA sequence variation.

The Angiosperm Phylogeny Group (APGII[410]), though recognizing the heterogeneity of the Agavaceae, reckoned that they did warrant their own family, and kept them in that family (in the order Asparagales), but an update of the system (APGIII[411]) included Agavaceae in the Asparagaceae, and invoked the sub-family Agavoideae. Agave then became one of 18 genera in this sub-family. It was recognized that the heterogeneous nature of the Agavaceae had led to it being apportioned in several ways in the past and had "further confused the issue".

Agave plants are monocarpic (the plant flowers after 20–30 years, and then dies) perennials, and are succulents with a large rosette of thick fleshy leaves (Figure 5.26) emanating from a short stout stem. Each leaf tip ends in a sharp point, and the leaf margins are serrated and spiny. The overall morphology gives the impression that the leaves and flower stalk arise directly from the root.

Good-Avila *et al.*[412] examined the timing and rate of speciation in *Agave* and found that it is a young genus, somewhere between 7.8 and 10.1 million years old. The estimated age of the family was between 20 and 26 million years, and the age of *Agave sensu lato* approximately 10 million years. They also showed that speciation rates were significantly higher between 8 and 6 Mya, and again between 3 and 2.5 Mya.

Figure 5.26   Habit of the *Agave tequilana* F.A.C.Weber.

The ability of *Agave* spp. to grow on poor soil makes them important drink-giving plants in areas where indigent people find food hard to come by. These plants play an important role in the culinary traditions of Mexicans and Amerindians of the southwest deserts. *A. americana* is the principal culinary plant and, when about to be eaten, the plant is dug up just before the flower stalks emerge and the leaves are cut at their base. This exposes the cylindrical, whitish, pulpy trunk, which is cooked in a fire pit for several days. When cooked, the flesh can be eaten off the fibrous material pervading the trunk. Leaves and centres of agaves which have been roasted can then be stored. The crown also provides food, and juice that can be made into a syrup, or fermented, and the most common use of the plant was to ferment the sweet liquid that accumulates inside to make *pulque*. At a later date, Spanish settlers distilled the fermented juice to produce *mescal* and tequila. Leaves have also been used for fuel and for shelter. According to Lappe-Oliveras *et al.*:[413] "The great variety of agaves and their multiple uses have played an important role in the cultural identification of Mexico. They have been exploited in many ways for over 10,000 years ..."

Iberian explorers brought specimens back to Europe where they proved popular garden plants. Those that have been grown on that continent for some time often show characteristics different from their wild counterparts. The most familiar species is *A. americana*, often called "American aloe" (although *Agave* is not a close relative of *Aloe*). The name "Centuary plant" emanates from the protracted time that it takes to flower, flowering involving great energy expenditure. Other trivial names include "blue agave" and "maguey". Four parts of the agave are edible, sap, flower, leaves and stalk, and evidence from the caves of the Tehuacán Valley suggests that agave was an important food plant by 6500 BC. A few people arrived in the valley around 11,000–10,000 BC, and the first evidence for plant cultivation occurred *ca.* 7000 BC. It is claimed that the Tehuacán sequence is one of the best examples of how, over thousands of years, man made the transition from a hunter-gatherer to an agricultural being. The Tehuacán Valley of Puebla, Mexico, lies in the southeastern corner of this state and extends into the northern edge of Oaxaca. The natural vegetation is thorn-scrub below 1800 m, and above this the natural cover is oak-pine forest, and agriculture is practised below the 1800 m contour.

Agriculture in the valley proper is more complex than that in the mountain fields. The level areas in the valley are among the most intensely cultivated in Mexico. As Smith[414] concluded, there are "at least eight species of maguey native to the Tehuacán area" and fibrous remains of leaves can be found "from all levels for which preservation was adequate to preserve any plant parts". The easiest possible site for crop planting would have been the more open stretches of alluvial soil. Of interest is the fact that New World agricultural pioneers in the Tehuacán Valley probably concentrated on planting succulent perennials, such as maguey, whereas their Old World counterparts started with the cultivation of annual crops. Regarding the natural vegetation of this area, it will be appreciated that it is not easy to reconstruct the wild vegetation of a valley whose floor has been disturbed by agriculture for more than 7000 years!

A well-known early site for maguey and domestication of other plants is the Guilá Naquitz cave[415] in the eastern range of mountains in the Oaxaca Valley. First studied by Kent Flannery in the 1960s, the cave was occupied seasonally by hunter-gatherers for over 2000 years (8000–6500 BC). There is some of the earliest evidence for the cultivation of maize (see p. 190), bottle gourd and squash.

5.4.1 Beverages from Agaves

Since pre-Hispanic times, Mesoamerican civilizations have produced alcoholic fermentations from a variety of native plants, and consumption of these products has played a vital role in the religious and secular lives of these peoples. In Mexico in particular, traditional alcoholic beverages, such as pulque, have long played an important role in the daily life of indigenous communities, and have become deeply embedded in various customs and rituals.

The most important and widely used plants in Mexico are the agaves or magueys (*Agave* spp.), and three main beverages are made from them: pulque, mezcal and tequila. The former is a pre-Conquest, non-distilled drink, and is regarded as a Mexican icon. Traditionally, pulque is obtained by spontaneous fermentation of the sap (or aguamiel) of certain *Agave* spp., but nowadays it is also produced on a small industrial scale through a pure, mixed starter culture fermentation, which aims to control the quality (and safety) of the product. Mezcal and tequila are also seen as national beverages, but they made their appearance after the Conquest, when Spaniards brought the knowledge of distillation that they had learned from the Moors. Mezcal is a liquor obtained by braising, fermenting and distilling maguey heads or yolks, whilst tequila is obtained by fermenting and then distilling cooked minced maguey. A number of *Agave* species may be used to make mezcal but, by definition, *Agave tequilana* var. *azul*, grown near the town of Tequila and other regions in the state of Jalisco, should be used for tequila. Thus, tequila is one category of mescal (a drink that can be produced in different Mexican regions, particularly Oaxaca state). Nowadays, distilled drinks are more popular than pulque. A summary of Mexican alcoholic non-distilled and distilled agave beverages is provided in Lappe-Oliveras *et al.*[413]

Prior to the Conquest, legitimate drinking of alcohol was mostly ceremonial, and confined to the upper strata of society, the elderly and the wise. Ceremonial drinking was mostly undertaken to honour (or placate) the gods, to bless collective tasks and to promote fertility of the land. Commoners were allowed to partake only in certain circumstances, women after giving birth, males after exhausting tasks, and drunkenness was looked upon as a grave misdeed. Habitual drunkards were considered unworthy neighbours and had their homes demolished. Legend has it that the Toltecs, one of Mexico's greatest civilizations, were destroyed through misuse of alcohol. In modern Mexico, knowing how to drink is a reason for admiration and respect, and whoever can sink the greatest quantity is regarded as the most macho. Not being able to drink is a "female trait"!

Since ancient times, *Agave* species have been the most important and widely used plants in Mexico. They belong to the family Agavaceae, which is endemic to America, and, because they are succulents and grow in similar environments, have often been misidentified as cacti. During ancient times, these plants were integral to Mexican society, being used for food, alcoholic beverages, a water "substitute", clothing and as herbal medicine. The distribution of the genus *Agave* is centred on Mexico, but the plants extend deep into the USA, and into northern South America.

Pulque (*octli*), the oldest and most traditional of Mexican alcoholic drinks, is a milky white, viscous, slightly acidic drink

(<6.0% ABV) produced from the sap that is extracted from several *Agave* species, mainly *A. salmiana*, *A. atrovirens* and *A. mapisaga*. It is not a beer, although it has sometimes been erroneously described as such. Because of its great historical, religious, social, medical and economical importance, pulque is the most widely studied and documented of the agave-based drinks. In ancient Mexico, it played a major role in daily life, and was a decisive influence on religious and war rituals. The Aztecs distinguished three different pulque-type beverages: "metocli" or agave wine, "iztacoctli" or white wine and "teoctli" or ceremonial wine. A sample that had become too aged, and had an unpleasant odour and flavour, was referred to as "poliuhquioctli", and it is from this word that the term "pulque" is derived, and was used by the Spaniards to designate the freshly prepared drink. When the Aztec empire disintegrated, pulque lost most of its religious significance, but retained relevance as a nutritional supplement and water substitute. Once the drink became profane, it was introduced to the country's large conurbations, particularly Mexico City. Today, it is the preferred drink for the low-income population, and forms an important part of their daily diet, because of its vitamin and essential amino acid content. Traditionally, pulque has only been sold in special taverns (*pulquerias*), but latterly it has been promoted in restaurants, with the intention of it becoming more popular with the younger generation. The drink is naturally unstable, with a shelf-life of a couple of days, but a means of preserving and canning it has been developed, and the product (usually 6% ABV) is now exported to the USA.

The drink is made from the sweet, yellowish, slightly cloudy sap, which is extracted from 8- to 10-year-old agave plants that are about to produce their (only) inflorescences. Anything from 200–1000 litres of sap can be obtained from a single plant. The principal sap components are water, sugars (glucose, fructose, sucrose), proteins, gums and mineral salts, and this provides a suitable medium for the spontaneous growth of desirable microbes. The principal carbohydrate in agaves is inulin, which can be hydrolysed to fermentable sugars (mainly fructose) with steam heating.[416]

Mexican regulations recognize two types of sap: Type I represents the highest quality (cleanest and highest sugar content), whilst Type II encompasses all other sap samples. Similarly, the regulations recognize two types of *pulque*: Type I includes *pulque de semilla* (starter) and *pulque pie de cuba*, and Type II refers to commercial *pulque*. It is now difficult for us to realize that, in ancient Mexico, *pulque* played a major role in everyday life.

The initial step in the artisanal production of *pulque* is the elimination of the floral bud (called "castration"), leaving a cavity (*cajete*) in the centre of the agave stem. The plant is then left to "age" for at least one month, which enables maturation of the central leaves, and a raising of the sap sugar content (to 7–14% w/v). The healed cavity is then scraped to permit the sap to flow again, and this accumulates in the natural recess. The sap is then removed (traditionally, by means of oral suction *via* a dry, hollowed gourd, called an *acocote*), and then transferred to a shed, called a *tinacal*, where fermentation takes place in open (*ca.* 700 L) containers. After the initial removal of sap, more will be produced by the plant, and this re-secretion will continue for about a month. The person harvesting the agave juice is known as a *tiachiquero.* Fermentation is either spontaneous, or a response to the addition of a fermented sap starter culture (a *semilla*). The latter is made initially from around 10–15 L of highest quality sap, which has been fermented in a small closed vessel. After several days, depending on environmental factors (temperature, *etc.*), the alcoholic fermentation ends, and an acetic fermentation commences. A thick, floating layer is then formed, indicating that the mixture is ready for a second fermentation. To effect this, another batch (600–900 L) of best-quality sap is then mixed with the starter at a rate of 1–3% v/v. This occurs in a tank called the "head fermenter", and fermentation will proceed until the sugars have been exhausted. A batch (300–400 L) of the supernatant (essentially *pulque*, and called *pie de cuba*) is removed, and poured into a second vessel, the "tail fermenter", and used as an inoculum (50% v/v) to ferment a batch of regular sap. After several days in the tail fermenter, the desired ABV, viscosity and sensorial characters are attained, and the product is barrelled. Essentially, this process is semi-continuous since, on withdrawal of batches of pulque from the head fermenter, the bulk of the microbial population is retained for subsequent sap additions.

Most of the microbes needed for sap fermentation can normally be obtained from the walls of the *cajete*, the utensils used for scraping and transferral, from the atmosphere or from insects (*e.g.* *Drosophila* spp.), but starter cultures increase the population of essential microbes, and ensure a more rapid and regulated initiation of fermentation. The most important organisms for sap fermentation are bacteria and yeasts, although some filamentous fungi are also present. In both sap and fresh young *pulque*, there is a predominance of bacteria over yeasts, caused by the neutrality (pH 7–7.4) of the medium. The yeast population increases gradually during fermentation, and these organisms eventually become dominant when the pH drops to around 4.5. There is a four-fold change in the type of microflora during fermentation, after which putrefying organisms take over. Differing groups of microbes elicit a variety of biochemical transformations to the sap, and the initial community is dominated by lactic-acid-producing bacteria (*Leuconostoc* and homo- and heterofermentative *Lactobacillus* spp.). During the very early stages of fermentation, the flora is dominated by two sub-species (*mesenteroides* and *dextranicum*) of *Leuconostoc mesenteroides*. They increase sap acidity, reduce oxygen tension and develop desirable viscosity, through the production of polysaccharides.[416,417] The synthesis of lactic acid, ethanol and dextrans through LAB fermentation, plus the generation of secondary metabolites such as fatty acids and their esters, organic acids and higher alcohols, creates an environment unsuitable for pathogenic bacteria and conducive for the growth of ethanologenic species: *Zymomonas mobilis* ssp. *mobilis* and yeasts. *Z. mobilis* is a crucial organism in the *pulque* fermentation, and produces some lactic acid, acetic acid, acetate, acetone, glycerol, acetaldehyde, acetylmethylcarbinol and gums.[418,419]

Fermentation was effected by a number of microbes, the early stages being dominated by non-*Saccharomyces* spp. As ethanol concentration increased, ethanologenic organisms (*S. cerevisiae* and strains of *K. marxianus*) proliferated. Yeasts encountered were: *S. cerevisiae*, *S. bayanus*, *S. paradoxus*, *Candida* spp., *C. parapsilopsis*, *C. lusitaniae*, *Kluyveromyces marxianus*, *K. lactis*, *Hanseniaspora uvarum*, *Pichia* spp., *P. guilliermondii* and *Torulaspora delbrueckii*. *Leuconostoc* spp. are important bacteria during *pulque* fermentation owing to their ability to lower pH and production of dextrans. Some of the above-mentioned yeasts are involved in the "ennoblement"[420] of *pulque*, providing amino acid, vitamin and protein additions. The amino acid and vitamin content of agave sap, and some *pulque* samples, is shown in

Table 5.8 Amino acid and vitamin content in agave sap, traditional, pilot plant and industrialized pulque (after Lappe-Oliveras et al.[423]).

Essential compounds	Pulque types			
	Agave sap	Traditional pulque	Traditional pulque	Pilot plant pulque	Industrialized pulque
Amino acids (mg 100 mL^{-1})					
Lysine		16.2	3.0	7.5	12.0
Tryptophane		2.7	2.5	9.0	6.0
Hystidine		4.7	4.0	1.0	–
Phenylalanine		11.2	6.5	7.5	–
Leucine		10.5	4.0	6.0	–
Threonine		6.4	1.5	5.0	–
Methionine		0.7	3.0	5.0	–
Valine		6.6	2.5	7.5	–
Arginine		10.9	2.5	3.2	–
Tyrosine			3.0	7.5	–
Vitamins (mg 100 mL^{-1})					
Thiamine	0.06	0.02	0.02	0.02	0.02
Riboflavin	0.02	0.025	0.02	0.03	0.03
Niacin	0.45	0.32	0.28	0.37	0.35
Pyridoxine	–	–	0.02	0.03	–
Biotine	–	–	0.02	0.02	–
Ascorbic acid	9.0	5.6	–	–	5.1

Table 5.8. Because of these micronutrients, *pulque* forms an important part of the daily diet of poorer folk, since it supplies some substances not obtainable from a corn diet. There is some evidence that *pulque*, owing to its partially fragmented fructan content, may have prebiotic activity[421] and probiotic activity because of the presence of LAB and yeasts.[417]

Microflora studies thus far showed that *pulque* contains the following yeasts: *Cryptococcus* spp., *Candida parapsilopsis*, *Clavispora lusitaniae*, *Debaryomyces carsonii*, *Hanseniaspora uvarum*, *Kluyveromyces lactis*, *Kl. marxianus*, *Geotrichum candidum*, *Pichia* spp., *P. guilliermondii*, *P. membranifaciens*, *Rhodotorula* spp., *R. mucilaginosa*, *Saccharomyces bayanus*, *S. cerevisiae*, *S. pastorianus* and *Torulaspora delbrueckii*. The bacterial flora is constituted by: *Acetobacter aceti*, *A. aceti* ssp. *xylinus*, *Bacillus simplex*, *B. subtilis*, *Cellulomonas* sp., *Escherichia* sp., *Kocuria rosea*, *Lactobacillus* spp., *L. delbrueckii*, *L. vermiforme*, *L. mesenteroides*, *Macrococcus caseolyticus*, *Micrococcus luteus*, *Sarcina* spp. and *Zymomonas mobilis* ssp. *mobilis*.[413]

Using 16S rRNA gene sequences, the bacterial diversity of *pulque* was determined by Escalante *et al.*,[422] who identified the following: *Lactobacillus* strain ASF360, *L. acidophilus*, *L. kefir*, *L. acetotolerans*, *L. hilgardii*, *L. plantarum*, *Leuconostoc pseudomesenteroides*, *Microbacterium arborescens*, *Flavobacterium johnsoniae*, *Acetobacter pomorium*, *Gluconobacter oxydans* and *Hafnia alvei*. *Lactobacillus* spp. comprised *ca.* 81% of all bacteria.

The main distilled beverages produced from *Agave* spp. are tequila, mescal, bacanora and raicilla. Juice preparation and fermentation techniques are slightly different. Lachance[423] reported on the yeast communities in a tequila distillery where spontaneous fermentation is practised. Fresh *Agave* material contained a diverse microflora dominated by *C. lusitaniae* and a new endemic species, *Metschnikowia agaves* (*agaveae*). Cooked agave, and its extract, contained many species, including *S. cerevisiae*, while fermenting juice exhibited a gradual decrease in floral components, with *T. delbrueckii*, *K. marxianus* and *Hanseniaspora* spp. giving way to *S. cerevisiae*, *Zygosaccharomyces bailii*, *Candida milleri* and *Brettanomyces* spp. *P. membranifaciens* was the only yeast common to all communities in the distillery. Fruit flies (*Drosophila* spp.) played a part in vectoring certain yeasts, notably *Hanseniaspora* spp.,

Pichia kluyveri and *Candida krusei*. Cooked agave molasses produced during processing harboured predominantly *Schizosaccharomyces pombe*, with lower numbers of *K. marxianus*. From evidence presented, it is evident that the tequila ecosystem studied comprised five principal yeast communities.

Lachance also studied killer yeasts, knowing that they are important in directing yeast communities in some "natural" fermentations. The most noticeable killer species encountered was *T. delbrueckii*, which was quickly overcome as fermentation progressed. The broadest spectrum killer activity was shown by *P. anomala*. Evidence suggested that the most important yeast, *S. cerevisiae*, perpetuates itself in the distillery, possibly with the aid of the fruit fly.

5.5 CASSAVA

Manioc (*Manihot esculenta* Crantz), or more accurately *M. esculenta* Crantz subsp. *esculenta*), is a root crop with great economic importance worldwide, yet evolutionary and geographical origins have remained unresolved and controversial. It is a staple for some 500 million people living throughout the tropics; it is the primary source of carbohydrates in sub-Saharan Africa, and ranks sixth among crops in terms of global production. Manioc ranks second among African staple crops and, while the crop is native to South America, over half of the world's production comes from sub-Saharan Africa. Despite its immense importance in the developing world, cassava has historically received less attention by researchers than have temperate crops, earning it the status of an "orphan crop". Unlike some other crops, manioc requires relatively little labour for cultivation. The root can be kept in the ground until required, and yields more calories per acre than any other food plant. On the down side, in terms of nutrition, manioc offers little else than a source of carbohydrate. In another world, manioc was classified as *M. utilissima*.

The genus *Manihot* (of the spurge family, Euphorbiaceae), comprising just under 100 species, is distributed throughout the New World tropics, from Mexico to Argentina. Most species (~80) occur in northern South America, and there is a secondary centre of species diversity in Central America and Mexico. All *Manihot* species studied thus far have a chromosome number of

$2n = 36$ and show regular bivalent pairing at meiosis. There are few reliable phenotypic characters for inferring evolutionary relationships within *Manihot*, as many species, including cassava, show tremendous intraspecific morphological variability. In addition, cassava has been cultivated throughout the Neotropics for several thousand years, making its site of earliest cultivation very difficult to infer from archaeological or anthropological data. In the absence of a single, clear-cut candidate for a progenitor species, a hybrid-origin hypothesis came to be the favoured explanation for cassava's origin.[424] In its most extreme form, this hypothesis has interpreted the crop as being a "compilospecies" derived from multiple hybridizing *Manihot* species complexes throughout the New World. In the last taxonomic revision,[425] for example, it was proposed that cassava probably had multiple origins in Central and South America, with Central American species being the most likely to have played a key role in the crop's evolution.

Commencing with work carried out during the 1980s, evidence started to accumulate that questioned the "compilospecies" domestication hypothesis. Firstly, the availability of molecular markers led to re-examinations of the crop's evolutionary relationship to South American *vs.* Central American *Manihot* species. Analyses of RFLPs, AFLPs and DNA sequences all indicated that South American and Central American members of the genus form two distinct lineages, and that cassava is more closely related to the former. This questioned cassava's putative origin from Central American *Manihot* species.

Secondly, extensive collecting of *Manihot* populations in South America by Antonio Allem revealed naturally occurring populations that are phenotypically similar enough to cassava to be considered potential direct wild progenitors of the crop.[426] These wild populations were referred to as *M. esculenta* ssp. *flabellifolia* (Pohl) Ciferri, which grows as a clambering understorey shrub, and differs from cultivated cassava primarily in root and stem traits. The candidate wild progenitor occurs in a zone of transitional forest that separates the lowland rainforest of the Amazon basin and the drier savanna of the Brazilian Shield plateau, to the south and east. Another wild species, *M. pruinosa* Pohl, was found overlapping in the eastern section of the *M. esculenta* ssp. *flabellifolia* range (Figure 5.27), and this plant had been grouped

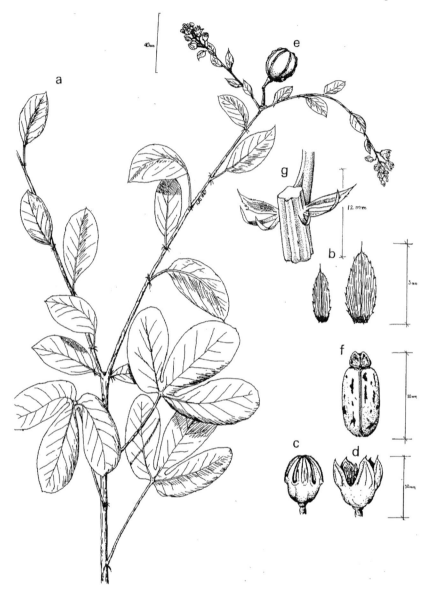

Figure 5.27 Manihot pruinosa.

by Allem[427] into cassava's "secondary gene pool" of potentially interfertile species. As the only such species to have an overlapping distribution with *M. esculenta* ssp. *flabellifolia*, *M. pruinosa* is arguably the most likely candidate for hybridization with *M. esculenta* ssp. *flabellifolia* and/or cassava.

In 2004, Kenneth Olsen, having used molecular methods, reported that cassava was most likely domesticated from a single wild *Manihot* species, rather than from multiple hybridizing species.[428] Olsen also intimated that cassava most likely originated somewhere in the southern Amazon basin. Methods used to trace the evolutionary and geographical origins of cassava were single nucleotide polymorphisms (SNPs) and simple sequence repeat (SSR) variation.

Manioc is known as *mandioca* in Brazil and Paraguay, and *yuca* (not to be confused with yucca) in most other parts of South America. The plant is also commonly referred to as cassava, although this term should strictly be applied to flat round cakes of bread made from manioc flour. Tapioca is a kind of manioc flour, hence another name for the plant. The word "manioc" is from *maniot* in the Tupi language of coastal Brazil, whilst "mandioca" is reputedly derived from *Mani-óca*, or "the house of Mani". According to legend, Mani was the Indian woman from whose body the manioc plant grew! The term "cassava" comes from the Arawak word *kasabi*.

Although a native of Brazil, manioc travelled by canoe with South American Indians and migrated northwards into the Caribbean and, later, *via* Portuguese ships to Africa. The plant was domesticated in Brazil before the beginning of the sixteenth century. When the Spanish reached Central America and the Caribbean, they found that the indigenous peoples were cultivating manioc, a plant they referred to as *yuca*, and, accordingly, the earliest European description of the plant dates from 1494. A description of the first voyage of Columbus refers to "venomous roots" being used by the natives to make bread. The Portuguese first came across manioc after 1500 on the coast of Brazil, and when the Dutch invaded that country during the seventeenth century, they too became familiar with manioc cultivation.

From Brazil, the Portuguese took manioc to their stations along the Upper Guinea coast in West Africa to the Kingdom of Kongo in northern Angola, where it was successfully introduced and adopted. The first documentation of manioc in Africa is to be found in a letter of 1593, and by the end of that century it was an accepted item of trade in Angola, and a food for slaves in transit to the Americas. Manioc would serve as a staple of the slave trade until its abolition in the mid-nineteenth century. The use of the plant spread slowly through West Africa, and extensive manioc cultivation did not develop until the nineteenth century, when liberated slaves returned to their homeland. Further diffusion of manioc in West Africa in the late-nineteenth and twentieth centuries was linked to European colonialism. It is not exactly known when colonists took manioc to East Africa, but we do know that the Portuguese did not introduce it into Mozambique until 1750. Manioc is now cultivated throughout Africa, except for desert areas in the north, and in the far south.

Europeans also transported manioc to Asia, although it is possible that Polynesians may have introduced it into the Pacific *via* Easter Island. From what we know, the first Asian region to receive the crop was the Philippines (*via* Spanish settlers). Again, it spread rather slowly, but by the seventeenth century it was being grown in the Moluccas, and by 1653 was found on one of the outer islands of Indonesia. By 1740, manioc was being cultivated on Java, and it is thought that the Portuguese took the crop to Goa in the early eighteenth century. Additional plants were taken to India from South America in 1794, and from the West Indies in 1840. Cassava has never supplanted rice as the main staple in Asia, but it has become an increasingly more important crop.

Manioc is a perennial woody shrub with umbrella-shaped leaves and starchy storage roots (Figure 5.28), and grows to a height of around 1.8–2.7 m. It has leaves with 5–7 lobes that grow toward the end of the branches. These are considered edible and may be cooked like spinach. The roots (tubers) radiate from the base of the stem rather like the spokes of a wheel, and on harvesting may reach 0.3–0.6 m in length, with a width of 0.05–0.15 m. Individual plants may yield up to 8 kg of root material. There are numerous varieties (cultigens) differing in plant height and appearance, and colour and starch content of the tuber.

As for its reproductive biology, the cassava plant produces separate male and female flowers on the same plant (monoecious), and male and female flowers are borne on the same branched inflorescence, with female flowers at the base and males toward the tip. Flowers are small, with males being around 0.5 cm in diameter and females slightly larger. In a given inflorescence, female flowers open first and the male flowers do so at least a week later, a trait

Figure 5.28   Fleshy roots of *Manihot esculenta* Crantz.

called "protogyny". By the time that male flowers have opened, the female flowers in close proximity will have been fertilized or have aborted. Because the flowering period on a single plant may last for more than two months, pollen from a male may fertilize females on the same plant, as well as flowers on surrounding plants. Flowering of clones is largely influenced by environmental factors, and plant habit. A flower bud typically forms when the plant branches, so that more highly branched genotypes flower more prolifically than those that are sparsely branched.

Cassava pollen grains are relatively large and are sticky, so anemophily can be ruled out. Wasps and honeybees seem to be the main pollinators. Pollen loses viability very soon after it is shed and, in practice, breeders perform their pollinations within one hour of pollen collection to enhance the chance of fertilization. Developing seeds are viable for around two months after pollination, and the fruit matures around a month after that. The fruit is a schizocarp, and dehiscence is thus explosive, with the large ovoid (100 mm × 4–6 mm) seeds falling fairly adjacent to the mother plant. From this position they are further distributed by ants, and so a seed may end up several metres from its birthplace. Cassava seeds are adapted to ant dispersal, with large energy reserves that permit deep burial in a nest, and a long dormancy period. Seed germination is heat-activated and this trait, together with extended dormancy and a tolerance of deep burial, suggests that the progenitor of cassava may have evolved under conditions of sporadic natural fire and was thus uniquely suited to domestication under slash and burn agriculture.[429]

Manioc has traditionally been divided into two main types: sweet and bitter. The former types have a shorter growing season, and can be harvested in 6–9 months and then simply peeled, and boiled or baked and eaten as a vegetable. If not harvested soon after maturity, sweet manioc will deteriorate rapidly. Bitter varieties require 12–18 months to mature, but will tolerate being left in the ground for many months before harvesting. On the down side, bitter forms are toxic and require elaborate processing before they are edible. The root of bitter manioc is only edible after washing, grinding, scraping and heating[430] and if untreated, or poorly processed, material is ingested it will cause health problems for people that subsist on cassava-based diets.

Various health disorders are associated with the consumption of cassava containing residual cyanogens, including hyperthyroidism, tropical ataxic neuropathy and konzo. Konzo is a neurological disorder characterized by the onset of sudden paralysis of the legs. Cyanide poisoning from high-cyanogenic cassava is typically associated with insufficient consumption of cysteine and methionine in the diet, for reduced sulfur-containing compounds are substrates for the detoxification of cyanide catalysed by the enzymes rhodanese and/or β-cyanoalanine synthase.

Cassava roots have low protein content (0.7–2%), and some amino-acids, such as lysine and methionine, are also notably low, the latter even being reported as absent by some workers. Breeding programmes have come up with an interspecific hybrid, ICB 300, which has ten times more lysine and three times more methionine than a "common" cassava cultivar.

Sweet and bitter types belong to the same species, but have differing quantities of cyanogenic glycosides in their tubers. These glycosides are a group of nitrile-containing secondary compounds in plants that yield poisonous prussic acid (HCN) on their enzymatic breakdown (cyanogenesis). The precise function of such compounds remains to be determined, but plants produce cyanide as a by-product of ethylene synthesis, and other likely explanations are that cyanoglycosides act as a deterrent to predatory herbivores and/or provide transportable forms of reduced nitrogen for plant metabolism. With reference to the latter, Siritunga and Sayre[431] proposed that cyanide is a major source of reduced nitrogen for cassava root protein synthesis.

Manioc is generally regarded as being a cyanogenic crop, because all tissues, with the exception of seeds, contain glycosides. Leaves have the highest levels (5.0 g linaramarin kg^{-1} fresh weight), whereas roots have approximately 20-fold lower levels, although the peel (cortex) has concentrations approaching those found in leaves. In addition to these tissue-specific differences, there are cultivar-dependent variations in root cyanogens levels. The sweet cultivars are incorrectly termed "non-cyanogenic" and their fresh roots contain less than 100 mg kg^{-1} cyanoglycoside. Bitter cultivars, on the other hand, contain up to 500 mg kg^{-1} fresh root. Present evidence suggests that the cyanoglycosides that accumulate in the roots are mostly synthesized in shoots and then transported to the roots.

The ability of plants to release HCN was first reported in 1803, and over the subsequent couple of centuries most higher plants have demonstrated some capacity for cyanogenesis. Only in relatively few species, however, has the actual source of HCN been identified. Most cyanogenesis in higher plants arises from the catabolism of cyanogenic glycosides, of which around 75 have been identified, and which are O-β-glycosidic derivatives of α-hydroxynitriles. Most cyanogenic glycosides are derived from the five hydrophobic protein amino acids tyrosine, phenylalanine, valine, leucine and isoleucine (Figure 5.29). The cyanogenic glycosides found in cassava are principally linamarin and lotaustralin, the

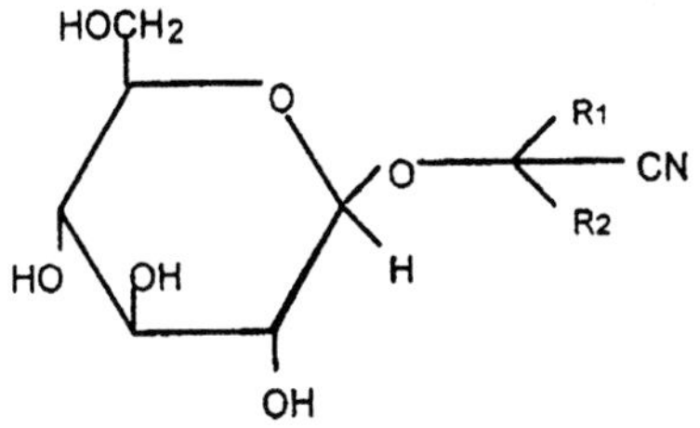

Figure 5.29 General structure of cyanogenic glycosides.

former accounting for 90–95% of the cyanoglycoside content of most varieties.

Biochemical studies have shown that the cyanogenic glycosides accumulate in cell vacuoles, while enzymes for their degradation are located in cell walls. When tissue is damaged (*i.e.* through processing and/or mastication) these compounds are brought together and hydrolysis occurs and acetone cyanohydrin and glucose are liberated. Acetone cyanohydrins is then broken down spontaneously (at pH >5, or temperatures >35 °C) or by hydroxynitrile lyase (HNL) to produce acetone and HCN.[432]

The cyanogenic content of cassava is under genetic control and is influenced by environmental conditions such as drought, varying between cultivars from 10 to 500 mg CN equivalents kg^{-1} dry weight in tubers.[433] Even the lower end of this range exceeds FAO recommended levels for food-derived cyanide exposure, and careful processing is essential if the above-mentioned disorders are to be avoided. Acyanogenic cassava is unknown from nature, and there is a desire within the scientific community to produce plants with a zero, or very low, level of cyanogenic material. By the end of the first decade of the twenty-first century, considerable progress had been made.

To some extent, it is surprising why ancient Amerindian peoples should have persevered with the domestication of a plant that had the potential to poison. The reason, of course, is that manioc has many attributes and, as Rosling[434] points out, the plant has been used as a human food for millennia and the benefits derived from its consumption far outweigh any detrimental effects. It grows well in the lowland tropics where there is a warm, moist climate and no frost (although there are cold-tolerant varieties in the Andes). In addition, manioc yields decent crops on soils of low fertility, and will tolerate acid soils. It can also survive drought conditions and in severe winds its roots are relatively unaffected, even if the leaves are damaged. For propagation purposes, seeds or stem cuttings are used – rather than (edible) root portions. Another bonus is that, once processed, manioc may be preserved in a tropical climate as *farinha* (meal) or as a bread (*pan de tierra caliente*, as it was called by late-colonial Mexicans[435]).

About 70% of wild species of *Manihot* are from Brazil. They are concentrated at the *Manihot* quadrilateral between 15° and 35° S and 35° and 55° W. They go from small shrubs measuring less than

0.1 m to tall trees or creepers reaching 12 m. Many species are threatened and a small number of plants are growing in this new environment. It is very difficult to keep the *in situ* populations growing far from official conservation areas. Many of these *Manihot* species, which can be used for cassava genetic enhancement, have important traits, such as tolerance and resistance to insects and diseases, and to environmental factors.

As reported, *M. flabellifolia* is regarded as the gene source of today's cassava cultivars and landraces. A dynamic collection strategy was therefore set up to conserve its germplasm. This collection shows links to natural processes associated to species evolution, thereby making the appearance of new genotypes on the field possible as a result of open pollinated (OP) seeds. Such seeds drop down and can germinate, which leads to new plants from the same species but adapted to the environment where they are growing. Dynamic conservation enables the progressive evolution of a new population. By handling the collection, it is possible to assess and use this germplasm in the field throughout the years. Plants identified as resistant to insects and diseases could be found more quickly by observing the plants in the field. The biodiversity and genetic resources of *M. flabellifolia* and other *Manihot* species should be preserved by *in situ* and *ex situ* methods.

As Taylor *et al.*[436] pertinently point out in their summary of transgenic technologies in cassava, lack of investment in research and development of the crop during the 1980s has resulted in a significant proportion of cassava's yield potential remaining untapped. It is a fact that most cassava farmers are resource-poor, and lack the ability to purchase and apply agrochemicals on a regular basis. Development and deployment of enhanced germplasm, therefore, remains the most important method for ensuring improved cassava production. Conventional breeding programmes have been successful in developing and delivering cassava varieties with enhanced disease and insect resistance, dry matter content and improved processing qualities in Africa, Asia and the Americas, but traditional breeding programmes alone will not be able to provide all the answers. In order to suit the varying requirements of all farmers, the ability to transfer new genetic material into the cassava genome is essential if the crop is to benefit from the major advances offered by this post-genomic era.

Transgenic technologies permit beneficial traits to be transferred from one cassava cultivar to another and from wild relatives to cultivated *Manihot*, circumventing species boundaries and the problems of outcrossing and inbreeding depression, inherent in this vegetatively propagated crop.

Cassava is normally propagated by means of stem cuttings, which are known horticulturally as "stakes". Stakes are typically at least 20 cm long, and have 4 to 5 nodes each with a viable bud. Stakes must be transported carefully to avoid damage and may be treated with agrochemicals to prevent pest or disease establishment in the new plants.[437]

According to Briggs,[438] numerous different [fermented] products are made using malted sorghum and millets, often supplemented with raw cereals or non-cereal starchy materials, such as cassava or bananas.

In 2007, a team from the University of Colorado at Boulder, led by professor Payson Sheets, discovered an ancient field of manioc that had been buried by a volcanic eruption some 1400 years ago. This was the earliest evidence for the ancient cultivation of the crop in the New World. The manioc field was located under some ten feet of volcanic ash during the excavation of the ancient village of Joya de Cerén,[439] an ancient Maya settlement in El Salvador (15 miles west of San Salvador). Considered the best-preserved ancient village in Latin America, Cerén's buildings, artefacts and land-scape were frozen in time by the sudden eruption of the nearby Loma Caldera volcano around 600 AD, thus providing a unique window on the everyday life of prehistoric Mayan farmers. It was known that these ancients cultivated maize and beans, but manioc cultivation was only suspected. Ash hollows in the planting beds left by decomposed plant material were cast with dental plaster to preserve their shapes and structures, and these could be positively identified as cassava tubers. Evidence indicated that the manioc had just been cut down, most of the tubers harvested and the beds replanted with manioc stalk cuttings placed horizontally in the ground. The manioc field, "at least one-third the size of a football field", had been intensively cultivated and Sheets calculated that the Cerén field "would have produced roughly 10 metric tons of manioc annually for the 100–200 villagers that lived there". The Cerén site was discovered by Sheets in 1978.

5.5.1 Alcohol from Cassava

Brewing was not necessarily confined to peoples from regions where cereal crops were the indigenous staples, and the South American beverage *chicha* has been brewed by Amazonian Indians for several millennia.[440] Many indigenous peoples of the tropical forests and savannahs of South America depend for their subsistence on manioc, which gives the highest yield of starch per hectare of any known crop; some 90% of the fabric of the crop can be regarded as potentially fermentable carbohydrate. This crop forms the basis of the beer known as *chicha* (see below). Not all manioc beer is referred to as *chicha*, and not all *chicha* is made from manioc, and there are as many terms as there are recipes. Whatever the recipe, fermentation of *chicha* from manioc is basically induced either by masticating cassava and letting the salivary enzymes do their work, or by allowing a mould culture to develop.

5.5.2 *Chicha*

"Aye? Well then, Don, refill my cup. Your chicha's very fine ..."
Hermann Melville, *Moby Dick*

As Hugh Cutler and Martin Cardenas[441] point out, "simple fermented beverages" enlivened the ceremonies of most Indian groups long before the Conquest. These mildly alcoholic brews were so common in some regions that they were regarded as providing substantial contributions to the diet. The word *"chicha"* was spread by the Spaniards, and now designates both alcoholic and non-alcoholic beverages made from a wide variety of plants and prepared in diverse fashions. Many of these were listed by Weston La Barre.[442]

Indigenous Andean peoples today, as well as at the time of contact, use the word *chicha* to denote all indigenous fermented beverages, but since the principal beverage in the Inca Empire was maize beer, the term became most commonly associated with this type of beverage. Remember that the Andean watershed is home to around 100 language groups, and each of these languages will include words to describe the nature of these beverages and their methods of preparation. In purely quantitative terms, tuberous crops were more important than maize in the highland Andean diet.

It is thought that the word derives from *chichal*, which translates as "with saliva", or "to spit". The name does not only apply to beverages from sweet and bitter manioc, but also to some of those made from maize, sugar cane and various fruits, such as the peanut. Mesquite, quinoa (*Chenopodium quinoa*), kiwicha (*Amaranthus caudatus*) and plantain have also been used as raw materials. In the regions conquered by the Incas, *chicha* was made from maize, their most important crop, in terms of quantity and prestige. The beverage is prepared in a variety of ways, but the two major variants in this maize beer preparation involve the source of the diastase. In much of the Americas a common source of diastase is saliva (as with *chicha* from manioc). The practice of pre-chewing grain for subsequent feeding to infants, the sick or the elderly developed many moons ago, and the use of saliva to hydrolyse starch was likely the original way of initiating fermentation.

Cutler and Cardenas[441] found the custom of masticating (better called "salivating", because the teeth are rarely used) roots, fruits and grains in the preparation of beverages to be widespread – among both "primitive" and "advanced" tribes. They list reported substrates as: mandioca in lowland Ecuador; mandioca and maize in central Brazil; maize and sweet potatoes in coastal Brazil; algarroba (*Prosopis* spp.), tusca (*Acacia aroma* Gill.) and chanar fruits (*Gourleia spinosa* (Mol.) Skeels) in the Chaco of Bolivia, Paraguay and Argentina near the Pilcomayo River; and maize in the highlands of Peru. They found malting to be common in most of Bolivia and Peru, especially in the highlands, but not common in other regions.

The manufacture of *chicha* was probably the most elaborate task in Andean cuisine, and the process took several weeks, involving grinding of dried maize kernels, chewing of maize flour, soaking of more flour (the mash), separating liquid from solid components, boiling, fermenting and decanting. It is evident that native Andeans had more accoutrements for preparing and storing *chicha* than for any other purpose. A simple alcoholic *chicha* can be made by mixing a substance containing starch or sugar with water and allowing the liquid to ferment, but most is made by methods that increase the alcoholic content and improve the flavour (*i.e.* to include malting).

Figure 5.30 *Chicha* flag and sign (from Cutler and Cardenas[441] - by kind permission).

Well-made *chicha* is an attractive drink, clear and sparkling, and its alcoholic content varies greatly, from about 2% ABV in new or in watered down samples to as high as 12% in a few prime samples. When *chicha* is ready for sale, a sign, the *aca llantu* (*"chicha* flag"), is displayed (Figure 5.30). *"Llantu"* is probably related to *llaitu*, the word for a fringe, which denotes nobility or rulership. The frequent occurrence of fringes and tassels in *chicha* signs suggests that there may be some basis to local belief that the better types of *chicha* were formerly reserved for the ruling classes.

In *chicha* manufacture, dried, ground corn is put into the mouth in slightly moist balls and worked with the tongue until it has absorbed saliva. The "gob" is then pressed and flattened against the roof of the mouth, and removed as a single mass. These lumps of "salivated" maize flour are known as *muko* and they are sun-dried and stored in stacks. *Muko* represents a valuable commodity, and is the starting point for the brewing of *chicha*, which commences by filling a wide-mouth, earthenware pot one-third full with dried, pulverized, salivated flour. Unsalivated flour and/ or sugar may also be added. The pot is then filled with water, to a level just below the jar rim, and heated. Alternatively, hot water (just below boiling point) can be added and mixed in. Either way, the temperature of the mixture needs to be around 75 °C. Boiling the mixture causes it to go very glutinous. The well-mixed pot contents are heated for about one hour, then cooled and settled, after which three layers are discernible: 1) the liquid top, called *upi*; 2) a jelly-like (semi-congealed) middle layer; and 3) the coarse particles at the bottom. The liquid upper layer is ladled out and placed in another wide-mouth pot, where it is allowed to stand. The jelly-like layer is then removed and placed in a shallow pan, where it is simmered and concentrated down to a caramel-like, sweet paste. Nowadays, this mass is reincorporated into the beer but, in earlier times, before the introduction of sugar, it was used as a sweetener. When the middle layer has been removed, more *muko* is added to the sediment in the first jar, and the process is repeated. As the top liquid layer forms in this "run", it too is removed and added to the *upi* already collected. Additional sweet jelly is also removed and added before fermentation is started. On the third day, the collected *upi* becomes rather bitter, and on day four, fermentation begins and the liquid bubbles vigorously. Fermentation is usually complete by day six or, at the latest, by day ten, depending upon ambient temperature. Some of the floating froth from the ferment is removed and may be used as an inoculum for a future brew, although this is rarely necessary because the earthenware pots used to hold *upi* are so impregnated with microbes from fermentation that no additional inoculum is required. Some consumers prefer to drink the beverage when it is still fermenting (it will be very cloudy), whilst others leave it for several days after fermentation has finished before consumption. Most forms of *chicha* are cloudy, but well-made samples are attractively clear and effervescent, resembling apple cider in flavour. The alcoholic content of indigenous *chichas* can vary from 2–12%, but most products contain 4–5% alcohol by volume.

An alternative method of enzyme production is to allow the maize to germinate, malted maize being known as *jora*. *Jora* is made by soaking maize kernels in water overnight in earthenware pots. The following day, the moist grains are placed in layers 2–3 inches deep to germinate in the dark (they are covered with leaves or straw). Optimum germination temperature has been shown to be 33 °C., and it is most important that germination throughout the sample is uniform. When the emerging shoots are about the same length as the grains, germination is deemed to be complete; the kernels will now have a sweet taste. The germinated maize is then heaped up and covered in burlap to keep in the heat. Within two days the kernels become white and parched, whence they are covered with a layer of ash. Kernels are then sun-dried for 2–5 days, after which time they may be called *jora*. When *jora* is milled, the resultant flour is called *pachucho*, and it is this that is mixed with water, and undergoes a series of boiling processes to separate the starch from the hulled material. Eventually, it is trained through cloth, and the liquid falls into pots that have been used previously for fermentation and, therefore, contain the necessary inoculum. Fermentation is thus spontaneous. When the *chicha* has lost its sweetness and has assumed a degree of sharpness, it is ready to drink. To increase the alcohol content of *chicha*, brown sugar or molasses can be added. In Bolivia, *jora* may be chewed in order to make *chicha*, whilst in Brazil a beer known as *kaschiri* is made from sweet cassava tubers, which are chewed and expectorated in order to initiate the brewing process. Finally, in Mozambique, women chew the yuca plant (= *Manihot esculenta*), spit it out and allow it to ferment into a beer known locally as *masata*.

While a large sector of the population would profess to be aware of the origins and significance of European-style beer, one really has to be an archaeologist or an ethnographer in order to appreciate the importance of *chicha* to past Andean cultures. A wide variety of plants, such as manioc and peanuts, were used to brew beer in the prehistoric Andes, but corn beer, known as *chicha* today, was perhaps the most common. Today, the alcoholic strength of *chicha* is generally low (certainly less than 5%, but can reach 12% ABV), and colour will vary according to the colour of the maize grain used. Our understanding of *chicha* is largely derived from contemporary ethnography and the experience of researchers working in regions where the beer is still brewed. In these communities, *chicha* and other alcoholic beverages are essential for acquiring labour and for use in rituals. As Goodman-Elgar[443] says: "Andean ethnography has the benefit of "direct historical analogy", because many contemporary Andeans are direct descendents of pre-Colombian peoples."

Although there are no written records in the Andes before the Spanish Conquest (1532 AD), pots, strainers and botanical remains found in archaeological contexts suggest that contemporary methods of preparing *chicha* were likely similar to methods used in the Andes since at least the Early Intermediate Period (200 BC– AD 750). We know most about the use of *chicha* during the Inca Empire (1430–1532), when the drink was consumed in vast quantities at their feasts. The Incas produced and consumed huge amounts of maize beer, and its role in shaping identity and ethnicity is well known from Inca chronicles. Historical accounts relate that *chicha* was made by "the chosen women", or *aqlla* (sometimes *acllacunas*), who were cloistered women that served the community by weaving cloth and brewing. These young women were reputedly sent to special, all-female schools to learn the art of brewing and other domestic arts.

An Inca king would offer *chicha* to his subjects in order to thank them for their loyal service throughout the year, and the quantity of *chicha* that was provided was seen as a reflection of power and prestige. It has been estimated that the Incas produced millions of litres of *chicha* each year at administrative centres across the Andes.

There is evidence that earlier cultures probably consumed similar quantities of the beer at their public celebratory events.[444,445] As Moore relates, *chicha* played a variety of roles in pre-Hispanic Peruvian societies. It was an everyday beverage, an essential element in social and ritual interactions, and part of the payment the ruling class were obliged to provide for their subjects. Moore's paper arose from an archaeological investigation of *chicha* production at a non-elite residential site at Manchan, the regional centre of the Chimu Empire (AD 900–1470) in the Casma Valley on the north coast of Peru. The assembled data indicated that: 1) the equipment for making *chicha* was generally available within the community, 2) at different times, certain households would brew large quantities of the drink, and 3) this production took place in a self-sufficient unit without much "state" involvement.

In modern times, our economies are based upon the market-exchange system where supply and demand are all-important. This is in contrast to the non-market Andean economy which was based on two related forms of exchange (referred to as "reciprocity" and "redistribution" by anthropologists – see Chapter 6), which are basically forms of exchange similar to gift-giving. Here, the social and economic aspects of the exchange are linked, and a crucial feature of reciprocity and redistribution is that the actual "value" of the goods exchanged is very closely allied to the socio-political status of the people involved in the transaction, *i.e.* the value of the gift depends not just on what the gift is, but who is giving it (a small token from the king would be worth more than a large gift from a commoner). Another characteristic of reciprocity is that the exchange itself is not usually completed in a single act; in reality, the gift represents an obligation that may be fulfilled at a later date. Throughout history, it would seem that food and drink have always been commodities that are particularly well suited to this sort of exchange but, in modern societies, products thus exchanged are no longer as important.

As Murra pointed out,[446] maize was, by far, the most highly esteemed crop in the pre-Columbian Andes, and one of its most important uses was for making *chicha*. Archaeological evidence from the central Andes of Peru indicates that the role of maize changed between AD 500 and 1500, and shifted from being a culinary item, which was simply prepared by boiling, to a more complex symbolic food, transformed through grinding and other processes (brewing) into beer. This drink had elaborate political significance, and the change in the way that maize was processed and consumed coincided with a period of increased political and social tensions.[447]

We have little information about the early stages of "salivary amylase" fermentations, certainly in relation to *chicha* production, but some related indigenous beverages have been studied. The Tapirapé people of the Tapi'itãwa tribe of Brazil produce several fermented foods and beverages, one of which is called *cauim*. This essentially non-alcoholic product, which can variously be made from cassava, maize, rice and peanuts, as well as some other plants, is often the main staple food for adults and children.[448] *Cauim* is fermented and cooked and is consumed daily and, although non-alcoholic, many of the stages of its manufacture are pertinent to indigenous alcoholic drinks. To commence preparation, water is boiled in large clay pots, and the warmed plant ingredients are then added and cooked for a period. Afterwards, the cooked material is removed and left to cool, before the inoculum is added. The latter is saliva obtained from women who have been chewing sweet potato, and this is the case whatever plant material is used as a base. The saliva-sweet potato mixture is added slowly to the cooled substrate.

The *cauim* used in the above-mentioned study was prepared by the Tapirapé near Urubu Branco (Mato Grosso, Brazil), and employed a mix of cassava and rice as the substrate. Cassava tubers were left in running water for 3–4 days to soften the skin, and then peeled, sun-dried, grated and ground. Rice was soaked in water for 15 minutes and then dried and ground to a flour. To commence the mash, 2 kg of rice flour was mixed with around 30 L water and cooked. After one hour at *c*. 85 °C, 1 kg of cassava flour was added, and the mixture was brought to the boil and cooked until a smooth consistency was attained. At this point, it was removed from the heat source and cooled for around 4–6 hours prior to being inoculated with a mixture of sweet potato and saliva. The inoculum is added slowly to the base mash. The fermentation period was two days.

Microbiological and biochemical parameters were measured during fermentation, and perhaps the most obvious change was a reduction in pH, from 4.2 (before inoculation) to 3.6 (after 48 h fermentation). This was likely attributable to the growth of lactic acid bacteria. Temperature of the ferment gradually increased from 22 to 30 °C. To assess microbial activity, carbohydrates and fermentation products were assayed. As was to be expected, starch levels gradually declined with time, whilst the fermentable sugars identified (maltose, glucose, fructose and sucrose) showed different profiles, with fructose and sucrose always being present in low concentrations (always <30 μg mL^{-1}). Maltose was continuously being liberated from starch, and reached peak levels after 36 h (Table 5.9), and the subsequent decrease in concentration of this disaccharide was probably due to microbial activity. The glucose peak observed after 12 h of fermentation was most likely the result of the microbial breakdown of maltose, and this was followed by a gradual depletion of the sugar. The major fermentation product encountered was lactic acid (lactate), and its concentration increased with time. The initial concentration of

Table 5.9 Main sugars, pH, soluble starch and organic acid contents of beverage cauim (after Schwan *et al.*[448]).

Fermentation time (h)	Starch (%)	Maltose (μg mL^{-1})	Glucose (μg mL^{-1})	Lactose (μg mL^{-1})	pH
WI	11.8 ± 0.8	242 ± 22	96 ± 4.5	ND	4.2 ± 0.11
0	9.5 ± 0.5	240 ± 22	89 ± 3.6	200 ± 15	4.9 ± 0.05
12	4.0 ± 0.4	286 ± 18	190 ± 3.8	450 ± 22	4.0 ± 0.05
24	1.8 ± 0.09	335 ± 25	95 ± 4.0	480 ± 22	3.8 ± 0.10
36	1.7 ± 0.09	450 ± 33	89 ± 3.5	550 ± 12	3.8 ± 0.10
48	1.65 ± 0.08	180 ± 11	42 ± 2.0	780 ± 13	3.6 ± 0.05

WI: without inoculum (sample before inoculation); ND: not detected. The data are means of four replicates ± SD.

lactate encountered (200 µg mL^{-1}) was probably due to the activity of lactic acid bacteria following the cooling of the mash. Small amounts of acetate (always <10 µg mL^{-1}) were detected, and low concentrations of ethanol (*ca.* 20 µg mL^{-1}) were manifest between 12 and 24 h fermentation.

The cooked *cauim* mash prior to inoculation with saliva had a yeast population of 3.2×10^2 CFU mL^{-1}, which after inoculation increased to 7.5×10^3 CFU mL^{-1}. After 48 h, final yeast concentration was 6.9×10^7 CFU mL^{-1}. Based on morphological and physiological characters, some 441 yeast strains were identified, and these were divided into 17 groups. By sequencing the D1/D2 domain of the 26S rRNA gene, and comparing with sequences given in GenBank, seven species emerged (homology with GenBank sequences was within 98–100%), and these, together with their patterns of occurrence, are shown in Table 5.10.

Candida tropicalis was the most common yeast species isolated and represented some 26% of the total found. Then followed *Pichia guilliermondii* (22%), *Candida intermedia* (17%), *Trichosporon asahii* (11%), *Saccharomyces cerevisiae* (11%) and *Candida parapsilosis* (11%). *Exophalia dermatidis*, a black, yeast-like mould, often associated with blastomycosis, was found in the mash before inoculation and during the first few hours of fermentation, before it disappeared. The presence of this species is probably attributable to a lack of hygiene during processing. As Table 5.10 shows, all species except for *S. cerevisiae* and *C. intermedia*, were present prior to inoculation, albeit in low numbers. Some isolates of *C. tropicalis* secreted their own amylase, but all other yeasts encountered had to rely on exogenous enzymes in the mash to break down starch to fermentable sugars.

It has been suggested that the development of yeasts in fermented foods and beverages is favoured by the acidic environment created by bacteria and that the growth of bacteria is then stimulated by the presence of yeasts, which may provide growth factors, such as vitamins and soluble nitrogen compounds.[449] The production of acids and other antimicrobial compounds during fermentation may promote or improve the microbiological safety and stability of a fermented food product.

First-hand knowledge of twenty-first century *chicha* brewing has recently been reported by Frances Hayashida,[450] who forcefully, and quite rightly, states: "Archaeological studies of alcohol have

Table 5.10 Distribution of isolated yeasts found during cauim fermentation (after Schwan et al.[448]).

Yeast species	Microbial counts of yeast species ($CFU\ mL^{-1}$) at					
	WI	0 h	12 h	24 h	36 h	48 h
Candida intermedia	ND	4.5×10^2	11.4×10^3	4.2×10^6	2.1×10^6	13.7×10^6
Candida parapsilosis	0.2×10^2	4.5×10^2	6.1×10^3	3.5×10^6	1.4×10^6	0.8×10^5
Candida tropicalis	0.2×10^2	9.1×10^2	19.1×10^3	6.6×10^6	3.6×10^6	19.8×10^6
Exophiala dermatidis	1.0×10^2	2.0×10^2	ND	ND	ND	ND
Pichia guilliermondii	0.8×10^2	13.6×10^2	14.5×10^3	1.2×10^6	2.9×10^6	1.1×10^5
Saccharomyces cerevisiae	ND	4.5×10^2	11.4×10^3	3.3×10^6	1.2×10^6	1.1×10^5
Trichosporon asahii	1.0×10^2	6.8×10^2	5.4×10^3	2.1×10^6	1.2×10^6	2.05×10^6

WI: without inoculum (sample before inoculation); ND: not detected.

tended to focus on consumption while production, particularly of beer, has been more difficult to recognize and interpret." In an exhaustive piece of ethnoarchaeological work on modern *chicha* production on Peru's north coast, Hayashida provided information on: (1) production steps and their material correlates, (2) labour and raw material inputs at different scales and possible labour bottleneck, and (3) variation in technology and organization that is linked to contexts of consumption and cultural differences through time across the Andes.

Chicha is still consumed in many areas as an everyday or ceremonial beverage, though the contexts, meanings and technologies of brewing have undergone several changes over time. These include the introduction, following the Spanish Conquest, of widespread commercial *chicha* production in small breweries (called *chicherías*), colonial efforts to eradicate *chicha* because of its central role in religious rites and the more recent adoption of *chicha* as a powerful symbol of indigenous identity.[451]

Chicha production on the Peruvian north coast goes back almost two millennia, with archaeological evidence of brewing activity having been found at the centre of the Moche polity (second–eighth century AD).[452] Subsequent pictorial and ethnographic accounts indicate that north coastal brewing techniques emanate from at least the eighteenth century. Given the similarity of some modern brewing vessels to those found at some late-pre-Hispanic north coastal sites, some practices must be even older. In some instances, of course, it is inevitable that clay vessels have been abandoned in favour of plastic bins and aluminium pots! It is also evident that the use of sugar, the reduction in the use of native maize varieties and the use of commercially milled corn are other post-Conquest changes. One facet of brewing has not changed, however, because most brewing is still carried out by women.

Hayashida's study[450] encompassed two different contexts of *chicha* production: "brewing for sale" – relative large-scale production in the Department of Piura (Chulucanas and surrounding towns and settlements) – and "domestic brewing", where the product was purely for home consumption. The latter took place in various settlements in the Department of Lambayeque, to the southeast of Piura. This work complemented that of Jennings,[453] who provided estimates of the resources required for *chicha* production on different scales. As Hayashida avers, her goal was not to create a homogeneous "*chicha* model" that could be projected into the past, but to provide analogues that can serve as points of comparison with the archaeological record.

In the Chulucanas area *chicha* is typically from a mix of ground malted maize (*pachucho*) and maize flour (*crudo*), which is purchased from vendors or directly from the mill. The most commonly used maize was a yellow hybrid, locally called *maiz perla* or *pato*, and it was evident that this had replaced local varieties. Maize was initially prepared for brewing (malted) *via* a four-step steeping regime, with soak times varying from 8 hours to 30 hours, and steeping being carried out in shallow cement, or cement-lined tanks. Moist kernels are then spread out in the shade for 4–5 days, and regularly sprinkled with water. The rootlets should now be around an inch in length, at which point the grains are made into mounds and covered with available material (plastic sheets or sacking). Coverings are weighted down and the grains are left to "bake" for 7–12 days. Mounds are referred to as "ovens" because of the heat generated therein. Finally, kernels are left exposed to dry in the sun for 1–2 days. There is a malting loss of around 10% when making *pachucho*, which is only prepared during the dry months to prevent damp spoilage. Grain is stored whole, and ground on demand, to reduce the likelihood of spoilage.

In the Lambayeque district, most women made *chicha* solely from malted kernels (*jora*), and did not mix in maize flour. Again, the yellow hybrid corn was mainly used, although a local red variety, *alazán*, was used occasionally. Ground *jora* is often obtained from suppliers, although some women grow, harvest and grind their own maize. For malting purposes, grain would be soaked in a tub of water, and then spread out in another, shallow container. This would then be covered by a sheet and, after 6–8 days germination, the grains would be cast out to dry in the sun, before being taken to a mill. Some makers placed the soaked grains on banana leaves, then covered them with more leaves and a cloth to facilitate germination.

Brewing protocol differed in the two study areas. In Piura, brewing is a four-day process, and the beer is then sold immediately (fifth and sixth days). Early on day 1, *crudo* and *pachucho* flours are mixed with water in cooking jars (called *ollas*) and boiled for anything from one to ten hours (depending on the brewer and the type of *olla* used). The mixture is stirred frequently with a wooden paddle, and aerated (*venteado*) using a dipper (*umaz*) made from a large hollowed-out gourd (sometimes nowadays a cut-out plastic container) attached to a stick. This activity helps to cool the mixture and prevent the thick scum that forms on top from spilling over. Pots are placed over a wood fire, and the spaces between pots may be covered with pieces of sheet metal or large broken pot fragments to help control the fire and reduce fuel consumption. To aid stability, *ollas* may be propped up with supports, or the bases may be partially buried.

In established brewing sites, pots are arranged in groups of two rows in permanent positions. They are only moved if broken. Vessel size is measured in *latas* (1 *lata* = 5 gallons), and volumes range from 2 to 4 *latas*. Brewing capacity of the sites examined varied from 8–32 *latas*. Vessel exteriors were often smeared with clay, and/or reinforced with wire to make them more heat-resistant and less likely to fracture through thermal shock. Even so, the life of an *olla* is short – around two to four weeks (around eight boilings). In many of the breweries visited, the hearth area was slightly raised due to accumulation of ash and other debris over a period of time.

After the first boiling, the developing *chicha* is cooled by transferring it from *ollas* to a series of wide-mouthed jars (*tinajóns*) situated nearby, transfer being *via* the afore-mentioned dippers. In addition to being cooled, the mash is "worked", and this is effected by placing a long, narrow log (called a *vieja*) across the mouth of the *tinajón*. Handfuls of sieved mash are then rubbed back and forth across the *vieja*, presumably to aid the extraction of starch (and sugars). As the mash is worked, small portions of it may be chewed, and then returned to the pot, thus promoting saccharification. Once it has been worked (the act of working is called "*trabajando*"), the mash is strained through a sieve (often a framed coarse cloth) and then left to cool completely. The sieved mash is aerated periodically with an *umaz*, while the sieved remains (*afrecho*) are used for animal feed.

Early on day 2, the *chicha* is boiled again for anything between 30 minutes and three hours. Sugar may be added at this point. When cooled, the *chicha* is strained through a cotton cloth held by two people before being transferred to *tinajóns*. The ends of the cloth are rapidly moved up and down to encourage the liquid to

Figure 5.31   Making *chicha* (from *Truijillo del Peru*, by Bishop Baltasar Jaime Martinez Compañón, 1783–1785).

pass through, and the sludgy residue is then pressed or twisted in the cloth to express the last drop of liquid (Figure 5.31). This second sieving through cloth is called to *taquear* the *chicha*, and the solid residue, or *taca*, is fed to animals or sold as feed.

The liquid is left in *tinajóns* to cool, and is aerated with an *umaz*. Once cooled, it is transferred to large, narrow-necked jars (*cántaros*) – or, more likely these days, a large plastic bin (+ lid). Whatever vessel is used it is capped and the *chicha* is left to ferment for two days. By day 5, the *chicha* is ready, and is transferred to smaller clay or metal pitchers (traditionally by use of an *umaz*). At ceremonies *etc.*, *chicha* is drunk from small gourd bowls called *potos* or *cojuditos*. Ideally, *chicha* should be consumed within three days of production.

In Lambayeque, *chicha* preparation takes only three days, and the product is consumed immediately, or may be left for a day or so prior to drinking. The women in this area use only malted grain and do not engage in salivary amylolysis. The mash is given only one boil, and the whole regime of cooking, cooling, *trabajando* and sieving takes one or sometimes two days. Cooking pots in this region are taller than in Piura, and have a slightly constricted neck. Cooking begins on the morning of day 1, and lasts for 7–15 hours, the mash being stirred with a stick. As the mash thickens, hot water is added to maintain consistency.

For everyday domestic use, a single cooking pot with a capacity of 3–7 *latas* is typically employed, and *chicha* would be brewed either weekly or fortnightly. Pots are normally partially buried for stability (and, maybe, to prevent charring at the base), and fuel for the fire is placed to one side. This feature, and the fact that *chicha* hearths are usually located outside, demands that pots are protected from wind and consequential heat loss, which is why the hearth is often situated next to a wall. Cooking pots are periodically removed for cleaning, rather than being left *in situ*. The type of wood used for fuel is important, because some types can impart an unwanted smokiness to the drink. If large quantities of drink are required for special celebrations, additional cooking pots are borrowed. Lambayeque women reckon that the life-span of their pots is about two years, with weekly or bi-weekly usage.

After cooking, the *chicha* can be left to cool in the same pot, or transferred to another clay cooking vessel, a large metal pot or a shallow, wide-mouthed *callana* (also called a *lebrillo*). Relatively few of the large *tinajóns* common in Piura were used by *chicha* brewers in Lambayeque, and the large gourd dippers (*umaces*) were not used at all, either because of their expense or because of smaller batch size. Smaller dippers (*chiculas*) made from split-necked gourds are used. *Chicha* is left to cool for a few hours, or left overnight, and, once cool it is sieved through cloth. Sieving usually involves two people, each one holding an end, and rapidly moving the corner up and down, although it can be accomplished by one, if one end of the cloth is attached to a fixed object. The sieved mash (*afrecho*) is then milled by a variety of methods (more than in Piura), one of which uses just hands, and another a grinding trough. The ground *afrecho* is usually mixed back into the *chicha* or, occasionally, may be mixed with boiling water to give a mildly flavoured drink, *chicha de afrecho*.

The *chicha* with the milled *afrecho* is stirred and sieved through cloth before being introduced into narrow-necked *cántaros* (*mulos*). Sometimes sugar may be added as each *cántaro* is filled, or just prior to serving. As in Piura, *chicha* is drunk from small gourd bowls. Many of the types of jar used for making *chicha* are also used for storing water, although never ones that are used for brewing. Presumably, this is to prevent the indigenous yeast flora, which will gradually build up in a *chicha* vessel, from being removed by rinsing.

The climate in the Piura area is generally warmer than in the Lambayeque region, and this may be one of the reasons that *chicha* produced in the former is less robust, and much more likely to sour – even though it usually has greater alcoholic strength. Other explanations might be that in Piura a higher ratio of grist to water is used, there is a longer period of fermentation and sugar is added during manufacture rather than just prior to consumption. It was also evident that brewing tended to be on a larger scale in Piura and was a weekly event. This meant that brewing areas were larger and more permanent, with individual spots being reserved for certain steps of the process (see layout in Figure 5.32). Thus, there is a fixed hearth, cooling areas and areas for fermentation. In Lambayeque, the hearth may be set up in the same place, but other areas may be moved around. Brewing in the latter region takes place outdoors, adjacent to a house, while in Piura (where most brewers studied lived in small towns) *chicha* was in roofed areas next to a kitchen. It seems reasonable to suggest that larger scale production is concomitant with more permanent facilities.

Because of the fragility of *chicha*, when feasts, or the like, are to be catered for a lot of labour is recruited to brew the large quantities required. As an example, Hayashida gives an account of the preparations for a three-day religious feast in Piura, where some 250 guests were invited (although many more arrived!). The grist consisted of 800–1000 lb of malted and unmalted grain, and this gave a *chicha* yield of 256–320 *latas* (*ca.* 4840–6050 litres). Thirty-four metal cooking *ollas* were used, mostly borrowed, along with *tinajóns* and fermentation vessels. Forty people were needed for the preparation of food and drink.

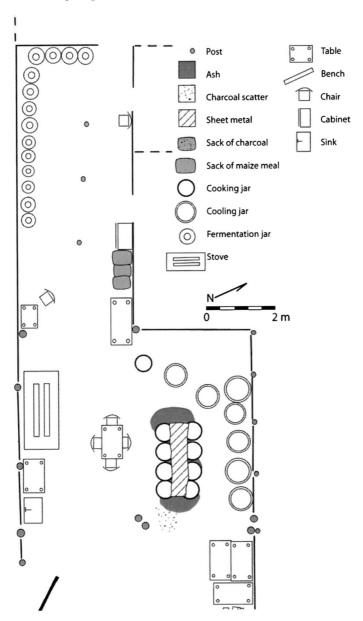

Figure 5.32   Layout of the chichería in Chulucanas (after Hayashida,[450] with permission).

After providing a very useful account of the inputs and scale of *chicha* brewing in these two areas, Hayashida looks at the evidence for stored *chicha*. The ability to conserve food products was one of the fundamental concerns for ancient societies, and a delicate product such as *chicha* would have presented considerable problems – especially in the absence of added herbal, or any other, preservative material. In Lambayeque, she found that if *chicha* was allowed to ferment until there was no further sign of "bubbling" (*i.e.* the presumed depletion of sugars), this could be put into small (1 or 1.5 *lata*) *cántaros*, which are sealed with clay (or plaster) and then buried. The *cántaros* are then exhumed when special occasion demands. Excavating the drink may, of itself, warrant a special ceremony, and different vintages are spoken of as "one month"; "one year", *etc*. *Chicha* so treated is said to be strong, dry and "similar to a port wine in flavour".

Chicha ageing has been reported from other regions, including around Titicaca. Bertonio's seventeenth-century Aymara dictionary[454] includes terms for *chicha* that is "valuable, or stored for a long time" (*chicha preciada, o guardada por mucho tiempo: Llutapu, yanu yakusa*), and another account from the same period talks of "*chicha* of many years" (*chicha de muchos años*) being specially made for a toast to the Sun. In his chronicle of Inca life, Guaman Poma refers several times to a one-month-old *chicha*, called *yamor aca*, which was made for the Inca king and for serving at special ceremonies.

5.6 PALMS AND PALM WINE

The palm family (Palmae, or more recently Areaceae) comprises 200 genera and around 2200 species that occur naturally in the tropics and sub-tropics, and many of which have long been of use to mankind. Palm wine, also known generally (incorrectly, according to some) as "toddy" in Europe, is an alcoholic drink produced by fermenting the sugary sap of certain of these palm plants, although most palm saps can yield a wine. The palms most often used are the coconut palm (*Cocus nucifera* L.), the oil palm tree (*Elaeis guineensis* Jacq.) and the raffia palms (*Raphia hookeri* Mann & Wendl, and *R. vinifera* P. Beauv.). The wild (or silver) date palm (*Phoenix sylvestris* (L.) Roxb.), the kithful palm, or "toddy palm" (*Caryota urens* L.), the nipa or attap palm (*Nypa fruticans* Wurmb.) and the Asian palmyra palm (Cambodian palm; sugar palm), *Borassus flabellifer* L., are also frequently used. Swings and DeLey[418] have provided a list of palms used to make wine. It should be pointed out that the term "toddy palm" describes a variety of species depending on where one is in the world. The increasing demands on the Earth's natural resources pose a serious threat to palm biodiversity, the main threats being over-exploitation and habitat destruction.

Apart from being an alcoholic drink, "toddy" also refers to the sweet (unfermented) sap from several Asian palm trees and, according to some authorities, the term "toddy" should be confined to the fermented flower sap from the coconut palm (*Cocus nucifera* L.), once described as "the most perfect tree that is found, to our knowledge".[455] Principally because heat and humidity mitigate against the preservation of fossils, we have little record of the coconut palm in prehistoric times. Add to this the fact that the plant grows and produces fruit all year round, thus obviating the need for crop storage, and the fact that the high water content of the nut makes storage impractical, and it is easy to see why there is so little for the archaeologist to play with. It is likely, however, that where it grew, *Cocos nucifera* played an important role in the diet of our ancestors. As Harries put it,[456] the coconut is "the milk bottle on the doorstep of mankind".

Toddy produced after overnight fermentation will contain around 3–4% ethanol (with 10% residual fermentable sugar) and, after around 36 hours, ABV will have risen to around 8%. Sweet (unfermented) sap would contain 16–30 mg ascorbic acid per 100 grams, and this is scarcely diminished during fermentation.

Palm wine is a popular drink in Nigeria (where over 450 million gallons are consumed annually) and West and Central Africa, where it plays an important role in popular culture, and is regarded as a food. It is also known from southern India (particularly the state of Kerala), Sri Lanka, Mexico, Philippines, Myanmar, Malaysia and other parts of Southeast Asia but, basically, palm wines are produced wherever palm trees grow. Dowson[457] reports that equivalents of palm wine are produced in Tunisia, Algeria and Libya. The drink is rich in nutrients, containing carbohydrates, proteins, amino acids and minerals, as well as alcohol. Fermentation increases the content of some vitamins in palm sap, most notably B_{12}, pyridoxine, riboflavin and thiamine.[458] As well as being fermented, palm sap can be concentrated, by boiling in shallow pans, to give a 12–15% yield of a crude, semi-crystalline, sugar called "jaggery" ("Javanese sugar"). Slightly less concentration produces a syrup.

Because palms are monocotyledons, there is no cambial layer beneath the bark to facilitate "easy" tapping of sap *via* the trunk (stem); vascular bundles are scattered throughout the stem tissue. For this reason, unopened inflorescences are usually the targets for sap removal, which is an art in itself. As would be expected, there are many variations of the product, and a variety of production methods. There are, however, two main ways of obtaining palm sap; either from a tree that is intact (*i.e.* standing upright and growing), or from a felled tree. From a growing tree sap can be won from an immature male flower head (inflorescence tapping), as is the practice with raffia palms, or from the stem (stem tapping), which applies to oil palms. Raffia palms usually yield more sap than oil palms, although the former can only be tapped once because the terminal inflorescence is destroyed.

When trees are felled, as practised in Ghana and Cameroon, the trunks are left for around a fortnight to permit the sap to concentrate. Sap is then withdrawn over a period of eight weeks, and the fermented product is called "down wine". Evidence exists[459] that tapping from a felled tree yields sap of a different composition to that from a live tree, and down wine contains methanol and propanol as well as ethanol. These chemical differences are linked to a variation in sap microfloras and, as Tuley[460] showed conclusively, the method of tapping affects the composition of the sap and the quality of the wine. The yield of sap from a single tree may be considerable; Simonart and Laudelout[461] give information about a felled tree that produced 150 litres of sap in 32 days.

According to Bassir[462] freshly taken sap is a clear, colourless liquid containing around 10–12% sugar; 4.29 ± 1.4% sucrose; 3.31 ± 0.95% glucose and with a pH of 7.4. The same author reported rapid natural inoculation by yeasts (mainly *S. cerevisiae* and *Schizosaccharomyces pombe*), and then growth of *Lactobacillus plantarum* and *Leuconostoc mesenteroides*, which lower the pH to 6.8. After 12 hours fermentation, pH has reached 6.5–5.5, and after 48 hours, 4.0. After 12 hours fermentation, alcohol by volume ranged from 1.5 to 2.1% and, after 72 hours this had increased to 4.5 to 5.2%. Bassir maintained that with a "natural" fermentation ABV would never rise above 7.0%.

Once sap has been removed, fermentation is practically unavoidable, and it is often necessary to take steps to slow it down! The traditional method of preventing autofermentation after tapping was to add pieces of bark from the tree *Sacoglottis gabonensis*,[463] about which more later.

Whatever the raw material, palm wine fermentation always follows a lactic-alcoholic-acetic path, although other microbes can be implicated as well. As a rule of thumb, without any extraneous treatment, palm wine is very drinkable for 24 hours after collection; post this time it is acceptable, and beyond 48 hours it is totally unacceptable and is normally distilled (originally, to prevent waste). The main distillation product of palm wine is "arrack", a popular spirit that can be refined, flavoured and made into a kind of gin.

Elaeis guineensis and the *Raphia hookeri* are native to West Africa, and both are believed to have originated from the swamps of that region. Characterized by their compound pinnate leaves, *Raphia* species are the largest palms in Africa and are one of the most economically useful, the stem producing sap and the leaves providing shelter. All parts of the plant have some use for humans, with the sap being the most important entity. The production of palm wine forms one of the most important occupations of the rural people in the Nigerian palm belt, and the palm wine industry is of considerable economic importance in West Africa generally.

5.6.1 Oil Palm Wine

Oil palm wine is mildly alcoholic (1.5–2.5% ABV), sweet, effervescent and milky (mainly due to yeast proliferation; the drink should contain 10^2–10^4 microbial cells per ml when consumed). When taken straight from the palm, sap generally has a slightly brownish hue. The product is highly prone to oxidation, and should be consumed within a day or so. When sap is collected from *Elaeis guineensis*, a microflora of *Saccharomyces cerevisiae*, *Zymomonas mobilis*, *Lactobacillus* and *Leuconostoc* spp. can be detected.[464]

An intensive study of the sugars of the unfermented sap and wine from *Elaeis guineensis* was carried out by Eze and Ogan,[465] whose results would seem to account for the quality variation in oil palm wine during storage. Using improved methodology, they found that sucrose levels in the pure, unfermented sap ranged from 9.59 to 10.59 (w/v), whereas glucose or fructose levels were less than 1% (w/v) (0.13–0.73%). Raffinose occurred only in trace amounts (0.13–0.35% w/v). With time, fermentation gradually diminished sucrose levels, but fructose levels increased somewhat – reaching a peak at the ninth hour, before starting to decline. Glucose and raffinose levels remained continuously low. All sugars had disappeared by the thirty-third hour. Sap pH was initially 6.60, and gradually dropped to 3.30 at 48 hours.

Investigations into the biochemical constituents of fresh (3 to 6 hours old) oil palm wine from trees around Ibadan were carried out by Ezeagu *et al.*,[466] who monitored the sugar, protein, amino acid, alcohol, mineral element and trace element content, and briefly addressed public health implications. The pH of samples varied between 7.2 and 7.4, and the mean specific gravity was 0.99 ± 0.01. Ethanol content averaged out at 3.4% ABV (varying from 2.8–5.2) and mean sucrose, glucose, fructose, raffinose and maltose levels were 2.67 ± 0.14, 0.65 ± 0.11, 0.90 ± 0.12, 0.32 ± 0.17 and 0.10 ± 0.02 mg 100 ml^{-1}, respectively. Total nitrogen varied between 85.12 and 119.70 mg 100 ml^{-1}, and protein varied from 34.60 to 47.90 mg 100 ml^{-1}, while the mean free amino acid and lipid content were shown to be 59.63 ± 8.15 and 62.65 ± 11.80 mg 100 ml^{-1}.

The microbiological and biochemical changes that occur in palm wine during the tapping of felled oil palm trees was studied by Amoa-Awua *et al.*[467] They found that the accumulation of ethanol in palm wine occurs in three stages, and that it runs parallel to lactic acid and acetic acid fermentations. Yeast growth was dominated by *S. cerevisiae*, and commenced immediately after tapping. A lactic acid bacterial flora, dominated by *Lactobacillus plantarum* and *Leuconostoc mesenteroides*, caused a rapid acidification of the wine during the first 24 hours of tapping, and the presence of acetic acid bacteria (*Acetobacter* and *Gluconobacter* spp.) increased as the level of ethanol built up by day three. After four days, acetification reached an unacceptable level (0.6% acetic acid v/v).

In view of conflicting reports of non-*Saccharomyces* fermentation in palm wines, oil palm wine samples (obtained by "inflorescence tapping") from three different sites in Nigeria were examined for the presence of *Zymomonas* species.[468] *Z. mobilis* was the only species encountered, with the initial population being roughly the same in samples from all sites (98–99 × 10^5 cfu ml^{-1}). Carbohydrate fermentation, ethanol tolerance and growth parameters (pH and temperature) were evaluated, and a comparison of *Zymomonas* and *Saccharomyces* fermentation rates was established. In terms of the latter, *Z. mobilis* was found to ferment at a faster rate than *S. cerevisiae*. It reached its maximum density (2.72 × 10^7 cfu ml^{-1}) between 12 and 15 hours after sap was tapped, and was acid- and ethanol-tolerant, being able to grow between pH 4.0 and 7.0, and tolerate ethanol concentrations between 2.5 and 15.0% ABV.

A number of aroma-active compounds have been identified from palm wine made from *Elaeis guineensis* including 3-isobutyl-2-methoxypyrazine ("earthy"), acetoin ("buttery"), ethyl hexanoate ("fruity") and 2-acetyl-1-pyrroline ("popcorn"), which help to impart the distinct aroma. 3-Methylbutyl acetate ("fruity") has also been identified.[469]

Using modern, culture-independent (molecular) identification techniques, as well as culture-dependent methods, Stringini *et al.*[470] investigated yeast diversity during the tapping and fermentation of oil palm trees in Cameroon, Central Africa. Fresh palm wine samples were collected at 24-hour intervals during the first five days following tapping. They were collected in sterile plastic containers through a bamboo tube that was inserted into an incision into the interior of a felled tree. Results of their analyses confirmed the broad, quantitative presence of yeasts, lactic acid bacteria and acetic acid bacteria during the palm wine tapping process, and clearly demonstrated a reduced diversity of yeast species as fermentation progressed and, by day three, *S. cerevisiae* was the only surviving yeast. At the onset, *Saccharomycodes ludwigii* and *Zygosaccharomyces bailii* were also detected, by both "new" and "old" methods. In addition, denaturing gradient gel electrophoresis (DGGE) analysis detected *Hanseniaspora uvarum*, *Candida parapsilosis*, *Candida fermentati* and *Pichia fermentans*, which could not be recovered by "traditional" methods. Thus, the tapping of *Elaeis guineensis* in the locations studied in this work indicates that a multi-starter, semi-continuous auto-fermentation takes place.

Further investigations into the physiology of ten oil palm wine yeast isolates from localities in southern Cameroon were conducted by Bechem *et al.*[471] Factors studied were: sensitivity to chloramphenicol, ethanol tolerance, acetic acid tolerance, osmotolerance and ADH and protein polymorphism. It was shown that eight strains could tolerate 15% ABV, and two could tolerate 40% (w/v) sucrose. None of the isolates was sensitive to 30 µg ml^{-1} chloramphenicol, and none could tolerate 1% acetic acid. The denatured protein pattern (SDS-PAGE) and the native protein pattern was similar for all strains, while the ADH pattern showed a high diversity, with three distinct groups being discernible.

5.6.2 Raffia Palm Wine

The *Raphia* palm is hapazanthic, *i.e.* after a period of vegetative growth, it produces flowers (and then fruits) only once before the plant dies when the seeds have matured. Since the palm is tapped when a flower bud appears, there are inherent problems if tapping is indiscriminate. The imminence of a flower is signalled when the crown of the tree throws up more than one extended spear leaf. Precise details on how to tap the Raffia palm have been given by Tuley,[460,472] but the time-honoured method basically involved making an incision into the young (unexpanded) spathe and tying a hollowed-out gourd to it. Sap trickles into the container and up to three litres per day can be collected.[473]

In a study of nine sugars identified in Raffia palm sap, Faparusi[474] showed that there was a variation in their levels over a period of time. Sucrose, glucose and fructose all decreased in concentration over the 22-day tapping period, while cellobiose, maltose and xylose showed slight increases in concentration over the same time. Rhamnose, arabinose and galacturonic acid showed no regular patterns of distribution.

Rokosu and Nwisienyi[475] investigated the variation in the components of *Raphia vinifera* palm wine during auto-fermentation. Over a four-day period sap pH fell from 7.0 to 3.5, and then rose to 4.5 on day five. Fresh, unfermented sap contained 3.92% (w/v) glucose, 10 mg ml^{-1} vitamin C and 4 mg ml^{-1} protein. Figure 5.33 illustrates the results of the work, which suggest that fresh *Raphia vinifera* palm sap is converted into palm wine in two stages, the first consisting of ethanol and organic (mainly lactic)

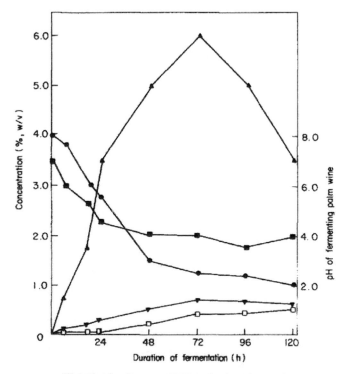

Variation in the components of palm wine during
fermentation. Freshly collected sap was allowed to autoferment to
palm wine for 120 h. During the fermentation aliquots (1 ml) were
taken out with sterile pipettes. These (1 ml) samples were used for
the various determination of the components as described in the
text at specified intervals. ▲, Alcohol; ■, pH; ●, glucose; ▼, titratable
acidity; □, volatile acidity

Figure 5.33   Autofermentation of *Raphia vinifera* P. Beauv. palm sap (from
Rokuso and Nwisienyi[475] - by kind permission).

acid production. The second stage involved the activity of
acetobacteria, thus lowering the pH of the medium.

In addition to naturally fermented palm wine, they also prepared
a "laboratory fermented" version, which, although more hygienic,
did not possess the same flavour attributes as wine fermented
naturally.

In a preliminary evaluation of the microbial quality of raphia
palm wine (called "ogoro" in western Nigeria), Olawale *et al.*[476]
analysed 150 samples collected at different locations and varying
stages of fermentation. Using the methods described by Lodder,[477]

which have been largely superseded, they found "three groups of yeast strains": *S. cerevisiae*, *S. chevalieri* (now *S. cerevisiae*) and *Kloerckera apiculata*, and using two somewhat selective media (and incubating at 37 °C!) identified bacteria that also fell into "three groups": *Bacillus cereus*, *B. firmus* and *Enterococcus faecalis* – all evidence of poor hygiene.

In a comparison of the microbiology, stability and shelf-life of palm wines obtained from *Elaeis guineensis* and *Raphia hookeri*, Ogbulie et al.[478] found that the former harboured more yeast species, while the latter contained a higher population of *E. coli* and various other heterotrophic bacteria. As Okafor[479] had previously found, highest bacterial numbers were found in sap straight from the trees (0 hours), but numbers decreased over time, indicating a loss of viability as ethanol content increased (and sugar levels decreased). Other physico-chemical parameters affecting the microflora would have been pH, water availability and oxygen tension. Five genera of bacteria were isolated and identified: *Staphylococcus* sp., *Micrococcus* sp., *Bacillus* sp., *Lactobacillus* sp. and *Brevibacterium* sp., plus *E. coli* and a yeast (*Saccharomyces* sp.). The presence of *E. coli* in the raffia palm sample and *Micrococcus* and *Staphylococcus* in samples from both palms suggests obvious public health issues, and could account for the alimentary tract problems exhibited by some palm wine drinkers.

5.6.3 Other Palm Wines

According to Batra and Millner,[480] there are three basic types of toddy in India: *sendi*, from palm, *tari*, from palmyra and date palm, and *nareli*, from coconut palm. Microbes implicated in the fermentation of these drinks include species of *Saccharomyces*, *Schizosaccharomyces* and *Geotrichum*.

Merican[481] reported that a single species of *Saccharomyces* had been recovered from Malaysian toddy samples, and that yeast counts in fresh toddy range from 10×10^8 to 3.1×10^9 ml^{-1}. The yeast was reported as being "similar to *S. cerevisiae* except that it does not ferment maltose". Bacterial numbers ranged from 1.50×10^8 to 3.6×10^9 ml^{-1}, with species of *Lactobacillus* and *Acetobacter* being identifiable.

Atputharajah et al.,[482] investigating the microbiology and biochemistry of the natural fermentation of coconut palm sap, found the usual lactic-alcoholic-acetic fermentation succession. They identified 166 yeast isolates (18 species; 8 genera), and 39 isolates of bacteria, mostly bacilli, but with evidence of *Leuconostoc*, *Lactobacillus*, *Micrococcus* and *Enterobacter* species. Some 72% of yeasts belonged to the genera *Saccharomyces*, *Pichia* and *Candida*, with 35% of the total population assigned to *Saccharomyces chevalieri*. The same group[483] demonstrated that these yeasts exhibited a high ethanol-yielding ability in laboratory batch cultures; being capable of producing 9% (v/v) ethanol from sap containing 15–18% (w/v) sucrose. These alcohol-tolerant *S. chevalieri* isolates were subsequently the subject of a study to assess their suitability for use in continuous fermentation processes,[484] one strain showing exceptional suitability.

According to Shamala and Sreekantiah,[485] who carried out microbiological and biochemical studies on traditional Indian wild date palm wine fermentations, microbes identified in the freshly tapped sap are: *Saccharomyces cerevisiae*, *Schizosaccharomyces pombe*, *Acetobacter aceti*, *A. rancens*, *A. suboxydans*, *Leuconostoc dextranicum*, *Micrococcus* sp., *Pediococcus* sp., *Bacillus* sp. and *Sarcina* sp. The majority of these survived into the fermented product. Ethanol production in this study was attributable solely to the two yeasts, with no evidence of participation by *Z. mobilis*. The maximum amount of ethanol by volume found in fresh toddy was 3.47% (at 72 h), and by 96 hours this had reduced to 1.80%, due to the activity of acetobacteria. Viability of yeast cells was greatly reduced after 78 hours fermentation.

Like many other alcoholic beverages, palm wine can serve as a carrier for medicinals and, in West Africa, for example, the bark or stem of plants such as *Alstonia boonei* De Wild. (anti-inflammatory, antipyretic, analgesic properties) and *Sacoglottis gabonensis* (Baill.) Urb. (antioxidant protective properties) are added to combat various ailments. The last named has been used for a variety of purposes, and imparts alkaloids and phenolics into palm sap, thus producing an antimicrobial effect in palm wine.

5.6.4 Shelf-life Studies

Because of the popularity of palm wine (over ten million people regularly consume the product in West Africa alone) it has been necessary to extend the shelf-life of the drink. One such method is bottling, which essentially involves filtration of the wine, diluting with water, bottling and pasteurization. Because of its perishable nature, much palm wine is wasted if not consumed straight away, and several workers have attempted to devise additional preservation regimes, with six months' shelf-life being seen as desirable. One of the first suggestions for preservation was to use sodium metabisulfite and benzoate,[486] and, a decade later, Chinerasa[487] used pasteurization for bottled wine.

As a result of some preliminary experiments on the preservation of oil palm wine, Okafor[488] found that a water extract of the bark from *Sacoglottis gabonensis* failed to inhibit several yeasts and bacteria. Sodium metabisulfite, diethyl pyrocarbonate (DEPC) and sorbic acid all showed varying degrees of inhibitory activity against palm wine microbes, but the latter was deemed to be the most suitable. Pasteurization at 70 °C for 30 minutes was more effective than any of the chemical methods, and it was proposed that this pasteurization regime, followed by treatment with sorbic acid, would be a useful means of preserving oil palm wine.

A couple of years later, Okafor[489] reported an improved method for successfully preserving palm wine for a period of at least six months. The method is based on the fact that when palm wine is centrifuged the clear supernatant has all the organoleptic properties of "raw" palm wine. Centrifugation greatly reduced the microbial load, and pasteurization of the supernatant with 0.05% sodium metabisulfite provided the enhanced protection. As Okafor related, the presence of the all-important palm wine opacity "was restored with a proprietary non-toxic substance, which was added before pasteurizing". A little later, Esechie[490] recommended the use of sodium metabisulfite and pasteurization at 70 °C for 40 minutes.

Sodium metabisulfite was also investigated as a preservation agent by Morah,[491,492] and an increase in acidity with age has been shown to be considerably reduced by the metabisulfite ion. Another positive result of metabisulfite addition is the fact that ethanol production does not peak and then decline after three or four days, but yeast activity continues, and wines of greater strength can be made.

Obahiagbon and Oviasogie[493] extended the shelf-life of Raphia palm sap to 24 months by pasteurizing at 75 °C for 45 minutes, without resorting to any chemical preservative. Essential sap nutrients all survived the heat treatment and were present at the end of the storage period. Raphia palm sap, whether fermented or unfermented, contributes around 20% of the palm-derived drink consumed in southern Nigeria. A review of the origin, morphology, cultivation, economic products and health and physiological implications of raphia palms has been provided by Obahiagbon.[494]

Using water and ethanol as solvents, Ogbulie *et al.*[478] used extracts from a variety of plants known to have preservative properties in an attempt to extend the shelf-life of palm wines. They found that wines treated with extracts of *Sacoglottis gabonensis* maintained their "condition" (carbonation) and organoleptic properties for an extended period (up to 96 hours after tapping).

The first investigations into the effects of *S. gabonensis* bark on the microflora of palm wine was carried out by Faparusi and Bassir,[495] who showed that a bark extract inhibited the growth of *Leuconostoc mesenteroides* and *Lactobacillus plantarum*. They showed that at least three different bark constituents were inhibitory to the growth of these two bacteria. Bergenin, a compound that had been shown to accelerate fermentation and inhibit acidification of palm wine, was not reported as one of the "active" constituents – something that has since been shown to be incorrect. Bergenin is a C-glucoside of 4-*O*-methylgallic acid that occurs in several plant genera, and a review of some its wide array of biological activities has been provided by Rastogi and Rawat.[496]

The bergenin molecule, which was first identified in *S. gabonensis* bark by Ogan,[463] is composed of three six-membered rings: an aromatic ring; a glucopyranose ring and an annellated δ-lactone ring (Figure 5.34). According to Rastogi and Rawat,[496] the following properties have attributed to bergenin: antiulcerogenic,

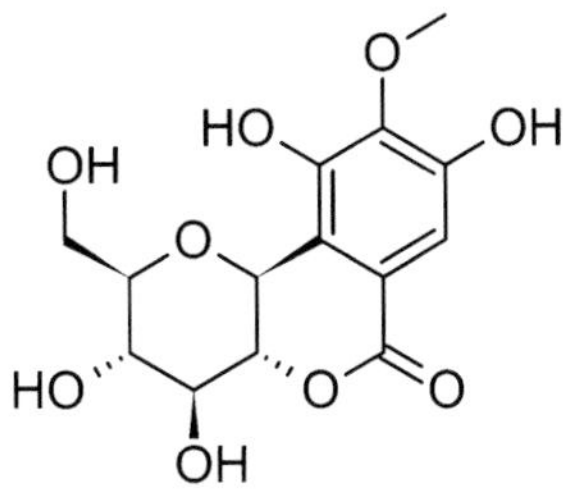

Figure 5.34 Structure of bergenin.

hepatoprotective, antiviral (against hepatitis C and HIV), antidia-
betic/antiobesity (acting as a PTP1B inhibitor), anti-arrhythmic,
antioxidant, antiarthritic, burn wound healer and trypanocide.

The antimicrobial activity of bergenin (albeit from *Endopleura
uchi*, another member of the Humiriaceae, widespread in the
Amazon basin) was studied by da Silva *et al.*,[497] who found that
the compound was active against a range of potentially pathogenic
fungi, including *Candida albicans*, but inactive against selected
Gram-negative and Gram-positive bacteria.

S. gabonensis (Humiriaceae), the bitter bark tree, is found in the
tropical rainforest of West Africa and America. Ekong and
Ejike[498] reported that addition of bark extract arrested the
lowering of pH during fermentation, suggesting a cytotoxic
activity, and the antioxidant activity of the extract was later
confirmed. Maduka *et al.*[499] compared the antioxidant effect of the
bark extract with that of vitamins C and E, and found a certain
amount of parity. The mechanism of inhibition of the observed
antioxidant action of the extract appeared to be by the inhibition
of propagation of lipid peroxidation. Maduka and Okoye,[500] using
the rat as an experimental animal, and 2,4-dinitrophenyl hydrazine
(DNPH) as the experimental oxidant, studied the effect of a
bergenin extract on membrane lipid peroxidation and tissue
ascorbic acid level. They found that pretreatment with bergenin
significantly reduced DNPH-induced lipid peroxidation in the
liver, brain and erythrocytes, and also protected against DHPH-
induced depletion of tissue ascorbic acid.

Over the years, there have been numerous health problems
arising from the consumption of bottled palm wine.
Unfortunately, there are still numerous small-scale bottling
enterprises using poor quality water in their processing, and this
has led to inevitable health problems for consumers. This situation
has been particularly prevalent in Benin City, Nigeria, which has a
high concentration of palm wine bottling outfits.[501] It has been
shown[502] that the ground water in Benin City was/is contaminated
with unacceptably high levels of some heavy metals (Pb, Cd, Cr,
Zn). Research showed[501] that zinc, chromium and nickel levels
were from two to fifteen times higher in Benin-bottled palm wine,
as compared to fresh samples. Such contamination was traced to
the dilution stage during bottling, and the problem was found to be
more severe in wine bottled by the smaller concerns.

With some palm wine yeasts showing some characters likely to be of industrial use, such as ethanol tolerance, Nwachukwu *et al.*[503] carried out investigations on yeasts isolated from palm wines from southeastern Nigeria. Using samples from both oil and raffia palms, ethanol tolerance and flocculation/sedimentation rates were calculated. Isolates (9 strains of *S. cerevisiae*; 2 strains of *S. globosus*; 2 strains of *Hanseniaspora uvarum*) were taken from 600-hour old wines. Ethanol tolerances varied between 10 and 20% ABV, and sedimentation rates varied from 55.5 to 93.1%, indicating promising industrial credentials for some strains. In some instances, ethanol tolerance could be increased by addition of supplements (such as soybean and palm kernel) to the growth medium.

At least one report[504] has been made of an attempt to create a synthetic palm wine. The basal solution consisted of: 12.5% fructose; 5.0% glucose; 12.5% sucrose; 2.5% raffinose, and 0.01% ammonium phosphate. Washed palm wine dregs were added and the mixture was fermented at 28 °C for 24 hours. A parallel system was set up consisting of the above with added 0.1% yeast extract. An acceptable "laboratory" palm wine would obviate the need for laborious sap collection methods, and conserve trees that might otherwise be felled.

As the authors state: "This innovation will definitely open a new chapter in the palm wine industry in Nigeria when production on a large scale commences."

REFERENCES

1. J. G. D. Clark, *Antiquity*, 1942, **16**, 208.
2. E. Crane, *The Archaeology of Beekeeping*, Duckworth, London, 1983.
3. J. A. Kelhoffer, *Greek, Roman, and Byzantine Studies*, 2005, **45**, 59.
4. E. Crane (ed.), *Honey: A Comprehensive Survey*, Heinneman Edition, London, 1975.
5. E. Neufeld, *Ugarit-Forschungen*, 1978, **10**, 219.
6. F. S. Bodenheimer, *Animal and Man in Bible Lands*, E. J. Brill, Leiden, 1960.
7. T. Forti, *Vetus Testamentum*, 2006, **56**, 327.

8. J. Neusner, *The Mishnah: A New Translation*, Yale University Press, New Haven, CT, c.1988.
9. W. S. Hett, *Aristotle: Minor Works*, Harvard University Press, Cambridge, MA, 1965.
10. A. Mayor, *Archaeology*, 1995, **48**, 32.
11. P. C. Plugge, *Arch. Pharm.*, 1891, **229**, 552.
12. J. Ott, *Econ. Bot.*, 1998, **52**, 260.
13. H. A. Hoffner, Jr., in *Law Collections from Mesopotamia and Asia Minor*, ed. M. T. Roth, Scholars Press, Atlanta, GA, 1995.
14. G. M. Beckman (ed.), *Hittite Myths*, Scholars Press, Atlanta, GA, 1990.
15. M. T. Roth (ed.), *Law Collections from Mesopotamia and Asia Minor*, Scholars Press, Atlanta, GA, 1995.
16. H. M. Ransome, *The Sacred Bee in Ancient Times and Folklore*, George Allen and Unwin, London, 1937.
17. R. J. Head, *The FARMS Review*, 2008, **20**, 57.
18. A. Mazar, D. Namdar, N. Paniz-Cohen, R. Neumann and S. Weiner, *Antiquity*, 2008, **82**, 629.
19. S. Gitin, *Eretz Israel*, 1989, **20**, 52.
20. A. Mazar and N. Paniz-Cohen, *Near Eastern Archaeology*, 2007, **70**, 202.
21. G. Bloch, T. M. Francoy, I. Wachtel, N. Panitz-Cohen, S. Fuchs and A. Mazar, *Proc. Nat. Acad. Sci. USA*, 2010, **107**, 11240.
22. S. Dalley, *Mari and Karana, Two Old Babylonian Cities*, Gorgias, London, 1984.
23. E. V. Kvavadze, *Paleontol. J.*, 2006, **40**, S595.
24. E. Crane, *The World History of Beekeeping and Honey Hunting*, Duckworth, London, 1999.
25. N de G. Davies, *The Tomb of Rekh-me-Re at Thebes*, Plantin Press, New York, 1943.
26. H. B. Tristram, *The Natural History of the Bible*, SPCK, London, 1867.
27. G. Kuény, *J. Near E. Stud.*, 1950, **9**, 84.
28. D. J. Brewer, D. B. Redford and S. Redford, *Domestic Plants and Animals: The Egyptian Origins*, Aris & Phillips, Warminster, 1994.
29. L. Manniche, *An Ancient Egyptian Herbal*, British Museum Press, London, 1989.

30. J. H. Breasted, *The Edwin Smith Surgical Papyrus*, 2 volumes, Chicago University Press, Chicago, 1930.
31. B. Flower and E. Rosenbaum, *The Roman Cookery Book*, Harrup, London, 1958.
32. A. Dalby and S. Grainger, *The Classical Cookbook*, British Museum Press, London, 1996, 2000.
33. A. G. Sherratt, *Economy and Society in Prehistoric Europe*, Princeton University Press, Princeton, NJ, 1997.
34. A. Hagan, *A Second Handbook of Anglo-Saxon Food and Drink: Production and Distribution*, Anglo-Saxon Books, Hockwold, 1995.
35. S. Pollington, *Leechcraft: Early English Charms, Plantlore and Healing*, Anglo-Saxon Books, Hockwold, 2000.
36. E. Crane, P. Walker and R. Day, *Directory of Important World Honey Sources*, Int. Bee Res. Assn., London, 1984.
37. L. Persano Oddo, L. Piana, S. Bogdanov, A. Bentabol, P. Gotsiou, J. Kerkvliet, P. Martin, M. Morlot, A. Ortiz Valbuena, K. Ruoff and K. von der Ohe, *Apidologie*, 2004, **35**(Special Issue), 82.
38. H. V. Minh, B. V. Mendoza, F. M. Laigo, R. A. Morse and K. H. Steinkraus, *J. Api. Res.*, 1971, **10**, 91.
39. I. R. Siddiqui and B. Furgala, *J. Apic. Res.*, 1967, **6**, 139.
40. I. R. Siddiqui and B. Furgala, *J. Apic. Res.*, 1968, **7**, 51.
41. J. F. Cotte, H. Casabianca, S. Chardon, J. Lheritier and M. F. Grenier-Loustalot, *Anal. Bioanal. Chem.*, 2004, **380**, 698.
42. J. W. White, Jr., *Adv. Food Res.*, 1978, **24**, 287.
43. E. E. Stinson, M. H. Subers, J. Petty and J. W. White, Jr., *Arch. Biochem. Biophys.*, 1960, **89**, 6.
44. T. Ruiz-Argüeso and A. Rodriguez-Navarro, *J. Gen. Microbiol.*, 1973, **76**, 211.
45. S. Suárez-Luque, I. Mato, J. F. Huidobro, J. Simal-Lozano and J. C. García-Monteagudo, *Food Chem.*, 2003, **80**, 215.
46. B. Talpay, *Dtsch. Lebensm.-Rundsch.*, 1988, **84**, 41.
47. K. V. Giri, *Madras Agricultural J.*, 1938, **26**, 68.
48. S. R. Alonso-Torre, M. M. Cavia, M. A. Fernández-Muiño, J. F. Fernández-Muiño, G. Moreno, J. F. Huidobro and M. T Sancho, *Food Chem.*, 2006, **97**, 750.
49. P. Andrade, F. Ferreres, M. I. Gil and F. A. Tomás-Barberán, *Food Chem.*, 1997, **60**, 79.

50. J. W. White, Jr., in *Honey: A Comprehensive Survey*, ed. E. Crane, Heinneman Edition, London, 1975, p. 157.
51. M. J. Feller-Demalsy, B. Vincent and F. Beaulieu, *Apidologie*, 1989, **20**, 77.
52. M. J. Nozal Nalda, J. L. B. Yague, J. C. D. Calva and M. T. M. Gomez, *Anal. Bioanal. Chem.*, 2005, **382**, 311.
53. M. J. Latorre, R. Pena, C. Pita, A. Botana, S. García and C. Herrero, *Food Chem.*, 1999, **66**, 263.
54. O. M. Hernández, J. M. G. Fraga, A. I. Jiménez, F. Jiménez and J. J. Arias, *Food Chem.*, 2005, **93**, 449.
55. S. Bogdanov, K. Ruoff and L. Persano Oddo, *Apidologie*, 2004, **35**(Special Issue), s4.
56. H. Dold, O. H. Du and S. T. Dzaio, *Z. Hyg. Infektionskr.*, 1937, **120**, 155.
57. H. Dold and R. Witzenhausen, *Z. Hyg. Infektionskr*, 1955, **141**, 333.
58. D. Adcock, *J. Apic. Res.*, 1962, **1**, 38.
59. J. E. Schade, G. L. Marsh and J. E. Eckert, *Food Res.*, 1958, **23**, 446.
60. F. J. Ingram, H. Ottoway and J. E. Coppock, *Chem. Ind.*, 1956, **42**, 1154.
61. B. J. Macris, *Appl. Micro.*, 1975, **30**, 503.
62. E. Freese, C. W. Sheu and E. Galliers, *Nature*, 1973, **241**, 321.
63. S. Bogdanov, *Lebensm. Wissc. Technol.*, 1983, **17**, 74.
64. P. C. Molan and K. M. Russell, *J. Apic. Res.*, 1988, **27**, 62.
65. F. Ferreres, F. A. Tomás-Barberán, C. Soler, C. García-Viguera, A. Ortiz and F. Tomás-Lorente, *Apidologie*, 1994, **25**, 21.
66. H. A. L. Wahdan, *Infection*, 1998, **26**, 26.
67. J. W. White, Jr., M. H. Subers and A. I. Schepartz, *Biochim. Biophys. Acta*, 1963, **73**, 57.
68. J. W. White, Jr., and M. H. Subers, *J. Apic. Res.*, 1963, **2**, 93.
69. A. I. Schepartz and M. H. Subers, *Biochim. Biophys. Acta*, 1964, **85**, 228.
70. A. I. Schepartz, *Biochim. Biophys. Acta*, 1965, **96**, 334.
71. A. I. Schepartz, *Biochim. Biophys. Acta*, 1965, **99**, 161.
72. J. W. White, Jr., and M. H. Subers, *J. Apic. Res.*, 1963, **2**, 93.
73. P. C. Molan, *Bee World*, 1992, **73**, 5.
74. P. C. Molan, *Bee World*, 1992, **73**, 59.

75. J. Bardy, N. J. Slevin, K. L. Mais and A. Molassiotis, *J. Clin. Nurs.*, 2008, **17**, 2604.
76. N. Orsolic, *J. of ApiProduct and ApiMedical Science*, 2009, **1**, 93.
77. N. Al Somal, K. E. Coley, P. C. Molan and B. M. Hancock, *J. Roy. Soc. Med.*, 1994, **87**, 9.
78. N. S. Al Waili, *J. Med. Food*, 2004, **7**, 100.
79. O. Sherlock, A. Dolan, R. Athman, A. Power, G. Gethin, S. Cowman and H. Humphreys, *BMC Compl. Alternative Med.*, 2010, **10**, 47.
80. V. Bankova, *J. of ApiProduct and ApiMedical Science*, 2009, **1**, 23.
81. M. Viuda-Martos, Y. Ruiz-Navajas, J. Fernández-López and J. A. Pérez-Alvarez, *J. Food Sci.*, 2008, **73**, R117.
82. J. Matjan, *eCAM*, 2010, 1.
83. R. A. Cooper, P. C. Molan and K. G. Harding, *J. Appl. Micro.*, 2002, **93**, 857.
84. R. A. Cooper, E. Halas and P. C. Molan, *J. Burn Care Rehabil.*, 2002, **23**, 366.
85. P. H. S. Kwakman, J. P. Van den Akker, A. Güçlü, H. Aslami, J. M. Binnekade, L. de Boer, L. Boszhard, F. Paulus, P. Middelhoek, A. A. te Velde, C. M. Vandenbroucke-Grauls, M. J. Schultz and S. A. Zaat, *Clin. Infect. Dis.*, 2008, **46**, 1677.
86. P. H. S. Kwakman, A. A. te Velde, L. de Boer, D. Speijer, C. M. J. E. Vandenbroucke-Grauls and S. A. J. Zaat, *FASEB J.*, 2010, **24**, 2576.
87. P. H. S. Kwakman, L. de Boer, C. P. Ruyter-Spira, T. Creemers-Molenaar, J. P. F. G. Helsper, C. M. J. E. Vandenbroucke-Grauls, S. A. J. Zaat and A. A. te Velde, *Eur. J. Clin. Microbiol. Infect. Dis.*, 2011, **30**, 251.
88. R. Jones, *J. of ApiProduct and ApiMedical Science*, 2009, **1**, 2.
89. A. E. Jeffrey and C. M. Echazaretta, *Rev. Biomed.*, 1996, **7**, 43.
90. H. I. Assil, R. Sterling and P. Sporns, *J. Food Sci.*, 1991, **56**, 1034.
91. R. A. Gleiter, H. Horn and H.-D. Isengard, *Food Chemi.*, 2006, **96**, 441.
92. J. A. Snowdon and D. O. Cliver, *Int. J. Food Microbiol.*, 1996, **31**, 1.

93. B. O. Omotuybe and O. O. Akanbi, *African J. Micro. Research*, 2009, **3**, 891.
94. C. D. Michener and D. A. Grimaldi, *Proc. Nat. Acad. Sci. USA*, 1988, **85**, 6424.
95. M. S. Engel, *Proc. Nat. Acad. Sci. USA*, 2001, **98**, 1661.
96. M. S. Engel, *American Museum Novitates*, 2000, **3296**, 1.
97. M. Beye, M. Hasselmann, M. K. Fondrk, R. E. Page and S. W. Omholt, *Cell*, 2003, **114**, 419.
98. D. Grimaldi and M. S. Engel, *The Evolution of Insects*, Cambridge University Press, Cambridge, 2005.
99. F. Ruttner, *Biogeography and Taxonomy of Honey Bees*, Springer-Verlag, Berlin, 1988.
100. L. Garnery, J.-M. Cornuet and M. Solignac, *Mol. Ecol.*, 1992, **1**, 145.
101. C. W. Whitfield, S. K. Behura, S. H. Berlocher, A. G. Clark, J. S. Johnston, W. S. Sheppard, D. R. Smith, A. V. Suarez, D. Weaver and N. D. Tsutsui, *Science*, 2006, **314**, 642.
102. F. Ruttner, L. Tassencourt and J. Louveaux, *Apidologie*, 1978, **9**, 363.
103. P. Franck, L. Garnery, A. Loiseau, B. P. Oldroyd, H. R. Hepburn, M. Solignac and J.-M. Cornuet, *Heredity*, 2001, **86**, 420.
104. M. C. Arias and W. S. Sheppard, *Mol. Phylogenet. Evol.*, 2005, **37**, 25.
105. M. S. Engel, *J. Hym. Res.*, 1999, **8**, 165.
106. F. Ruttner and V. Maul, *Apidologie*, 1983, **14**, 309.
107. W. S. Sheppard, *American Bee Journal*, 1989a, **129**, 617.
108. W. S. Sheppard, *American Bee Journal*, 1989b, **129**, 664.
109. N. M. Schiff and W. S. Sheppard, *Experientia*, 1993, **49**, 530.
110. W. S. Sheppard and D. R. Smith, *Ann. Entomol. Soc. Am.*, 2000, **93**, 159.
111. S. S. Schneider, G. De Grandi-Hoffman and D. R. Smith, *Ann. Rev. Entomology*, 2004, **49**, 351.
112. A. Zayed and C. W. Whitfield, *Proc. Nat. Acad. Sci. USA*, 2008, **105**, 3421.
113. E. O. Wilson and B. Holldobler, *Proc. Nat. Acad. Sci. USA*, 2005, **102**, 13367.
114. F. G. Barth, *Insects and Flowers: The Biology of a Partnership*, Princeton University Press, Princeton, 1985.

115. C. O'Toole, and A. Raw, *Bees of the World*, Blandford, London, 1991.
116. D. Grimaldi, *Ann. Mo. Bot. Garden*, 1999, **86**, 373.
117. C. D. Michener, *The Bees of the World*, Johns Hopkins University Press, Baltimore, 2000.
118. Honeybee Genome Sequencing Consortium, *Nature*, 2006, **443**, 931.
119. S. K. Behura, *Mol. Biol. Evol.*, 2007, **24**, 1492.
120. D. Bensasson, D. X. Zhang, D. L. Hartl and G. M. Hewitt, *Trends Ecol. Evol.*, 2001, **16**, 314.
121. S. K. Behura, *Mol. Ecol.*, 2006, **15**, 3087.
122. G. R. Gayre, *Wassail! In Mazers of Mead*, Phillimore & Co., London, 1948.
123. R. Evershed, S. J. Vaughan, S. N. Dudd and J. S. Soles, *Antiquity*, 1997, **71**, 979.
124. M. Rösch, *Veg. Hist. Archaeobotany*, 2005, **14**, 179.
125. P. E. McGovern, J. Zhang, J. Tang, Z. Zhang, G. R. Hall, R. A. Moreau, A. Nuñez, E. D. Butrym, M. P. Richards, C.-s. Wang, G. Cheng, Z. Zhao and C. Wang, *Proc. Nat. Acad. Sci. USA*, 2004, **101**, 17593.
126. P. E. McGovern, A. P. Underhill, H. Fang, F. Luan, G. R. Hall, H. Yu, C.-s. Wang, F. Cai, Z. Zhao and G. M. Feinman, *Asian Perspect.*, 2005, **44**, 249.
127. H. T. Huang, in *Science and Civilisation in China Volume 6, Biology and Biological Technology*, ed. J. Needham, Cambridge University Press, Cambridge, 2000.
128. M. Loewe and E. L. Shaughnessy (ed.), *The Cambridge History of Ancient China*, Cambridge University Press, Cambridge, 1999.
129. P. B. Ebrey, *China: A Cultural, Social and Political History*, Houghton Mifflin, Boston, c.2006.
130. J. Zafiropulo, *Mead and Wine: A History of the Bronze Age in Greece*, Sidgwick & Jackson, London, 1966.
131. A. G. Sherratt, in *Bell Beakers of the Western Mediterranean*, ed. W. Waldren and R. C. Kennard, British Archaeological Reports, International Series, 287, BAR, Oxford, 1987.
132. A. S. Henshall, *Proc. Soc. Antiq. Scot.*, 1964, **97**, 166.
133. J. H. Dickson, *Antiquity*, 1978, **52**, 108.
134. A. S. C. Deans, *Bee Research Assn. Report*, 142, 1957.
135. T. Thomsen, *Nordisk Fortidsminder*, 1929, **2**, 165.

136. M. Rösch, *Veg. Hist. Archaeobotany*, 1999, **8**, 105.
137. U. Körber-Grohne, in *Der Keltenfürst von Hochdorf, Methoden und Ergebnisse der Landesarchäologie in Baden-Würtemberg*, ed. D. Planck, J. Biel, G. Süsskind and A. Wais, Stuttgart, 1985, p. 116.
138. P. E. McGovern, *Uncorking the Past*, University of California Press, Berkeley, CA, 2009.
139. F. J. Furnival, *A Compendyous Regyment or a Dyetary of Helth*, N. Trübner & Co., London, 1870, p. 257.
140. K. Digby, *The Closet of the Eminently Learned Sir Kenelme Digbie, Kt. Opened*, H. Brome, London. 1669.
141. W. Harrison, *The Description of Britain*: The first volume of the *Chronicles of England, Scotland and Irlande* by Raphaell Holinshed, John Harrison, London, 1577.
142. C. Renfrow, *A Sip Through Time*, Published by the author, 1995.
143. S. Norton and J. C. Vuillemard, *Crit. Rev. Biotechnol.*, 1994, **14**, 193.
144. N. Qureshi and D. V. Tamhane, *Appl. Microbiol. Biotechnol.*, 1985, **21**, 280.
145. N. Qureshi and D. V. Tamhane, *Appl. Microbiol. Biotechnol.*, 1986, **23**, 438.
146. B. R. Thakur, R. K. Singh and A. K. Handa, *Crit. Rev. Food Sci. Nutr.*, 1997, **37**, 47.
147. M. Navrátil, E. Šturdík and P. Gemeiner, *Biotechnol. Lett.*, 2001, **23**, 977.
148. R. A. Morse and K. H. Steinkraus, in *Honey: A Comprehensive Survey*, ed. E. Crane, Heinneman Edition, London, 1975, p. 392.
149. K. Schramm, *The Compleat Meadmaker*, Brewers Publications, Boulder, CO, 2003.
150. E. Sarin, *Z. Unters. Nahr. –u Genussmittel.*, 1921, **42**, 90.
151. Y. Teramoto, R. Sato and S. Ueda, *Afr. J. Biotechnol.*, 2005, **4**, 160.
152. O. Gonçalves de Lima, J. Francisco de Mello, I. L. D'Albuquerque, F. D. Monache, G. B. Marini-Bettolo and M. Sousa, *Lloydia*, 1977, **40**, 195.
153. J. E. S. Thompson, *A Catalog of Maya Hieroglyphs*, University of Oklahoma Press, Norman, 1962.

154. D. de Landa, *Relacion de las cosas de Yucatan*, 1566. [translation by Jean Genet, 1928]
155. J. E. S. Thompson, *Ethnology of the Maya of Southern and Central British Honduras*, Field History Museum of Natural History, Chicago, 1930.
156. F. Delle Monache, F. Marletti, G. B. Marini-Bettolo, J. F. De Mello and O. Goncalvez de Lima, *Lloydia*, 1977, **40**, 201.
157. R. G. Wasson, *Botanical Leaflets Harvard University*, 1963, **20**, 161.
158. F. J. Lipp, *The Mixe of Oaxaca: Religion, Ritual, and Healing*, University of Texas Press, Austin, TX, 1991.
159. A. Hofmann, *Botanical Leaflets Harvard University*, 1963, **20**, 194.
160. A. T. Weil and W. Davis, *J. Ethnopharmacol.*, 1994, **41**, 1.
161. S. Vogel and A. Gobezie, in *Symposium on Indigenous Fermented Foods*, Bangkok, Thailand, 1977.
162. B. Dester, *Ethiopian Med. J.*, 1977, **15**, 65.
163. B. Bahiru, T. Mehari and M. Ashenafi, *Food Microbiol.*, 2006, **23**, 277.
164. I. Pardo, M. Jose-Garcia, M. Suniga and F. Uruburu, *Appl. Environ. Microbiol.*, 1989, **55**, 539.
165. M. Toussaint-Samat, *A History of Food, New Expanded 2nd edn*, Wiley-Blackwell, Chichester, 2009.
166. V. I. Zapriagaeva, *Wild Growing Fruits in Tadzikistan* [in Russian, with English summary], Nauka, Moscow, 1964.
167. E. Maul, R. Topfer and R. Eibach, *Vitis International Variety Catalogue 2008*, http://www.vivc.bafz.de.
168. V. H. Heywood and D. Zohary, *Flora Mediterranea*, 1991, **5**, 375.
169. L. Levadoux, *Annales de la' Amélioration des Plantes, Ser. B*, 1956, **1**, 59.
170. D. Zohary and M. Hopf, *Domestication of Plants in the Old World: The Origin and Spread of Cultivated Plants in West Asia, Europe and the Nile Valley*, Clarendon Press, Oxford, 3rd edn, 2000.
171. A. M. Negrul, *Atti. Acad. Ital. Vine e Vino*, 1960, **12**, 113.
172. R. K. Jansen, C. Kaittanis, C. Saski, S.-B. Lee, J. Tomkins, A. J. Alverson and H. Daniell, *BMC Evol. Biol.*, 2006, **6**, 32.
173. J. E. Planchon, *Monographia Phanerogamerum*, 1887, **5**, 305.

174. R. S. Jackson, *Wine Science: Principles and Applications*, Elsevier, Amsterdam, 3rd edn, 2008.
175. A. C. Renfrew, *Sci. Am.*, October 1989, 106.
176. A. C. Stevenson, *J. Biogeogr.*, 1985, **12**, 293.
177. H. J. De Blij, *Wine: A Geographic Appreciation*, Rowman & Allanheld, Totowa, NJ, 1983.
178. R. Arroyo-García, F. Lefort, M. T. de Andrés, J. Ibáñez, J. Borrego, N. Jouve, F. Cabello and J. M. Martinéz-Zapater, *Genome*, 2002, **45**, 1142.
179. R. Arroyo-García, L. Ruiz- García, L. Bolling, R. Ocete, M. A. López, C. Arnold, A. Ergul, G. SöylemezőLu, H. I. Uzun, F. Cabello, J. Ibáñez, M. K. Aradhya, A. Atanassov, I. Atanassov, S. Balint, J. L. Cenis, L. Costantini, S. Gorislavets, M. S. Grando, B. Y. Klein, P. E. McGovern, D. Merdinoglu, I. Pejic, F. Pelsy, N. Primikirios, V. Risovannaya, K. A. Roubelakis-Angelakis, H. Snoussi, P. Sotiri, S. Tamhankar, P. This, L. Troshin, J. M. Malpica, F. Lefort and J. M. Martinez-Zapater, *Mol. Ecol.*, 2006, **15**, 3707.
180. P. E. McGovern, D. L. Glusker, L. J. Exner and M. M. Voigt, *Nature*, 1996, **381**, 480.
181. P. E. McGovern, *Archaeology*, 1998, **51**, 32.
182. V. R. Badler, P. E. McGovern and R. H. Michel, *MASCA Research Papers in Science and Archaeology*, 1990, **7**, 25.
183. T. G. H. James, in *The Origins and Ancient History of Wine*, reprint, ed. P. E. McGovern, S. J. Fleming and S. H. Katz, Routledge, London, 2004, p. 197.
184. D. Zohary and P. Spiegel-Roy, *Science*, 1975, **187**, 319.
185. P. E. McGovern, U. Hartung, V. Badler, D. L. Glusker and L. J. Exner, *Expedition*, 1997, **39**, 3.
186. P. Kaplony, *Die Inschriften der ägyptischen Frühzeit*, 3 volumes, O. Harrassowitz, Wiesbaden, 1963–64.
187. P. E. Newberry, *Beni Hasan I*, Egyptian Exploration Fund, London, 1893.
188. M. A. Powell, in *The Origins and Ancient History of Wine*, reprint, ed. P. E. McGovern, S. J. Fleming and S. H. Katz, Routledge, London, 2004, p. 97.
189. J. Janssen, *Commodity Prices from the Ramesside Period: An Economic Study of the Village Necropolis Workmen at Thebes*, E. J. Brill, Leiden, 1975.

190. R. Palmer, in *Wine in The Mycenean Palace Economy*, ed. R. Palmer, University of Liège, Aegaeum 10, Liège, 1994.

191. M. A. Murray, in *Ancient Egyptian Materials and Technology*, ed. P. T. Nicholson and I. Shaw, Cambridge University Press, Cambridge, 2000, p. 577.

192. H. Ren and J. Wen, in *Flora of China*, ed. Z. Y. Wu and P. Raven, Science Press and Missouri Botanical Garden Press, Beijing and St Louis, 2007, vol. 12, p. 210.

193. B. Laufer, *Sino-Iranica: Chinese Contributions to the History of Civilization in Ancient Iran, with Special Reference to the History of Cultivated Plants and Products*, Field Museum of Natural History, Publication No. 201, Chicago, 1919.

194. X. Liu, *The Silk Road*, 2005, **3**, 23.

195. H.-E. Jiang, Y.-B. Zhang, X. Li, Y.-F. Yao, D. K. Ferguson, E.-G. Lü and C.-S. Li, *J. Archaeol. Sci.*, 2009, **36**, 1458.

196. Y. Wan, H. Schwaninger, D. Li, C.J. Simon, Y. Wang and C. Zhang, *Vitis*, 2008, **47**, 81.

197. F. Radler and D. H. S. Horn, *Aust. J. Chem.*, 1965, **18**, 1059.

198. C. Srinivasan and M. G. Mullins, *Ann. Bot.*, 1976, **38**, 1079.

199. S. S. Negi and H. P. Olmo, *Vitis*, 1971, **9**, 265.

200. H. Caspari, A. Lang and P. Alspach, *American Journal of Enology and Viticulture*, 1998, **49**, 359.

201. M. C. Candolfi-Vasconcelos and W. Koblet, *Vitis*, 1990, **29**, 199.

202. M. C. Candolfi-Vasconcelos, M. P. Candolfi and W. Koblet, *Planta*, 1994, **192**, 567.

203. M. Chaumont, J.-F. Morot-Gaudry and C. F. Foyer, *J. Exp. Botany*, 1994, **45**, 1235.

204. T. Roitsch, *Curr. Opin. Plant Biol.*, 1999, **2**, 198.

205. A. Lindenmayer, *J. Theor. Biol.*, 1968, **18**, 280.

206. P. Prusinkiewicz and A. Lindenmayer, *The Algorithmic Beauty of Plants*, Springer-Verlag, New York, 1990.

207. M. T. Allen, P. Prusinkiewicz and T. M. DeJong, *New Phytologist*, 2005, **166**, 869.

208. A. Quereix, R. C. Dewar, J.-P. Gaudillere, S. Dayau and C. Valancogne, *J. Exp. Botany*, 2001, **52**, 2313.

209. S. I. Gibson, *Curr. Opin. Plant Biol.*, 2005, **8**, 93.

210. H. Wada, K. A. Shackel and M. A. Matthews, *Planta*, 2008, **227**, 1351.

211. T. Roitsch and M. C. Gonzalez, *Trends Plant Sci.*, 2004, **9**, 606.

212. G. Lebon, G. Wojnarowiez, B. Holzapfel, F. Fontaine, N. Vaillant-Gaveau and C. Clément, *J. Exp. Botany*, 2008, **59**, 2565.

213. A. Martini, M. Ciani and G. Scorzetti, *American Journal of Enology and Viticulture*, 1996, **47**, 435.

214. D. Wibowo, R. Eschenbruch, C. R. Davis, G. H. Fleet and T. H. Lee, *American Journal of Enology and Viticulture*, 1985, **36**, 302.

215. B. G. Coombe, *American Journal of Enology and Viticulture*, 1992, **43**, 101.

216. E. Longo, J. Cansado, D. Agrelo and G. Villa, *American Journal of Enology and Viticulture*, 1991, **42**, 141.

217. J. Sabate, J. Cano, B. Esteve-Zarzoso and J. M. Guillamón, *Microbiol. Res.*, 2002, **157**, 267.

218. G. Muyzer, *Curr. Opin. Microbiol.*, 1999, **2**, 317.

219. F. J. Las Heras-Vazquez, L. Mingorance-Cazorla, J. M. Clemente-Jimenez and F. Rodriguez-Vico, *FEMS Yeast Res.*, 2003, **3**, 3.

220. V. Renouf and A. Lonvaud-Funel, *Journal International des Sciences de la Vigne et du Vin*, 2004, **38**, 219.

221. V. Renouf, O. Claisse and A. Lonvaud-Funel, *Australian Journal of Grape and Wine Research*, 2005, **11**, 316.

222. V. Renouf, M. Falcou, C. Miot-Sertier, M. C. Perello, G. De Revel and A. Lonvaud-Funel, *J. Appl. Microbiol.*, 2006, **100**, 1208.

223. O. Jaillon, J.-M. Aury, B. Noel, A. Policriti, C. Clepet, A. Casagrande, N. Choisne, S. Aubourg, N. Vitulo, C. Jubin, A. Vezzi, F. Legeai, P. Hugueney, C. Dasilva, D. Horner, E. Mica, D. Jublot, J. Poulain, C. Bruyère, A. Billault, B. Segurens, M. Gouyvenoux, E. Ugarte, F. Cattonaro, V. Anthouard, V. Vico, C. Del Fabbro, M. Alaux, G. Di Gaspero, V. Dumas, N. Felice, S. Paillard, I. Juman, M. Moroldo, S. Scalabrin, A. Canaguier, I. Le Clainche, G. Malacrida, E. Durand, G. Pesole, V. Laucou, P. Chatelet, D. Merdinoglu, M. Delledonne, M. Pezzotti, A. Lecharny, C. Scarpelli, F. Artiguenave, M. E. Pè, G. Valle, M. Morgante, M. Caboche, A.-F. Adam-Blondon, J. Weissenbach, F. Quétier and P. Wincker, *Nature*, 2007, **449**, 463.

224. M. K. Aradhya, G. S. Dangl, B. H. Prins, J. M. Boursiquot, M. A. Walker, C. P. Meredith and C. J. Simon, *Genet. Res.*, 2003, **81**, 179.

225. J. Bowers, J. M. Boursiquot, P. This, K. Chu, H. Johansson and C. Meredith, *Science*, 1999, **285**, 1562.

226. H. Karataş and Y. S. Ağaoğlu, *Hereditas*, 2008, **145**, 58.

227. M. S. Lopes, D. Mendonça, M. Rodrigues dos Santos, J. E. Eiras-Dias and A. da Câmara Machado, *Genome*, 2009, **52**, 790.

228. B. Schaal, J. Beck, S.-C. Hsu, T. Beridze, M. Gamkrelidze, M. Gogniashvili, I. Pipia, V. Tabidze, P. This, R. Bacilieri, V. Gotsiridze and M. Glonti, Plastid DNA sequence diversity in a worldwide set of grapevine cultivars (*Vitis vinifera* L. subsp. *vinifera*), Oral Presentation OR.1.02, 33rd World Congress of Vine and Wine, Tblisi, Georgia, June 20–27, 2010. www.oiv2010 ge/index

229. A. Soejima and J. Wen, *Am. J. Bot.*, 2006, **93**, 278.

230. J. Wen, Z.-L. Nie, A. Soejima and Y. Meng, *Can. J. Bot.*, 2007, **85**, 731.

231. I. Pipia, M. Gamkrelidze, M. Gogniashvili and V. Tabidze, Genetic diversity of Georgian varieties of *Vitis vinifera* subsp. *sylvestris*, Poster Presentation P.1.10, 33rd World Congress of Vine and Wine, Tblisi, Georgia, June 20–27, 2010. www.oiv2010 ge/index

232. J. F. Vouillamoz, P. E. McGovern, A. Ergul, G. Söylemezoğlu, G. Tevzadze, C. P. Meredith and M. S. Grando, *Plant Genet. Resour.*, 2006, **4**, 144.

233. K. M. Sefc, H. Steinkellner, J. Gloessl, S. Kampfer and F. Regner, *Theor. Appl. Gen.*, 1998, **97**, 227.

234. E. M. Sanchez-Escribano, J. R. Martin, J. Carrero and J. L. Cenis, *Genome*, 1999, **42**, 87.

235. J. F. Vouillamoz, D. Maigre and C. P. Meredith, *Theor. Appl. Gen.*, 2003, **107**, 448.

236. P. This, A. Jung, P. Boccacci, J. Borrego, R. Botta, L. Costantini, M. Crespan, G. S. Dangl, C. Eisenheld, F. Ferreira-Monteiro, S. Grando, J. Ibáñez, T. Lacombe, V. Laucou, R. Magalhães, C. P. Meredith, N. Milani, E. Peterlunger, F. Regner, L. Zulini and E. Maul, *Theor. Appl. Gen.*, 2004, **109**, 1448.

237. K. M. Sefc, I. Pejic, E. Maletic, M. R. Thomas and F. Lefort, in *Grapevine Molecular Physiology and Biotechnology*, ed. K. A. Roubelakis-Angelakis, Springer, London & New York, 2009, p. 565.

238. P. This, T. Lacombe and M. R. Thomas, *Trends Genet.*, 2006, **22**, 511.

239. J. F. Manen, L. Bouby, L. Dalnoki, P. Marinval, M. Turgay and A. Schlumbaum, *J. Archaeol. Sci.*, 2003, **30**, 721.

240. J.-F. Terral, E. Tabard, L. Bouby, Ivorra S, Pastor T, Figueiral I, Picq S, Chevance JB, Jung C, Fabre L, Tardy C, M. Compan, R. Bacilieri, T. Lacombe and P. This, *Ann. Bot.*, 2010, **105**, 443.

241. J.-P. Péros, G. Berger, A. Portemont, J.-M. Boursiquot and T. Lacombe, *J. Biogeogr.*, 2011, **38**, 471.

242. L. Riahi, N. Zoghlami, K. El-Heit, V. Laucou, L. Cunff, J. Boursiquot, T. Lacombe, A. Mliki, A. Ghorbel and This, P, *Genet. Resour. Crop Evol.*, 2010, **57**, 255.

243. E. Cappellini, M. R. Thomas, M. T. P. Gilbert, F. Geuna, G. Fiorentino, A. Hall, J. Thomas-Oates, P. D. Ashton, D. A. Ashford, P. Arthur, P. F. Campos, J. Kool, E. Willerslev, and M. J. Collins, *Naturwissenschaften*, 2010, **97**, 205.

244. C. Arnold, A. Schnitzler, A. Douard, R. Peter and F. Gillet, *Biodiversity and Conservation*, 2005, **14**, 1507.

245. M. Ben Slimane Harbi, H. Snoussi, R. Bouhlal and H. Nahdi, *Afr. J. Plant Sci. Biotechnol.*, 2010, **4**, 17.

246. R. Bacilieri, D. Maghradze, S. Grando, I. Pejic, E. Maul, G. Munoz, J. Eiras-Dias, A. Schneider, M. Boselli and P. This, Conservation, characterisation and management of grapevine genetic resources: The European Project Grapegen06, Oral Presentation OR.1.01, 33rd World Congress of Vine and Wine, Tblisi, Georgia, June 20–27, 2010. www.oiv2010 ge/index

247. I. S. Hornsey, *The Chemistry and Biology of Winemaking*, Royal Society of Chemistry, Cambridge, 2007.

248. U. Hendrick (ed.), *Sturtevant's Notes on Edible Plants*, Lyon, Albany, NY, 1919.

249. J. C. Loudon, *An Encyclopædia of Gardening, A New Edition*, Longman, Rees, Orme, Brown, Green and Longman, London, 1835.

250. S. S. Korban and H. Chen, in *Biotechnology of Perennial Fruit Crops*, ed. R. Litz and F. A. Hammerschlag, CAB International, Wallingford, UK, 1992, p. 203.

251. D. Potter, T. Eriksson, R. C. Evans, S. Oh, J. E. E. Smedmark, D. R. Morgan, M. Kerr, K. R. Robertson, M. Arsenault, T. A. Dickinson and C. S. Campbell, *Plant Syst. Evol.*, 2007, **266**, 5.

252. *The Plant List* (2010) Version 1. Published on the Internet; http://www.theplantlist.org/

253. S. A. Harris, J. P. Robinson and B. E. Juniper, *Trends Genet.*, 2002, **18**, 426.

254. C. S. Campbell, C. W. Greene and T. A. Dickinson, *Syst. Bot.*, 1991, **16**, 333.

255. D. J. Mabberley, C. E. Jarvis and B. E. Juniper, *Telopea*, 2001, **9**, 421.

256. B. E. Juniper and D. J. Mabberley, *The Story of the Apple*, Timber Press, Portland, Oregon, 2006.

257. J. L. Luby, in *Apples: Botany, Production and Uses*, ed. D. C. Ferree and I. J. Warrington, CABI Publishing, Cambridge, MA, 2003, p. 1.

258. S. S. Korban and R. M. Skirvin, *HortScience*, 1984, **19**, 177.

259. J. B. Phipps, K. R. Robertson, P. G. Smith and J. R. Rohrer, *Can. J. Bot.*, 1990, **68**, 2209.

260. R. Velasco, A. Zharkikh, J. Affourtit, A. Dhingra, A. Cestaro, A. Kalyanaraman, P. Fontana, S. K. Bhatnagar, M. Troggio, D. Pruss, S. Salvi, M. Pindo, P. Baldi, S. Castelletti, M. Cavaiuolo, G. Coppola, F. Costa, V. Cova, A. Dal Ri, V. Goremykin, M. Komjanc, S. Longhi, P. Magnago, G. Malacarne, M. Malnoy, D. Micheletti, M. Moretto, M. Perazzolli, A. Si-Ammour, S. Vezzulli, E. Zini, G. Eldredge, L. M. Fitzgerald, N. Gutin, J. Lanchbury, T. Macalma, J. T. Mitchell, J. Reid, B. Wardell, C. Kodira, Z. Chen, B. Desany, F. Niazi, M. Palmer, T. Koepke, D. Jiwan, S. Schaeffer, V. Krishnan, C. Wu, V. T. Chu, S. T. King, J. Vick, Q. Tao, A. Mraz, A. Stormo, K. Stormo, R. Bogden, D. Ederle, A. Stella, A. Vecchietti, M. M. Kater, S. Masiero, P. Lasserre, Y. Lespinasse, A. C. Allan, V. Bus, D. Chagné, R. N. Crowhurst, A. P. Gleave, E. Lavezzo, J. A. Fawcett, S. Proost, P. Rouzé, L. Sterck, S. Toppo, B. Lazzari, R. P. Hellens, C.-E. Durel, A. Gutin, R. E. Bumgarner, S. E.

Gardiner, M. Skolnick, M. Egholm, Y. Van de Peer, F. Salamini and R. Viola, *Nat. Genet.*, 2010, **42**, 833.

261. C. S. Campbell, R. C. Evans, D. R. Morgan, T. A. Dickinson and M. P. Arsenault, *Plant Syst. Evol.*, 2007, **266**, 119.

262. R. C. Evans and C. S. Campbell, *Am. J. Bot.*, 2002, **89**, 1478.

263. N. I. Vavilov, *International Horticultural Congress*, 1930, Group B, 271.

264. P. L. Forsline, E. E. Dickson and A. D. Dzhangaliev, *HortScience*, 1994, **29**, 433.

265. J. Janick, J. N. Cummins, S. K. Brown and M. Hemmat, in *Fruit Breeding*, ed. J. Janick and J. N. Moore, John Wiley & Sons, New York, 1996, vol. 1, *Tree and Tropical Fruits*.

266. C. Vila, J. A. Leonard, A. Götherström, S. Marklund, K. Sandberg, K. Lidén, R. K. Wayne and H. Ellegren, *Science*, 2001, **291**, 474.

267. M. Archetti, *Proc. Roy. Soc.*, *Series B*, 2009, **276**, 2575.

268. M. Archetti, *J. Theor. Biol.*, 2000, **205**, 625.

269. G. M. Volk, C. M. Richards, A. D. Henk, A. Reilley, D. D. Miller and P. L. Forsline, *HortScience*, 2009, **44**, 516.

270. Z.-Q. Zhou, *Genet. Resour. Crop Evol.*, 1999, **46**, 599.

271. P. L. Forsline, H. S. Aldwinckle, E. S. Dickson, J. J. Luby and S. C. Hokanson, *Horticultural Reviews*, 2003, **29**, 2.

272. B. Lion, *Mémoires de Nouvelles Assyriologies Brèves et Utilitaires*, 1992, **1**, 107.

273. K. Mudge, J. Janick, S. Schofield and E. E. Goldschmidt, *Horticultural Reviews*, 2009, **35**, 437.

274. R. Ellison, J. Renfrew, D. Brothwell and N. Seeley, *J. Archaeol. Sci.*, 1978, **5**, 167.

275. H. Helbaek, *Acta Archaeologica*, 1952, **23**, 108.

276. J. M. Renfrew, *Palaeoethnobotany*, Methuen, London, 1973.

277. I .M. Lonie, *The Hippocratic Treatises. "On Generation", "On the Nature of the Child", "Diseases", IV*, de Gruyter, Berlin, 1981.

278. A. M. Watson, *Agricultural Innovation in the Early Islamic World*, Cambridge University Press, Cambridge, 1983.

279. J. Evelyn, *Pomona*, John Martyn and James Allestry for the Royal Society, London, 1664.

280. J. Beale, *Herefordshire Orchards, a Pattern for All England*, Roger Daniel, London, 1657.

281. J. Worlidge, *Vinetum Britannicum*, Tho.Dring, and Tho.Burrel, London, 1676.
282. R. J. Garmer, *The Grafter's Handbook*, Oxford University Press, Oxford, 5th edn, 1988.
283. A. du Breuil, *The Scientific and Profitable Culture of Fruit Trees: Including Choice of Trees, Planting, Grafting*, Bibliolife Reproduction Series, 2009.
284. R. Rosenfeld, *The Gardener's Guide to Propagation*, Lorenz Books, London, 2011.
285. P.-E. Lauri, K. Maguylo and C. Trottier, *Am. J. Bot.*, 2006, **93**, 357.
286. A. C. Hulme, *J. Exp. Botany*, 1951, **2**, 298.
287. J. D. Phillips, A. Pollard and G. C. Whiting, *J. Sci. Food Agric.*, 1956, **7**, 31.
288. H. Maarse, *Volatile Compounds in Food and Beverages*, Marcel Dekker, New York, 1991.
289. P. S. Dimick and J. C. Hoskin, *Crit. Rev. Food Sci. Nut.*, 1983, **4**, 387.
290. N. M. M. Paillard, in *The Flavour of Fruits*, ed. I. D. Morton and A. J. MacLeod, Elsevier Scientific, Amsterdam, 1990, p. 1.
291. D. R. Rudell, D. S. Mattinson, J. P. Mattheis, S. G. Wyllie and J. K. Fellman, *J. Agric. Food Chem.*, 2002, **50**, 2627.
292. J. Dixon and E. W. Hewett, *New Zealand Journal of Crop and Horticultural Science*, 2000, **28**, 155.
293. K. Lee, Y. Kim, D. Kim, H. Lee and C. Lee, *J. Agri. Food Chem.*, 2003, **51**, 6516.
294. M. B. Davies, J. Austin and D. A. Partridge, *Vitamin C: Its Chemistry and Biochemistry*, Royal Society of Chemistry, Cambridge, 1991.
295. L. W. Mapson, in *The Biochemistry of Fruits and Their Products*, ed. A. C. Hulme, Academic Press, London, 1970, vol. 1, p. 369.
296. J. Boyer and R. H. Liu, *Nutr. J.*, 2004, **3**, 5.
297. K. Aso and K. Matsuda, *J. of Agr. Research*, 1951, **2**, 135.
298. D. Blanco Gomis, M. D. Gutierrez Alvarez, J. J. Mangas Alonso and A. Noval Vallina, *Chromatographia*, 1988, **25**, 701.
299. W. H. Loescher and J. D. Everard, in *Photoassimilate Distribution in Plants and Crops: Source-Sink Relationships*,

ed. E. Zamski and A. A. Schaffer, Marcel Dekker, New York, 1996, p. 185.

300. N. Noiraud, L. Maurousset and R. Lemoine, *Plant Physiol. Biochem.*, 2001, **39**, 717.
301. D. H. Lewis and D. C. Smith, *New Phytologist*, 1967, **66**, 143.
302. A. Moing, *Dev. Crop Sci.*, 2000, **26**, 337.
303. K. L. Webb and J. W. A. Burley, *Science*, 1962, **137**, 766.
304. K. Klages, H. Donnison, J. Wünsche and H. Boldingh, *Aust. J. Plant Physiol.*, 2001, **28**, 131.
305. C. A. Priestley, *J. Exp. Botany*, 1983, **34**, 1740.
306. R. L. Bieleski, *Aust. J. Plant Physiol.*, 1977, **4**, 11.
307. C. A. Priestley, *Ann. Bot.*, 1980, **46**, 77.
308. J. C. McQueen, P. E. H. Minchin, W. B. Silvester and T. G. A. Green, *Acta Horticulturae*, 2004, **636**, 267.
309. J. W. Patrick, *Annu. Rev. Plant Physiol. Plant Mol. Biol.*, 1997, **48**, 191.
310. J. W. Patrick and C. E. Offler, *J. Exp. Bot.*, 2004, **52**, 551.
311. R. L. Bieleski, in *Encyclopedia of Plant Physiology New Series*, ed. F. A. Loewus and W. Tanner, Springer-Verlag, Berlin, 1982, vol. 13A, *Plant Carbohydrates*, vol. 1, *Intracellular Carbohydrates*, p. 158.
312. M. Nozarzewski, A. M. Clements, A. B. Downie and D. D. Archbold, *Physiologia Plantarum*, 2004, **121**, 391.
313. S. Yamaki and K. Ishiwaka, *J. Amer. Soc. Hort. Sci.*, 1986, **111**, 134.
314. R. L. Bielski and R. J. Redgwell, *Aust. J. Plant Physiol.*, 1985, **12**, 657.
315. R. Zhou, R. Sicher and B. Quebedeaux, *Aust. J. Plant Physiol.*, 2001, **28**, 1143.
316. W. H. Loescher, G. C. Marlow and R. A. Kennedy, *Plant Physiol.*, 1982, **70**, 335.
317. J. Berüter, *J. Exp. Bot.*, 1993, **44**, 519.
318. J. Berüter and M. E. S. Feusi, *J. Plant Physiol.*, 1995, **146**, 95.
319. Z. Gao, L. Maurousset, R. Lemoine, S. D. Yoo, S. van Nocker and W. Loescher, *Plant Physiol.*, 2003, **131**, 1566.
320. J. Watari, Y. Kobae, S. Yamaki, K. Yamada, K. Toyofuku, T. Tabuchi and K. Shiratake, *Plant Cell Physiol.*, 2004, **45**, 1032.

321. R.-C. Fan, C.-C. Peng, Y.-H. Xu, X. F. Wang, Y. Li, Y. Shang, S. Y. Du, R. Zhao, X. Y. Zhang, L. Y. Zhang and D. P. Zhang, *Plant Physiol.*, 2009, **150**, 1880.
322. J. W. Riesmeier, L. Willmitzer and W. B. Frommer, *EMBO J.*, 1992, **11**, 4705.
323. N. Sauer and R. Stadler, *Plant J.*, 1993, **4**, 601.
324. A. C. Hulme and M. J. C. Rhodes, in *The Biochemistry of Fruits and Their Products*, ed. A. C. Hulme, Academic Press, London, 1971, vol. 2, p. 333.
325. A. G. Brown and D. M. Harvey, *Euphytica*, 1971, **20**, 68.
326. T. Visser, A. A. Schaap and D. P. de Vries, *Euphytica*, 1968, **17**, 153.
327. Y.-X. Yao, M. Li, H. Zhai, C.-X. You and Y.-J. Hao, *J. Plant Physiol.*, 2011, **168**, 474.
328. Y.-X. Yao, Q.-L. Dong, H. Zhai, C.-X. You and Y.-J. Hao, *Plant Physiol. Biochem.*, 2011, **49**, 257.
329. Y.-X. Yao, M. Li, C.-X. You, Z. Liu, D.-M. Wang and Y.-J. Hao, *Acta Horticulturae Sinica*, 2010, **37**, 1.
330. J. M. Bain and R. N. Robertson, *Aust. J. Sci. Res.*, 1951, **4**, 75.
331. A. N. Lakso, L. Grappadelli and M. C. Goffinet, *J. Hort. Sci.*, 1995, **70**, 389.
332. M. C. Goffinet, T. L. Robinson and A. N. Lakso, *J. Hort. Sci.*, 1995, **70**, 375.
333. A. N. Lakso, L. Grappadelli, J. Barnard and M. C. Goffinet, *Acta Horticulturae*, 1998, **466**, 13.
334. J. Berüter, *J. Exp. Bot.*, 1983, **34**, 737.
335. C. Zhang, K. Tanabe, F. Tamura, K. Matsumoto and A. Yoshida, *J. Exp. Bot.*, 2005, **56**, 2713.
336. S. W. Park, K. J. Song, M. Y. Kima, J.-H. Hwanga, Y. Uk Shina, W.-C. Kima and W.-II Chung, *Plant Sci.*, 2002, **162**, 513.
337. D. D. Archbold, M. Nozarzewski, A. M. Clements and A. B. Downie, *Acta Horticulturae*, 2004, **636**, 443.
338. L.-Y. Zhang, Y.-B. Peng, S. Pelleschi-Travier, Y. Fan, Y.-F. Lu, Y.-M. Lu, X.-P. Gao, Y.-Y. Shen, S. Delrot and D.-P. Zhang, *Plant Physiol.*, 2004, **135**, 574.
339. L. Cheng, R. Zhou, E. J. Reidel, T. D. Sharkey and A. M. Dandekar, *Planta*, 2005, **220**, 767.

340. G. Teo, Y. Suzuki, S. L. Uratsu, B. Lampinen, N. Ormonde, W. K. Hu, T. M. DeJong, and A. M. Dandekar, *Proc. Nat. Acad. Sci. USA*, 2006, **103**, 18842.
341. E. J. Reidel, E. A. Rennie, V. Amiard, L. Cheng and R. Turgeon, *Plant Physiol.*, 2009, **149**, 1601.
342. E. A. Rennie and R. Turgeon, *Proc. Nat. Acad. Sci. USA*, 2009, **106**, 14162.
343. S. Lalonde, D. Wipf and W. B. Frommer, *Ann. Rev. Plant Biol.*, 2004, **55**, 341.
344. K. Shiratake, *Genes, Genomes and Genomics*, 2007, **1**, 73.
345. J.-C. Jang and J. Sheen, *Plant Physiol.*, 1994, **6**, 1665.
346. A. Malladi and L. K. Johnson, *J. Exp. Bot.*, 2011, **62**, 205.
347. A. V. Forte, A. N. Ignatov, V. V. Ponomarenko, D. B. Dorokhov and N. I. Savelyev, *Russ. J. Genet.*, 2002, **38**, 1150.
348. E. Coart, S. Van Glabeke, M. De Loose, A. S. Larsen and I. Roldán-Ruiz, *Mol. Ecol.*, 2006, **15**, 2171.
349. M. Ng and M. F. Yanofsky, *Plant Cell*, 2001, **13**, 739.
350. H. Shan, L. Zahn, S. Guindon, P. K. Wall, H. Kong, H. Ma, C. W. dePamphilis and J. Leebens-Mack, *Mol. Biol. Evol.*, 2009, **26**, 2229.
351. B. J. Janssen, K. Thodey, R. J. Schaffer, R. Alba, L. Balakrishnan, R. Bishop, J. H. Bowen, R. N. Crowhurst, A. P. Gleave, S. Ledger, S. McArtney, F. B. Pichler, K. C. Snowden and S. Ward, *BMC Plant Biol.*, 2008, **8**, 16.
352. T. Jack, *Plant Cell*, 2004, **16**(Supplement), S1.
353. R. D. Newcomb, R. N. Crowhurst, A. P. Gleave, E. H. A. Rikkerink, A. C. Allan, L. L. Beuning, J. H. Bowen, E. Gera, K. R. Jamieson, B. J. Janssen, W. A. Laing, S. McArtney, B. Nain, G. S. Ross, K. C. Snowden, E. J. F. Souleyre, E. F. Walton and Y.-K. Yauk, *Plant Physiol.*, 2006, **141**, 147.
354. M. Höfer and A. Meister, *J. Bot.*, 2010, **2010**, article 480873, 8pp.
355. R. K. French, *The History and Virtues of Cyder*, Robert Hale, London, 1982.
356. M. Nelson, *The Barbarian's Beverage: A History of Beer in Ancient Europe*, Routledge, London, 2005.
357. F. A. Roach, *Cultivated Fruits in Britain – Their Origin and History*, Basil Blackwell, Oxford, 1985.
358. I. S. Hornsey, *A History of Beer and Brewing*, Royal Society of Chemistry, Cambridge, 2003.

359. S. Davies, in *Liquid Nourishment*, ed. C. A. Wilson, Edinburgh University Press, Edinburgh, 1993, p. 79.

360. W. Harrison, *Description of Britain*, first part of *Chronicles, The Historie of Englande, Scotlande, Irelande*, by Raphael Holinshed, John Hunne, London, 1577.

361. N. Kent, *A General View of the County of Norfolk*, George Nicol, London, 1796.

362. F. W. Beech, *J. Inst. Brew.*, 1972, **78**, 477.

363. J. Morgan and A. Richards, *The Book of Apples*, Brogdale Horticultural Trust in assoc. with the Ebury Press, London, 1993.

364. A. G. H. Lea and G. D. Ford, *Fruit Processing*, 1991, **1**, 29.

365. B. Jarvis, M. J. Forster and W. P. Kinsella, *J. Appl. Bacteriol. Symp. Suppl.*, 1995, **79**, 5S.

366. S. C. Marks, W. Mullen and A. Crozier, *J. Agric. Food Chem.*, 2007, **55**, 8723.

367. J. L. Toribio and J. E. Lozano, *Lebensm. Wiss. Technol.*, 1987, **20**, 59.

368. M. Revier (ed.), *Le Cidre – Hier et Aujord'hui*, La Nouvelle Librarie, Paris, 1987.

369. L. F. Burroughs, *J. Sci. Food Agric.*, 1957, **8**, 122.

370. L. F. Burroughs, in *The Biochemistry of Fruits and Their Products*, ed. A. C. Hulme, Academic Press, London, 1970, vol. 1, p. 119.

371. A. G. H. Lea and J. R. Piggott (ed.), *Fermented Beverage Production*, Kluwer Academic, New York, 2nd edn, 2003.

372. B. L. Wedzicha, *Chemistry of Sulphur Dioxide in Foods*, Elsevier Applied Science, Amsterdam, 1984.

373. F. W. Beech, in *The Yeasts*, ed. A. H. Rose and J. S. Harrison, Academic Press, London, 2nd edn, 1993, vol. 5, *Yeast Technology*, p. 163.

374. I. Masneuf, J. Hansen, C. Groth, J. Piškur and D. Dubourdieu, *Appl. Environ. Microbiol.*, 1998, **64**, 3887.

375. S. S. González, E. Barrio, J. Gafner and A. Querol, *FEMS Yeast Res.*, 2006, **6**, 1221.

376. S. Torriani, G. Zapparoli, P. Malacrino, G. Suzzi and F. Dellaglio, *Lett. Appl. Microbiol.*, 2004, **38**, 239.

377. J. G. Carr, A. Pollard, G. C. Whiting and A. H. Williams, *Biochem. J.*, 1957, **66**, 283.

378. T. Henick-Kling, *J. Appl. Bact. Symp. Suppl.*, 1995, **79**, 29S.

379. A. G. Salih, J. F. Drilleau, F. F. Cavin, C. Divies and C. M. Bourgeois, *J. Inst. Brew.*, 1988, **94**, 5.

380. J. G. Carr and P. A. Davies, *J. Appl. Bact.*, 1970, **33**, 768.

381. J. G. Carr, J. D. Phillips, A. Pollard, G. C. Whiting and A. H. Williams, *Chem. Ind.*, 1954, 1515.

382. J. G. Carr and P. A. Davies, *J. Appl. Bact.*, 1972, **35**, 463.

383. C. E. Arthurs and D. Lloyd, *Appl. Environ. Microbiol.*, 1999, **65**, 3360.

384. J. F. Bowen and F. W. Beech, *J. Appl. Bact.*, 1964, **27**, 333.

385. J. F. Bowen and F. W. Beech, *J. Appl. Bact.*, 1967, **30**, 475.

386. J. M. Laplace, S. Apery, J. Frère and Y. Auffray, *J. Inst. Brew.*, 1998, **104**, 71.

387. F. W. Beech, *J. Appl. Bact.*, 1958, **21**, 257.

388. J. G. Carr, *J. Appl. Bact.*, 1983, **55**, 383.

389. J. M. Laplace, A. Jacquet, I. Travers, J. P. Simon and Y. Auffray, *J. Inst. Brew.*, 2001, **107**, 227.

390. I. S. Pretorius, *Yeast*, 2000, **16**, 675.

391. M. Dueñas, A. Irastorza, C. Fernandez, A. Bilbao and A. Huerta, *J. Food Sci.*, 1994, **59**, 1060.

392. G. Del Campo, J. L. Santos, I. Berregi, S. Velasco, I. Ibarburu, M. T.Dueñas and A. Irastorza, *J. Inst. Brew.*, 2003, **109**, 342.

393. C. Cabranes, J. Moreno and J. J. Mangas, *Appl. Environ. Microbiol.*, 1990, **56**, 3881.

394. F. Ness, F. Lavellee, D. Dubourdieu, M. Aigle and L. Dulau, *J. Sci. Food Agric.*, 1993, **62**, 89.

395. D. Paffetti, C. Barberio, E. Casalone, D. Cavalieri, R. Fani, G. Fia, E. Mori and M. Polsinelli, *Microbiology*, 1995, **146**, 587.

396. B. Suárez Valles, R. Pando Bedriñana, A. González Garcia and A. Querol Simón, *J. Appl. Microbiol.*, 2007, **103**, 778.

397. L. Castellari and G. Pacchioli, *J. Food Sci.*, 1992, **3**, 179.

398. G. I. Naumov, *J. Ind. Appl. Microbiol.*, 1996, **17**, 295.

399. F. W. Beech and R. R. Davenport, in *The Yeasts*, ed. A. H. Rose & J. S. Harrison, Academic Press, New York & London, 1970, vol. 3, p. 73.

400. M. Combina, A. Elia, *et al.*, *Int. J. Food Microbiol.*, 2004, **99**, 237.

401. A. Picinelli, B. Suárez, J. Moreno, R. Rodríguez, L. M. Caso-García and J. J. Mangas, *J. Agric. Food Chem.*, 2000, **48**, 3997.
402. G. I. Naumov, H. V. Nguyen, E. S. Naumova, A. Michel, M. Aigle and C. Gaillardin, *Int. J. Food Microbiol.*, 2001, **65**, 163.
403. W. F. Morrissey, B. Davenport, A. Querol and A. D. W. Dobson, *J. Appl. Microbiol.*, 2004, **91**, 647.
404. J. Hutchinson, *The Families of Flowering Plants II. Monocotyledons*, Macmillan & Co. Ltd, London, 1934.
405. P. S. Nobel, *Remarkable Agaves and Cacti*, Oxford University Press, Oxford, 1994.
406. A. Cronquist, *An Integrated System of Classification of Flowering Plants*, Columbia University Press, New York, 1981.
407. R. M. T. Dahlgren, H. T. Clifford and P. F. Yeo, *The Families of Monocotyledons*, Springer-Verlag, Berlin, 1985.
408. D. J. Bogler and B. B. Simpson, *Syst. Bot.*, 1995, **20**, 191.
409. D. J. Bogler and B. B. Simpson, *Am. J. Bot.*, 1996, **83**, 1225.
410. Angiosperm Phylogeny Group II, *Bot. J. Linn. Soc.*, 2003, **141**, 399.
411. Angiosperm Phylogeny Group III, *Bot. J. Linn. Soc.*, 2009, **161**, 105.
412. S. V. Good-Avila, V. Souza, B. S. Gaut and L. E. Eguiarte, *Proc. Nat. Acad. Sci. USA*, 2006, **103**, 9124.
413. P. Lappe-Oliveras, R. Moreno-Terrazas, J. Arrizón-Gaviño, T. Herrera-Suárez, A. García-Mendoza and A. Gschaedler-Mathis, *FEMS Yeast Res.*, 2008, **8**, 1037.
414. C. E. Smith, Jr., *Fieldiana*, 1965, **31**, 55.
415. K. V. Flannery (ed.), *Guilá Naquitz: Archaic Foraging and Early Agriculture in Oaxaca, Mexico* (Studies in Archaeology), Academic Press, Orlando, FL, 1986.
416. A. Sánchez-Marroquín and P. H. Hope, *J. Agric. Food Chem.*, 1953, **1**, 246.
417. K. H. Steinkraus (ed.), *Handbook of Indigenous Fermented Foods*, Marcel Dekker, New York, 2nd edn, 1996.
418. J. Swings and J. DeLey, *Bact. Rev.*, 1977, **41**, 1.
419. J. F. Ramirez, A. Sánchez-Marroquín, M. M. Álvarez and R. Valyasebi, in *Industrialization of Indigenous Fermented Foods*,

ed. K. Steinkraus, Marcel Dekker, New York, 2nd edn, 2004, p. 547.

420. B. S. Platt, *Food Tech.*, 1964, **18**, 662.

421. R. I. Ortiz-Basurto, G. Pourcelly, T. Doco, P. Williams, M. Dornier and M. P. Belleville, *J. Agric. Food Chem.*, 2008, **56**, 3682.

422. A. Escalante, M. E. Rodriguez, A. Martinez, A. López-Munguía, F. Bolívar and G. Gosset, *FEMS Microbiol. Lett.*, 2004, **235**, 273.

423. M.-A. Lachance, *Antonie van Leeuwenhoek*, 1995, **68**, 151.

424. J. D. Sauer, *Historical Geography of Crop Plants*, CRC Press, Boca Raton, FL, 1993.

425. D. J. Rogers and S. G. Appan, Manihot manihotoides (Euphorbiaceae), *Flora Neotropica*, Monograph 13, Hafner Press, New York, 1973.

426. A. C. Allem, *Euphytica*, 1999, **107**, 123.

427. A. C. Allem, in *Report of the First Meeting of the International Network for Cassava Genetic Resources*, ed. W. M. Roca and A. M. Thro, Centro Internacional de Agricultura Tropical (CIAT), Cali, Colombia, 1992, p. 87.

428. K. M. Olsen, *Plant Mol. Biol.*, 2004, **56**, 517.

429. B. Pujol, G. Gigot, G. Laurent, M. Pinheiro-Kluppel, M. Elias, H. M. McKey and D. McKey, *Econ. Bot.*, 2002, **56**, 366.

430. R. Beier, *Rev. Environ. Contam. Toxicol.*, 1990, **113**, 47.

431. D. Siritunga and R. Sayre, *Plant Mol. Biol.*, 2004, **56**, 661.

432. W. L. B. White, D. I. Arias-Garzon, J. M. McMahon and R. T. Sayre, *Plant Physiol.*, 1998, **116**, 1219.

433. G. M. O'Brien, A. J. Taylor and N. H. Poulter, *J. Sci. Food Agric.*, 1991, **56**, 277.

434. H. Rosling, in *The Impact of Plant Molecular Genetics*, ed. W. B. S. Sobral, Birkhaeuser Boston, New York, 1996, p. 315.

435. A. von Humboldt, *Political Essay on the Kingdom of New Spain*, 4 volumes, Longman, Hurst, Rees, Orme and Brown, London, 1811.

436. N. J. Taylor, P. Chavarriaga, K. Raemakers, D. Siritunga and P. Zhang, *Plant Mol. Biol.*, 2004, **56**, 671.

437. D. Leihner, in *Cassava: Biology, Production, and Utilization*, ed. R. J. Hillock, J. M. Thresh and A. Belotti, CAB International, Wallingford, 2002, p. 91.

438. D. E. Briggs, *Malts and Malting*, Blackie, London, 1998.
439. P. D. Sheets, *The Cerén Site: An Ancient Village Buried by Volcanic Ash in Central America*, Case Studies in Archaeology, Cengage Learning, Florence, KY, 2nd edn, 2005.
440. L. Mowat, *Cassava and Chicha*, Shire Publications, Princes Risborough, 1989.
441. H. C. Cutler and M. Cárdenas, *Botanical Museum Leaflets* (Harvard University), 1947, **13**, 33.
442. W. La Barre, *Am. Anthropol.*, 1938, **40**, 224.
443. M. Goodman-Elgar, in *Drink, Power, and Society in the Andes*, ed. J. Jennings and B. J. Bowser, University Press of Florida, Gainesville, FL, 2009, p. 75.
444. J. Moore, *Am. Anthropol.*, 1989, **91**, 682.
445. M. Moseley, D. J. Nash, P. R. Williams, , S. D. deFrance, A. Miranda and M. Ruales, *Proc. Nat. Acad. Sci. USA*, 2005, **102**, 17264.
446. J. V. Murra, in *Culture and History: Essays in Honour of Paul Radin*, ed. S. Diamond, Columbia University Press, New York, p. 393.
447. C. A. Hastorf and S. Johannessen, *Am. Anthropol.*, 1993, **95**, 115.
448. R. F. Schwan, E. G. Almeida, M. A. Souza-Dias and L. Jespersen, *FEMS Yeast Res.*, 2007, **7**, 966.
449. M. J. R. Nout and P. K. Sarkar, *Antonie van Leewenhoek*, 1999, **76**, 395.
450. F. M. Hayashida, *J. Anthropol. Archaeol.*, 2008, **27**, 161.
451. M. Weismantel, in *Drink, Power, and Society in the Andes*, ed. J. Jennings and B. J. Bowser, University Press of Florida, Gainesville, FL, 2009, p. 257.
452. I. Shimada, *Pampa Grande and the Mochica Culture*, University of Texas Press, Austin, TX, 1994.
453. J. Jennings, in *Foundations of Power in the Prehispanic Andes*, ed. C. Conlee, D. Ogburn and K. Vaughn, American Anthropological Association, Washington, DC, 2005, p. 241.
454. L. Bertonio, *Vocabulario de la Lengua Ayamara*, F. del Canto, Chucuito, Peru, 1612.
455. H. C. Harries, *Bot. Rev.*, 1978, **44**, 265.
456. H. C. Harries, *Ann. Bot.*, 1981, **48**, 873.
457. V. H. W. Dowson, *Trop. Agr.*, 1953, **34**, 295.

458. O. Bassir, *West Afr. J. Biol. Appl. Chem.*, 1968, **10**, 42.
459. G. K. S. Ayernor and J. S. Matthews, *Trop. Sci.*, 1971, **13**, 71.
460. P. Tuley, *J. Nigerian Inst. Oil Palm Res.*, 1965, **4**, 283.
461. P. Simonart and H. Laudelout, *Inst. R. Colon. Belge Bull. Seances*, 1951, **22**, 385.
462. O. Bassir, *W. Afr. J. Appl. Sci.*, 1967, **10**, 42.
463. A. U. Ogan, *Phytochemistry*, 1971, **10**, 2832.
464. S. I. Faparusi, *J. Appl. Bacteriol.*, 1973, **36**, 559.
465. M. O. Eze and A. U. Ogan, *Plant Foods Hum. Nutr.*, 1988, **38**, 121.
466. I. E. Ezeagu, M. A. Fafunso and F. E. Ejesie, *Ecol. Food Nutr.*, 2003, **42**, 213.
467. W. K. Amoa-Awua, E. Sampson and K. Tano-Debrah, *J. Appl. Microbiol.*, 2007, **102**, 599.
468. O. Obire, *J. Appl. Sci. Environ. Manag.*, 2005, **9**, 25.
469. O. Lasekan, A. Buettner and M. Christlbauer, *Food Chem.*, 2007, **105**, 15.
470. M. Stringini, F. Comitini, M. Taccari and M. Ciani, *Food Microbiol.*, 2009, **26**, 415.
471. E. E. T. Bechem, C. Omoloko, D. Nwaga and V. P. K. Titanji, *Afr. J. Biotechnol.*, 2007, **6**, 1715.
472. P. Tuley, *Niger. Field*, 1964, **30**, 120.
473. K. H. Steinkraus, in *Fermented Food Beverages in Nutrition*, ed. C. F. Gastineau, W. J. Darby and T. B. Turner, Academic Press, New York, 1979, p. 35.
474. S. I. Faparusi, *Food Chem.*, 1981, **7**, 81.
475. A. A. Rokosu and J. Nwisienyi, *Enzym. Microb. Tech.*, 1980, **2**, 63.
476. A. K. Olawale, A. A. Olubiyi and D. O. Moses, *New York Science J.*, 2010, **3**, 35.
477. J. Lodder (ed.), *The Yeasts: A Taxonomic Study*, North Holland Publishing Co., Amsterdam, 1970.
478. T. E. Ogbulie, J. N. Ogbulie and H. O. Njoku, *Afr. J. Tech.*, 2007, **6**, 914.
479. N. Okafor, *J. Sci. Food Agr.*, 1972, **23**, 1399.
480. L. R. Batra and P. D. Millner, *Mycologia*, 1974, **66**, 942.
481. Z. Merican, Malaysian coconut palm toddy, Symposium on Indigenous Fermented Foods Bangkok, Thailand, 1977.
482. J. D. Atputharajah, S. Widanapathirana and U. Samarajeewa, *Food Microbiol.*, 1986, **3**, 273.

483. J. D. Atputharajah, U. Samarajeewa and S. Widanapathirana, *J. Food Sci. Tech.*, 1986, **23**, 5.
484. D. G. N. G. Wijesinge and U. Samarajeewa, *Food Microbiol.*, 1988, **5**, 119.
485. T. R. Shamala and K. R. Sreekantiah, *Food Microbiol.*, 1988, **5**, 157.
486. S. S. Levi and C. B. Oruche, *The Preservation and Bottling of Palm Wine*, Research Report No. 1 Federal Ministry of Industry, Lagos, 1957.
487. E. Chinarasa, *The Preservation and Bottling of Palm Wine*, Research Report No. 38, Federal Ministry of Industry, Lagos, 1968.
488. N. Okafor, *J. Appl. Bact.*, 1975, **38**, 1.
489. N. Okafor, *J. Appl. Bact.*, 1977, **43**, 159.
490. H. A. Esechie, *Niger. Agr. J.*, 1978, **15**, 158.
491. F. N. I. Morah, *J. Sci. Educ. (Nigeria)*, 1986, **2**, 100.
492. F. N. I. Morah, *Food Chem.*, 1995, **53**, 153.
493. F. I. Obahiagbon and P. Oviasogie, *Am. J. Food Tech.*, 2007, **2**, 323.
494. F. I. Obahiagbon, *Afr. J. Food Sci.*, 2009, **3**, 447.
495. S. I. Faparusi and O. Bassir, *Appl. Microbiol.*, 1972, **24**, 853.
496. S. Rastogi and A. K. S. Rawat, *Herba Polonica*, 2008, **54**, 66.
497. S. L. da Silva, V. G. de Oliveira, T. Yano and R. de C. Nunomura, *Acta Amazonica*, 2009, **39**, 187.
498. D. E. U. Ekong and C. Ejike, *J. West African Sci. Assn*, 1974, **19**, 63.
499. H. C. C. Maduka, H. M. Uhwache and Z. S. C. Okoye, *Pakistan J. Biol. Sci.*, 2003, **6**, 202.
500. H. C. C. Maduka and Z. S. C. Okoye, *Int. J. Toxicol.*, 2006, **3**, 1.
501. M. E. Okhun, N. P. Okolie and A. O. Oyerinde, *Afr. J. Biotechnol.*, 2005, **4**, 829.
502. P. O. Erah, C. N. Akujieze and G. E. Oteze, *Trop. J. Pharm. Res.*, 2002, **1**, 75.
503. I. N. Nwachukwa, V. I. Ibekwe, R. N. Nwabueze and B. N. Anyanwu, *Afr. J. Biotechnol.*, 2006, **5**, 1725.
504. N. Uraih, Y. S. Izuagbe and P. Ivbievbiokun, *Appl. Microbiol. Biotechnol.*, 1985, **23**, 77.
505. R. J. Forbes, *Studies in Ancient Technology, Volume 5*, E. J. Brill, Leiden, 1957.

CHAPTER 6

Anthropological and Archaeological Aspects

6.1 THE DRUNKEN MONKEY

Scientists view ethanol in a different light from that of anthropologists, and the latter's attention has focused largely on health and nutrition, the generation and reinforcement of individual and group identities and rituals related to religion and status.[1] In an anthropological context, fermentation can be conveniently viewed as controlled spoilage of food. The development of anaerobic conditions and the presence of ethanol prevent the growth of microbes that might otherwise oxidize a food substrate and, as food scientists will tell you, fermentations were used long before refrigeration and chemical preservatives came along. The corpus of anthropological and ethnological work published over the past couple of decades has clearly shown that the production and consumption of alcohol are central to the creation of identity, the instigation and maintenance of power, the functioning of social networks and the practice of religions.[2,3]

Before embarking upon an appraisal of some of the possible reasons for the importance of alcohol to mankind, it should be made clear that yeasts, the principal microbes responsible for alcoholic fermentation, are all around us in the biosphere and need no invitation to take up glucose (or any other fermentable sugar) and set in motion ethanologenic metabolism. The competition for

Alcohol and its Role in the Evolution of Human Society
Ian S. Hornsey
© Ian S. Hornsey 2012
Published by the Royal Society of Chemistry, www.rsc.org

fermentable sugars in nature is immense, and mechanisms by which one organism might gain an advantage over another for the "alcoholic prize" almost certainly arose during the Cretaceous period when angiosperm fleshy fruits evolved. The adaptive radiation and eventual dominance of angiosperms during the Cretaceous opened up a whole new world of dietary opportunities for animals, which could now forage on leaves, flowers, pollen, nectar, seeds and fruits.

There is no reason to suppose that ethanol was anything but relatively plentiful in the environment during the early evolution of mankind and it is likely that the penchant of modern man for enthusiastic alcohol consumption may derive from our fruit-eating ancestors – what Robert Dudley has called an "evolutionary hangover".

The "Drunken Monkey Hypothesis", a stimulating, and somewhat controversial, proposition formulated by Dudley early in the millennium[4,5] and fully discussed in a joint paper with Dustin Stephens in 2004,[6] is a convenient topic with which to begin this chapter. The pair suggested that alcoholism in humans can be traced back to primate evolutionary history, reasoning that, because ethanol is a characteristic of ripe fruit, frugivorous primates evolved mechanisms for tolerating the compound in their diets. Goldman *et al.*[7] showed that ethanol use (and abuse) by humans can be partly linked to genetically inheritable traits. Most current theories on alcoholism postulate that man and his ancestors were either used to being without alcohol at all or, at most, receiving low doses from fermented fruits – before man commenced brewing and winemaking. Availability of ethanol at concentrations above those attainable solely by yeast activity (*i.e.* distillation) is, of course, a very recent event in terms of human evolution.

Aware of the genetic connection, Dudley realized that alcohol present in ripening fruits provided a cue to ancient frugivores that valuable nutrients were at hand, and that these foraging animals would have been exposed to alcohol for a long period of time. The levels of ethanol generally available were/are too low to permit consuming animals to receive harmful doses of the compound and genetic traits for increased alcohol intake became positively selected for. There was no similar selection pressure working on genes protecting against the harmful effects of ethanol. When man started to produce alcoholic beverages (say 9000–10,000 years ago), the levels of ethanol that they contained would have been harmful (intoxicating). In theory, this could have caused a reversal of selection pressures, but the period of time that has elapsed is too short to induce sufficient evolutionary responses.

As Katharine Milton[8] says, although primates are viewed as omnivorous, "they are omnivores of a very particular type in that the great majority of their foods each day come from plants". Much evidence suggests that contemporary primate diets are predominantly plant-based and, in many, fruits form an important part of the plant material consumed. This is nothing new, for primate diets have included substantial amounts of fruit for over 40 million years and, in tropical regions in particular, yeast activity results in the formation of ethanol, which can accumulate to detectable levels. In species of the palm genus, *Astrocaryum* (particularly *A. standleyanum*), for example, unripe fruit is devoid of ethanol, but when ripe and still attached to the tree levels of 0.6% ethanol by weight have been observed. When fruit becomes overripe, and often falls to the ground, levels of 4.0% can be attained [the howler monkey that Stephens observed on Barro Colorado Island was feasting on fruit near to peak ripeness – with ethanol content at 1.0%].

Sugars naturally present in fruit pulp act as a nutritional reward for frugivorous animals, which will ultimately disperse the seeds within that fruit and reproduce the species. The main agents of seed dispersal are birds and mammals and they are generally attracted to fruits only when seeds are ripe, since attracting them to immature seed accomplishes nothing in terms of reproduction. Plants use a variety of chemical and physical defence mechanisms to prevent premature consumption of fruit by potentially dispersing species and, upon ripening, these mechanisms cease to operate and fruit provides a completely different milieu. Unfortunately, this means that fruits are most susceptible to microbial attack once they have attained ripeness, and the more time that elapses before they are consumed by frugivores, the more vulnerable they become. Fruit ripening involves a number of biochemical processes including pulp texture softening, colour change, conversion of starch to sugars and production of attractive volatile compounds.[9,10]

There is evidence that decaying fruit discourages uptake by frugivorous vertebrates[11] and, as Janzen[12] made us aware, fruit decomposition can be regarded as a race between microbes and dispersal agents to gain nutritional advantage. Fruit-bearing plants often resort to the production of antifungal agents[11,13–15] when fruits ripen in order to retard mycologically induced decay (and fermentation), and thus tilt the "race" in favour of dispersal agents. Through their potential effects on frugivores, the anti-fungal defences of fruits have implications for fruit dispersal[13,16] and the study of fruit-fungi interactions might be of great use for studying fruit-frugivore interactions. The antifungal compounds in fruit are secondary metabolites and, according to Cipollini and Stiles,[11] their primary function is to protect against "destructive fungi that compete with seed dispersers for the pulp". In some members of the Solanaceae, competing fungi are inhibited by two glycoalkaloids (α-solasoline and α-solamargine), but these compounds are also somewhat toxic to (wanted) seed-dispersing vertebrates.[15,17] As one would expect, as a rule of thumb, all secondary metabolites are more toxic to ravaging fungi than to seed-dispersing animals. Jordano[18] has written an interesting chapter on fruits and frugivory, and Fenner and Thompson[19] have documented the ecology of seeds.

Championing the microbial world, Janzen[12] has even suggested (in the absence of much substantiating evidence) that ethanologenesis by fruit yeasts is part of a strategy that has evolved in order to make ripe fruit unattractive to vertebrates; as he says: "fruits rot, seeds mold, and meat spoils because that is the way microbes compete with bigger organisms." It is certainly the case that when an animal eats spoiled or rotten food, it has a fair chance of being "upset" by toxins or other microbially produced compounds (including antibiotics) and being infected by microbes.

Bearing in mind the toxicity of ethanol to many microbes, ethanologenesis by one organism is certainly a means of controlling the microflora of fruit. Auto-fermentation of fruit will naturally be more pronounced in warm and humid environments, of which tropical forests, a happy hunting ground for frugivorous primates, are a prime example. Associated with fermentation, ripe fruits emit odour plumes that contain other volatile, low-molecular-weight compounds apart from ethanol. These plumes serve notice that ripe fruit is available and can be detected by animals over long distances.

Because of the fierce competition for ripe fruit, a scarce commodity in tropical forests, some frugivores have evolved highly efficient foraging schemes in order to satisfy their nutritional requirements, and Katharine Milton is of the opinion that arboreal primates developed spatial memory and cognitive ability to aid foraging efficiency.[8,20,21] Some primate species are still heavily dependent on fruit and, as Milton has shown, howler monkeys (*Alouatta palliata*) and capuchin monkeys (*Cebus capucinus*) living on Barro Colorado Island, Panama, regularly use over 40% of fruit in their annual diets.[22] Primates evolved in the tropical forest canopy, and it is postulated that the strategies that the early primates adopted to cope with life in that environment influenced the way in which they, particularly the anthropoids, evolved. As Milton[8] put it, "the evolutionary trajectory of hominoids is intimately bound up with the exploitation of ripe, fleshy fruits." Thus, natural selection favoured traits that enhanced the efficiency of canopy foraging, such as the ability to move deftly in trees, and enhanced visual (including colour vision for fruit detection) and odour-detecting apparatus. Dominy *et al.*[22] have made a contribution to our knowledge of the evolution of primate colour vision. Adaptive radiation of primates, especially the higher primates (anthropoids – monkeys and apes), was attributable to their ability to fill vacant niches in the canopy and to dominate the energy-rich foodstuffs found there.[20]

While the tropical forest canopy is relatively rich in leaves, flowers and fruit, arboreal life has its own problems, mainly because most of the foods available are inadequate in one way or another. For example, fruits are high in readily available calories, but have only moderate fibre content and low protein levels. In addition, they are not generally available all year round. Mature leaves, on the other hand, are low in calories, and have moderate protein and high fibre. Clearly, the dietary situation had to change before all evolutionary pathways could be explored and, by the time *Australopithecus*, the first genus in our family, emerged in Africa during the Pliocene, it is clear from their huge molar teeth that their diet consisted of much more substantial ("tougher") plant material. The *Australopithecus* brain was not much larger than today's apes, but the animal was now bipedal, and no longer an arboreal dweller. With climate changes toward the end of the Pliocene and during the next epoch, the Pleistocene, tropical forest acreage shrank and the habitats were largely replaced by savanna woodlands with fewer tree species.

Australopithecus became extinct during the Pleistocene, but the genus *Homo* had evolved during the Pliocene, the first species being *H. habilis*, which was similar in size but had an appreciably larger brain. *H. habilis* became replaced by an even larger-brained species, *H. erectus*, during the Pliocene, and this species was to be ousted by *H. sapiens* during the Pleistocene. Our species had an even greater brain size, and this was linked to an increased stature and a decrease in the size of molar and premolar teeth – the latter reflecting diet (teeth have increasingly "less work to do").

Levey,[24] while agreeing with some of Dudley's arguments, wonders exactly how man became exposed to sufficiently high concentrations of ethanol to allow frequent/continual drunkenness, and he argues that the regular consumption of ethanol-rich fruits by our ancestors is not a prerequisite for excessive alcohol consumption by *Homo sapiens*. He reckons that humans "discovered" alcohol while they were using fermentation as a means of preserving their food.

6.1.1 Food Preferences

In relation to the above, Galef[25] has reported that human beings show scant evidence of nutritional wisdom, and he questioned the suggestion that relatively well-fed omnivorous animals are able to consistently choose a nutritionally adequate diet from an array of foodstuffs. Until relatively recently, most human communities consumed age-old diets that emanated from their ancestors, and the raw materials for them could be gathered close by. Today, in industrialized urban societies, people are often far removed from their sources of food and, therefore, out of touch with ancient dietary tradition.

Until the contribution by Berbesque and Marlowe,[26] no studies of systematically collected data existed on human forager food preferences. They studied the Hadza of Tanzania, full-time foragers in an area where the hominid record stretches back some 3–4 million years. Where foraging is concerned, there is a known sexual division of labour, but this work showed that there were sexual differences in food preference as well.

According to Monica Smith,[27] "food preference" is a socially constructed concept in which both consumers and producers define what is "good to eat". She believes that, while the food choices of today are conditioned by the capitalist or nationalist goals of food purveyors, what actually gets eaten is the result of individual decisions made within a complex social context. In the context of humans, food preference predates our early complex societies, and was an extremely important factor in deciding which wild plants were to participate in domestication at the end of the Holocene. Whether domestication was a "Neolithic Revolution", as Childe[28] thought, or whether the process was rather more a series of gradual transitions, humans had to make conscious decisions (edibility, ease of harvest, storability, *etc.*) about which plants to manipulate. The plant's propensity for being able to provide a substrate for alcoholic fermentation seems to have been one of the more important characters to have come under consideration. For example, in the pre-Colombian Andes, *chicha* (see p. 484) could be made from a variety of plants, but that made from maize (corn) was particularly highly regarded and was probably one of the main reasons why an uninteresting plant like teosinte was persevered with and managed to end up as corn. A similar situation might be claimed for barley domestication in the ancient Near East.[29]

Sclafani[30] has written about the psychobiology of food preferences and the role played by experimentation with animals in obtaining information about the basic psychobiological determinants of food preference. Of all food flavour elements (taste, texture, *etc.*) the sweet taste of sugar has an inherent attraction to many animals, including humans. Rats able to access sucrose (or glucose) as well as their laboratory food will consume 60% of their total calories as sugar and increase their energy consumption by around 20%, and this leads to obesity. When fed maltodextrin solution (bland-tasting to humans), these produced as much overeating in rate as did the simple sugars. This suggested that taste and palatability were not important determinants of carbohydrate choice, and that rats, unlike humans, are very attracted to the taste of maltodextrins (which are rapidly digested to glucose). One of the main findings of the work was that food preferences result from complex interactions between the orosensory and nutritional properties of food. Thus, great care should be taken when assessing the reasons for an animal choosing one particular foodstuff over another – and this applies to assessments about the possible reasons for the diet choices of our ancestors. Rozin[31] has written a readable contribution on psychobiological perspectives of food preferences and avoidances, and Garber[32] has written about the foraging strategies of living primates.

6.1.2 ... But a Drunken Shrew?

When considering human alcohol use and the genetic traits associated with it, one should realize that it has still not been unequivocally proven whether these have been shaped by natural selection directly connected to ethanol intake.[33] In an attempt to throw light on the subject, Wiens *et al.*[34] suggested that "moderate to high human alcohol consumption could be evolutionarily maintained by positive effects on evolutionary fitness that outweigh any associated health risk in most current settings, despite negative net effects in some others". For supporting evidence they turned to extant non-primate animals that naturally consume ethanol.

While there are several reports of wild animals (including mammals) exhibiting signs of intoxication as a result of activities in the natural environment, evidence for daily consumption of inebriants, as seen in some humans, is lacking. Weins's team studied pen-tailed tree-shrews (*Ptiloceros lowii*) and six other mammals that consume alcoholic nectar daily from the bertam palm (*Eugeissona tristis*), a stemless plant that grows in dense clusters in many rainforests of western Malaysia. With its unique floral structure (the large inflorescences contain around 1000 flowers; half being male, half hermaphrodite) and mode of pollination, the palm produces nectar from its hermaphrodite flower buds that harbour a community of yeasts, some of which are fermentative. During nectar production, the inflorescences produce a strong, "brewery-like" smell, which acts as a cue for foragers. The flowers in an inflorescence develop synchronously and fermented nectar, which has been called "nectar beer", is produced for a period of around six weeks whilst in the bud stage (before stigma and pollen are exposed), thus serving to attract potential pollinators. Once pollen is mature nectar production ceases, but because the palm (unusually) flowers almost throughout the year there is nearly always fermented nectar available in the vicinity of a cluster of plants.

The maximum alcohol content recorded from the palm nectar is 3.8% ABV (mean: 0.6%; median 0.5%), the highest level reported from a natural food, and the tree-shrews regularly consume alcohol doses that would intoxicate humans and be positively dangerous to other mammals. Analysis of an ethanol metabolite, ethyl glucoronide, in *P. lowii* hair samples indicated imbibed alcohol concentrations far higher (bodyweight for bodyweight) than humans with high alcohol intakes but, amazingly, these animals did not show any co-ordination problems, or other signs of being drunk.

Interestingly, *P. lowii* may be ecologically and behaviourally close to extinct species, ancestral to primates, that lived around 55 Mya.[35,36] Weins[34] believes that there are positive psychological effects of the tree-shrews' alcohol consumption, and that the exposure to higher than anticipated (even harmful) ethanol levels was probably significant early on in the evolution of several mammalian lineages, tree-shrews and primates included. From this it would seem that the trait of alcohol consumption has been actively maintained during evolution – therefore there must be some benefit. Add to this the fact that one of the most convincing pieces of evidence presented in favour of the "evolutionary hangover" is the fact that over 10% of the total soluble protein in the human liver is in the form of alcohol dehydrogenase, and it seems as though Dudley's drunken monkey may have lived after all! The past ancestral history of humans as fruit-eating primates is not in doubt; what is debatable is that they possess an evolutionarily based affinity for ethanol.

With primates being a predominantly frugivorous lineage, Dudley's hypothesis stimulated much thought, and in 2004 a symposium entitled "*In Vino Veritas*" was held in New Orleans,[37] at which an attempt was made to place alcohol consumption within historical and comparative contexts. The historical origins of human alcoholism were discussed by Dudley,[38] who noted that addictions in general, and alcoholism in particular, have usually been viewed as novel afflictions devoid of evolutionary context. Wolf and Heberlein[39] have shown that behavioural responses to ethanol and molecular pathways of inebriation are shared among many species, and in the symposium Heberlein *et al.*[40] outlined the molecular genetic analysis of ethanol intoxication of the fruit fly (*Drosophila melanogaster*). The fruit fly is one of the most thoroughly studied of organisms and has provided insights into cellular and developmental processes that are conserved in mammals, including man. The genome sequence of *D. melanogaster*[41] revealed a high level of molecular similarity between flies and mammals (see also Rubin *et al.*[42]). Genetic screens for mutants with altered responsiveness to ethanol have been made and some of the disrupted genes have been identified, and in their own words Heberlein *et al.*[40] said: "The availability of powerful tools for genetic manipulation in *Drosophila*, together with the high degree of conservation at the genomic level, make *Drosophila* a promising model organism to study the mechanism by which ethanol regulates behaviour and the mechanisms underlying the organism's adaptation to long-term ethanol exposure."

It is known that fruit flies collected in brewery and winery environments often show higher ethanol tolerance than specimens collected nearby,[43] which indicates localized adaptation to high ethanol levels. On a larger scale, *D. melanogaster* populations from high latitudes (northern or southern hemisphere) show greater ethanol tolerance than tropical populations.[44] Although this species is largely ethanol-tolerant and attracted to the substance, the sibling species *D. simulans* (from which it diverged within the last 3 million years[45]) is attracted more moderately to ethanol and its responses to it are not uniform with *D. melanogaster* (*e.g. D. simulans* populations show no high latitude/tropical differences). The adaptation of *D. melanogaster* to high levels of ethanol on differing geographic scales provides us with a very useful system for studying the genetics of ecophysiological adaptation.

The contribution of Sanchez *et al.*[46] showed how yeasts, thanks to their ethanol-producing ability, can influence the interaction between frugivores and fleshy-fruited plants. Using Egyptian fruit bats (*Rousettus aegyptiacus*), they found that moderate alcohol consumption (*i.e.* at the levels close to those found in ripe fruit) may have beneficial short- and long-term effects for the dispersing animal. Ethanol may enable the animals to assess the quality of fruit, and the team suggested that it may serve as an appetitive stimulant. The possibility that ethanol may be used as an energy source by frugivorous vertebrates was also discussed. Overripe fruit, with an ethanol content of over 1% (v/v), was avoided by the bats, which would reduce their food consumption if continually presented with such fruit. Thus, if ethanol levels are below 1% (v/v) the compound will act as a cue and feeding will ensue. This will aid seed dispersal. Conversely, with ethanol over 1% (v/v) feeding (and seed dispersal) will decrease. Dominy's contribution[47] indicated that ethanol has played "a significant and underestimated role in the regulation of primate foraging behaviour".

6.2   SOME EARLY WORK ON THE ANTHROPOLOGY AND ETHNOLOGY OF ALCOHOL STUDIES

Although humans have experimented with a myriad of psychoactive substances, alcohol is the most widely used agent in this category used by man; nothing has been used for longer or more widely for mind-altering purposes, and complex and diverse models of belief and behaviour. With the passage of time and the adoption of various "fad" substances, alcoholic beverages remain the best known and most widely used means of altering human consciousness. Since ancient times, alcohol has been widely valued for defining societal roles and for rituals and, with the exception of most of North America and Oceania (until contact with European settlers), peoples from most parts of the world were familiar with some kind of alcoholic drink.

The psychopharmacological properties of ethanol have persuaded humans throughout the world to encircle alcoholic beverages with a plethora of diverse rules and regulations. So embedded was alcohol in the culture of these peoples that protocols for its use and proscription are usually documented in fine detail, and there can be huge variations in cultural use of the substance – from enthusiastic imbibing to total abstinence. A classic example of the former attitude is exemplified by the Kofyar of northern Nigeria who "make, drink, talk, and think about beer ... they certainly believe that man's way to god is with beer in hand".[48] Compare this to the Hopi and other Pueblo Indian tribes of the American southwest who see drinking alcohol as a threat to their way of life, and who managed to ban the substance from their settlements for many years.[49] Yet again, another southwestern Indian tribe, the Navajo, seemingly expect their young males to be heavy drinkers and are loath to isolate them from society.[50] Interestingly, cirrhosis mortality rates are higher among the Hopi, which is seen as a result of imbibers being rejected from their communities. Douglas[51] expands this subject a little when she relates some information relating to Jewish attitudes to drinking. Traditional attitudes toward alcohol recommend sobriety, whereas alcoholism is quite widespread among New York Jews, something that she does not simply attribute to the assimilation of these immigrants to the American way of life. She believes that such behaviour results from the people being "far from home" and therefore "distanced from traditional orthodoxy and from all that the ceremonial participation does to enable Jews to re-enact solidarity and to renew the moral authority of their faith". She says that a similar situation appertains to the New York Irish, but that "Catholicism offers nothing like the same ritual control of personal behaviour as Judaism".

With its pivotal role in many societies, anthropologists have necessarily had a distinctive perspective on alcohol consumption. Before the 1970s, relatively few anthropologists in the field regarded the use of alcohol as a problem, and so their data interpretations are not problem-oriented. Clinicians, on the other hand, mainly focus(ed) on pathology and the serious problems of excessive drinking. As Mary Douglas[51] wrote:

"Anthropologists bring several challenges to the assumptions of other writers on alcohol. They challenge the common view that some races are, because of their biological inheritance, peculiarly vulnerable to ill effects from alcohol. They challenge the view that alcohol leads to anomie. They would be more satisfied with the notion that a state of anomie leads to alcoholism and they are prepared to face the theoretical problems involved in defining anomie. They find no clear relation between the use of alcohol and a tendency to aggressive or criminal behaviour. They dispute that drunken behaviour exemplifies a relaxation of cultural constraints before the levelling effects of nature ... Drinking is essentially a social act, performed in a recognized social context. If the focus is to be on alcohol abuse, then the anthropologists' work suggests that that the most effective way of controlling it will be through socialization."

Collections of articles devoted to cross-cultural literature on alcohol use were a feature of the early years of the second half of the last century. Edited contributions by McCarthy,[52] Pittman and Snyder,[53] Everett *et al.*[54] and Marshall[55] enriched the field of alcohol studies. Prior to this, relatively few anthropologists engaged the field, confirming Joffe's opinion[1] that "an under-appreciated feature of complex societies is the production and consumption of alcoholic beverages, in particular wines and beers". Virtually nothing appeared in print prior to 1940 and the first major contribution came from Ruth Bunzel[56] in that year. Her work sought to explain why people in a particular culture drink and behave in a particular way. She made a detailed study of the sociological and psychological significances of the use of alcohol by the Indians of Chichicastenango in Guatemala and the Chalmula Indians of Mexico, and presented the first "controlled comparison" between similar cultures that used alcohol in very different ways.

The second major work, published three years later, was by Donald Horton,[57] who conducted a statistical, cross-cultural comparative study of alcohol usage. His aim was to find out whether there was any connection between societal characteristics and the differing ways in which members of the societies studied used and reacted to alcohol. He tested a number of hypotheses based on the assumption that man's main reason for consuming alcohol was to reduce anxiety brought about by either a struggle to exist or acculturation pressures. By the 1970s, the number of anthropologists studying the beliefs and behaviours surrounding the use of alcohol had grown considerably.

One aspect of study was "alcohol and culture", and Daniel Mandelbaum produced a seminal paper on this theme in 1964.[58] Noting the fact that cultural expectations define the ways in which both "normal" and "abnormal" drinking is carried out in a society, he provided an excellent introduction to some socio-cultural aspects of alcohol, and he was able to bring the subject to the attention of a broad section of the professional community. His comments on the cultural variation in the use of alcohol stressed anthropological points that had been forgotten for too long and neglected by the medical fraternity. For example, he notes that drinking in a society may be either a sacred or a profane act, depending on the context, and the people partaking may not necessarily be aware of the meanings involved. He also notes that addiction to alcohol and hangovers are both heavily influenced by cultural interpretations.

Perhaps the most significant piece of early work in this sphere was Dwight Heath's ethnographic study of ritual drunkenness among the Camba of eastern Bolivia.[59–61] The original work was carried out at a small town called Montero, in a part of Bolivia where the Amazon basin meets the vast hot, semi-arid Gran Chaco plain (around 500 miles east of La Paz). The Camba are a mestizo people descended from the indigenous Indian population and Spanish settlers. Having persuaded the locals that he and his family were not missionaries, Heath was able to observe, at first hand, their daily lives. It was soon obvious that they drank as heavily and as frequently as any known group of people. Heath noted that Camba society was marked by a singular lack of "communal expression", since they were essentially itinerant farm-workers and their long hours of daily labour tended to be solitary. Such a lifestyle precluded the development of kinship ties, and so there were no "neighbourhoods" as such. Life revolved around weekly drinking parties, which might last for a whole weekend. The participants drank rum from one of the local sugar refineries, in Heath's words, "an undiluted distillate of sugar cane that contains 89% ethyl alcohol", which caused irritation to their mouths but, seemingly, no other ill effects! During the course of a session, a drinker might pass out and, on recovering, start all over again, and this may happen several times. Again, as Heath attested: "hangovers and hallucinations are unknown among these people, as is addiction to alcohol."

The structure of these drinking parties was heavily ritualized, and the invited participants would sit in a circle and a bottle and a glass were placed on a table. The host would stand, fill the glass and walk over to someone in the circle (the "toastee"). Standing before the toastee, the host nods and the toastee smiles and reciprocates the nod. The host then drinks half of the glass and hands the rest to the toastee – who drinks it. The toastee now refills the glass and passes toward another guest in the circle and the ritual is repeated. This happens again and again. If a guest becomes incapable, he curls up and sleeps (without leaving the circle) and, on recovering, he will rejoin the drinking party. The Camba never drink alone or on work nights – only within the context of this ritual. Incredibly, the majority of drinkers do not

even like the "rum" that they drink, and Heath certainly didn't –
although he accepted it out of respect! Heath's work in this part of
South America was the forerunner of other studies and the nearby
Andes became one of the first regions in the world to be studied
specifically for the role of alcohol in the society structure of their
peoples.

In contrast to the Camba, the Navajo were one of the few
cultures that did not develop a native alcoholic beverage. That
said, the people have a longstanding alcohol problem despite the
fact that in 1802 the US government gave the president powers to
ban all liquor traffic to Indians, and in 1832 a law was enacted that
made the sale of liquor to all Indians a criminal offence. The law
was repealed in 1953, and the amount of alcohol-related crime
increased considerably, even though the law had always been
stretched – in the 1880s, for example, a trading post agent
complained that traders were selling "whiskey of the vilest
description" (Navajo Yearbook, 1961). On the repealing of the
liquor law, it was naively, and erroneously, supposed that the tribe
would self-regulate the supply of alcohol, but from a situation in
1940, when around 50% of all crimes were alcohol-related, by the
early 1960s the rate had risen to 84%.

A number of factors have been held responsible for the increase
in serious drinking in the Navajo (and other indigenous North
American peoples). Horton[57] found that subsistence anxiety was a
common problem in cultures in which excessive drinking was rife,
and this was certainly applicable to the Navajo. It was also found
that economic anxiety was exacerbated by the influence of white
culture. This leads directly to the second point: acculturation. The
Navajo have always been willing to adopt material aspects of
cultures with which they came into contact, but contact with the
"white man" was a step too far. "White ways" tend to be adopted
by new generations rather than the "old ways", but while
embracing white culture they were treated as social inferiors.
Thirdly, there is aggression anxiety, and there is much evidence of
repressed hostility in the Navajo. Ancient protocol does not allow
any expression of hostility toward family members and, through
unpleasant experiences, it has been learned that it is pointless
behaving aggressively to whites.[62] Some other factors affecting
binge drinking in the Navajo have been discussed by Ferguson,[63]

and a study of the clinical course of alcoholism in some Mission Indians has been carried out by Ehlers *et al.*[64]

6.3 THE ANCIENT NEAR EAST

As Joffe[1] has said: "The question of alcoholic beverages in Mesopotamia is more complex than that of Egypt or the Levant ... Ceramic evidence suggests that production and consumption of alcoholic beverages increased throughout the 3rd and 4th millennia." I have outlined the role of this region in the history of brewing elsewhere,[29] and few would dispute that the "Fertile Crescent" was a seminal location for domestication of some of the important crops of today.[65,66] Mesopotamian economies were strongly agrarian, with the emphasis on barley and wheat cultivation. It would appear that, as with ancient Egypt, a large percentage of the people obtained the bulk of their calories from grain products – whether in solid or liquid form.

States emerged during the fourth millennium BC (the Uruk period) as a result of increased settlement rates and the growth of towns and cities. By the Early Dynastic, southern Mesopotamia was composed of numerous city-states, each made up of one, or a few, major cities and a rural hinterland. It seems as though only by the end of this period most of the population was urbanized. Records attest to the fact that feasting played an important role in determining and maintaining social privilege and power. Feasts occurred on many occasions and for a variety of reasons.[67] Oppenheim[68] and Postgate[69] have given good accounts of the region, and Joffe[1] has written about alcohol and social structure in Western Asia.

With its reliance on suitable grain crops, beer produced in this region was an ancestor of "Western-style" beer, although, judging by the number of depictions of it being drunk through straws, it would not have been the lucid drink that we know today. Beer figures prominently in ancient Near Eastern life and it is no coincidence that the earliest recipe that we have for the drink is the celebrated Hymn to Ninkasi (Figure 6.1), the Sumerian goddess of brewing.[70] Much has been written about this aspect of brewing history and, despite conjecture about interpretation of the tablets, the fact that beer-production warranted its own deity indicates the importance of beer to these ancient people. We know from

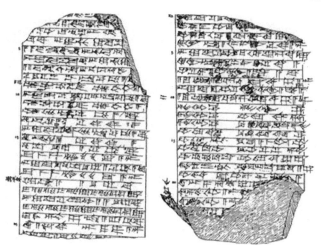

Figure 6.1   Hymn to Ninkasi.

numerous texts that beer was THE drink of the masses in lowland Mesopotamia and, as in ancient Egypt, the higher echelons of society would also have normally drunk beer.

At one time it is reckoned that 40% of all cereals grown were used for brewing[71] and, apart from being used in a religious context, the drink was available in beer shops and taverns. Taverns helped to maintain a complex, civilized society (think Dr Johnson), and to fulfil a useful function they needed to be regulated. Thus, we have the first regulations associated with beer: the law code of King Hammurabi.[72] Hammuarabi (*c.* 1792–1750 BC), when king of Babylon, raised that town to eminence in Mesopotamia, and his codes laid down strict regulations for tavern-keepers and tavern servants in an effort to ensure that they operated sensibly. His code specified the quality, the price, and even terms of credit for beer.

From numerous extant cylinder seals we get an idea of how beer was consumed in Mesopotamia. Drink is contained in large jars and drinkers invariably have straws. They are often depicted quaffing with at least one partner – suggesting that this was a communal activity. The earliest known example of this motif is a clay sealing recovered from Tepe Gawra in the Zagros Mountains (Figure 6.2), although it is possible that the two figures are stirring the contents of the jar rather than drinking from it! The Late Uruk (Early Period V) site at Godin Tepe, high in the central Zagros Mountains of western Iran, has provided us with the earliest

Figure 6.2 Clay seal from Tepe Gawra excavations (reproduced by kind permission of The University of Pennsylvania Museum of Archaeology and Anthropology).

chemical evidence of wine (determined as tartaric acid from pottery sherds[73,74] and, later, of barley beer (detected as calcium oxalate, "beerstone"[29,75]). A detailed account of these landmarks in beverage science can be found in McGovern.[76]

The numerous banquet scenes show seated individuals drinking from cups, or other vessels, and through straws (Figure 6.3). The underlying theme of these depictions seems to be of the distribution of the beverage in a ceremonial or celebratory context. Pollock has suggested[77] that seals depicting a banquet are mostly associated with female burials, while those showing some kind of contest are connected with the burial of males. She concludes that "while the ritual activity and representation of the banquet

Figure 6.3 Drinking beer through straws. From an Early Dynastic (mid-3rd millennium BC) cylinder seal from Mesopotamia.

involved males and females, the rituals that reinforced male authority and solidarity were limited to males". It is interesting to note that most of the participants at these banquets were sitting – rather than reclining on couches as they did in ancient Greece. Having said that, the habit of reclining during the ritualized banquet is not Greek in origin, but is apparently first attested in the ancient Near East.

The contents of graves give us a good insight into the importance of food and drink in the burial ceremony and/or as an accompaniment for the dead in the afterlife. In Mesopotamian cosmology the afterlife was an awful place to be, as can be gleaned from the *Epic of Gilgamesh* (II, 123–126), where "dirt is their drink [and] their food is of clay".[78] Supplying the deceased with food and drink was deemed essential in order to prevent them from acting in a hostile fashion toward the living.[79]

Schmandt-Besserat[80] argued that the feast was a significant factor in the Mesopotamian redistribution economy, and that the economic function of feasting was maximized by the fact that festivals were religious and socio-political events. She concludes her contribution with: "The combined information derived from the art and economic texts of the ancient Near East suggests that, in the third millennium BC, the religious, social, and political aspects of feasting were manipulated to amass palace wealth. (1) The feasts honoured deities and therefore giving was obligatory.

Failing to provide the gods spelled divine rage, famine, epidemics, and doom for the community. (2) The procession of offerings was a social event when, in the midst of pageantry, before the entire citizenry, worshippers rivalled each other in presenting the most generous gift. (3) The queen, acting as an intermediary between people and the pantheon, gave the palace economic control. Finally, the king was the central figure of the ceremonies and the ultimate recipient of the offerings. The delivery of "gifts to the gods" set in motion the wheels of the Mesopotamian redistribution economy. The contributions forced each household to produce a surplus. Of course, all the monthly deliveries of animals and merchandise were not consumed at the feast, but increased the palace herds and granaries until their redistribution. In Mesopotamia, the feast was a fulcrum of the state redistribution economy."

A milestone in the study of the significance of ancient Near Eastern alcoholic beverages was Lucio Milano's edited volume,[81] a synopsis of a symposium held in Rome in 1990. The chapter by Michalowski[82] on alcohol in Mesopotamian ritual and mythology, and Pinnock's[83] contribution on the figurative art relating to banqueting in Mesopotamia and Syria, are particularly pertinent. Two contributions by Martin Stol,[84,85] as one would expect, impart much information, and Henry Lutz's *Viticulture and Brewing in the Ancient Orient*,[86] now reprinted, still has much to offer.

6.4   ANCIENT EGYPT

I have made an attempt to cover much of the literature on ancient Egyptian brewing,[29] and Delwen Samuel's painstaking work around the turn of the millennium has contributed much to our understanding of the subject (culminating with Samuel, 2000[87]). Space does not allow me to cover this work again. Suffice to say that we have evidence for the production and consumption of beer in Egypt going back to the Predynastic era (5500–3100 BC). The Greek Geographer Strabo, presumably a wine drinker, commented that: "Barley beer is a preparation peculiar to the Egyptians, it is common to many tribes, but the mode of preparing it differs in each." He also noted that in many parts of the land (Alexandria, for example) beer was the principal beverage.

Figure 6.4   Grape harvest and treading scene in the Theban tomb-chapel of
Nakht.

Wine-drinking (Figure 6.4) became more popular from the New
Kingdom period (1550–1069 BC) onwards, and even more so
during the subsequent Roman Period (30 BC–AD395). Murray
et al.[88] sum up the situation concisely when they write:
   "While there is archaeobotanical (and possibly chemical) evidence
for the grape from the earliest periods in both Egypt and
Mesopotamia, beer was made from barley, the ubiquitous cereal
staple throughout the archaeological and cultural records of these
two regions." Cereal production and cereal processing in ancient
Egypt are the subject of an excellent chapter by Mary Anne Murray.[89]
In Egypt, the consumption of wine became more widespread during
the Ptolemaic period due to the influx of a large Greek population,
and improvements in irrigation techniques at that time.
   But in ancient Egypt, beer was certainly the defining beverage,
an indispensable staple, and it would have been drunk daily as a
highly refreshing, nourishing and reliable potable substitute for
water. Beers brewed for everyday drinking would not have been
very alcoholic *per* se and would have therefore had a very short
shelf-life, necessitating their being brewed daily (for immediate
consumption). Although there was no distinct division between
beer-drinking and wine-drinking regions in ancient Egypt, beer
was especially important in regions where the vine would not grow.
   All sections of the community drank beer, from the Pharaoh
downwards, and it was a product that was inextricably woven into

the fabric of daily existence, as well as being a feature of religious festivals and state occasions (when "special" brews were produced). Most Egyptologists are of the opinion that grain production and distribution, for brewing and baking purposes, underpinned the ancient Egyptian economy and the political organization of that ancient society, and that a study of beer production can provide an insight into the structure of ancient Egypt itself.

Some authorities maintain that beer was the usual drink of the commoners, while wine was the drink of the rich.[90] This may well be the case from Greco-Roman sources, but is not a true reflection of Dynastic Egypt, when, being a common offering to the gods, it is highly unlikely that beer would have been spurned by the well-to-do. It has been proposed that the relative dearth of information about ancient Egyptian beer, as opposed to wine, during the post-Pharaonic period, is due to the customs and habits of the poorer classes being inadequately documented by chroniclers of the time. Indeed, Greek and then Roman travellers to Egypt equated beer with poverty, and wine with wealth.

Ashour[91] maintains that "Egyptian drinking is unique", and says that the drinking pattern and overall consumption have remained stable for over 4000 years. He names "culture and religion" as being the major determinants of this. From the Predynastic period onwards, Egyptians were convinced that life continued after death, and that the afterlife was very similar to life prior to death. Thus, the deceased were provisioned with victuals for continued existence. Beer was a common divine offering and a mortuary offering, and so we find that references to beer are prevalent in the Pyramid Texts,[92] which are the oldest Egyptian funerary texts. Funerary temples were built to accommodate large amounts of food and beer. Bread and beer were inseparable commodities, and the latter was perceived more as a food than as a drink. Apparently, this concept still survives. If bread and beer were seen to be essential for one's existence, wines and liqueurs (made from grapes, dates and figs) were regarded, by some, as being "dangerous". Over-indulgence, however, was not necessarily frowned upon (Figure 6.5.)

As we know, by Old Kingdom times Egypt had developed into a highly organized and centralized theocracy, and ethical codes were quickly developed. Religious and cultural rituals (involving beer)

Men carried home from a drinking party.   *Beni-Hassan.*

A servant called to support her mistress.   *Thebes.*

Figure 6.5   Over-indulgence (from Wilkinson, *The Manners and Customs of the Ancient Egyptians*, 1878).

were not gestures of piety toward the deities, but a self-imposed duty and a behavioural norm. To illustrate the central role beer played in those days, Ashour[91] recalls that "when guests arrived at the door of a host holding a party, the host would greet them with the phrase 'bread and beer' (the equivalent of 'hello')".

6.5 THE AMERICAS

As Beauvais[93] admits, before European colonization, the native population of what would become the United States and Canada was "relatively naïve to alcohol's effects". Some tribes produced weak beers or other fermented beverages, but these were generally only used for ceremonial occasions. Even when forest clearances around AD 800 led to maize plantings, the corn was not used for extensive *chicha* production. When these peoples has access to distilled, and other more abusable, forms of alcohol – kindly provided by colonists – they had virtually no time to develop social, legal or moral guidelines to regulate its use. Early traders soon established a demand for alcohol, often using it in exchanges, and they found that providing it free during negotiations proved to be highly beneficial.[94]

Dr Havard, in his late-nineteenth-century paper *"Drink Plants of the North American Indians"*,[95] sums it up nicely: "All authorities substantially agree that American Indians, north of Mexico, had not acquired the knowledge of preparing alcoholic drinks at the time of landing of Columbus, and that, whatever their vices may have been, they were free from that of drunkenness." A similar sentiment had been expressed by French missionary Father Gabriel Sagard-Théodat, who, in his 1636 *Histoire du Canada*[96] says: "Our savages, in their feasts, are, thank God, free from such misfortune, for they use neither wine, beer nor cider; if any one among them asks for a drink, which very rarely happens, he is offered fresh water, not in a glass, but in a dipper or off the kettle itself from which he freely drinks and is thus saved from drunkenness, a great blessing to the body and soul, for it is likely that if they had wine they would become as intemperate as ourselves."

Since those times, alcohol consumption has caused problems for native Americans, although the level of use among adults is difficult to estimate. Drinking practices vary from tribe to tribe according to cultural, economic and lifestyle differences, and Levy and Kunitz (1971) attributed the variability between tribes to differences in their tolerance to deviant behaviour, which in turn lead to differing levels of acceptable drinking. The prevalence of alcohol abuse in these peoples has been studied by Hisnanick,[97] while May[98] suggested that they have at least two main patterns of alcohol abuse: "anxiety drinkers" and "recreational drinkers".

6.5.1 The Andes

For millennia, the people of the Andes have made alcoholic beverages from a wide variety of cereals, roots, fruits and seeds and these have often been categorized under the generic name of *chicha* (see p. 483). Since the pioneering work of Bunzel and Horton, an increasing volume of fieldwork has thrown light on the role of alcoholic drinks in the creation of identity, the construction and maintenance of power, the functioning of social networks and religious practice. Notable contributions over the years include the

learning of drinking behaviour,[99] the social uses of alcoholic
beverages in an Andean community,[100] Morris's survey of the role
of maize beer in the economics, politics and religion of the Inca
empire,[101] the *chicha* economy and the expansion of the Tiwanaku
civilization[102] and the production of maize beer in the Peruvian
Middle Horizon.[103] Books relating to the subject include: Isbell,[104]
Bastian,[105] Weismantel[106] and Allen.[107] The chronology of the
central Andes is shown in Figure 6.6.

The recent edited book by Jennings and Bowser[108] contains a
variety of contributions based on three main themes: "reciprocity
and power", "identity and society" and "continuity and change",
each documenting an aspect of the role of alcohol in the Andes
over a considerable length of time. There is a consideration of how
alcohol has maintained gender roles, kinship bonds, status
hierarchies (Figure 6.7), exchange relationships and production

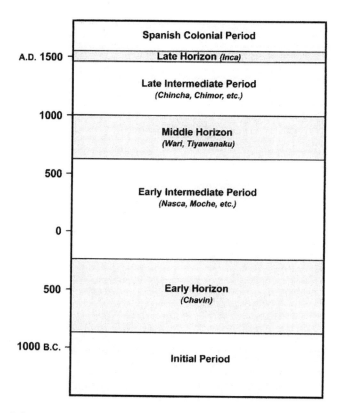

Figure 6.6   Chronology for central Andes.

Figure 6.7   Festival of the Sun, 1615: Woman serving *chicha* to the Inca king.

regimes that form communities. There is also evidence to show how social change within Andean cultures has been brought about by shifts in alcohol production and consumption. In their introduction, four books are mentioned as being of particular note: Cavero Carrasco,[109] Camino,[110] Saigres[111] and Butler[112] – the latter is in English and documents how the role of beer in Ecuadorian society has changed since the Spanish Conquest.

   Reciprocity is the backbone of the traditional Andean economy[113,114] and reciprocal obligations are put into effect during hospitable occasions, the central theme of which is alcohol. Nowadays, the most frequent manifestation of exchange is called *ayni* in the language of the Aymara and *mita* in Quechua. Aymara (*Aymar aru*) is spoken in the Altiplano region of western Bolivia, southern Peru and northern Chile and the people were conquered by the Inca and subjected. Quechua is the most widely spoken language family of the indigenous Andean peoples.[115] Aymara, together with Quechua and Spanish, is an official language of Peru and Bolivia. One should note here that *chicha* is a Spanish word, and the Quechua word for this beer is *aqa* (*akka*), with it being *kusa* in Aymara.

Ayni (*mita*) represent a delayed exchange of labour between small networks of family or friends and, whatever the project is, it is incumbent on the perpetrator (in effect, the "host") to provide food and drink for the invited labourers. As Mayer[116] has attested, labourers come partly for the food and drink and partly for the camaraderie, and their presence at the "do" makes it inevitable that the host will return the favour sometime in the future. When larger projects are undertaken, such as work for the state, there is no onus on the organizer ("host") to repay his workers with a bout of labour at a later date. Instead, the host must provide a feast for them. These asymmetrical exchanges, where there is no labour returned, are called *minka* (Aymara) or *minga* (Quechua), and they have long been important in the political economy of the Andes, with *chicha* usually being the prescribed beverage.[107,117] As Mary Weismantel[114] and others have shown, reciprocity was the structuring principle of the Inca, and the passing of a drink from host to guest was the most widely recognized metaphor for this relationship. Accordingly, *chicha* has become one of the universal emblems of indigenous culture in Bolivia, Ecuador and Peru.

Minka exchanges, and the food and drink accompanying them, are integral to the establishment of power (as defined by the ability to coerce, or persuade, people to behave in a certain way). The creation of and maintenance of power is *via* social relationships and the feasts that accompany *minka* exchanges are integral to such relationships in a number of ways, principally by firm establishment of the host's high social position in the community. The most overt manifestations of high standing are the ways in which the food and drinks are served (to whom, and how: Colloredo-Mansfeld,[118] Jennings and Chatsfield[119]) and the degree of ostentation shown through the goods on offer and the way they are served.[120] By means of generous offerings, the host's power is continually being restated – especially if the recipient ("labourer") knows that he/she is unable to reciprocate. Another point, picked up by Mayer[116] and Jennings[121], is that the recipient will normally provide more work than the labour-equivalent cost of the food and drink that he/she consumes. This is of ultimate benefit to the host family, but it should be remembered that the burden of under-writing the cost of a large feast can drain resources and, arguably, curtail the power of the host. There is the additional problem that, if the feast recipients are not satisfied with their fill, they might not work as hard as they could and, if really dissatisfied, may not work at all.[108]

Where huge power was involved (*i.e.* the state or the church), enormous amounts of *chicha* would be provided at *minka* exchanges, and recipients were encouraged to drink to excess.[101,117,122,123] As the Jesuit scholar Bernabé Cobo (1580–1657) reported in his 1653 *History of the Inca Empire*,[124] "the principal activity is to drink until they cannot stand up". His observations on ritual, often obligatory, drunkenness are interesting: "The public banquets lasted a long time and participants would drink heavily at these feasts until they became inebriated … They would take turns offering each other *chicha* in the following way: the one who was offering would get up and go over to a member of the other group carrying two glasses of *chicha* in his hands, giving one glass to his counterpart and keeping the other himself, they would drink together."

For someone like an Inca emperor, who might have to provide huge feasts quite regularly in order to keep the labour force in his debt, it became necessary to expand and control maize production and centralize *chicha* brewing. Thus, new areas opened up for maize cultivation, and *chicha* production was ascribed to women chosen specially for the purpose, called "chosen women", or *mamacuna*.[101]

Whether alcoholic beverages have been a vital part of life in Andean cultures as they have been in other parts of the world is a moot point. There are not the ancient written records that we have for Egypt, the Near East and China, and there are no records of *chicha* production and use that precede the Spanish Conquest. To date, we have no archaeological vessel-residue evidence of ancient alcoholic drinks from the region, although work in this area is on-going. In the absence of direct evidence, workers (*e.g.* Moore,[125]

Goldstein *et al.*[126]) have searched for archaeological correlates to indirectly identify *chicha* production and consumption in the prehistoric Andes.

Morris[101] was the first archaeologist to make an in-depth study of beer-making as part of a project in the Inca administrative city of Huánuico Pampa. His work provided evidence that confirmed the importance of *chicha* in diet and ritual, and showed that it was brewed on a huge scale, "apparently under state management". He elaborates by saying: "the degree to which the high level of the state administrative apparatus was involved with beer production and consumption, I believe, was truly extraordinary. Beer was not necessarily the secret ingredient that enabled the Inca to establish the largest native empire of the New World, and one of the largest in history, but it certainly was one of a series of interrelated key features that enabled the Cuzco rulers to extend and maintain their power over a vast region." The evidence for *chicha* in the city took the form of aggregations of sherds from large brewing jars and large flattening stones used to crack the malt (*jora*).

Another innovator was Christine Hastorf who used stable carbon isotopic (SCI) analysis of human bone to track the changes in maize consumption during the Late Intermediate to Late Horizon transition in the Upper Mantaro Valley of southern Peru.[127,128] She showed that maize consumption increased after the Inca Conquest of the region, and that men were consuming more of the crop than women during the Late Horizon period. The latter she attributed to the fact that males had select access to Inca feasts where huge volumes of *chicha* were consumed. A similar trend of high maize consumption was shown by males of the Tiwanaku,[102] but these findings are at variance with Burger and van der Merwe's[129] findings at the Early Horizon site at Chavín de Huántar, where maize consumption was much lower. While these isotope studies can measure maize consumption, they do not distinguish between whether the staple was taken as solids (meal) or drunk as *chicha*.

SCI analysis of fossils is an important method for gathering dietary and environmental information in terrestrial and aquatic ecosystems, and carbon isotope variations in an animal's diet leave a geochemical fingerprint in the inorganic (bioapatite) and organic (*e.g.* collagen) components of bones and teeth.[130] Carbon isotope compositions vary among primary producers, and can be used to assess dietary preferences.[131,132] The basis of the ability to differentiate maize *via* this method[133] emanates from the extensive work of van der Merwe[134,135] and van der Merwe and Vogel[136] after it was realized[137] that plants undergoing C_3 photosynthesis have different SCI ratios (av. $\delta^{13}C = -26.5‰$) from C_4 plants (av. $\delta^{13}C = -12.5‰$). The latter group, of course, include maize and other sub-tropical grasses. For details of isotope analyses and the history of maize, see the chapter by Robert Tykot.[138]

With the religious, ceremonial and social importance of food and drink in early cultures, it should not surprise us that vessels and pots (the "ceramic assemblage") assumed enormous significance and, as Bray said,[139] studying pottery "offers a window into the ways in which food, feasting, and gender figured in the negotiation of state power and imperial expansion". She centres on Inca polychrome pottery and discusses its functional significance and contemplates the role of such objects in the context of elite identity and empire building, averring that: "the meaning and importance of imperial Inca polychrome pottery is bound up with its involvement in the political practices of the Inca state." While there are several records of the importance and ubiquity of ceramic containers in the Andes made by early Spanish settlers (for example, Cobo[124] wrote that the average Indian's household furnishings consisted primarily of "pots, large jars, pitchers, and cups"), there are very few references to the forms of these vessels and what they were used for. As elsewhere, these were everyday objects and, thus, of little documentary value, but, fortunately, the culinary practices and means of subsistence were worth noting, and there are some good early records of food preparation patterns, storage methods and details of consumption, and these give insights into ceramic vessel requirements.

An anonymous source from 1573,[140] a typical entry for the time, described the basic Andean diet as follows: "Their usual sustenance is wine made of maize ... and some herbs which they call *yuyo* and potatoes, and beans, and cooked maize; their daily bread is any of these cooked with a little salt, and what they consider as a good seasoning to put in their stewed foods is red pepper." As we have said, corn was by far the most highly prized crop in the Andes, and it is a major item in nearly every account of pre-Colombian native subsistence. Most maize seems to have been used for *chicha* production and, again, quoting Cobo: "... they keep in their house no spirits, not even water, other than their wine or *chicha*, and this lasts for such a short period, they make it quite frequently in quantities of four or six *arrobas* each time." [Nowadays, an *arroba* can vary from 2.5–6.5 imp. gal.]

Cobo also noticed that the natives possessed more *chicha*-related equipment than for any other food-related activity: "To produce, store, and drink this beverage, they had more instruments and vessels than they did for their foods. They use clay jars, the largest being four and six *arrobas*, as well as other smaller ones; they use a large quantity of large and small jugs, and three or four types of cups and glasses." Another early chronicler remarked upon the fact that different vessels were used for making and for storing *chicha*, with the former being wide-mouthed and smaller (*olla*), and the storage jar having a narrow mouth and longer neck.

Cobo's observations on *chicha* manufacture are worth noting here, for he notes that the drink can be made in a number of ways: "One very strong type, called *sora*, is made by first burying the maize underground until it sprouts; another type of *chicha* is made from toasted maize … But the most common type that the Indians of Peru drink is made from masticated maize … Women chew a portion of the maize to be used and spit the masticated mash into jars of warm water. More grain is added to the jars and the mixture is then allowed to ferment to the desired strength."

As Tamara Bray's excellent chapter intimates, prehistoric pottery has not generally been rigorously analysed, in terms of its functional and technological properties, by archaeologists, as it has in terms of its ethnicity, and she feels that this might relate to "generic associations of pottery with women, cooking, and the domestic sphere of activity". Studies, she says, have tended to centre on larger social processes, and this presupposes that, in those far-off times, there was a conscious distinction between "public" and "domestic" spheres of life (as there is now in Western ideology). Bray assessed Cuzco pottery as culinary equipment and as material symbols of the state, and highlighted the close links between food, gender and politics. She developed the idea that the Inca worked out a specific elite, or "haute", cuisine and a distinctive, and distinguishing, collection of ceramic cooking, service and storage vessels as an intentional strategy aimed at creating visible differences between social classes. Incan pottery and culinary practices are discussed.

From the point of view of *chicha*, perhaps the most important drinking apparatus is the *kero*, a distinctive wooden vessel that has been made in Peru both before and after the Spanish Conquest.[141] A close relative is the *paccha*, but this is much less common. Wood is the commonest material for both types, which are usually decorated by painted motifs, but they can be made from stone, pottery (Figure 6.8) or metal. *Kero* is a Quechua word, defined by de Torres Rubio[142] as "vase of wood from which they drink *chicha*", and Cobo, in an exposition on drinking cups, said: "the most common of these are of wood, of the shape of our glass tumblers, wider at the top than the bottom. They hold a pint of wine. They are painted outside with a kind of lacquer, very

Figure 6.8   *Kero.*

relucent in various colours, and with different raised figures and paintings. These wooden vessels are called *queros.*"

A wide variety of subjects are painted on keros, with processions, wars, hunting scenes and drinking scenes being the commonest motifs. Other themes include trees, flowers, birds and insects, and textile-like patterns are also important. Post-conquest-dated keros tend to have depictions of Spanish clothing, guitars, harps, angels, *etc.* Most of the keros Lothrop examined had long vertical cuts on their inner walls – presumably cut with a metal instrument.

First described and studied by Joyce,[143] the *paccha* is basically a vessel with a drinking outlet at the base (Figure 6.9) and, in Lothrop's words, "liquid might issue through a tube across a long tongue with open zigzag channels on the upper surface leading to a conical mouthpiece". *Pacchas* with short extensions could probably be used unaided, but the more elaborate forms (heavy, and over 2 feet long) almost certainly required an assistant. There are three main types of these curious vessels, those made from: 1) human skulls, 2) pottery, 3) metal. There are no surviving specimens of the first category! Pottery *pacchas* are pre-Spanish and of various types, but all have a tube extending from the base. They are the commonest form of the vessel. Wooden *pacchas* are quite rare and Lothrop said: "it is evident that such intricate instruments were not easy to manufacture and could not well have been devised for general use", and Joyce[143] has suggested that they might have been designed for serving *chicha* to the dead.

Orlove and Schmidt[144] studied the relationship between food and society identity *via* an analysis of consumption in Bolivia and Peru. They looked at eleven sites from six Andean regions and

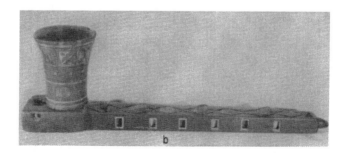

Figure 6.9   Wooden *paccha* (after Lothrop[141] - by kind permission).

focused on *chicha* and Western bottled beer. The latter product was first introduced after independence as a luxury item from Europe. They found that beer was more widely consumed in cities, *chicha* in the countryside, but the divisions among settings could not be reduced to a simple urban/rural polarity, and there was no correlation between levels of urbanism and regional consumption levels. They finish by saying: "The weak association between beer-drinking and Spanish monolinguals, who presumably are mestizos, is undercut by the strong association between *chicha* and Spanish-Quechua bilinguals, suggesting a complex relation between ethnicity and consumption."

After the reports of *chicha* and drunken revelry, it is time for a sobering thought. The Andean tradition of drinking to excess, with its roots in the great festivals held in the Inca capital prior to the Conquest, did not necessarily conflict with the principles of the Catholic Church and was thus largely tolerated. Catholic rituals were used by indigenous communities to keep alive pre-Columbian religious beliefs and practices, but since the introduction and spread of Protestantism, numerous Quechua- and Aymara-speakers have converted to sects that regard drink as abominable.[145] As Weismantel reports, ritual toasting is still carried out – but with *chicha* brewed to be non-alcoholic! She finishes with: "As the economy of the Andes changes, those aspects of Andean beliefs about food and drink that express extinct economic relationships are disappearing. But as *chicha* ceased to be the drink of farmers, it has nevertheless, as it did for the Inca, become a liquid of potent power in the politics of the state and a rousing symbol of the strength of the Andean people."

Enlarging the field of study somewhat, Duncan *et al.*[146] have recovered evidence of what was consumed at feasts at a central Peruvian preceramic site dating back to around 2200 BC. Through studying gourd and squash residues, they identified starch grains from manioc, potato, chilli pepper (*Capsicum* spp.), arrowroot (*Maranta arundinacea*) and algarrobo (*Prosopis* sp.). One should not underestimate the significance of squash and bottle gourds to man, and these artefacts are plentiful in preceramic contexts.[147] They had a sacred symbolism as containers for serving ritual libations. The history and distribution of the cultivated cucurbits in the Americas has been well documented.[148]

6.5.2 The Tahahumara

As has been reported (*e.g.* Lumholtz[149]), in some cultures drinking
alcohol is integrated into society for reasons other than solely as
part of religious ceremony or reinforcement of status, and in the
case of the Tarahumara Indians of Chihuahua, northern Mexico,
drink is seen to be essential for overcoming innate shyness.[150] The
people engage in a beer-drinking party, the *tesguinada*, which
seems to be of fundamental importance – especially to "un-
Christianized" (un-baptized) peoples called *cimarones*, who are the
most conservative and isolated members of the tribe. Kennedy[151]
studied the role of beer-drinking and how it is integrated into the
culture and social organization of the Tarahumara, and found it to
be institutionalized. He found the main drink of the Indians to be
tesguino, a maize beer common to most of the tribes in northern
Mexico, although mestizos in the area tended towards distilled
beverages.

The timidity and taciturnity of the Tarahumara has often been
interpreted, often erroneously, as unsociability. Many students of
the tribe have noted that is only after participating in a *tesguinada*
that they are released from their inhibitions (one commentator
writing that it is "only amidst the fumes of alcohol that the
Tarahumara becomes human"). Lumholtz,[149] who pioneered
research in this area, felt that it was mainly through *tesguino* that
the tribe was able to perpetuate itself, since it was only while drunk
that a male would be vigorous enough to enforce his conjugal
rights. The lack of confidence of these people has not made them
an easy topic for study.[152]

Tesguino is a thick, opaque, nutritious brew with a short shelf-
life. It is made from a maize base with a local grass seed
(*basiáhuari*) added. The Tarahumara have two native names for it:
batári and *suguí*. The beer is made in large earthenware vessels and
takes around a week to make, including 3 days for the corn to
sprout, and must be consumed within a day (*tesguino* is at its peak
12–24 hours after preparation). Once a social gathering has started
to imbibe, every effort is made to finish the drink before it
deteriorates and, if the participants cannot drink the batch in one
sitting, they will sleep for a period and start again. If any of them
become too sick or satiated, they will drink a mixture of parched,
ground corn and water (called *esquiate*) as an antidote so that they

can start afresh. It is serious business, and Kennedy (1963) likens the sacred character of *tesguino* to the Tarahumara people to *kava* in the Pacific (*e.g.* Lemert[153]), and documents some of the ritual practices involving the drink.

As we have seen before, *tesguino* is used to attract cooperative labour, and if there is a substantial task to be undertaken (*i.e.* more than can be dealt with by family members alone – such as housebuilding) then a "working *tesguinada*" is instigated. Rarely does a man make *tesguino* with no practical motive or excuse. The prestige of a Tarahumara man is gauged by his ability to get as many people as possible drunk for as long as possible! Kennedy reckoned that "the Mexican Tarahumara spend about one-third of the year directly concerned with *tesguino* and much of the time under its influence". Robert Zingg[154] has documented some of the values of Tarahumara culture.

6.5.3 The Wari at Cerro Baúl

Prior to the reign of the Inca, two great empires ruled the central Andes from AD 600–1000. To the north the Wari ruled much of mountainous central Peru, while to the south were the Tiwanaku from the windswept shores of Bolivia's Lake Titicaca. The two peoples faced each other when they both colonized the Moquegua Valley sierra in southern Peru. The Wari had for ages occupied the high mesa of Cerro Baúl, and two adjacent hills, at the southern edge of their empire and had built monumental buildings atop the mesa as part of an embassy-like delegation. In essence a small city, with palace complexes, temples and ritual feasting facilities for entertaining dignitaries, it was manifestly a combination of fortress, provincial capital, imperial embassy and sacred city. It is reckoned that the remote Cerro Baúl stronghold supported a population of over 1000.

The summit of Cerro Baúl is sheer-sided, and because of its inaccessibility was uninhabited prior to Wari times. All daily needs had to be hauled up, and this required great effort. Temples around the base of the mesa strongly suggest that Cerro Baúl was a sacred mountain, or *apu*. Moseley *et al.*[155] think that the colony was "established atop a sacred natural bastion to emphasize political prowess regardless of economic impracticalities". The enclave was established just prior to AD 600 in what was essentially

Tiwanaku territory and it was one of the few places that the two peoples came into direct contact (they were normally separated by a buffer zone). These ancient states never conquered each other, and the Moquegua frontier is a unique place for studying the dynamics of imperial confrontation and ethnic diversity. The relationship that developed between the two ultimately contributed to the rise of the Inca Empire. Not long after AD 1000, both peoples abandoned their Moquegua settlements, and by 1100 their states had collapsed. A number of small regional kingdoms developed, which built on the achievements of the Wari and Tiwanaku. One of these states was to evolve into the Inca Empire.

Although the Wari and Tiwanaku shared similar iconography and religious practices, they were economically and politically quite different, with the Wari basically secular and militaristic, and the Tiwanaku more ecclesiastical and mercantile. Both societies were evidently stratified with commoners (principally farmers and herders), artisans, technicians, religious specialists and a class of governing aristocracy. Distinctions in class and rank were demarcated by differences in location and quality of living accommodation, and degree of access to food and durable goods.

In what seems like a response to Cerro Baúl, the Tiwanaku built a huge temple complex downstream at Omo, which was the sole such example outside of their altiplano heartland. A small brewing and drinking facility was built at Omo. It had one boiling vat, three fermentation vessels and a brew length estimated at 360 litres. Grain milling, if performed, must have occurred elsewhere, but drinking from *keros* took place on site. It is possible that, after the building of high contour canals, the Cerro Baúl settlement was built to exploit the agricultural potential of the immediate area.

The monumental architecture was built on an artificially levelled area of around three hectares on the mesa top. Stone blocks were cut from the mesa (leaving quarry pits) and these were held together by earthen mortar. Water was hauled up from below the mesa since the quarry pits would not hold water. All other essential building materials had to be hauled up from below. A palace complex represented the most lavish residential quarters on the site, with one building being an elite palace. There was also a temple annexe, but most of the hilltop residents lived in basic accommodation.

All archaeological evidence suggests that the enclave was swiftly abandoned along with lavish termination ceremonies which involved brewing, drinking, feasting, vessel smashing and the burning of buildings. The most elaborate of these ceremonies was at the site of a monumental brewery, which still bore evidence of brewing equipment and final offerings. The brewery was trapezoidal in plan (Figure 6.10), with individual compartments for milling, boiling and fermentation. The milling room was thatched, and its floor divulged high phosphate levels consistent with incessant grinding of germinating grain (maize?) kernels. Grinding stones were also present. The boiling room contained at least seven fire pits, each one associated with a pair of opposed pedestals (Figure 6.11), which, it is assumed, acted as supports for the ceramic boiling vessels. Ash, of which there was much deposition, also helped to stabilize boiling vessels. Copious quantities of *S. molle* seeds were also recovered amid the floor ash, but it was not possible to determine whether *chicha de molle* and *chicha de jora* were both being brewed or whether *S. molle* seeds were being added to the *chicha de jora* recipe to create a special brew (maybe with some religious significance). After

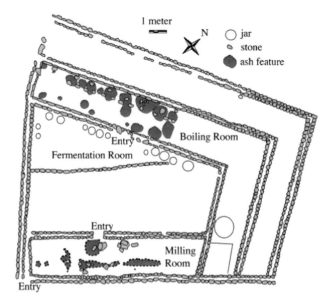

Figure 6.10 Plan of the brewery at Cerro Baúl (after Moseley *et al.*[155] - by kind permission).

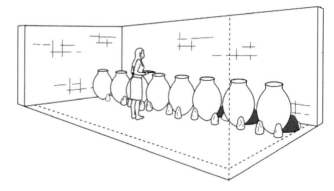

Figure 6.11 Reconstruction of the boiling room in the brewery complex atop Cerro Baúl. Reproduced with permission from Jill Seagard (copyright 2005, The Field Museum).

boiling, the mixture was transported to the fermentation area where it would be distributed into large vessels lined against the wall of the fermentation area. There were 12 of these vessels, each one being able to accommodate 150 litres of beer. After fermentation, beer was ladled into smaller vessels prior to serving. The brew length of the plant was estimated at around 1800 litres,[155] which makes Cerro Baúl the largest known production site for alcoholic beverages in the ancient Americas. Native Andean women fastened their shawls with beautiful clasps called *tupus*, which usually identified the wearer's area of residence and status. The usual form of the *tupu* was a straight pin with a spatulate head. Within the brewery complex, numerous *tupu* pins were recovered, suggesting a prominent role for elite women in *chicha* production.

The role of brewsters in this region was to persist, for when the royal Inca rulers drank *chicha*, it was no ordinary brew that touched their lips, but rather a liquid prepared by women consecrated to the task. These *acllacuna* or "chosen ones" lived in houses inhabited only by women like themselves, and into which men dare not venture. Their lives were dedicated to sacred and royal duties, including the making of fine textiles, the tending of sacred fields of maize and the making of special foods and drinks out of the harvested ears.[145]

Assuming the brew length mentioned above, it would be necessary to stockpile raw material, and the most likely storage facility was a row of three rectangular rooms 15 m north of the brewery. In addition to evidence of maize and S. *molle*, evidence of squash seeds and peanut husks was obtained, as well as some animal bones.

For undetermined reasons, both the Wari and Tiwanaku declined around AD 1000, and it has been suggested that, if the latter entered decline first then the Wari had no justification for their "embassy", hence their evacuation. But why such rapidity?

Of the end of the citadel, based on ceramic archaeological evidence, Moseley *et al.* offer the following:

"If the brewery was the central facility sustaining libation rites throughout the monumental compounds, then it was essential to colonial political economy and probably the last building accorded closure ceremonies. The excavated evidence supports a scenario of planned closure that began with the brewing of a final batch of *chicha*. One week later, it was served in pitchers to nobles assembled in the court in front of the fermentation patio. There were 28 lords if each drank individually from one *kero* produced in seven distinct four-vessel sets. Sets were ranked by ornamentation and size, with simpler ones holding 12 oz. (1 oz. = 29.57 ml). Volume graded up through a penultimate group of black-and-white panelled vessels identical to high status beakers used at the Wari capital. Holding 64 oz. of brew, the largest, most ornate *keros* depicted the head of the front-facing deity, an icon rarely depicted in the colony. Presumably, as ceremonies neared completion and the *chicha* batch neared depletion, the edifice was ceremonially torched. Serving pitchers were thrown in as flames consumed the thatched milling room, and each attendant lord cast his emblematic *kero* into the conflagration in an act of sacrifice."

The sacrifice of heraldic beakers by the elite was not an act evident in other ritual contexts, and emphasized the significance of the brewery site, and the fact that the brewery was the last building to be abandoned illustrates how important *chicha* must have been in all of the preceding closure ceremonies.

For some reason the Wari suddenly vacated their stronghold in the Andes after having what can only be described as a "huge booze-up". Ritual mugs were then smashed and the whole complex was razed to the ground! When the embers had cooled, six necklaces of shell and stone and a bracelet were placed on top of the ashes in a final act of reverence.

The study of the importance of alcohol in the Andean world has burgeoned over the last couple of decades and time and space only permit an overview of some other work. Perlov,[156] in a study of women brewing *chicha* (*chicheras*) in highland Bolivia, has shown how some gained social and economic power through the production and distribution of *chicha*. She also remarked on the fact that, with the decline of *chicha* drinking, there has been a concomitant reduction in the influence of women. Goodman-Elgar[123] has focused on maize cultivation among the Inca, and has suggested that the terraced fields used to grow the crop were an important site of Inca rituals that linked maize, *chicha* and the Inca elite to the seasonal productive cycle. For example, it is common for Indians to worship fertile soil by spilling *chicha* on it. She suggests that roles for terraced fields go beyond mere food production and that, in the Andean region, agricultural terraces "not only support cultivation but may also represent the power to placate deities, control land and labour, and reinforce a stratified social order".

6.5.4 The Maya and Their Periphery

The Mesoamerican Mayan culture can be recognized from *ca.* AD 150, and probably evolved from earlier agricultural settlements in the area. They are noted for having the only fully developed written language in the pre-Colombian Americas, and for their art, architecture and mathematical and astronomical systems, and their Classic period, when many cities reached their developmental zenith, lasted from AD 300–800. The life-cycle of maize was the cornerstone of Mayan belief, and so one can appreciate the importance of beverages made from that crop.

It has been said that the Maya have always lived in the sacred, as opposed to the profane, world. They seek the mystery of things, and all their relations with nature, work or sexuality have a sacred meaning, and even the act of feeding is considered thus. It is no wonder, therefore, that food and drinks were symbolic objects, especially those relating to their ceremonies. A glance at the *Popol Vuh*, the sacred book of the Maya,[157,158] will illustrate these points. In most Mayan ceremonies the special drink *balché* played a part, it being made with an extract of the bark of the tree *Lonchocarpus longistilus* Pitter. The word "balché" arises from "baal" (to hide) and "che" meaning tree. The bark is first boiled to remove most of the bitterness, and then boiled again in fresh water. The extract is cooled, honey is added and the mixture fermented. Other Mayan alcoholic beverages were made from corn and cacao.

Throughout the Maya region, feasting was a crucial component of rituals for the elite and non-elite alike. Mayan rulers participated in a form of competitive feasting with defined requirements for repayment, and from Tozzer's[159] translation of Bishop Diego de Landa's major work (see below), it is evident that the obligations incurred from feasting could be passed on to the next generation:

"And they[†] often spend on one banquet what they have earned by trading and bargaining many days. And they have two ways of celebrating these feasts: the first, which is that of the nobles and of the principal people, obliges each one of the invited guests to give another similar feast. And to each guest they give a roasted fowl, bread, and drink of cacao in abundance; and at the end of the repast, they were accustomed to give a *manta* [cloth] to each to wear, and a little stand and vessel, as beautiful as possible. And if one of the guests should die, his household or his relations are obliged to repay the invitation. The second way of giving feasts was used among kinfolk when they marry their children or celebrate the memory of the deeds of their ancestors, and this does not oblige the guests to give a feast in return, except if a hundred persons have invited an Indian to a feast, he also invites them all when he gives a banquet or marries his children. They have strong friendship and they remember for a long time these invitations, although they are far apart from one another."

Diego de Landa (1524–1579), Spanish Bishop of Yucatán, left a mixed legacy. He destroyed much Mayan literature (he thought it was inspired by the devil!), and later (1566) wrote *Relación de las costas de Yucatán*, which documents Maya religion, history, language, writing and mathematics.

We have already mentioned the Mayan farming village of Cerén in El Salvador (see p. 482 and Sheets[160]) but, apart from its information about ancient manioc farming, it holds unique information about rural village life and ritual feasting. Brown[161] has produced a summary of some aspects of the feasting and festivals at the Cerén site, and we know that the inhabitants had been able to exploit the rich soil there and, manioc apart, grew numerous plants including maize, beans, squash, maguey, cacao

---

[†]The sixteenth-century Yucatec Maya.

and chilli, as well as medicinal/intoxicating plants.[162] From modern-day states in ancient Mayan territory, we have synopses of drinking in Guatemala[163] and Honduras,[164] and a similar treatment has been accorded for Mexico.[165]

6.5.5 Cacao Beverages

The chocolate tree, *Theobroma cacao* L., is an understorey tree native to the upper Amazon, which is now cultivated widely in the humid tropics, and is a major cash crop in several West African countries.[166] *Theobroma* literally translates as "food of the gods". The plant provides sustainable economic and environmental benefits to some of the poorest and most ecologically sensitive areas of the world.[167] It is a diploid ($2n = 20$) plant of the mallow family (Malvaceae) with a genome size of *ca.* 400 Mb.[168] The flowers and fruits (pods) (Figure 6.12) are produced directly on woody stems and branches (called "cauliflorous"), and it is the pods that contain the seeds (beans). When the Spanish conquistadors arrived in the New World, they found native peoples using cacao beans as currency.

Traditionally, two main genetic groups, "Criollo" and "Forastero", have been defined based on morphological traits and geographical origins.[169] A third group, "Trinitario" has been recognized and these are Criollo × Forastero hybrids. Botanists recognize two sub-species of *Cacao*: *cacao* (Criollo) and *sphaer-*

Figure 6.12   Cacao pod.

ocarpum (Forastero), which, according to some authorities, evolved in South and Central America, respectively.[170] *T. cacao* flowers are hermaphrodite, but wild specimens possess self-compatibility mechanisms, which promote out-crossing. Cultivated plants are generally self-compatible. The genetic diversity of cacao has been covered by Bartley.[171] Breeding and management of genetic resources of the plant have been hindered by a lack of a sound classification of *T. cacao* based on genetic data, but this situation is gradually being addressed and the genome is being sequenced. To expand the utility of the genome sequence to the improvement of cacao, the genome of the closely related *T. grandiflorum* is also being sequenced.[172] Motamayor *et al.*[173] have recently proposed that *T. cacao* has ten phylogenetic clusters, one of which is Criollo.

For a variety of reasons, cacao is not an ideal experimental organism. Its life-cycle takes a minimum of 2–3 years from seed to seed; self-incompatibility makes genetic analysis and breeding labour intensive, and seed recalcitrance means that germplasm must be conserved as living specimens. Despite this, breeding programmes commenced from the 1920s onwards, and today's genetic resources research is highly collaborative.[174] The latest general literary treatment of the crop is by Smith and Le Ber.[175]

Criollo cacao was domesticated and utilized by Amerindian civilizations in Mesoamerica including the Olmecs (400–1200 BC) and the Mayas (AD 300–900). Some controversy surrounds the origin and domestication of cacao and, although the initial home of domestication was said to be Central America, Van Hall[176] thought that "the most probable home is the region of the Orinoco and the Amazon basins, with the valleys of their tributary streams". Cheeseman[169] considered the site to be in the Upper Amazon near the Columbian-Ecuadorian border on the eastern flanks of the Andes – where the greatest morphological diversity occurs. He also argued that, although cacao had been cultivated in Mexico and Central America for over 2000 years, no truly wild populations were present in this region, suggesting that the plant was introduced into these regions.

Controversy persists, but Motomayor *et al.*[177] reckoned that *T. cacao* ssp. *cacao* was cultivated by the Mayas over 1500 years ago. Using RFLP and microsatellite analyses, they attempted to throw new light onto the origin and genetic basis of Criollo cacao from

Central America. During their procedures, they avoided mixing plants classified as Criollo (which might have been introgressed with Forastero genes) with pure Criollo plants. These two types of plant were described as "modern" and "ancient", respectively, and, in contrast to previous studies, "ancient" individuals formerly classified as "wild" were found to form a closely related group together with ancient Criollo plants from South America. The "ancient" Criollo trees were found to be closer to Colombian-Ecuadorian Forastero plants than the latter were to other South American Forastero plants. Results showed that a high level of homozygocity and a very low genetic diversity were displayed by the "ancient" Criollo group. The results also suggested that the "ancient" Criollo plants represent the original Criollo group, and that this group does not represent a separate sub-species and that it probably originated from a few individual plants in South America that might have been spread by man's activity in Central America. A year later, a similar team,[178] using the same techniques, reported on the genetic structure of the traditional Trinitario cultivar and attempted to identify the source of the putative Forastero parents, thus contributing to the knowledge of cacao domestication.

Whatever the exact origins of the crop, it is evident that the Olmec, Mayas and Aztecs all enjoyed the cocoa bean many moons before it was sampled in Europe. The Maya called the tree *cacahuaquchtl* and by the sixth century AD were making a cold, bitter drink out of it, called *xocolatl* ("bitter water"). The drink was imbued (rightly, as it turns out) with health-giving properties. Later on, the Aztec hierarchy used the same drink (pronounced "shoco-latle") and thought that it conferred power and wisdom and had aphrodisiac attributes. As Bennett beautifully put it:[174] "From its divine origins in Mesoamerica, cacao was introduced to the Spanish royal court in the mid-1550s and products derived from cacao continue to hold a special appeal in modern culture." One of the first documents to mention "cacao" or "chocolate" in a European language was a letter written by Hernando Cortés to the Emperor of Spain (dated 30th October, 1520).

In 2007, Henderson *et al.*[179] reported on the chemical analysis of residues recovered from pottery vessels from Puerto Escondido, on the Ulúa River in northern Honduras. Dating to around 1400 BC, these pottery samples were pre-Olmec and some of the earliest found in Mesoamerica. Eleven out of the thirteen pottery sherds examined tested positive for theobromine and/or caffeine. In this location, theobromine could only have come from *Theobroma* spp. Two distinctive types of vessel were recovered and analysed, one of which was shaped like a cacao pod, with a distinctive long neck, and was clearly designed to contain liquid (Figure 6.13). This type dated from the site's earliest phase (the Ocotillo, *ca.* 1400–1100

Figure 6.13   Barraca Brown bottle from northern Honduras (after Henderson *et al.*[179] - with permission).

Figure 6.14   Bodega Brown bottle from northern Honduras (after Henderson
            et al.[179] - with permission).

BC). The second major class of vessel was of the much more
widespread "teapot"-type, with a small spout (Figure 6.14), which
dated to the more recent Playa phase (900–200 BC). The chemical
evidence obtained did not enable the team to distinguish between a
beverage made from cacao seed or one made from the pulp around
the seed. In relation to the latter, it is known that the fruity pulp
was fermented by indigenous Mesoamericans to produce a form of
chicha. Henderson had already advocated that the earliest use of
cacao in Mesoamerica was likely to have been for production of an
alcoholic drink from fermented seed-enclosing pulp.[180]

   In a highly readable account of the cultural history of the
medicinal and ritual use of chocolate, Dillinger *et al.*[181] tell us that
the medicinal use of cacao, both as a primary remedy and as a
vehicle to deliver other medicines, originated in the New World
and diffused into Europe in the mid-sixteenth century. The practice
originated among the Olmec, Maya and Aztec, and chocolate paste
was a favoured medium for administering drugs – many of which
had a bitter taste. Treatments are detailed in some of the early
colonial medical literature, and the Badianus Codex of 1552
recommended cacao flowers to combat fatigue. The Florentine
Codex (1590) offers a prescription of cacao seeds, maize and the
herb *tlacoxochitl* (*Calliandra anomala*) to "alleviate fever and

$_2$ka w(a)

Figure 6.15 Maya hieroglyph, *"kakawa"*.

panting of breath and to treat the faint of heart". The sap from *C. anomala*, a leguminous plant known also as "Cabeza de Angel", was used as a stimulant and the resin was also collected.

The word "cacao" is derived from the Mayan word for chocolate, *kakaw* (or *kakawa*) (Figure 6.15), itself with roots in Olmec. The chocolate-related term, *cacahuatl*, is from an Aztec language (Nahuatl) again derived ultimately from Olmec. Two early English texts dealing with cacao are *The American Physitian*[182] and Henry Stubbe's *The Indian Nectar* of 1662.[183] Additionally, the clergyman Thomas Gage's classic travel account, *The English-American: His Travail by Sea* (1648),[184] contains a chapter devoted to New World beverages, including chocolate. Hughes[182] was quite enlightening and enthusiastic about chocolate: "Chocolate is most excellent, it nourishing and preserving health entire, purging by expectorations, and especially the sweat-vents of the body, preventing unnatural fumes ascending to the head, yet causing a pleasant and natural sleep and rest ... eaten twice a day a man may very well subsist therewith, not taking anything else at all."

6.5.6 Peruvian Pepper

This plant, *Schinus molle* L. (Anacardiaceae – the cashew family) (Figure 6.16), is an evergreen tree (up to 15 m) native to the Peruvian Andes. It has a number of other trivial names, including

Figure 6.16 *Schinus molle* L.

"false pepper", "pink peppercorn" and "Peruvian peppercorn tree". It is now widely distributed around the world, being mainly planted as an ornamental tree and for provision of a spice. With its ability to flower and seed throughout the year, drought tolerance and longevity it has assumed weed status in some areas. Barkley[185] has written a monograph on the genus *Schinus*, while Kramer[186] has documented some of the biology of *S. molle*.

The specific name "molle" derives from the Quechua word, *mulli*, and Spanish chroniclers noted that the tree had a variety of uses, including a type of *chicha* made by fermenting the fruits in water, and which they described as being "sweet and agreeable". It was also noted that, after being processed for *chicha*, the spent seeds were ground and used as a "pepper". Ominously, it was recorded that seeds could not be used in this way for extended periods without adverse health effects.

The essential oils, which are produced by leaves and branches, as well as fruits, have an aroma reminiscent of black pepper, and have long been used as an insect repellent – relevant parts being strewn in, or around, middens. Various parts of the tree have been used for medicinal purposes, from being anti-rheumatic to having diuretic properties. It is evident that indigenous peoples within its native range have benefited from *S. molle* for millennia.

Recent archaeological investigations in the Central Andes have indicated that *S. molle* was used extensively during the Middle Horizon period. Excavations have revealed varying concentrations (from a few tens to thousands) of seeds, which Goldstein and Coleman[187] feel are mostly related to *chicha de molle* production and consumption. From Middle Horizon deposits, *S. molle* seeds occur in deposits throughout the Colonial periods of Andean history. There are two distinct patterns of deposition, which probably reflect different uses of the seed. Some are found in pits, containing "tens of thousands of seeds deposited in single episodes of disposal", while others are found as a result of "low-density garbage (1–10 seeds m^{-2}) on house floors throughout the excavation areas".[187] The former pattern of deposition "corresponds to the expected pattern of *chicha de molle* production detritus", while the latter can be described as a "household" context. In an expanded coverage of this topic.[126] further distinctions between household sites are made. The deposition of really large quantities of seeds was restricted to certain archaeological contexts, which suggests that production (and consumption) of this drink was a communal enterprise.

Still produced in the Andes today, although now quite rare, *chicha de molle* was apparently prepared by dissolving sugars from *S. molle* fruit (a drupe) in hot water, cooling the extract and then allowing ethanologenic bacteria to ferment the liquid in semi-closed ceramic vessels. The excavation work was associated with the Cerro Baúl Archaeological Project, and investigations were made further afield in order to find out how the drink was surviving. It was evident[187] that while many people knew about the drink, very few knew exactly how to make it. They did learn, however, that only the fruit was used in *chicha de molle* production (no leaves or sap), and that 250 g (*ca.* 4000) of *S. molle* drupes yielded roughly 20 L *chicha*. The team then decided to couple experimental work with archaeological data and attempted to "brew" *chicha de molle* (see below), which is, in essence, a wine and totally different from "beer-like" *chicha de jora*. On ethnographic and ethnohistorical evidence, *chicha de molle*, at least over the last 500 years, was consumed in a similar way to *chicha de jora*.

When mature, the drupe has a papery outer layer (exocarp) that surrounds the actual seed. The seed wall is much invaginated, and in the recesses (pits) there are pockets of resin. The fruit is ideal for making *chicha de molle* just as it reaches maturity (the exocarp starts to become papery and the resin pockets are still slightly fleshy). At this stage, the fruit contains high levels of fermentable sugar but, if they are allowed to age further, the fruits become dry and develop an unwanted bitterness. Peppery flavours are also transmitted from seed to pulp. Under-ripe fruit will also be too bitter. Although fruit can be found on the tree all year round, it is most abundant during January and February (early Andean summer) when trees buckle under the weight of fruit.

6.5.7 Production of *Chicha de Molle*

The Spanish colonialists did not record any methods for obtaining alcohol from *S. molle*, but there was sufficient ethnographic detail to construct a brewing process experimentally.[187] The first stage, fruit harvesting, would have been no different from any other everyday chore if small (domestic) quantities were being produced, but for ceremonial occasions the huge quantity of fruit needed would have posed considerable labour logistic problems. The second stage is, in effect, a winnowing process whereby the fruit is separated from its papery scales and stalks. Goldstein and Coleman[187] used wind-driven and cloth sieve methods. Winnowing would have resulted in accumulations of stalks and paper coats. Step three involves boiling the fruit to dissolve out the sugar, a process requiring much hot water and numerous large ceramic pots. On the recommendation of a "local", a couple of sticks of cinnamon (*Cinnamomum* sp. L.), and a few cloves (*Syzygium aromaticum* (L.) Merrill and Perry) were added to the boil, as this was said to reduce the likelihood of stomach disorder after consuming the drink. After boiling, the liquid was strained before being distributed into ceramic vessels (*tinaja*) for fermentation. A damp cloth was placed over the *tinaja* neck to create an anaerobic atmosphere. Jugs were stored in a cool room.

During Inca and colonial times, what one drank played an important role in making ethnic distinctions, so whether one drank a beverage made from corn, manioc, potato or the pepper tree had real significance. The role of alcoholic beverages in shaping identity and ethnicity is well documented in Inca writings, and it is evident that these people made and consumed enormous quantities of maize beer. According to Goodman-Elgar,[123] the political and religious status of maize beer in Inca society was due to the role maize played in the expansion of the state. The Inca drew upon the symbolic associations of maize and *chicha* in agricultural ceremonies to provide a basis for legitimizing the expanding Inca state. To the Incas, maize was a divine gift, and its consumption as a fermented drink formed a communion between the drinkers, their ancestors, the land and the entire Inca cosmology.

Hastorf and Johannessen[117] saw the transformation of raw materials from natural to cultural, especially as offerings, as being of great importance in Andean thought. Manufactured items (including *chicha*) are obviously not the same as their raw materials, and during processing they take on a new dimension; to some, they "now have a soul". To the Incas, objects transformed by human activity were especially worthy of being offered – either to a chief, or to the gods. But one can also make a soup out of maize, but this does not have the same religious or political significance. The difference is obviously ethanol, and *chicha* is not only the result of a transformation, it is also, itself, capable of causing significant transformations to the imbiber.

Maize intended for *chicha* production may have assumed its symbolic importance because it undergoes a number of transformations before it becomes *chicha* (during malting, mashing, boiling and fermentation, for example). Once *chicha* is consumed, it produces another transformation (intoxication) whereby "channels of communication" are opened, both with the gods and with fellow beings of a different class.

In the modern Andes, alcohol still plays an "identity-constructing" role,[51] because what one drinks still serves as a marker of one's origins and ethnicity.[188] Lowland indigenous peoples can sometimes be identified by what they drink, as for example in Peru where they tend to drink *masato*, a drink made from manioc. This habit makes them distinct from other Andean peoples and they are known as "*masato* drinkers". In the Amazon region, *masato* can be made from a variety of tubers and fruits, as well as manioc, and some of the drinks there are made from corn (*chicha*), so the situation can become complicated. In a more modern assessment of the social uses of alcoholic beverages in Peru, Doughty[100] examined the extent of alcohol consumption and the ritual functions of drinking. He found that the use of alcoholic drinks is highly patterned and integral to normal social interaction. Drinking served to promote certain community activities, such as public work projects. "Disfunctional aspects of drinking" were found to be minimal in the community studied (in the north central highlands).

6.6 THE CELTS

With their reputation for being prodigious drinkers, the Celts of Western Europe have been the subject of much study from archaeologists and, thanks to their "furious passion for drinking", we have plenty of attestations of Celtic drinking habits from Greek and Roman authors. Dietler[189] feels that, while the Celts were fond of drink, their reputation was muddied by a misunderstanding of the nature of their drinking and feasting customs. As he said: "Drinking was extremely important to Greeks and Romans as well; they just did it differently and were appalled by practices that did not conform to their own." Plato, for example, included the Celts among "six barbarian peoples noted for drunkenness, as opposed to Spartan restraint" (*Laws* 1:637), and Diodorus Siculus recorded their "furious passion for drinking" and noted that they would enthusiastically drink imported wine and then "fall into a stupor or into a maniacal disposition" (V: 26, 2–3). Other writers, such as Plutarch and Polybius, commented on the fact that their tendency to overindulge after battle was ultimately self-defeating, the most notorious example of this being in the early fourth century BC when, after sacking Rome, they were easy prey as they slept after their customary drunken celebrations.

The Celt's predominant native drink was beer made from cereals, described by Dionysus of Halicarnassus (XIII: 10) as "a foul smelling liquor made from barley rotted in water", but they developed a passion for imported wine, which, horrifyingly for the Greeks and Romans, they drank undiluted. An attestation of the dreadful habit of drinking neat wine was noted by Pausanius (X: 23.12), who reported that Brennus, the leader of the Celtic army which attacked Delphi in 278 BC, "committed suicide" by doing so.

During the last couple of centuries BC, Celtic imports from Rome reached massive proportions, and one conservative estimate puts the figure at forty million amphoras being imported over the last 100 years of the Iron Age (more than 50 shipwrecks laden with Roman wine have been found off the coast of southern France). The above estimate works out at some 2.65 million gallons imported annually. Smaller volumes of wine had been imported earlier, from the Etruscans (seventh century BC) and from Massalia (now Marseille) around 600 BC. Michael Dietler has extensively studied this ancient wine trade and rues the fact that there are no records of the activity. One of his early contributions[190] contended that many ancient societies may have been "driven by drink". He also makes the point that the French words for "brewing" (*brasser*), "barrel" (*tonneau*) and "un-hopped beer" (*cervoise*) are all of Celtic origin (quoting Moulin[191]).

A fascinating aspect of this early trade in wine is the resulting effect of it on Celtic art. Dietler believes that for centuries "Celtic thirst was the main feature linking the Mediterranean states and the Celtic world". As an example, he cites[189] the fact that the Late Iron Age art style known as La Tène was heavily influenced by the motifs on imported wine-drinking paraphernalia. Arnold[192] says that during the Hallstatt period Celtic drinking behaviour and equipment can be linked to the Greek colonies of the Mediterranean coast. An earlier scholar of this subject, J. M. de Navarro,[193,194] went further and reckoned that "Celtic art owed its existence to Celtic thirst". André Tchernia[195] has also gone into this subject in great detail.

In an important contribution to Celtic drinking, Bettina Arnold, recognizing that alcoholic beverages were important consumable status items in prehistoric Europe, and served as a social lubricant as well as a social barrier, carried out a cognitive analysis of the material culture of Iron Age drinking and feasting by integrating archaeological and documentary evidence. She also dealt with the impact of contact with the Mediterranean world, gender config- urations and the ideology of power and patronage in relation to changing material culture assemblages. Her literature review and documentation of archaeological evidence is particularly useful and commences with the Late Hallstatt period in West Central Europe (600–400 BC) – particularly from the British Isles and the Continent. As Arnold says:[192] "It is at this time that the peoples known as Celts begin to appear in recognizable form in the material record, but the contemporary Mediterranean written sources are meagre and problematic." Pitfalls relating to the union of archaeological evidence and the written word are fully discussed.

Not all workers have been totally happy with the marriage of archaeological finds and Classical texts, however. With the profusion of studies exploring various aspects of prehistoric eating, drinking and feasting over the past two decades (*e.g.* Parker-Pearson[196] and Poux[197]), Loughton[198] questioned the degree of reliance on the Classical accounts of Celtic drinking, and asked: "Do we really need to give them so much prominence given the wealth of archaeological evidence available? The most obvious drawback lies in the flexibility of these passages to interpret archaeological evidence from the Hallstatt to the late La Tène periods. Geographically these accounts have been used to interpret finds from Germany, Britain, Switzerland and France, yet most of these texts refer to events that took place in southern and central France during the Late La Tène period." The obvious problem here is that one tends to get a uniform and unchanging picture of drinking and feasting behaviours at all sites in the European Iron Age. Loughton continues: "It is questionable how relevant many of these Classical accounts are towards understanding drinking and feasting behaviours in non-Mediterranean France during the Late Iron Age, let alone during the preceding Hallstatt and early La Tène periods."

Mediterranean ceramics and metal vessels have generally been regarded as luxury objects with their use being confined to the elite. Similarly, wine consumption has also been seen as a privilege of elites, including the warrior class, and great feasts involving the drink were organized in enclosed places ("sanctuaries"). Concentrating on the abundant evidence relating to the Arverni, a Gallic tribe living in what is now the Auvergne region of France during the last centuries BC, Loughton challenged some of the existing interpretations of the trade in Republican amphorae, and looked at who exactly was consuming wine and where it was being consumed. He questioned whether sites where large numbers of amphorae have been recovered had higher status (*i.e.* were "better") than those with no, or fewer, amphorae. He goes on: "Yet, from several perspectives, there is nothing privileged or special about this. It seems strange to link, what are after all, higher rates of alcohol consumption with being a good thing! Discussions on the

Republican amphora trade often fail to remember that we are dealing with a large-scale trade in alcohol. The consequences of the volume of wine being consumed during the late La Tène period and its effects on society are rarely considered."

In a brief consideration of how frequent intoxication (even addiction) might have been perceived and interpreted by Iron Age peoples, Loughton moots that alcoholic beverages might have been "infused with special powers and with the spirits of the ancestors". Thus, the state of intoxication linked the drinker with his gods and forefathers. Being "under the influence" really did mean something; under the influence of outside forces, and temporarily being able to access the "otherworld". He also suggests that, far from being solely for the higher echelons of society, wine was likely used by artisans to enable them to access the special powers of the otherworld to aid them in their work (*i.e.* inspiration). As he put it: "Artisans may have equated the physical and mental transformation of materials experienced during intoxication with the magical transformation of materials during the manufacturing process. An individual's craving for wine was probably seen as a sign that the gods wished to commune with them."

It is probably not simply coincidence that large assemblages of Republican amphorae are found on urban, or semi-urban, sites during the Late Iron Age, and that such vessels are virtually absent from rural sites. The connotation is that high levels of wine consumption are linked to the stresses associated with living in the urban environment (rather than any connection with "eliteness"), and that it was probably the "working classes" drinking most of it. Urbanization has been related to heavy wine consumption throughout historical time, from Mesopotamia[199] to early modern Europe.[200] Loughton draws an analogy with the phenomenal consumption of gin in eighteenth-century London,[201] where one of the main problems was that gin was drunk in the same quantities as if it were (the much weaker) beer. The converse can be said to be true of life in the countryside, where the stresses are very different and there is little mass overcrowding.

The possibility exists that the elite avoided drinking wine during much of the late La Tène period because it was now widely available and was enjoyed by the "working classes" (who had, hitherto, been obliged to consume the drinks of their ancestors, beer and mead). Loughton ends his discourse with: "Finally, with

Figure 6.17 Bell beaker.

the decline in large-scale importation of Italian amphorae during the course of the first century BC, wine may have been taken up by the nouveaux riche and its consumption confined to more restricted social contexts, such as funerary feasts and rites." Loughton's PhD thesis (2000) is worth reading and is available on-line.[202]

Drinking vessels have always been integral items in determining status, and they have always been useful archaeological markers. I have always been fascinated by bell beakers (Figure 6.17, and see page 361), and the late Andrew Sherratt's "Cups that cheered" is a fascinating read.[203] Accounts of ancient Greek and Roman drinking vessels abound in archaeological literature.

6.7 SUB-SAHARAN AFRICA

The consumption of indigenous beer is a widespread and long-standing characteristic of many African societies, and a practice of historical and contemporary importance. It is only in the past couple of decades that in-depth studies of beer-drinking in Africa have been carried out. It was with great foresight that Ivan Karp[204] suggested that historians of "anthropological analyses of cultural forms in African societies may place beer drinking alongside the more standard forms of exotica such as witchcraft and sorcery, spirit possession and divine kinship". Karp went further when summarizing his contribution to an analysis of beer-drinking among the Kenyan Iteso, when he proffered that "were he to analyse the role of beer drinking in the context of the formal structure of Iteso symbolism, he might focus on the role of beer as a mediating symbol".

With Africa's pivotal role in the ascent of man, it is perhaps not surprising that the continent has been credited with the earliest fermentations. In his valuable contribution to bee culture and the use of honey in Africa, Carl Seyffert[205] maintained that apiculture was transmitted to Africa by the Hamites (a people believed to have been descended from Ham, one of Noah's sons), and that fermented honey drinks were probably the earliest alcoholic beverages known to man. He also remarked that, of all the regions in Africa, "the West African tropical rainforest area is the only one that does not make much use of honey". As we find elsewhere, honey requires only rainwater to initiate yeast activity.

The remarkable role of alcohol, and its production, in Africa was remarked upon by many of the early adventurers, as can be seen from this early-seventeenth-century account from Ethiopia (as cited in Beckingham and Huntingford[204]):

"Five or six parts of water and one of honey are put into a pot with a handful of toasted barley which makes it ferment. Afterwards they add some morsels of a certain wood called Sado‡, which draws out the sweetness and in five or six days it is ready to drink. It had not the taste of our grape wine, but its wholesomeness undoubtedly gives it an advantage. These people have one great vice, which is that they are much addicted to wine. They do not drink while eating, but when the table is cleared the *caloes*§ are brought into the house and neighbours or friends assemble. In noblemen's houses and in the Emperor's many guests

---

‡A local name for a buckthorn, *Rhamnus staddo* A. Rich.
§Large earthenware pots.

are always invited. There is no conversation without the wine circulating and being drunk in turns until either they are laid out on the spot or else withdraw so much warmed up that they cannot easily find the door. They achieve this by drinking to such excess that no one knows how they can hold so much. The wine is really very mild and if it were drunk in moderation no harm could be done."

There are many such insightful accounts of alcohol and the Africa of yore, and many are well worth searching for.

Africa is such a large continent that it is impossible to generalize but, in most African countries, alcoholic drinks are considered to be a food, and drinking has always played an important role in traditional social life. For convenience, Adomakoh[207] divides Africa up into three groups: the Arab group occupying northern Africa, the Black African group, which is south of the Sahara, and the White African countries in central and southern Africa. Of the Black African countries, he ascertains that "the local and traditional drinks are basically similar", and divides them into: "beer brewed from millet; palm wine, and a distillate obtained from palm wine, which may contain up to 85% alcohol and is rich in impurities, for example aldehydes, methyl and butyl alcohol and esters".

While it is dangerous to make generalized statements about drinking in Africa, they have been uttered in the past, and often lead to misconceptions; Carothers[208] is an example, when he stated that alcohol consumption in Africa was "in the form of palm wine and beer from millet, that spirits are not drunk to any extent, except in the townships, and that although the wine and beer are of low alcoholic content, sufficient quantities are consumed to induce intoxication that produced results more often criminal than psychiatric".

Beer-brewing in Africa, as elsewhere, requires a considerable amount of grain, and it is a time-consuming process demanding a substantial level of labour (mainly female). By processing grain into beer (rather than bread), one is emphasizing one of its major roles; namely communal consumption. From information we have, brewing among the Mabaso women of South Africa takes 10–12 days,[209] while beer made by the Maale of southern Ethiopia takes five days to produce.[210] In many processes the first step is to allow grains to dampen in a moist environment. After a time, these are dried and mixed with more grain prior to being stone-ground (very time-consuming). The resulting grist is mixed with hot water (sometimes involving huge quantities of labour) and the mixture is boiled, cooled and fermented in large pots.

The value of alcohol, particularly the locally made beer or palm wine, seems to depend on factors other than its ability to intoxicate. This was certainly the impression that Netting got when describing the Kofyar of West Africa:[48] "The Kofyar crave beer as much for its food qualities of taste and nutrition as for the psychological effects of alcohol." He also relates that men "make, drink, talk, and think about beer [all the time]. It is a feature of cultural concern and activity in much the same way as are cows in pastoral African societies". Drunkenness is frowned upon, indeed, Netting said: "I never saw a person intoxicated enough to lose motor control to the point of staggering or to become sick ... Stupor is neither the desired goal nor the usual outcome of Kofyar drinking."

The Kofyar are the so-called "hill pagan" populations found in the broken hills and high plains of the Jos Plateau in northern Nigeria, and they have adapted to their rugged environment with much success. They are sedentary subsistence farmers and survive by intensive agriculture on small farms. They have little political centralization, and are largely self-sufficient and lacking in occupational specialization.

In East Africa, there has always been a plentiful supply of raw material for making alcoholic drinks, a fact remarked upon by some of the early colonists. Wheat, barley, sorghum, maize, millet, cassava and bananas are all oft-used sources of fermentable substrate, and alcohol has been used for centuries to enhance a wide variety of social functions. After the arrival of the Europeans (who introduced distilled beverages), alcohol became part of the package of trade that formed the basic tie between Africa and Europe,[211] and alcohol was one of the main articles of the barter system through which European goods were exchanged for African slaves. The situation regarding alcohol consumption got out of hand and, in 1889, a group of European powers met in Brussels (!) and drew up an International Treaty under which they agreed to observe certain moral principles with regard to their African territories. A number of articles on the traditional use of alcohol in East Africa have been published.[212,213]

Mushanga,[214] in a study of the traditional use of alcohol in western Uganda, wrote: "Alcohol played a very important social function in African cultures. Alcohol was always present in meetings when elders settled disputes, after a successful hunt or harvest, and when marriages were arranged. The bride price usually paid in cows was traditionally accompanied by several pots of beer."

Platt[213] contends that there are "five main groups of African alcoholic beverages: fermented honey water; fermented fruits and juices; fermented sap from various species of palm and bamboo; drinks made from milk, and the beers. All of these have more or less familiar equivalents in mead, cider, fermented birch sap, fermented mare's milk (koumiss) and old-fashioned ales". He saw beer as being "the commonest alcoholic beverage in Africa", and that "various materials – sorghum, maize, millets, cassava flour with additions of cereals, and plantains are used in the preparation of these beers". Platt could not find any details on the alcoholic strength.

6.7.1 An Early Problem

Despite its central role in society, alcoholic drink has long been a problem in Africa, much of it, apparently, attributable to distilled beverages introduced *via* the influence of Europeans. Late-nineteenth-century traveller and anthropologist Joseph Thomson,[215] on an exploratory trip, offered: "Stated briefly, I have to confess, with shame and reluctance, that the opinion I have formed is, that contact with the European in West Africa, has been attended with almost unmixed evil to the natives." He continues: "Along the whole line of that unhappy continent there rises the cry for more drink – more drink – give us tobacco, gunpowder and guns. Those are the wants roused by a hundred years of European trade, till now the proudest boast of a native village is the size of its pyramid of empty gin bottles as showing how much spirit they can afford to drink."

On a journey up the Niger River, Thomson searched for evidence that the influence of imported European drink was not all bad, and forwarded the following: "In steaming up the river, I saw little in the first two hundred miles to alter my views, for there luxuriated in congenial union fetishism, cannibalism, and the gin trade. But as I left behind me the low-lying coast region, and found myself near the southern boundary of what is called the Central Sudan, I observed an ever-increasing improvement in the appearance and character of the native; cannibalism disappeared, fetishism followed in its wake, the gin trade largely disappeared." He goes on to describe domestic matters and arrives at a significant conclusion – something I wish we had more time to discuss: "Everything indicated a leavening of some higher element, an element that was clearly taking a deep hold on the negro nature and making him a new man. That element you will perhaps be surprised to learn is Mohammedanism."

A somewhat different view of drink, albeit beer, as opposed to gin, was given by a man of the cloth at the turn of the twentieth century (as reported by Bryant[216]), who made a plea for a sensible attitude toward beer: "It is a great error, in my opinion, to assume, as many, even 'Old Colonists', are apt to do, that this Kaffir-beer is simply and solely an intoxicating drink. It is, in my view, much more than a luxurious and supererogatory beverage. It is rather a very admirable, very beneficial, even, perhaps, very necessary form of food; and Governments in their legislation thereanent should recognize this fact, and aim rather at preventing its abuse than preventing its use."

6.7.2 Xhosa Beer Drinking

Patrick McAllister has spent over 20 years in South Africa studying ritual, oratory and beer drinking, and much of his work is presented in his 2006, *Xhosa Beer Drinking Rituals*,[217] a compilation of participant-observation ethnologies that demonstrates the social significance inherent in Xhosa beer drinking rituals. The Xhosa people now live primarily in the Eastern Cape areas called Transkei Ciskei, although they are found all over South Africa. They were part of the gradual Bantu migration from southern Zaire (the most southern group), and are descended from a clan of the Nguni. Historically, they have been herders and farmers, but now engage in many more activities. As Shaw noted some years ago,[218] beer is universally made by the southern Bantu, and historical sources suggest that a variety of raw materials, such as millets, sorghum and maize, were used (the latter being an introduction from America which rapidly became a staple crop).

The first records of indigenous beer from southern Africa come from a miscellany of European travellers and shipwrecked sailors,[219] and we read of "millet wine, consumed with great enjoyment". One drink of special note was *pombe* "which is very strong and intoxicating". Historical sources note that there were regional variations in the mode of brewing, and there was awe at the enormous quantity of beer available. Note was also made of the fact that beer appeared in folktales, songs, myths of origin and proverbs, and was of political importance.

In Africa, the nourishing character of beer has always been an important factor and in many parts of the continent it is still a major foodstuff. In one very early study,[220] involving "Kafir beer", solids were recorded at 5–7%, ethanol at 2–4%, and there were significant levels of vitamins B and C, protein and minerals. Also, as Hugh Ashton[221] picked up, brewing was a useful outlet for grain that was unsuitable for eating.

With increasing urbanization in southern Africa, and greater capitalist influence, the Xhosa have endured considerable pressure on their way of life. These changes came on top of the changes wrought by colonialism and apartheid. McAllister argues that the transition from a rural agrarian economy to one based on waged labour has resulted in a transition from a kin-based economy to one that is community based. These economic changes were reflected in cultural practices and it was noticeable that, since the beginning of colonization, large communal beer-drinking rituals started to prevail over smaller family-based rituals. These rituals were not, therefore, static events and are/were capable of responding to changing conditions. One conclusion made was that "beer enables rural people to turn assets (including grain, cattle, and money) into symbolic capital *via* ritual and communal activities. In the rural context, the social legitimacy of migrants' wages was dependent on the earnings being transformed into symbolic capital of some kind."

One particular beer-drinking ritual, the *umsindleko*, which is a "large beer drink" associated with labour migration, is noteworthy. McAllister considered the *umsindleko* (literally means "food prepared for a traveller by his wife to await his arrival" – see Kropf[222]) warranted extended study because "it allows one to muster the contemporary symbolic evidence relating to the series of interconnected historical changes that continue to govern local economic activities". The ritual is particularly pertinent in modern times, for many men are constrained to find work away from home. The name *umsindleko* is typically descriptive, and informs about gender and kin relationships and the fact it is celebrating a person's return.

Xhosa ritual beer-drinking is not simply about downing liquid, oratory is involved, and this has been likened to a public performance. At the *umsindleko*, speakers extol the virtues of good behaviour and castigate bad behaviour and, in terms of the latter, anything that prevents money going into the homestead is deemed "bad behaviour". A migrant worker who has returned with money saved for his family is much celebrated. A more profligate migrant worker is encouraged "to do better next time", and ancestor spirits are often invoked to aid with this encouragement. Beer is closely associated with the ancestors, who liked to drink it when they were alive and should be offered it by the living. When labour migration became more prevalent during the twentieth century, many families found that they were left with insufficient manpower to run their households. Accordingly, such households relied on neighbours to provide essential labour for routine tasks, and this has spawned another drinking ritual.

McAllister uses the term "beer drink" for such occasions and describes it as: "a public event, complete in itself, held in an individual homestead and for a particular purpose, at which beer is the central element and is distributed and consumed in a formally prescribed manner according to a set of social conventions." He found that beer drinks are open to all adult members of the local community as well as to visitors. He continues: "children, youths, newly married women and young men who have not yet been admitted to beer drinks, may not attend, though they are present and involved during the preparation, and receive an allocation [of beer] once the brewing is complete. Being public, beer drinks attract relatively large numbers of people – never less than about 30, sometimes hundreds. Generally, no one is specifically invited to attend and everyone in the vicinity is free to do so; the long brewing process is advertisement enough, and people learn by word of mouth where beer is available and when it will be ready."

Beer is metaphorically spoken of as *ubulawu*, or "medicine of the home", a liquid with sacred properties. All beer drinks are religious inasmuch as "the ancestors are thought to want their descendents to brew and to partake of beer 'in spirit'". Medicine of the home is used in expiation rituals, where the sinful person washes with it in order to obtain ancestral blessings and atonement.

With the undoubted social, economic and political value of beer in many African societies, it is perhaps strange that very few archaeological studies have attempted to study beer production and consumption, Edwards,[223] Samuel[224] and Arthur[225] being three notable exceptions. Edwards has worked on the social and political organization of the Kushite (Meroitic) period (late-first millennium BC–early-first millennium AD). Although strictly not sub-Saharan, Kush was an ancient Nubian state centred on the confluences of the Blue Nile, White Nile and Atbara rivers (now Sudan). For information about the archaeology of the area see Edwards.[226] It is highly likely that, from an early date, alcoholic beverages were an important way of utilizing sorghum in the Middle Nile basin, and that such usage had a notable social and nutritional significance. The importance of alcohol in the study area, in Edwards's words, is "further suggested by the very long history of wine importation into the region from Egypt and later the Mediterranean world". As we have noted elsewhere, grain beers play an important role in mobilizing labour, and in the social organization of production and reproduction. Archaeological evidence appears to suggest that sorghum beers were important within mortuary rites, both in terms of ritual alcohol consumption and, possibly, for mobilizing labour to construct funerary monuments. In addition inscriptional evidence from Kushite royal monuments points to substantial ritual links between the Meroitic monarch and sorghum, and particularly the monarch's ritual roles in ensuring land fertility, rain and a good harvest.

6.7.3 The Ethiopian Gamo

In his study of the Gamo people of Ethiopia, Arthur[225] found that the size and morphology of ceramic vessels, their context and the presence of a certain type of surface abrasion on the interior walls of these pots provide useful signatures for the recognition of beer production (and storage). The Gamo inhabit the highlands of southwestern Ethiopia along the Rift Valley and brew beer for ceremonial and subsistence use. They use different grains, mainly millet, wheat, maize or barley, depending on where they live (ecologically). The highland Gamo brew by boiling water in a large

Figure 6.18 Gamo beer jar (by permission of John W. Arthur).

ceramic cooking pot (*otto*) and then pouring it over flour and stirring it in a large serving bowl (*shele*), where it will cool. The liquid is then poured into a large beer jar (Figure 6.18) to ferment for five days. In lowland areas, water and flour are boiled together and, when cooled, the mixture is placed in beer jars to ferment.

Ethnographic evidence points to there being a strong connection between large jars and brewing,[210,227] where they are used for fermentation and storage. Large jars can also be used for other purposes, and archaeologists employ use-alteration analyses to distinguish between brewing and "other uses". Such analyses include carbon deposition, residue analysis and wall surface abrasion, with the condition of the inner wall being particularly useful for identifying brewing activity. Arthur found that there was an inequality within a population, in terms of feasting and brewing. Wealthier (high-caste) groups are able to attend political and social feasts, while poorer, artisanal (low-caste) groups are excluded. Similarly, wealthy households will have greater access to beer, because they have greater access to grain (*via* a larger land ownership). There was also a relationship between status and large (brewing) vessel ownership.

As part of his conclusion, Arthur offers: "The Gamo case study indicates that the size and morphology of ceramic vessels, their context and especially the presence of a certain type of surface attrition on the inner walls of these vessels provide valuable signatures for the identification of beer production. The consumption of beer often takes place outside the household, such as at specific feasting areas, but in this case study the majority of beer production took place within the household. Among the Gamo there is a strict social hierarchy and beer production and consumption is associated with the wealthier households."

6.7.4 Sorghum Beer and the Iraqw of Tanzania

The symbolic qualities of sorghum beer and milk among the Iraqw of northern Tanzania were the subject of a study by Rekdal.[228] Both milk and beer have deep historical roots among the Iraqw and Cushitic peoples, who entered the area 3000 to 5000 years ago. They were probably the first food-producing inhabitants of East Africa, and it is presumed that sorghum was one of the first crops to be cultivated in the region. Archaeological and linguistic evidence indicates that Cushites were probably the first East Africans to keep cattle and make use of their milk.[229] The drinks obtained from sorghum and cattle became associated with qualities understood to be positive and desirable.

The product known as *buura* refers to beer brewed either from sorghum (*mangwaré*) or from sorghum mixed with maize flour (*buusa*), although it may represent a range of alcoholic drinks. Both beer (over-indulgence apart) and milk have long been associated with attributes that are understood by these people to be positive and desirable, but with the advent of the market economy and of money as an exchange medium, the symbolic content of both drinks has become diluted somewhat. But, as Rekdal says: "As products in demand, they may today circulate in impersonal relations which lack the social and religious qualities that they traditionally communicated. The monetisation of sorghum beer and milk has not, however, caused a breakdown in established practices, nor in the structures of meaning in which such practices are embedded."

Rekdal explains that handling sorghum beer is a highly ritualized performance. Beer is poured from the pot in two or four stages, never in one or three, and it should never be drunk while standing, nor should anyone sit down while holding a gourd full of beer. Various vessels may be used for drinking, but they must be large enough to contain sufficient beer for several people (never one person). The gourd should be passed continually back and forth between persons of the same social standing.

Interestingly, Rekdal also speaks of the use of sorghum beer as a remedy for, or protection against, arguments and/or conflict. For example, hardwood sticks (themselves potential weapons) are smeared with *buura* if there is a suspicion of conflict between persons at a social gathering. He adds: "Sorghum beer, together with honey or honey mead, is also a central element in rituals seeking to withdraw and neutralise curses." *Buura* is also used to mediate with the "spirits of the dead" (*gi'i*), and so as to maintain, or restore, good relationships with *gi'i*, offerings of the drink (sometimes milk) are presented, usually at the threshold of the departed's dwelling or at his/her grave. The paper contains interesting information about the introduction of money into African societies, and the huge changes that have resulted. That is another story, and those interested should consult Parry and Block.[230]

6.7.5 Banana Beer

The Haya of northwest Tanzania use a mixture of sorghum and bananas to make their beer, with the latter being seen as an additional source of sugar.[231] The Haya are an interlacustrine Bantu-speaking people whose staple food is banana, and Carlson describes and analyses the traditional symbolic mediating role of banana beer, describes the effect of beer on the consciousness and illuminates the ways by which banana beer is utilized as an offering to "mediate the domains of everyday life and the ancestors". He goes on: "the value of describing the meanings associated with the use of banana beer, not merely in consumption but also in exchanges, tribute to the king, and in ancestor propitiation, lies in clarification of the role of alcoholic beverages 'as part of a larger configuration'."

Carlson[231] gives a recipe for this banana/sorghum beverage, which the Haya always refer to as "beer". The preferred bananas (called *embiile/enkundi*) are different from those used for cooking, and they are ripened by either hanging them over a hearth or by burying them in a pit for several days. This expedited ripening stimulates starch amylolysis, thus increasing fermentability. Ripened fruit (either peeled or unpeeled) is then placed in dug-out wooden troughs and broken up by foot maceration – dried grass being added to promote tissue breakdown. Gross material is removed from the resulting juice by using dried grass sieves and the cleaned juice is diluted (1 : 1) with water. Dried sorghum is added, the container covered with banana leaves and the mixture fermented for 24 hours minimum. No heat is applied to the juice. Brewing here is a male occupation, although females are allowed to perform certain tasks such as grass collecting and water collecting. Women may never crush bananas! Tradition has it that men and women must drink this beer from bottle gourd (*Lagenaria* spp.) containers of different shape. Males use a calabash (*ekilele*) with a long narrow neck, and sip through a hollow reed (*orusheke*). Females cannot use such a vessel, and must use either a short-necked gourd (*orushuba/endee*) without a straw or a banana leaf cup (*ekitegera*).

Carlson[231] said that "Banana beer is a secular refreshment, a necessary component of all celebrations and life-cycle rituals, including negotiations between clans for marriages, a traditional form of tribute to the king, and an essential substance for ancestor propitiation." He then continues with the effects of beer consumption, saying:

"Haya interpret the effects of beer consumption on a person's perception and behaviour in terms of a hierarchical series of levels. The first level satisfies a basic need: to refresh oneself (*okwehoreleza*) and to quench a thirst (*okutamba eiliho*). This stage is marked by the absence of altered perception or behaviour. Especially in the case of adult men, the light consumption of banana beer throughout the day is considered an essential feature of the diet. Moreover, the consumption of banana beer is believed to contribute to a man's strength generally, including his sexual potency."

The second stage is expressed primarily through the use of two verbs, *okuhaaga* and *okushemera*. The first means "to be full of banana beer", while *okushemera* means "to be high, happy, or hilarious". The third stage can be expressed by the phrase *okushaagwa amarwa* – "to be overcome by banana beer; to begin to lose control of oneself because of drinking". The verb *okushaaga* literally means "to go beyond". The fourth stage of altered consciousness is called *okutamiila and okusirana* – "to be very drunk" – and, in this state, one's speech is slurred, rapid, loud and incomprehensible. It is associated with staggering and fighting and, according to tradition, the proper use of banana beer should only result in the first two stages of altered consciousness.

Investigating the use and significance of alcohol in a rural parish in Uganda, Robbins[232] attempted to confirm the anthropological hypothesis that when alcohol is well integrated into a socio-cultural system its positive social and physiological functions will tend to override its role as a means of appeasing personal psychological problems. Conducted in a parish in the Buganda region, the work highlighted the differential functions of the banana beer, *mwenge*, a well-integrated, traditional beverage, and *nguli*, a distillate of *mwenge*, which is a recent development and a far less integrated product. The hypothesis under scrutiny was that people who drink mainly *nguli* do so more for psychological problem-solving reasons than those who mainly drink *mwenge*.

While admitting some other issues that impinge upon the central theme, such as the fact that *nguli* drinkers were rather younger than *mwenge* drinkers, it was found that *mwenge* is a more integrated drink than *nguli* and, as predicted, a sample of *mwenge* drinkers drank less for personal psychological reasons than a comparable sample of *nguli* drinkers.

6.7.6 Feasting, with Special Reference to Africa (*à la* Dietler)

The study of food and eating is important not only because of essential body maintenance, but because it has proved to be a fertile field for the germination of many archaeological and anthropological theories. Food studies have illuminated broad societal processes such as politico-economic value creation, and the social construction of memory in anthropology. Practices associated with food and eating support numerous meanings and play a pivotal role in the construction and function of identities.

The subject of "feasting" has been one of much study by Michael Dietler and the edited volume *Feasts*[233] is a must for anyone interested in this matter. Over the past three decades, Dietler[227] has made a concerted effort to develop the theoretical understanding of drinking as a social practice and, to some extent, this has encouraged the development of new techniques for detecting alcohol production and consumption at archaeological sites.[234]

Dietler made his case for the importance of studying "the feast" as follows: "There has been much written recently about the need to develop a practice-oriented approach in archaeology, but rather few coherent suggestions or effective demonstrations of how this can be accomplished. This is one of the principal attractions of a focus on feasts ... the "commensal politics" of feasting is a domain of political action that is both extremely important on a worldwide scale and potentially accessible to archaeological analysis." Dietler defines feasts as "a form of public ritual activity centred around the communal consumption of food and drink". This means that they differ in some way from acts of daily consumption, but it does not mean that they are necessarily highly elaborate ritual occasions, or even sacred in character. As Dietler points out, "a ritual act can be as simple as making the sign of the cross upon entering a church, pouring a few drops of beer on the threshold of a house as a libation, or throwing a small wine and cheese reception for a visiting anthropologist". The defining feature of rituals is that they are symbolically different from everyday activities; as Kertzer[235] said, they are "action wrapped in a web of symbolism". Hayden[236] uses a broader definition of the word, specifying that any unusual occasion accompanied by a shared meal should be called "a feast".

Feasts are an especially potent form of ritual activity because they are potentially visible in the archaeological record, especially in terms of the containers used for their preparation and consumption. Containers made of ceramic or metal are likely to be useful markers, even when broken. The food and drinks associated with feasts are, of course, highly perishable. To anthropologists and ethnologists the word "commensal" has a subtly different meaning from that understood by biologists and, in context of the former disciplines, commensality is eating with other people, especially at the same table, and commensal eating patterns reflect the social relationships of individuals.[237] By extension, the concept of commensal politics accrues from the above and, again, it is pertinent to quote Dietler:[238] "the potential of feasting and hospitality to be manipulated as a tool in defining social relations lies at the crux of the notion of commensal politics."

In a useful footnote, Dietler[233] puts the record straight about the word "commensal" as he uses it: "To avoid possible confusion, let me emphasize that I use the word *commensal* in its original sense, rather than its peculiar biological adaptation. The word derives from the Latin *com mensalis*, indicating the sharing of a table – hence, eating together. Needless to say, many people around the world manage to eat together quite well without using a table. Moreover, in a number of cases the sharing of food is accomplished without the host and guests actually eating in the same space – in some contexts it is actually considered impolite for the host to be present when his/her guests consume their food."

Having spent several years undertaking ethnographic research in Africa, familiarity with that continent, and the fact that it "gave birth to political anthropology as a field", made it a fertile place for Dietler's studies. Additionally, Africa has long been seen as a land mass with some remarkable traits in cuisine (despite Goody's[188] assessment that it was the "prototypic land without cuisine") and politics. As elsewhere, feasts in Africa serve a wide variety of important structural roles in the overall political economy. In the context of Africa, Dietler[233] defined three different patterns of commensal politics, the first being directed toward the acquisition, or creation, of social (and economic) power, and named "empowering feasts" (or sometimes "entrepreneurial feasts"). These involve "the manipulation of commensal hospitality toward the acquisition and maintenance of certain forms of capital, and sometimes economic capital as well". The term covers a range of symbolic consumption practices that help to negotiate social positioning. According to Dietler, "the empowering feast pattern operates on a variety of scales and in numerous contexts within a given society. It may extend from the private hosting of a pot of beer among a small group of friends, to the hosting of trade partners from another community, to the sponsorship of major community life-crisis ceremonies and religious festivals".

Patterns two and three are directed toward the maintenance of existing unevenness in power relations, and these two operate in different ways. One works *via* the expression of donor/receiver, superiority/subordination relationships (as in the above mode), while the other works through the idiom of distinctive exclusion in an insider/outsider relationship. Dietler's second major pattern of commensal politics he calls the "patron-role feast", and this involves the "formalized use of commensal hospitality to symbolically reiterate and legitimize institutionalized relations of asymmetrical power". This corresponds to a specific form of what has traditionally been called "redistribution" in the literature of economic anthropology" (*e.g.* Sahlins[239]). The main attribute of this mode of commensal politics is the establishment of reciprocal obligation produced through hospitality, as above, except that here there is no expectation that a state of equal reciprocation will be maintained. *Au contraire*, the acceptance of a continuous unequal pattern of hospitality "symbolically expresses the formalization of unequal relations of status and power and ideologically naturalizes it through repetition of an event that induces sentiments of social debt". Thus, we find that in Africa especially, but in many other parts of the world, the king, or chief, is expected to instigate regular lavish hospitality – and guests are never likely to reciprocate.

The third major mode, named the "diacritical feast", employs differing cuisines and styles of consumption as a distinguishing symbolic device to establish ranked differences in the status of social orders or classes. This concept differs from the last mode in a number of ways, the main one being the replacement of the emphasis on quantity to details of style and taste. Dietler adds: "Moreover, the emphasis shifts from an asymmetrical commensal bond between unequal partners to a statement of exclusive and unequal commensal circles: obligations of reciprocal hospitality are no longer the basis of status claims and power." Because this type of feasting relies upon style and taste for its *raison d'être*, it is quite possible for those wishing to attain higher status to imitate it.

Dietler goes on to discuss "feasts and social boundaries", "feasts and gender" and the "different modes of commensal politics", before describing aspects of feasting among the Luo of western Kenya. The Luo people live in the basin that surrounds the Winam Gulf of Lake Victoria in western Kenya. Their livelihood depends on small-scale agriculture, animal husbandry and fishing. They grow grain (sorghum, millet, maize), root crops (sweet potatoes, cassava) and other food crops, such as beans, bananas, lentils and peanuts, which vary locally in importance.[240] Agriculture is the domain of women and is carried out in scattered plots. Little food has to be imported into their area, and the crops grown provide a variety of dishes. The staple food, a thick porridge called *kuon*, made from boiled sorghum and maize flour, is seen as being symbolic to the Luo and some other dishes are derived from it. The main alcoholic beverage is a beer (*kong'o*) made from millet and/or maize but, in recent decades, a distillate known as *chang'aa* has become popular. Both of these drinks are now essential components of feasts, which are an important element of Luo life. Most Luo feasts can be placed into the aforementioned "empowering" category, with the largest examples being associated with funerals. They occur at the homestead of the deceased and involve huge quantities of beer (and beef), plus more "standard" fare such as *kuon*. As well as victuals, there are such events as parades of cattle, dancing, singing, oratory and the recitation of eulogies telling of the achievements of the deceased and the speakers. Such events can last for several days, and the prestige of the deceased person (and his/her family) is judged by the size of the feast and the length of time that it lasts. As Dietler says: "Influential men have the most ostentatiously lavish funerals, but every Luo is concerned about having an impressive funeral mounted for him/herself. This concern is often voiced by older widows as a major reason for joining religious groups, as these assure their followers of a proper funeral."

More modest feasts are held for marriages, crop gatherings, *etc.*, and small-scale gatherings between friends are usually signed by sharing a pot of beer. Whatever the size of these events, they will be distinguished from daily meals in some way. The most oft-used distinctions are the consumption of beer (or *chang'aa*) and beef, which are everyday foods. In some parts of Luo territory, homesteads have a specific shaded area (*siwanda*) where senior men can sit and feast. At the more important celebrations, special

Figure 6.19 Communal drinking vessels: *Thago* (left), *dakong'o* (right) (after Dietler[227] - by kind permission).

paraphernalia is used, the most evident being a large pot called a *thago* (Figure 6.19), which is partially buried in the ground to keep it upright. Drinkers sit around the *thago* and drink from it *via* large straws (Figure 6.20) constructed of hollow vine stems with a

Figure 6.20 Beer being consumed *via* straws from a *thago* (after Dietler[227] - by kind permission).

woven filter on one end. To have one's own straw (in a bamboo case) is an indication of senior status. The pot in which the beer has been brewed, the *dakong'o*, invariably stands near the *thago*, and beer is taken from it and mixed with hot water for consumption from the *thago*. Being a large, expensive pot, the *thago* is somewhat limited to wealthier households, and others will need to borrow one if they are holding an important feast. Conversely, every household will possess a smaller pot, the *mbiru*, which usually contains filtered beer. From this, drinkers remove beer, traditionally with a small dipping cup (*agwata*) made from half of a hollowed gourd.

Much labour is required to grow, store and process enough grain to enable a feast, and this responsibility is met by women who grow the crops, harvest and process them and carry out the cooking, brewing and serving. Thus, for a power-seeking male, having several wives is not just a sign of wealth, but a practical necessity if he is to host large feasts. To some extent, wives share in the benefits accruing from a successful feast. Dietler's observation of the extent of polygyny in Luo society is interesting: "It is noticeable that successful chiefs today are still conspicuously more polygynous than the rest of their people. For example, one chief in our research area had 45 wives when we arrived and over 50 when we left three years later."

Regarding other feast categories Dietler says: "Obviously, the Luo do not have diacritical feasts in the sense defined earlier (at least those in the countryside – the situation of the Luo who have moved to Nairobi is somewhat different). The situation regarding patron-role feasts is a little more complicated." He feels that, through Luo examples, "one can begin to understand the way in which the hospitality used in empowering feasts to acquire and maintain symbolic capital can become transformed into the institutionalized expectation of the patron-role feast; and one can see how a failure to meet those expectations can seriously weaken credibility and undercut authority ... Feasting is by no means the only arena of political action, but is very frequently an extremely important, if not crucial, one".

6.8   THE SYMPOSIUM

Much has been written about the role of the ancient Greeks and the Romans as civilizing influences during the history of man, and there is no reason to repeat it all here. Both civilizations were liberal in their use of alcohol, and it is somewhat strange that neither ancient language has a word meaning "alcohol". Instead, they use "intoxicant", a nomen that includes drugs, such as narcotics, as well as wine, beer, mead, *etc.* Distinctions were made, as witnessed by Aristotle, who recognized two categories of "intoxicant", those that cause "heavy-headedness" (*i.e.* soporific) and those that are "stupefying". In the first category, Aristotle includes wine, mandrake and opium poppy, while beer appears to be the only substance classed as stupefying. As Max Nelson avers:[241] "It is clear that Aristotle does not classify wine and beer together as alcohol-based intoxicants, but rather classes wine with opium, mandrake, and lolium, and beer in a separate category. Aristotle's false categorization certainly had considerable influence. A number of his students followed him in writing works on intoxication, including his greatest pupil, Theophrastus." Aristotle's *On Intoxication* confirms his distinction between beer and other intoxicants, even to the point of how one falls down: "But a peculiar thing happens in the case of the [drink made] of barley, the so-called *pinon.* Under the influence of all other intoxicants, those who become intoxicated fall in all directions, sometimes to the left, or to the right, or onto their fronts, or onto their backs. But those who become intoxicated from pinon only fall onto their backs and lie flat" (quoted by Athenaeus). A second quotation of Athenaeus from Aristotle, along the same line, appears to be somewhat contradictory: "Those who become intoxicated from wine fall onto their faces, but those who have drunk the barleyed turn upside down on their heads, because, on one hand, wine makes one heavy-headed, while, on the other, the barleyed is stupefying." Thus, when Aristotle did write about beer he was not particularly complimentary about it. *On Intoxication* (or *Symposium*) dealt with aspects of the Athenian drinking party and was one of the earliest Western texts to make comparisons between wine and beer.

To some Greek writers, most notably Aeschylus, beer was an effeminate drink. The cereal from which it was made was not thought of as such, so presumably something "happened" during the conversion of grain into beer that produced undesirable qualities. Maybe this is why beer was almost entirely overlooked by Hippocrates in his medical works. Wine and mead are discussed, but nothing much about beer.

In the classical Greek world, the word "symposium" ("to drink together") is used to describe a drinking party held to enable the participants to engage in intellectual discussion. They were much more complex affairs than the occasional parties thrown by the Athens elite. Aristotle's *Symposium* is thought to have been inspired by his teacher, Plato, and his experiences of intoxication recounted in his *Symposium*, a philosophical text dated *ca.* 385–380 BC. Like every Platonic dialogue, it comprises long, and often complex, arguments embedded in the dramatic form of a philosophical debate or conversation. Held at the house of the tragedian Agathon in Athens, the theme is love and each of the seven participants (*symposiasts*), one of whom is Socrates, must deliver an *encomium*, a speech in praise of Eros.

The Greek symposium, a male preserve, commenced with a banquet, which was eaten without wine, and this was followed by drinking, which followed a strict protocol. The first step was to elect a *symposiarch* from among the *symposiasts*, whose duty it was to arrange how they would be entertained (*i.e.* whether serious or sensual) while they drank. Then, under the auspices of the *symposiarch*, the guests would determine how many *kraters* of wine they would consume, and in what proportion the wine would be diluted with water (to drink undiluted wine was uncivilized). The *krater* (Figure 6.21) was a large mixing vessel designed to be carried by two men, and from which wine was distributed into drinking pitchers (*oenochoe*) or cups, which were often adorned with Bacchic or other drink-related scenes. The *krater* was the focal point of the symposium, on account of it being the source of wine! Mixing of wine and water in the *krater* was undertaken by an *oinochoos*, a young male slave (Figure 6.21). The *symposiarch* was also responsible for arranging suitable furniture (*e.g.* the high-legged couch, or *kline*) and drinking equipment to be in the special room (the *andron*) in which the symposium was held.[242] There was an accepted drinking protocol and over-indulgence was frowned upon. Some formalities were observed, however, the most important of which were libations – the pouring of a small quantity of wine in honour of the gods, or the mourned dead.

Traditionally, three toasts were drunk before a symposium; the third one usually being drunk to Hygeia, the goddess of health.

Figure 6.21   *Oinochoos* serving wine from *krater* (left).

Reasons for holding a symposium were varied, perhaps the most common being to celebrate some sort of victory (*e.g.* in athletics, or winning a literary competition). Introduction of young males into Athenian society (a sort of "coming out ball") was another. Entertainment was mainly in the form of music and poetry, but women were hired to converse, consort and perform with the guests, as were boys and slaves. Women of standing were precluded from symposia. Two musical instruments were closely associated with the symposium: the *barbitos*, an ancient stringed instrument related to the lyre, and the *aulos* (Figure 6.22), a "flute-like" wind instrument. Among the games played was *kottabos*, in which stale wine and/or wine lees was thrown from a target from a kylix (similar to dwyle-flunking, where stale beer is hurled!!). Another common activity was the *skolion*, when participants would, extempore, extol the virtues of gods and men. Some of these symposium details are described in Pomeroy *et al.*[243]

In a comic fragment (Fragment 93) of Eubulus, a Greek comic poet of the fourth century BC, Dionysus (speaking in the role of *symposiarch*) limits those who are healthy in mind to three *kraters*

Figure 6.22 *Aulos.*

of wine: the first for health; the second for Eros; the third for sleep, and then wise guests make for home. With the fourth *krater*, he says, one leaves the realm of Dionysus and enters that of hubris. Drinking even more, the fifth is for shouting; the sixth is for revel; the seventh for fights (black eyes); the eighth is to a summons; the ninth is for bile; the tenth for madness, tossing the furniture around and unconsciousness.[244] The title of Eubulus's play is traditionally given as *Semele* or *Dionysus*. According to Greek mythology, Semele was the human mother of Dionysus, the Greek god of wine – his father was Zeus.

For the lower echelons of society, wine was sold in taverns in large quantity for consumption at home, or else in smaller and diluted amounts on the premises. Drinking in a tavern was not as orderly as it was at the symposium. While Plato's account of the symposium was somewhat "idealized", in that afterwards the guests ambled back to their homes in varying degrees of insobriety, it was not unknown for the participants to leave and "go out on the town" on a drunken revelry, called a *komos*. Rollaston[245] has written an excellent account of alcohol in classical antiquity. Etruscan art shows similar scenes as Greek work, and the essence of the symposium lasted into Roman times whence it was known as the *convivium*.[246]

REFERENCES

1. A. H. Joffe, *Curr. Anthropol.*, 1998, **39**, 297.
2. D. B. Heath, *Drinking Occasions: Comparative Perspectives on Alcohol and Culture*, Brunner/Mazel, Philadelphia, 2000.
3. M. Dietler, *Annu. Rev. Anthropol.*, 2006, **35**, 229.
4. R. Dudley, *Q. Rev. Biol.*, 2000, **75**, 3.
5. R. Dudley, *Addiction*, 2002, **97**, 381.
6. D. Stephens and R. Dudley, *Nat. Hist.*, 2004, **113**, 40.
7. D. Goldman, G. Oroszi and F. Ducci, *Nat. Rev. Genet.*, 2005, **6**, 521.
8. K. Milton, *Integr. Comp. Biol.*, 2004, **44**, 304.
9. C. J. Brady, *Ann. Rev. Plant Physiol.*, 1987, **38**, 155.
10. G. B. Seymour, J. E. Taylor and G. A. Tucker (ed.), *Biochemistry of Fruit Ripening*, Chapman & Hall, London, 1993.
11. M. L. Cipollini and E. W. Stiles, *Ecology*, 1993a, **74**, 751.
12. D. H. Janzen, *Am. Nat.*, 1977, **111**, 691.
13. M. L. Cipollini and E. W. Stiles, *Adv. Ecol. Res.*, 1992, **23**, 31.
14. M. L. Cipollini and E. W. Stiles, *Am. Nat.*, 1993b, **141**, 663.
15. M. L. Cipollini and D. J. Levey, *Ecology*, 1997, **78**, 799.
16. C. M. Herrera, *Am. Nat.*, 1982, **120**, 218.
17. G. Kusano, A. Takhashi, K. Sugiyama and S. Nozoe, *Chem. Pharmacol. Bull.*, 1987, **35**, 4862.
18. P. Jordano, in *The Ecology of Regeneration in Natural Plant Communities*, ed. M. Fenner, CAB International, Wallingford, 2nd edn, 2000, p. 105.
19. M. Fenner and K. Thompson, *Ecology of Seeds*, Cambridge University Press, Cambridge, 2005.
20. K. Milton, *Am. Anthropol.*, 1981, **83**, 534.
21. K. Milton, *Sci. Am.*, August 1993, 70.
22. K. Milton, J. Giacalone, S. J. Wright and G. Stockmeyer, in *Tropical Fruits and Frugivores: The Search for Strong*

Interactors, ed. J. L. Dew and J. P. Boubli, Springer-Verlag, Berlin, 2005, p. 5.

23. N. J. Dominy, J.-C. Svenning and W.-H. Li, *J. Hum. Evol.*, 2003, **44**, 25.
24. D. J. Levey, *Integr. Comp. Biol.*, 2004, **44**, 284.
25. B. G. Galef, Jr., *Psychol. Rev.*, 1991, **98**, 218.
26. J. C. Berbesque and F. W. Marlowe, *Evol. Psychol.*, 2009, **7**, 601.
27. M. L. Smith, *Am. Anthropol.*, 2006, **108**, 480.
28. V. G. Childe, *Man Makes Himself*, Watts & Co., London, 1936.
29. I. S. Hornsey, *A History of Beer and Brewing*, Royal Society of Chemistry, Cambridge, 2003.
30. A. Sclafani, *Int. J. Obes.*, 2001, **25**(5), S13.
31. P. Rozin, in *Food and Evolution*, ed. M. Harris and E. B. Ross, Temple University Press, Philadelphia, 1987, p. 181.
32. P. A. Garber, *Ann. Rev. Anthropol.*, 1987, **16**, 339.
33. Y. Han, S. Gu, H. Oota, M. V. Osier, A. J. Pakstis, W. C. Speed, J. R. Kidd and K. K. Kidd, *Am. J. Hum. Genet.*, 2007, **80**, 441.
34. F. Wiens, A. Zitzmann, M.-A. Lachance, M. Yegles, F. Pragst, F. M. Wurst, D. von Holst, S. L. Guan and Rainer Spanagel, *Proc. Nat. Acad. Sci. USA*, 2008, **105**, 10426.
35. J. I. Bloch, M. T. Silcox, D. M. Boyer and E. J. Sargis, *Proc. Nat. Acad. Sci. USA*, 2007, **104**, 1159.
36. J. E. Janečka, W. Miller, T. H. Pringle, F. Wiens, A. Zitzmann, K. M. Helgen, M. S. Springer and W. J. Murphy, *Science*, 2007, **318**, 792.
37. R. Dudley and M. Dickinson, *Integr. Comp. Biol.*, 2004, **44**, 267.
38. R. Dudley, *Integr. Comp. Biol.*, 2004, **44**, 315.
39. F. W. Wolf and U. Heberlein, *J. Neurobiol.*, 2003, **54**, 161.
40. U. Heberlein, F. W. Wolf, A. Rothenfluh and D. J. Guarnieri, *Integr. Comp. Biol.*, 2004, **44**, 269.
41. M. D. Adams, S. E. Celniker, R. A. Holt, C. A. Evans, J. D. Gocayne, P. G. Amanatides, S. E. Scherer, P. W. Li, R. A. Hoskins, R. F. Galle, R. A. George, S. E. Lewis, S. Richards, M. Ashburner, S. N. Henderson, G. G. Sutton, J. R. Wortman, M. D. Yandell, Q. Zhang, L. X. Chen, R. C. Brandon, Y. H. Rogers, R. G. Blazej, M. Champe, B. D.

Pfeiffer, K. H. Wan, C. Doyle, E. G. Baxter, G. Helt, C. R. Nelson, G. L. Gabor, J. F. Abril, A .Agbayani, H. J. An, C. Andrews-Pfannkoch, D. Baldwin, R. M. Ballew, A. Basu, J. Baxendale, L. Bayraktaroglu, E. M. Beasley, K. Y. Beeson, P. V. Benos, B. P. Berman, D. Bhandari, S. Bolshakov, D. Borkova, M. R. Botchan, J. Bouck, P. Brokstein, P. Brottier, K. C. Burtis, D. A. Busam, H. Butler, E. Cadieu, A. Center, I. Chandra, J. M. Cherry, S. Cawley, C. Dahlke, L. B. Davenport, P. Davies, B. de Pablos, A. Delcher, Z. Deng, A. D. Mays, I. Dew, S. M. Dietz, K. Dodson, L. E. Doup, M. Downes, S. Dugan-Rocha, B. C. Dunkov, P. Dunn, K. J. Durbin, C. C. Evangelista, C. Ferraz, S. Ferriera, W. Fleischmann, C. Fosler, A. E. Gabrielian, N. S. Garg, W. M. Gelbart, K. Glasser, A. Glodek, F. Gong, J. H. Gorrell, Z. Gu, P. Guan, M. Harris, N. L. Harris, D. Harvey, T. J. Heiman, J. R. Hernandez, J. Houck, D. Hostin, K. A. Houston, T. J. Howland, M. H. Wei, C. Ibegwam, M. Jalali, F. Kalush, G. H. Karpen, Z. Ke, J. A. Kennison, K. A. Ketchum, B. E. Kimmel, C. D. Kodira, C. Kraft, S. Kravitz, D. Kulp, Z. Lai, P. Lasko, Y. Lei, A. A. Levitsky, J. Li, Z. Li, Y. Liang, X. Lin, X. Liu, B. Mattei, T. C. McIntosh, M. P. McLeod, D. McPherson, G. Merkulov, N. V. Milshina, C. Mobarry, J. Morris, A. Moshrefi, S. M. Mount, M. Moy, B. Murphy, L. Murphy, D. M. Muzny, D. L. Nelson, D. R. Nelson, K. A. Nelson, K. Nixon, D. R. Nusskern, J. M. Pacleb, M. Palazzolo, G. S. Pittman, S. Pan, J. Pollard, V. Puri, M. G. Reese, K. Reinert, K. Remington, R. D. Saunders, F. Scheeler, H. Shen, B. C. Shue, I. Sidén-Kiamos, M. Simpson, M. P. Skupski, T. Smith, E. Spier, A. C. Spradling, M. Stapleton, R. Strong, E. Sun, R. Svirskas, C. Tector, R. Turner, E. Venter, A. H. Wang, X. Wang, Z. Y. Wang, D. A. Wassarman, G. M. Weinstock, J. Weissenbach, S. M. Williams, T. Woodage, K. C. Worley, D. Wu, S. Yang, Q. A. Yao, J. Ye, R. F. Yeh, J. S. Zaveri, M. Zhan, G. Zhang, Q. Zhao, L. Zheng, X. H. Zheng, F. N. Zhong, W. Zhong, X. Zhou, S. Zhu, X. Zhu, H. O. Smith, R. A. Gibbs, E. W. Myers, G. M. Rubin and J. C. Venter, *Science*, 2000, **287**, 2185.
42. G. M. Rubin, M. D. Yandell, J. R. Wortman, G. L. Gabor Miklos, C. R. Nelson, I. K. Hariharan, M. E. Fortini, P. W.

Li, R. Apweiler, W. Fleischmann, J. M. Cherry, S. Henikoff, M. P. Skupski, S. Misra, M. Ashburner, E. Birney, M. S. Boguski, T. Brody, P. Brokstein, S. E. Celniker, S. A. Chervitz, D. Coates, A. Cravchik, A. Gabrielian, R. F. Galle, W. M. Gelbart, R. A. George, L. S. Goldstein, F. Gong, P. Guan, N. L. Harris, B. A. Hay, R. A. Hoskins, J. Li, Z. Li, R. O. Hynes, S. J. Jones, P. M. Kuehl, B. Lemaitre, J. T. Littleton, D. K. Morrison, C. Mungall, P. H. O'Farrell, O. K. Pickeral, C. Shue, L. B. Vosshall, J. Zhang, Q. Zhao, X. H. Zheng and S. Lewis, *Science*, 2000, **287**, 2204.

43. S. W. McKechnie and B. W. Geer, *Genetica*, 1993, **90**, 205.

44. J. R. David and C. Bocquet, *Nature*, 1975, **257**, 588.

45. J. R. Powell, *Progress and Prospects in Evolutionary Biology: The Drosophila Model*, Oxford University Press, Oxford, 1997.

46. F. Sanchez, C. Korine, B. Pinshow and R. Dudley, *Integr. Comp. Biol.*, 2004, **44**, 290.

47. N. J. Dominy, *Integr. Comp. Biol.*, 2004, **44**, 295.

48. R. McC. Netting, *Am. Anthropol.*, 1964, **66**, 375.

49. E. C. Parsons, *Pueblo Indian Religion*, 2 volumes, University of Chicago Press, Chicago, 1939.

50. J. E. Levy and S. J. Kunitz, *SW J. Anthropol.*, 1971, **27**, 97.

51. M. Douglas (ed.), *Constructive Drinking: Perspectives on Drink from Anthropology*, Cambridge University Press, Cambridge, 1987.

52. R. G. McCarthy (ed.), *Drinking and Intoxication*, Selected Readings in Social Attitudes and Controls, College and University Press, New Haven, CT, 1959.

53. D. J. Pittman and C. R. Snyder (ed.) *Society, Culture, and Drinking Patterns*, John Wiley & Sons, New York, 1962.

54. M. W. Everett, J. O. Waddell and D. B. Heath (ed.), *Cross-cultural Approaches to the Study of Alcohol: An Interdisciplinary Perspective*, Mouton, The Hague, 1976.

55. M. Marshall (ed.), *Beliefs, Behaviours, and Alcoholic Beverages: A Cross-cultural Survey*, University of Michigan Press, Ann Arbour, 1979.

56. R. Bunzel, *Psychiatry*, 1940, **3**, 361.

57. D. Horton, *Q. J. Stud. Alcohol*, 1943, **4**, 199.

58. D. G. Mandelbaum, *Curr. Anthropol.*, 1964, **6**, 281.

59. D. B. Heath, *Q. J. Stud. Alcohol*, 1958, **19**, 491.

60. D. B. Heath, in *Society, Culture, and Drinking Patterns*, ed. D. J. Pittman and C. R. Snyder, John Wiley & Sons, New York, 1962.
61. D. B. Heath, *Hum. Organ.*, 1994, **53**, 357.
62. C. Kluckhohn and D. Leighton, *The Navaho*, Harvard University Press, Cambridge, MA, 1951.
63. F. N. Ferguson, *Hum. Org.*, 1968, **27**, 159.
64. C. L. Ehlers, T. L. Wall, B. S. Betancourt and D. A. Gilder, *Am. J. Psychiatry*, 2004, **161**, 1204.
65. J. Diamond, *Nature*, 2002, **418**, 700.
66. I. S. Hornsey, *Brewer & Distiller Int'l*, December 2010, 41.
67. S. Pollock, in *The Archaeology and Politics of Food and Feasting in Early States and Empires*, ed. T. L. Bray, Kluwer Academic/Plenum Publishers, New York, 2003, p. 17.
68. A. Leo Oppenheim, *Ancient Mesopotamia: Portrait of a Dead Civilization*, University of Chicago Press, Chicago, 1964.
69. J. N. Postgate, *Early Mesopotamia: Society and Economy at the Dawn of History*, Routledge, London, 1992.
70. M. Civil, in *Studies Presented to A. Leo Oppenheim*, ed. R. D. Biggs and J. A. Brinkman, The Oriental Institute of the University of Chicago, Chicago, 1964, p. 67.
71. R. J. Forbes, *Studies in Ancient Technology*, E. J. Brill, Leiden, 1955, vol. III.
72. M. T. Roth, *Law Collections from Mesopotamia and Asia Minor*, Scholar's Press, Atlanta, GA, 2nd revised edn, 2000.
73. V. R. Badler, P. E. McGovern and R. H. Michel, *MASCA Research Papers in Science and Archaeology*, 1990, **7**, 25.
74. I. S. Hornsey, *The Chemistry and Biology of Winemaking*, Royal Society of Chemistry, Cambridge, 2007.
75. R. H. Michel, P. E. McGovern and V. R. Badler, *Nature*, 1992, **360**, 24.
76. P. E. McGovern, *Uncorking the Past*, University of California Press, Berkeley, 2009.
77. S. Pollock, in *Engendering Archaeology: Women and Prehistory*, ed. J. M. Gero and M. W. Conkey, Blackwell, Oxford, 1991, p. 366.
78. A. George, *The Epic of Gilgamesh*, Penguin, Harmondsworth, 2000.
79. P. Weadock, *Iraq*, 1975, **37**, 101.

80. D. Schmandt-Besserat, in *Feasts: Archaeological and Anthropological Perspectives*, ed. M. Dietler and B. Hayden, Smithsonian Institute, Washington, DC, 2001, p. 391.
81. L. Milano, *Drinking in Ancient Societies: History and Culture of Drinks in the Ancient Near East*, Sargon srl, Padua, 1994.
82. M. Michalowski, in *Drinking in Ancient Societies. History and Culture of Drinks in the Ancient Near East*, ed. L. Milano, Sargon srl, Padua, 1994.
83. F. Pinnock, in *Drinking in Ancient Societies*, ed. L. Milano, Sargon srl, Padua, 1994, p. 15.
84. M. Stol, in *Drinking in Ancient Societies. History and Culture of Drinks in the Ancient Near East*, ed. L. Milano, Sargon srl, Padua, 1994.
85. M. Stol, in *Civilisations of the Ancient Near East*, ed. J. M. Sasson, Simon & Schuster, New York, 1995, vol. 1.
86. H. F. Lutz, *Viticulture and Brewing in the Ancient Orient*, J. C. Hinrichs, Leipzig, 1922.
87. D. Samuel, in *Ancient Egyptian Materials and Technology*, ed. P. T. Nicholson and I. Shaw, Cambridge University Press, Cambridge, 2000, p. 537.
88. M. A. Murray, N. Boulton and C. Heron, in *Ancient Egyptian Materials and Technology*, ed. P. T. Nicholson and I. Shaw, Cambridge University Press, Cambridge, 2000, p. 577.
89. M. A. Murray, in *Ancient Egyptian Materials and Technology*, ed. P. T. Nicholson and I. Shaw, Cambridge University Press, Cambridge, 2000, p. 505.
90. W. J. Darby, P. Ghalioungi and L. Grivetti, *Food: The Gift of Osiris*, 2 volumes, Academic Press, London, 1977.
91. A. M. Ashour, in *International Handbook on Alcohol and Culture*, ed. D. B. Heath, Greenwood Press, Westport, CT, 1995, p. 63.
92. R. O. Faulkner (translated by), *Ancient Egyptian Pyramid Texts*, 2 volumes, Clarendon Press, Oxford, 1969.
93. F. Beauvais, *Alcohol Health and Research World*, 1998, **22**, 253.
94. R. Smart and A. Ogborne, *Northern Spirits: A Social History of Alcohol in Canada*, Addiction Research Foundation, Toronto, Ontario, 1996.

95. V. Havard, *Bulletin of the Torrey Botanical Club*, 1896, **23**, 33.
96. G. Sagard-Théodat, *Histoire du Canada*, Claude Sonnius, Paris, 1636.
97. J. Hisnanick, *Health Val.*, 1992, **16**, 32.
98. P. May, in *The Challenge of Participatory Research: Preventing Alcohol-Related Problems in Ethnic Minority Communities*, ed. P. Langton, DHHS Pub. No. (SMA) 95-3042, Center for Substance Abuse Prevention, Washington, DC, 1995, p. 183.
99. O. G. Simmons, in *Culture, and Drinking Patterns*, ed. D. J. Pittman and C. R. Snyder, John Wiley & Sons, New York, 1962, p. 37.
100. P. L. Doughty, *Hu. Org.*, 1971, **30**, 187.
101. C. Morris, in *Fermented Foods in Nutrition*, ed. C. Gastineau, W. Darby and T. Turner, Academic Press, New York, 1979, p. 21.
102. P. S. Goldstein, in *The Archaeology and Politics of Food and Feasting in Early States and Empires*, ed. T. L. Bray, Kluwer Academic/Plenum Publishers, New York, 2003, p. 143.
103. L. M. Valdez, *J. Anthropol. Res.*, 2006, **62**, 53.
104. B. J. Isbell, *To Defend Ourselves: Ecology and Ritual in an Andean Village*, Waveland Press, Prospect Heights, IL, 1978.
105. J. W. Bastian, *Mountain of the Condor: Metaphor and Ritual in an Andean Ayllu*, Waveland Press, Prospect Heights, 1978.
106. M. J. Weismantel, *Cholas and Pishtacos: Stories of Race and Sex in the Andes*, University of Chicago Press, Chicago, 2001.
107. C. J. Allen, *The Hold Life Has: Coca and Cultural Identity in an Andean Community*, Smithsonian Institute Press, Washington, DC, 2nd edn, 2002.
108. J. Jennings, and B. J. Bowser, *Drink, Power, and Society in the Andes*, University Press of Florida, Gainesville, FL, 2009.
109. R. Cavero Carrasco, *Maíz, chicha y religiosidad andina*, Universidad Nacional de San Cristóbal de Huamanga, Ayacucho, 1986.
110. L. Camino, *Chicha de maíz: Bebida y vida del pueblo Catacaos*, Centro de Investigación y Promoción del Campesinado-Piura, Piura, 1987.

111. T. Saignes (ed.), *Borrachera y memoria: La experience de lo sagrado en los Andes*, Instituto Francés de Estudios Andinos, Lima, Peru, 1993.

112. B. Butler, *Holy Intoxication to Drunken Dissipation: Alcohol among Quichua Speakers on Otavalo, Ecuador*, University of New Mexico Press, Albuquerque, 2006.

113. C. J. Allen, in *Drink, Power, and Society in the Andes*, ed. J. Jennings and B. J. Bowser, University Press of Florida, Gainesville, FL, 2009, p. 28.

114. M. J. Weismantel, in *Drink, Power, and Society in the Andes*, ed. J. Jennings and B. J. Bowser, University Press of Florida, Gainesville, FL, 2009, p. 257.

115. W. F. H. Adelaar and P. C. Muysken, *The Languages of the Andes*, Cambridge University Press, Cambridge, 2007.

116. E. Mayer, *The Articulated Peasant: Household Economies in the Andes*, Westview Press, Boulder, 2002.

117. C. A. Hastorf and S. Johannessen, *Am. Anthropol.*, 1993, **95**, 115.

118. R. Colloredo-Mansfield, *The Native Leisure Class: Consumption and Cultural Creativity in the Andes*, University of Chicago Press, Chicago, 1999.

119. J. Jennings and M. Chatfield, in *Drink, Power, and Society in the Andes*, ed. J. Jennings and B. J. Bowser, University Press of Florida, Gainesville, FL, 2009, p. 200.

120. M. Mauss, *The Gift: The Form and Reason for Exchange in Archaic Societies*, W. W. Norton, New York, 1990.

121. J. Jennings, in *Foundations of Power in the Prehispanic Andes*, ed. C. A. Conlee, D. Ogburn and K. Vaughn, Publications of the American Anthropological Association, American Anthropological Association, Washington, DC, 2005, vol. 14, p. 241.

122. T. L. Bray, in *Drink, Power, and Society in the Andes*, ed. J. Jennings and B. J. Bowser, University Press of Florida, Gainesville, FL, 2009, p. 108.

123. M. Goodman-Elgar, in *Drink, Power, and Society in the Andes*, ed. J. Jennings and B. J. Bowser, University Press of Florida, Gainesville, FL, 2009, p. 75.

124. B. Cobo, *History of the Inca Empire* – translated and edited by Roland Hamilton, University of Texas Press, Austin, TX, 1979 [original text 1653].

125. J. Moore, *Am. Anthropol.*, 1989, **91**, 682.
126. D. J. Goldstein, R. C. Coleman Goldstein and P. R. Williams, in *Drink, Power, and Society in the Andes*, ed. J. Jennings and B. J. Bowser, University Press of Florida, Gainesville, FL, 2009, p. 133.
127. C. A. Hastorf, in *Engendering Archaeology: Women and Prehistory*, ed. J. Gero and M. Conkey, Blackwell, Oxford, 1991, p. 132.
128. C. A. Hastorf, in *Empire and Domestic Economy*, ed. T. N. D'Altroy and C. A. Hastorf, Kluwer Academic, New York, 2001, p. 155.
129. R. L. Burger and N. J. van der Merwe, *Am. Anthropol.*, 1990, **92**, 85.
130. S. H. Ambrose and J. Krigbaum, *J. Anthropol. Archaeol.*, 2003, **22**, 193.
131. T. H. E. Heaton, *J. Archaeol. Sci.*, 1999, **26**, 637.
132. P. L. Koch, in *Stable Isotopes in Ecology and Environmental Science*, ed. R. Michener and K. Lajtha, Blackwell, Oxford, 2nd edn, 2007, p. 94.
133. J. C. Vogel and N. J. van der Merwe, *Am. Antiq.*, 1977, **42**, 238.
134. N. J. van der Merwe, *Am. Sci.*, 1982, **70**, 596.
135. N. J. van der Merwe, *Proc. Br. Acad.*, 1992, **77**, 247.
136. N. J. van der Merwe and J. C. Vogel, *Nature*, 1978, **276**, 815.
137. B. N. Smith and S. Epstein, *Plant Physiol.*, 1970, **47**, 380.
138. R. H. Tykot, in *Histories of Maize: Multidisciplinary Approaches to the Prehistory, Linguistics, Biogeography, Domestication, and Evolution of Maize*, ed. J. E. Staller, R. H. Tykot and B. F. Benz, Academic Press (Elsevier), Amsterdam, 2006, p. 131.
139. T. L. Bray, in *The Archaeology and Politics of Food and Feasting in Early States and Empires*, ed. T. L. Bray, Kluwer Academic/Plenum Publishers, New York, 2003, p. 93.
140. F. Salomon, *Native Lords of the Quito in the Age of the Incas*, Cambridge University Press, Cambridge, 1986.
141. S. K. Lothrop, *Am. Antiq.*, 1956, **21**, 233.
142. D. de Torres Rubio, *Arte de la Lenga Aymara*, Francisco del Canto, Lima, 1616.
143. T. A. Joyce, *J. Roy. Anthropol. Inst. Great Brit. Ireland*, 1922, **52**, 141.

144. B. Orlove and E. Schmidt, *Theor. Soc.*, 1995, **24**, 271.

145. M. J. Weismantel, *Mod. Lang. Notes*, 1991, **106**, 861.

146. N. A. Duncan, D. M. Pearsall and R. A. Benfer, Jr., *Proc. Nat. Acad. Sci. USA*, 2009, **106**, 13202.

147. J. Haas, W. Creamer and A. Ruiz, *Archaeology*, 2003, **56**, 7.

148. H. C. Cutler and T. W. Whitaker, *Am. Antiq.*, 1961, **26**, 469.

149. C. S. Lumholtz, *Unknown Mexico*, Macmillan, London, 1903, vol. 1.

150. W. Bennett and R. Zingg, *The Tarahumara*, University of Chicago Press, Chicago, IL, 1935.

151. J. G. Kennedy, *Am. Anthropol.* 1963, **65**, 620.

152. H. Passin, *Am. Anthropol.*, 1942, **44**, 235.

153. E. M. Lemert, in *Beliefs, Behaviours, and Alcoholic Beverages: A Cross-cultural Survey*, ed. M. Marshall, University of Michigan Press, Ann Arbour, 1979, p. 192.

154. R. M. Zingg, *Am. Anthropol.*, 1942, **44**, 78.

155. M. Moseley, D. J. Nash, P. R. Williams, S. D. deFrance, A. Miranda and M. Ruales, *Proc. Nat. Acad. Sci. USA*, 2005, **102**, 17264.

156. D. C. Perlov, in *Drink, Power, and Society in the Andes*, ed. J. Jennings and B. J. Bowser, University Press of Florida, Gainesville, FL, 2009, p. 49.

157. D. Tedlock, *Popol Vuh* (revised translation), Touchstone Books, New York. 1996.

158. J. Christensen, *Popol Vuh*, University of Oklahoma Press, Oklahoma City, 2007.

159. A. M. Tozzer, *Landa's Relacion de las Costas de Yucatan, a translation*, Peabody Museum, Cambridge, MA, 1941.

160. P. D. Sheets, *The Cerén Site: An Ancient Village Buried by Volcanic Ash in Central America*, Cengage Learning, Florence, KY, 2nd edn, 2005.

161. L. A. Brown, in *Feasts: Archaeological and Ethnographic Perspectives on Food, Politics, and Power*, ed. M. Dietler and B. Hayden, Smithsonian Institute Press, Washington & London, 2001, p. 368.

162. D. L. Lentz, M. P. Beaudry-Corbett, M. L. Reyna de Aquilar and L. Kaplan, *Lat. Am. Antiq.*, 1996, **7**, 247.

163. W. R. Adams, in *International Handbook on Alcohol and Culture*, ed. D. B. Heath, Greenwood Press, Westport, CT, 1995, p. 99.

164. K. W. V. Bustillo, in *International Handbook on Alcohol and Culture*, ed. D. B. Heath, Greenwood Press, Westport, CT, 1995, p. 111.

165. G. N. Rey, in *International Handbook on Alcohol and Culture*, ed. D. B. Heath, Greenwood Press, Westport, CT, 1995.

166. G. A. R. Wood and R. A. Lass, *Cocoa*, Longman, London, 4th edn, 1985.

167. R. A. Rice and R. Greenberg, *Nat. Hist.*, 2003, **112**, 36.

168. A. Figueira, J. Janick and P. Goldsborough, *J. Am. Soc. Hort. Sci.*, 1992, **117**, 673.

169. E. E. Cheeseman, *Trop. Agr.*, 1944, **21**, 144.

170. M. de la Cruz, R. Whitkus, A. Gomez-Pompa and L. Mota-Bravo, *Nature*, 1995, **375**, 542.

171. B. G. D. Bartley, *The Genetic Diversity of Cacao and Its Utilisation*, CABI Publishing, Wallingford, 2005.

172. D. N. Kuhn, A. Figueira, U. Lopes, J. C. Motamayor, A. W. Meerow , K. Cariaga, B. Freeman, D. S. Livingstone and R. J. Schnell, *Tree Genetics & Genomes*, 2010, **6**, 783.

173. J. C. Motamayor, P. Lachenaud, J. W. da Silva e Mota, R. Loor, D. N. Kuhn, J. S. Brown and R. J. Schnell, *PLoS ONE*, 2008, **3**, e3311.

174. A. B. Bennett, *Trends Plant Sci.*, 2003, **8**, 561.

175. S. Smith and D. Le Ber, *Cocoa*, Wiley, Chichester, 2010.

176. C. J. J. Van Hall, *Cocoa*, Macmillan, London, 1914.

177. J. C. Motamayor, A. M. Risterucci, P. A. Lopez, C. F. Ortiz, A. Moreno and C. Lanaud, *Heredity*, 2002, **89**, 380.

178. J. C. Motamayor, A. M. Risterucci, M. Heath and C. Lanaud, *Heredity*, 2003, **91**, 322.

179. J. S. Henderson, R. A. Joyce, G. R. Hall, W. J. Hurst and P. E. McGovern, *Proc. Nat. Acad. Sci. USA*, 2007, **104**, 18937.

180. J. S. Henderson and R. A. Joyce, in *Chocolate in Meroamerica: A Cultural History of Cacao*, ed. C. L. McNeil, University Press of Florida, Gainesville, FL, 2006, p. 140.

181. T. L. Dillinger, P. Barriga, S. Escárcega, M. Jimenez, D. Salazar Lowe and L. E. Grivetti, *J. Nutr.*, 2000, **130**(Supplement), 2057S.

182. W. Hughes, *The American Physitian, or a Treatise of the Roots, Plants, Trees, Shrubs, Fruit, Herbs etc. Growing in the*

English Plantations in America: Describing the Place, Time, Names, Kindes, Printed by J. C. for William Crook, London, 1672.

183. H. Stubbe, *The Indian Nectar, or, a Discourse on Chocolata*, Printed by J. C. for Andrew Crook, London, 1662.

184. T. Gage, *The English-American: His Travail by Sea*, R. Cotes, London, 1648.

185. F. A. Barkley, *Brittonia*, 1944, **5**, 160.

186. F. L. Kramer, *Econ. Bot.*, 1957, **11**, 322.

187. D. J. Goldstein and R. C. Coleman, *Econ. Bot.*, 2004, **58**, 523.

188. J. Goody, *Cooking, Cuisine, and Class: A Study in Comparative Sociology*, Cambridge University Press, Cambridge, 1982.

189. M. Dietler, *Archaeology*, 1994, **47**, 44.

190. M. Dietler, *J. Anthropol. Archaeol.*, 1990, **9**, 352.

191. L. Moulin, in *Manger et Boire au Moyen Age*, ed. D. Menjot, Publications de la Faculté des Lettres et Sciences Humaines de Nice, Les Belles Lettres, Paris, 1984, p. 13.

192. B. Arnold, *Camb. Archaeol. J.*, 1999, **9**, 71.

193. J. M. de Navarro, *Antiquity*, 1928, **2**, 423.

194. J. M. de Navarro, *Finds From the Site of La Tène*, Oxford University Press, Oxford, 1972, vol. 1.

195. A. Tchernia, in *Trade in the Ancient Economy*, ed. P. Garnsey and C. R. Whittaker, Chatto & Windus, London, 1983, p. 87.

196. M. Parker-Pearson (ed.), *Food, Culture and Identity in the Neolithic and Early Bronze Age*, BAR Int. Series 1117, Archaeopress, Oxford, 2003.

197. M. Poux, *L'Âge du Vin. Rites de boisson, festins et libations en Gaule indépendant*, M. Mergoil, Montagnac, 2004.

198. M. E. Loughton, *Oxf. J. Archeol.*, 2009, **28**, 77.

199. A. G. Sherratt, in *Consuming Habits. Drugs in History and Anthropology*, ed. J. Goodman, P. E. Lovejoy and A. Sherratt, Routledge, London, 1995, p. 11.

200. A. L. Martin, in *Food, Drink and Identity. Cooking, Eating and Drinking in Europe Since the Middle Ages*, ed. P. Scholliers, Berg, Oxford, 2001, p. 119.

201. J. Warner, *Craze: Gin and Debauchery in an Age of Reason*, Profile Books, London, 2003.

202. M. E. Loughton, *Republican Amphorae in the Auvergne Central France: an Archaeological and Petrological Study*, 2 volumes, PhD thesis, Bournemouth University, June 2000.
203. A. G. Sherratt, in *Bell Beakers of the Western Mediterranean*, ed. W. Waldren and R. Kennard, BAR Int. Series, 331, Archaeopress, Oxford , 1987, p. 81.
204. I. Karp, in *Explorations in African Systems of Thought*, ed. I. Karp and C. S. Bird, University of Indiana Press, Bloomington, IN, 1980, p. 83.
205. C. Seyffert, *Biene und Honig im Volksleben der Afrikaner mit besonderer Berücksichtigung der Bienenzucht ihrer Entstehung und Verbreitung*, R. Voigtländers Verlag, Leipzig, 1930.
206. C. F. Beckingham and G. W. B. Huntingford, *Some Records of Ethiopia: 1593–1646*, Haklyut Society, London, 1954.
207. C. C. Adomakoh, *Ann. New York Acad. Sci.*, 1976, **273**, 39.
208. J. C. Carothers, *J. Ment. Sci.*, 1948, **93**, 548.
209. D. Reusch, in *Ubumba: Aspects of Indigenous Ceramics in Kwazulu-Natal*, ed. B. Bell and I. Calder, Tatham Art Gallery, Pietermaritzburg, 1998, p. 19.
210. D. L. Donham, *History, Power, Ideology: Central Issues in Marxism and Anthropology*, University of California Press, Berkeley, CA, 1999.
211. L. Pan, *Alcohol in Colonial Africa*, The Scandinavian Institute of African Studies, Uppsala, 1975.
212. G. W. B. Huntingford, *Colonial Research Studies No.4*, HMSO, London, 1950.
213. B. S. Platt, *Proc. Nutr. Soc.*, 1955, **4**, 132.
214. T. M. Mushanga, *Crime and Deviance*, East African Literature Bureau, Kampala, Nairobi and Dar-es-Salaam, 1976.
215. J. Thomson, *J. Roy. Anthropol. Inst.*, 1887, **16**, 182.
216. A. T. Bryant, *A Description of Native Foodstuffs and Their Preparation*, Government Printer, Pretoria, 1939.
217. P. A. McAllister, *Xhosa Beer Drinking Rituals: Power, Practice and Performance in the South African Rural Periphery*, Carolina Academic Press, Durham, NC, 2006.
218. M. Shaw, in *The Bantu-speaking Peoples of Southern Africa*, ed. W. D. Hammond-Tooke, Routledge and Kegan Paul, London, 1974, p. 85.

219. G. M. Theal, *Records of South-eastern Africa*, Government of the Cape Colony, Cape Town, 1898, vol. I–IX.

220. F. W. Fox, *J. S. Afr. Chem. Inst.*, 1938, 21.

221. H. Ashton, *The Basuto*, Oxford University Press, Oxford, 1952.

222. A. Kropf, *A Kafir-English Dictionary*, ed. R. Godfrey, Lovedale Press, Alice, Cape Province, 2nd edn, 1915.

223. D. N. Edwards, *Norweg. Archaeol. Rev.*, 1996, **29**, 65.

224. D. Samuel, *J. Am. Soc. Brewing Chemists*, 1996, **54**, 3.

225. J. W. Arthur, *World Archaeol.*, 2003, **34**, 516.

226. D. N. Edwards, *Ann. Rev. Anthropol.*, 2007, **36**, 351.

227. M. Dietler, in *Feasts: Archaeological and Ethnographic Perspectives on Food, Politics, and Power*, ed. M. Dietler and B. Hayden, Smithsonian Institute Press, Washington & London, 2001, p. 65.

228. O. B. Rekdal, *Africa*, 1996, **66**, 367.

229. C. Ehret, *J. Afr. Hist.*, 1967, **8**, 1.

230. J. Parry and M. Bloch, *Money and the Morality of Exchange*, Cambridge University Press, Cambridge, 1989.

231. R. G. Carlson, *Ethnology*, 1990, **29**, 297.

232. M. C. Robbins, in *Beliefs, Behaviours, and Alcoholic Beverages: A Cross-cultural Survey*, ed. M. Marshall, University of Michigan Press, Ann Arbour, 1979, p. 362.

233. M. Dietler and B. Hayden (ed.), *Feasts: Archaeological and Ethnographic Perspectives on Food, Politics, and Power*, Smithsonian Institute Press, Washington & London, 2001.

234. W. R. Biers and P. E. McGovern (ed.), *Organic Contents of Ancient Vessels: Materials Analysis and Archaeological Investigation*, MASCA Research Pap. Sci. Archaeometry, Philadelphia, 1990.

235. D. I. Kertzer, *Ritual, Politics, and Power*, Yale University Press, New Haven, CT, 1988.

236. B. Hayden, in *Feasts: Archaeological and Ethnographic Perspectives on Food, Politics, and Power*, Smithsonian Institute Press, Washington & London, 2001, p. 23.

237. J. Sobal and M. K. Nelson, *Appetite*, 2003, **41**, 181.

238. M. Dietler, in *Food and the Status Quest: An Interdisciplinary Perspective*, ed. P. W. Wiessner and W. Shiefenhövel, Berghahn Books, Providence, RI, 1996, p. 87.

239. M. Sahlins, *Stone Age Economics*, Tavistock Publications, London, 1974.
240. M. Dietler and I. Herbich, *World Archaeol.*, 1993, **25**, 248.
241. M. Nelson, *The Barbarian's Beverage: A History of Beer in Ancient Europe*, Routledge, London, 2005.
242. O. Murray (ed.), *Sympotica: A Symposium on the Symposion*, Clarendon Press, Oxford, 1990.
243. S. Pomeroy, S. Burstein, W. Donlan and J. Tolbert Roberts, *Ancient Greece: A Political, Social, and Cultural History*, Oxford University Press, Oxford, 1999.
244. R. L. Hunter, *Eubulus: The Fragments*, Cambridge University Press, Cambridge, 1983.
245. J. D. Rollaston, *British J. Inebriety*, 1927, **24**, 101.
246. N. F. Hudson, *Am. J. Archaeol.*, 2010, **114**, 663.

Glossary

Abiogenesis Sometimes called "biopoesis", this is the study of how biological life arises from inorganic matter through natural processes, or, more specifically, the method by which life on Earth arose

Allele An allele is an alternative form of a gene (*i.e.* one of a pair) that is located at a specific position on a named chromosome. Dominant alleles are marked with a capital letter (*e.g.* Vrs1), while recessive forms are marked in lower case (vrs.1)

Alloploidy Relates to a hybrid organism or cell with two, or more, sets of chromosomes derived from two different ancestral species. Depending upon the number of multiples of the haploid set of chromosomes, alloploids are described as "allodiploids", "allotetraploids", *etc.*

Allozyme A variant form of an enzyme coded by different alleles at the same locus – as opposed to isozymes, which are enzymes that perform the same function, but which are coded by genes located at different loci

Ampelography The identification and classification of grapevines (*Vitis* spp.). Prior to the onset of DNA fingerprinting, this was effected by using leaves and berries as the main criteria

Anemophily Pollination by wind

Angiosperm Phylogeny Group (APG) An informal international group of systematic botanists who review plant taxonomy in the light of modern research findings. They have reinvigorated

Alcohol and its Role in the Evolution of Human Society
Ian S. Hornsey
© Ian S. Hornsey 2012
Published by the Royal Society of Chemistry, www.rsc.org

botanical taxonomic research, and have produced updated classification systems (in 1998, 2003 and 2009)

Apomyxis Development of an embryo without fertilization having occurred (*i.e.* a form of asexual reproduction)

Apoplast Is the free diffusional space outside the plant cell membrane. It is important to the plant because it is responsible for interactions with the environment. Carbon dioxide, for example, needs to be solubilized here before it can be taken up by chloroplasts (for photosynthesis)

Bacterial artificial chromosome (BAC) An artificially constructed chromosome in which medium-sized DNA segments (usually from 100,000–300,000 bases) from one species are cloned into a bacterium (often *E. coli*). Useful for sequencing in genome projects

Bottleneck Generally referred to as "population bottleneck", or "genetic bottleneck", it is an evolutionary event in which a substantial percentage of the population, or species, is killed or otherwise prevented from reproducing

C-value The amount of DNA contained in a haploid (gametic) nucleus – or one half the amount found in a diploid, somatic, eukaryotic cell. Expressed in pictograms (1 pg $= 10^{-12}$ g)

Clade A group of taxa that includes all descendants of one common ancestor. Members of such a group share homologous features

Cladistic methods These are methods of phylogenetic analysis that are based on the explicit assumption that a set of sequences evolved from a common ancestor by a process of mutation and selection without mixing (hybridization, or other horizontal gene transfer)

Complementary DNA (cDNA) Derived from a messenger RNA (mRNA) template in a reaction catalysed by the enzyme reverse transcriptase, and the enzyme DNA polymerase. cDNA is often used to clone eukaryotic genes in prokaryotic cells

Cultigen A cultivated organism (particularly a plant) not known to have a wild, or uncultivated, counterpart

Expressed sequence tag A unique DNA sequence derived from a complementary DNA library. It can be mapped to a particular locus in the genome and can serve to identify that locus

Fluorescence *in situ* hybridization (FISH) A cytogenetic technique used to detect and localize the presence or absence of specific

DNA sequences on chromosomes. FISH uses fluorescent probes that bind solely to those parts of the chromosome with which they show a high degree of sequence similarity

Flow cytometry (FCM) A technique for counting and examining microscopic particles, such as cells and chromosomes, by suspending them in a stream of liquid and passing through an electronic detector. Most devices can analyse thousands of particles per second

Genome The entire hereditary information of an organism that is encoded by its DNA, or RNA in some viruses

Genomics The study of the structure and function of genes

Genomovar A term used mainly by bacteriologists to denote strains that can be phylogenetically differentiated, but which are phenotypically indistinguishable

Germplasm A collection of genetic resources for an organism. In many plants, the seed represents a source of germplasm

Graminoid Grasses (Gramineae, or Poaceae) and grass-like plants such as rushes and sedges

Haplotype Is literally a contraction of "haploid genotype", and describes a set of closely linked markers that tend to be inherited together. Can be identified by patterns of SNPs

Holocene Literally means "entirely recent", this is the geological epoch that began at the end of the Pleistocene (*ca.* 10,000 ^{14}C years ago) and continues to the present

Homeobox Is a DNA sequence found within genes that are involved in the morphogenesis in animals, fungi and plants. A homeobox is around 180 bases pairs long and encodes a protein domain (the homeodomain), which, when expressed, can bind DNA

Homoeologous Of similar genetic constitution; the term is used of chromosomes believed to have been completely homologous in an ancestral form. Homoeologue would be the applicable noun

Homonymy A situation whereby a taxonomic name is applied (to a taxon) that already exists for another (different) taxon. It is thus unacceptable

Hypanthium A cup-shaped floral structure comprised of the bases of sepals, petals and stamens all fused together – as found in Rosaceae. Fusion of dissimilar floral parts is called "adnation". In the apple (a pome), the thickened, fleshy hypanthium is fused with the ovary wall (seed-bearing core). Most of an apple consists of hypanthium tissue

Introgression Infiltration of the genes of one species in the gene pool of another, through repeated backcrossing of an inter-specific hybrid with one of its parents

Karyotype The number, and appearance, of chromosomes in the nucleus of a eukaryotic cell

Landrace An old cultivar, usually only of local importance

Marker assisted selection (MAS) Is the process of using molecular markers as genetic tags to select indirectly for major genes or QTLs. Theoretical studies have demonstrated that the deployment of MAS in a plant breeding programme can be more effective than conventional phenotypic selection (PS) based on observations of performance in field trials

Maslin crop A mixture of different types of grain, usually wheat and rye – especially when sown together. The practice is seen as a precaution against failure of the wheat crop. Also applies to grain/pulse mixtures

Metabolic uncoupling The dissociation between catabolism and anabolism in an organism

Microarray A two-dimensional array on a solid surface (*e.g.* a glass slide) that will assay a large amount of biological material. Widely used is the DNA microarray (also known as the "biochip", "gene chip" or "DNA chip"), which consists of a large number of DNA spots attached to a glass, or silicon, slide

Microsatellites Also known as SSRs, or short tandem repeats (STRs), they are repeating sequences of 1–6 base pairs of DNA. Used as molecular markers

Monophyly Is a taxon which forms a clade, *i.e.* it contains all the descendants of the (maybe hypothetical) closest common ancestor of the members of the group

Orphan crop "Minor" crops, such as finger millet, yams, tef and various tubers, that tend to be regionally important, rather than being traded worldwide

Phenotypic selection (PS) Natural selection acts on the phenotype (observable characters) of an organism, and the genetic basis of any phenotype that gives a reproductive advantage to that organism will become increasingly common in a population. Over a period of time, this process can result in populations suited for particular niches and may eventually result in the emergence of new species.

Plesiomorphic Applied to features that are shared by different groups of biological organisms and are inherited from a common ancestor. The term means "old-featured" and the features to which it is applied were formerly called "primitive"

Ploidy The number of sets of chromosomes in a cell, or organism. Each set is designated "n", one set being haploid(n), two sets being diploid (2n), *etc.*

Polyphyly A polyphyletic taxon is composed of unrelated organisms descended from more than one ancestor

Prototrophic Describing organisms feeding solely on inorganic matter

Quantitative trait locus (QTL) Most critical characters in crops, such as quality and yield, are the result of a number of genes giving rise to continuous variation for such characters in which individual effects cannot be distinguished phenotypically. Given an extensive genetic map and appropriate phenotypic data, however, it is possible to identify regions of the genome that affect such characters. These regions are the QTLs and can be potentially tagged with molecular markers and manipulated in the same way as major genes

Randomly amplified polymorphic DNA (RAPD) markers These markers have several advantages over traditional RFLPs, such as the requirement for less sample DNA. These have been used successfully in a wide variety of plants

Restriction fragment length polymorphism (RFLP) This is a technique that exploits variation in homologous DNA sequences. In RFLP analysis, a DNA sample is broken down into pieces ("digested") by restriction enzymes, and the resulting 'restriction fragments' are separated, according to their lengths, by gel electrophoresis.

Rosids A name used in the APG systems for one of the principal clades of flowering plants, encompassing around 70,000 species

Satellite; minisatellite; microsatellite Tandemly repetitive sequences, commonly known as 'satellite DNAs', are put into three major groups: a) satellites are highly repetitive with repeat lengths of one to several thousand base pairs; b) minisatellites are moderately repetitive and moderately sized (9–100 base pairs); c) microsatellites are also moderately repetitive and found in arrays of short (2–6 base pairs) repeats. Tandemly repeated DNA sequences are interspersed throughout the genomes of all eukaryotes, including humans.

Simple sequence repeat (SSR) Stretches of 1–6 nucleotide units repeated in tandem and randomly spread in eukaryotic genomes. SSRs are extremely polymorphic owing to high mutation rate affecting the number of repeat units. Proved to be very useful markers for the analysis of the apple genome. Also known as microsatellites

Single nucleotide polymorphism (SNP) The difference in a single DNA nucleotide; for example, an SNP may replace cytosine (C) with thymine (T) in a certain stretch of DNA

Spathe A large bract that forms a sheathe to enclose a cluster of flowers (*e.g.* in palms)

Sympatry Describes two species (or populations) that occur in the same geographical area and are able to encounter and interact (but not breed) with each other. Antonym is allopatry

Synteny The physical co-localization of genetic loci on the same chromosome in a species. The general consensus of opinion suggests that organisms of relatively recent divergence show similar blocks of genes in the same relative position in the genome

Taphonomy The study of the conditions and processes involved in the fossilization of organisms

Tepal Describes the perianth of a flower in which there is virtually no distinction between petals and sepals (*e.g.* tulip)

Transposable element (TE) Also known as a transposon, this is a segment of DNA that can move around to different positions in the genome of a cell. Transposition can cause mutations and alter the genomic size in a cell. Sometimes called "jumping genes"

Subject Index